Air Pollution Modeling and Its Application I

NATO • Challenges of Modern Society

A series of edited volumes comprising multifaceted studies of contemporary problems facing our society, assembled in cooperation with NATO Committee on the Challenges of Modern Society.

Volume 1   AIR POLLUTION MODELING AND ITS APPLICATION I
Edited by C. De Wispelaere

Air Pollution Modeling and Its Application I

Edited by
C. De Wispelaere

Prime Ministers Office for Science Policy
National Research and Development Program on Environment
Brussels, Belgium

Published in cooperation with
NATO Committee on the Challenges of Modern Society

PLENUM PRESS • NEW YORK AND LONDON

Library of Congress Cataloging in Publication Data

International Technical Meeting on Air Pollution Modeling and Its Application (11th : 1980 : Amsterdam, Netherlands)
Air pollution modeling and its application I.

(NATO challenges of modern society series ; v. 1)
"Published in cooperation with NATO Committee on the Challenges of Modern Society."
"A report of the Air Pollution Pilot Study, NATO Committee on the Challenges to [i.e. of] Modern Society"—
Bibliography: p.
Includes index.
1. Air – Pollution – Mathematical models – Congresses. I. de Wispelaere, C. II. North Atlantic Treaty Organization. Committee on the Challenges of Modern Society. III. Title. IV. Series.

TD881.I59   1980   628.5'3'0724   81-12020

ISBN-13: 978-1-4613-3346-3   e-ISBN-13: 978-1-4613-3344-9
DOI: 10.1007/978-1-4613-3344-9   AACR2

Proceedings of the Eleventh International Technical Meeting on
Air Pollution Modeling and Its Application, held November 24-27, 1980,
in Amsterdam, The Netherlands

© 1981 Plenum Press, New York
Softcover reprint of the hardcover 1st edition 1981
A Division of Plenum Publishing Corporation
233 Spring Street, New York, N. Y. 10013

Foreword

This is the first in a new series of publications arising out
of the work of the Committee on Challenges of Modern Society of the
North Atlantic Treaty Organization.

The CCMS was established in 1969 with a mandate to examine
practical ways of improving the exchange of experience among member nations of the Alliance in the task of creating a better environment for their societies. It was charged with considering "specific problems of the human environment with the deliberate objective of stimulating action by member governments".

It may come as a surprise to some that NATO - generally
thought of as being an organization devoted solely to matters of
defence - should concern itself with the environment at all. But
this is to overlook Article 2 of the North Atlantic Treaty of 1949,
which expressly provides that member countries should contribute
towards the further development of peaceful and friendly international relations by promoting conditions of stability and well-
being. This concern is reflected in many non-military areas, in
addition to the environmental one.

I wish the present volume, which has been edited by the Belgian Prime Minister's Office for Science Policy Programming, every
success.

                              Robert Chabbal
                              Assistant Secretary
                              General for Scientific and
                              Environmental Affairs

Preface

In 1969 the North Atlantic Treaty Organisation established the Committee on the Challenges of Modern Society. Air Pollution was from the start one of the priority problems under study within the framework of the pilot studies undertaken by this Committee. The organisation of a yearly symposium dealing with air pollution modelling and its application is one of the main activities within the pilot study in relation to air pollution.

After being organised for five years by the United States and for five years by the Federal Republic of Germany, Belgium, represented by the Prime Minister's Office for Science Policy Programming, became responsible in 1980 for the organisation of this symposium.

This volume contains the papers presented at the 11th International Technical Meeting on Air Pollution Modeling and its Application held at Amsterdam, The Netherlands from 24th to 27th November 1980. The meeting was jointly organized by the Prime Minister's Office for Science Policy Programming, Belgium and the Ministry of Health and Environmental Protection, The Netherlands. The conference was attended by 139 participants and 45 papers have been presented. The members of the selection committee of the 11th I.T.M. were A. Berger (Chairman, Belgium), W. Klug (Federal Republic of Germany), L.E. Niemeyer (United States of America), L. Santomauro (Italy), J. Tikvart (United States of America), M.L. Williams (United Kingdom), S. Zwerver (The Netherlands), C. De Wispelaere (Coordinator, Belgium).

The main topic of this 11th I.T.M. was Interregional Transport of Air Pollution up to several hundreds of kilometers. On this topic three review papers were presented: one paper dealing with model types and results by W.B. Johnson, SRI International, USA, another paper, "Probability prediction of wet deposition of airborne pollution," by F.B. Smith, Meteorological Office, United Kingdom and finally a paper in relation to the transport of air pollutants over Western Europe, in particular the Netherlands-German case by N. van Egmond, Dutch National Institute of Public Health, The Netherlands.

Other topics of the conference were: meterological parameters for
use in advanced diffusion models, advanced techniques in air pollution modelling to take into account complex terrain, heavy gasses,
light wind conditions, forecasting of pollutant concentration under
episodic conditions and finally regulatory applications.

   On behalf of the selection committee and as organizer and
editor I should like to record my gratitude to all participants who
made the meeting so stimulating and the book possible. Among them
I particularly mention the chairmen and rapporteurs of the different sessions. Thanks also to the local organizing committee, especially Mr. S. Zwerver and Mr. J. van Ham, and Mrs. P.W.A.M. Venis-
pols who was the Conference Secretary. Finally it is a pleasure to
record my thanks to Miss A. Vandeputte for preparing the papers,
and Miss C. Bonnewijn and Miss L. Vandersmissen for typing the
papers.

                              C. De Wispelaere
                              Operational Director
                              of the National R-D
                              Program on Environment

Contents

1 : INTERREGIONAL TRANSPORT OF AIR POLLUTION
(UP TO SEVERAL HUNDREDS OF KILOMETERS)

1. Interregional exchanges of air pollution : Model types and
applications   3
    Warren B. Johnson

2. Trajectories as two-dimensional probability fields   43
    Perry J. Samson and Jennie L. Moody

3. The use of a regional-scale numerical model in addressing
certain key air quality issues anticipated in the 1980s. .   55
    Mei-Kao Liu and Phillip M. Roth

4. Probability prediction of the wet deposition of airborne
pollution.   67
    F.B. Smith

5. The statistics of precipitation scavenging during long
range transport.   99
    B.E.A. Fisher

6. Air pollution transport over Western Europe; exchanges between Germany and the Netherlands. 111
    N.D. Van Egmond, H. Kesseboom

7. A characterization of interregional transport of ozone and
precursors into an urban area 133
    Michael W. Chan, Douglas W. Allard

8. Air quality projections for the Ohio river basin 147
    M.T. Mills, E.Y. Tong, A. Hirata, A. Van Horn and
    L.F. Smith

2 : METEOROLOGICAL PARAMETERS FOR USE IN ADVANCED
     AIR DIFFUSION MODELS

9. Estimation of turbulence velocity scales in the stable and
    the unstable boundary layer for dispersion applications. .   169
         A. Venkatram

10. Meteorological input for a three dimensional medium range
    air quality model.   181
         H. Van Dop, B.J. de Haan and G.J. Cats

11. Interest of an atmospheric meso-scale model for air pollu-
    tion transport studies over medium distances   191
         Christian Blondin

12. A mass consistent wind field model over the Mid-Rhine
    Valley   201
         Patrick Racher, Robert Rosset and Yves Caneill

13. A mesoscale numerical model of atmospheric flow over the
    Alsace plain   213
         A.E. Saab, C. Rolin and V. Villouvier

14. The application of a stochastic wind model to the meteorol-
    ogy of North-West England   223
         J.W. Bacon, B. Henderson-Sellers and A. Henderson-
         Sellers

15. Atmospheric circulation on the regional scale and isentro-
    pic trajectories as support to the long range transport
    (LRT) of air pollution   235
         Sergio Borghi

16. Diffusivity profiles deduced from synoptic data.   251
         G. Schayes and M. Cravatte

17. Wind velocity variances in the atmospheric boundary layer.   267
         R. Berkowicz and L.P. Prahm

18. Estimation of mesoscale and local-scale roughness for at-
    mospheric transport modeling.   279
         Jon Wieringa

19. A convective plume model for PBL dispersion   297
         John D. Reid

20. Numerical modeling of stack plumes within a city environ-
    ment at distances of several kilometres downstream.   309
         Ann Henderson-Sellers, Brian Henderson-Sellers

21. Results of Lidar measurements of atmospheric barrier
    layers. 317
        Josef Giebel

22. Effects of release height on σ_y and σ_z in daytime condi-
    tions 337
        S.R. Hanna

23. Dispersion near to a tall stack 357
        R. Steenkist and F.T.M. Nieuwstadt

24. A statistical approach for estimating atmospheric stabili-
    ty classes from near-ground observations. 369
        S. Cieslik, H. Bultynck and J.G. Kretzschmar

25. Net radiation estimated from standard meteorological data . 385
        L.B. Nielsen, L.P. Prahm, R. Berkowicz and K. Con-
        radsen

26. Estimation of the sensible heat flux from standard meteo-
    rological data for stability calculations during daytime. . 401
        A.A.M. Holtslag, H.A.R. de Bruin and A.P. Van Ulden

3 : ADVANCED TECHNIQUES IN AIR POLLUTION MODELING TO
TAKE INTO ACCOUNT COMPLEX TERRAIN, HEAVY GASES AND
LIGHT WIND CONDITIONS

27. A comparison of finite difference schemes, describing the   411
    two-dimensional advection equation. . . . .
        B.J. De Haan

28. A comparison of some plume dispersion predictions with
    field measurements. 417
        G.A. Davidson and P.R. Slawson

29. Model investigations of spreading of heavy gases released
    from an instantaneous volume source at the ground 433
        A. Lohmeyer, R.N. Meroney, and E.J. Plate

30. Physical modeling of forty cubic meter LNG spills at China
    Lake, California. 449
        Robert N. Meroney and David E. Neff

31. The accidental release of dense flammable and toxic gases
    from pressurized containment - transition from pressure
    driven to gravity driven phase 463
        S.F. Jagger and G.D. Kaiser

32. Entrainment through the top of a heavy gas cloud. 477
        Niels Otto Jensen

33. Physical simulation of dispersion in complex terrain and
    valley drainage flow situations 489
        Robert N. Meroney

34. Analysis and simulation of local circulations and air pol-
    lution over a coastal, complex site 509
        E. Runca, G. Bonino, L. Briatore, G. Elisei and A.
        Longhetto

35. A new gaussian puff algorithm for non-homogeneous, non-
    stationary dispersion in complex terrain. 537
        Paolo Zannetti

36. Impact study in complex terrain 551
        K.E. Grønskei, B. Sivertsen

37. Conversion rate of nitrogen oxides in a polluted atmo-
    sphere 575
        R. Guicherit, K.D. van den Hout, C. Huygen,
        H. van Duuren, F.G. Römer and J.W. Viljeer

        4 : FORECASTING OF POLLUANT CONCENTRATION UNDER
        EPISODIC CONDITIONS

38. Forecasting of fumigation episodes in the Po Valley 595
        P. Bacci, A. Longhetto and D. Anfossi

39. Numerical computation of high air pollution levels. 609
        Cl. Demuth, G. Schayes, P. Hecq and M. Cravatte

40. Application of a photochemical dispersion model to the
    Netherlands and its surroundings 621
        P.J.H. Builtjes, K.D. van den Hout, C. Veldt,
        H.J. Huldy, J.Hulshoff, W. Basting and R. van Aalst

41. An application of a pollution episode predictor derived
    from a K-theory model. 639
        Pietro Melli and Giorgio Fronza

42. Forecasting of pollutant concentration under episode con-
    ditions. 653
        J.M. Fage, G. Gallay and J. Moussafir

5 : REGULATORY APPLICATIONS

43. An application of the empirical kinetic modeling approach
   (EKMA) to the Cologne area 681
         Rainer Stern and Bernhard Scherer

44. Air pollution impact calculated and measured during licensing procedures 703
         Lothar Kropp

45. The regulatory implications of using airport meteorological data instead of onsite data in air quality modeling. . 717
         Patrick T. Brennan and Mark L. Kramer

Participants   729

Authors Index   743

Subject Index   745

1: INTERREGIONAL TRANSPORT OF AIR POLLUTION (UP TO SEVERAL HUNDREDS OF KILOMETERS)

Chairman: W. Klug   Rapporteur: J. Tikvart

INTERREGIONAL EXCHANGES OF AIR POLLUTION:

MODEL TYPES AND APPLICATIONS

Warren B. Johnson

Atmospheric Science Center
SRI International
Menlo Park, California, USA 94025

INTRODUCTION

During the last two decades there has been a simultaneous growth in awareness of the quality of the environment, in the ability to measure its chemical constituents with greater accuracy and precision, and in the widespread use of mathematical models for estimating the transport and diffusion of air pollutants. In recent years, special attention has been given to problems resulting from long-range transport of air pollutants over large regional- or continental-scale areas. Air concentrations and ground deposition of sulfur compounds have been of particular concern.

The OECD-sponsored study of Long-Range Transport of Air Pollutants (LRTAP) conducted in Europe (Ottar, 1978; OECD, 1977) and the Sulfate Regional Experiment (SURE) in the United States (Perhac, 1978), typify recent major environmental assessments of the problems posed by aggregated emissions over large areas. In many of these studies, theoretical models have been developed and applied to simulate the transport, diffusion, transformation, and removal of SO_x released into the atmosphere from the multitude of sources located within a given region.

The development of suitable models for this application is a difficult task, considering the very large geographical areas involved. This is because of the need to strike an appropriate compromise between two equally desirable but conflicting model characteristics--accuracy and practicality. The achievement of additional accuracy in a model requires additional sophistication in the physical formulations, as well as additional detail in the controlling input variables. These in turn require additional

computer time for running the model and additional effort for preparing the input data, which together result in a less economical
and thus less practical model.

Air quality modeling specialists have made substantial progress
in recent years in developing new techniques to meet the challenges
of simulating long-range pollutant transport and deposition. A
number of excellent comprehensive reviews of the modeling work in
this field have been prepared recently, such as those of Eliassen
(1980), Bass (1980), Smith and Hunt (1979), Niemann et al. (1980),
Hosker (1980), MacCracken (1979), Fisher (1978), Demerjian
(1980).

In view of the above, it was not intended that this paper be
another comprehensive review of long-range transport (LRT) modeling.
Rather, the purpose of this paper is to briefly review some of the
general characteristics and problems associated with air quality
models on this scale, particularly interregional-type models (to be
defined below), and to illustrate the various kinds of results that
can be obtained by application of a sample model of this type.

CAPABILITIES AND CHARACTERISTICS OF INTERREGIONAL MODELS

The focus of this paper is on interregional air quality simulation models, which are defined as those long-range transport/
deposition models having the following characteristics:

- Multiple-source
- Scale: hundred to thousands of kilometers (mov-
  ing pollutants undergo at least one diurnal cycle)
- Averaging time: hourly to yearly
- Can determine contributions from individual
  sources or source areas.

The most important factor in this definition of interregional models
is the last one above, namely the ability of interregional models
to quantitatively assess the pollutant exchanges between individual
source-receptor pairs. Therefore, interregional models should
consider all (or as many as possible) of the pollutant emissions
from both area and point sources in the entire geographical area
of interest.

The capabilities and applications of models in this general
category have been summarized well by Venkatram (1980):

- Determine contributions of various sources to
  receptors of interest

- Estimate consequences of projected emissions changes

- Fill in gaps between observations

- Help plan field studies

  -- which variables to measure?
  -- where to site stations?

- Help interpret data

  -- e.g., infer transformation or deposition rates
     (this will be discussed further in a later
     section)

Note that most of these useful tasks can only be accomplished through the use of models--measurements by themselves are not adequate to provide the necessary information.

MAJOR TYPES OF LRT/INTERREGIONAL MODELS

The following general features of LRT/interregional models are to be considered highly desirable:

- Easy to understand and use

- Suitable for both short and long terms

- Spatial resolution 100 km or better

- Require only readily available input data

- Well evaluated against measurements

- Physically realistic and accurate

- Well documented

- Easily adapted for running on typical computers

- Efficient in computer running time

- Treat wet and dry deposition separately for different chemical species.

Unfortunately, some of these features are mutually contradictory, and thus the design of every model involves a number of compromises. Modeling skill becomes most important here, in determining which portions or features of a model are least important, and thus can be compromised with the least degradation in the usefulness of the model for its intended application.

As listed in Table 1, the LRT/interregional models that have been developed to date can be classed into three main types:

Eulerian grid, statistical trajectory, and Lagrangian trajectory.
Each of these model types and their advantages and disadvantages
will be briefly discussed next.

Table 1. Major Types of LRT/Interregional Models, and Some of the
          Investigators Who Have Developed Models of These Types

EULERIAN GRID	STATISTICAL TRAJECTORY	LAGRANGIAN TRAJECTORY
A. MOMENT METHOD Carmichael and Peters (1979) Egan et al. (1976) Pedersen and Prahm (1974) Nordø (1974, 1976) Lavery et al. (1980)	Rodhe (1972, 1974) Bolin and Persson (1975) McMahon et al. (1976) Fisher (1975, 1978) Scriven and Fisher (1975) Shannon (1979) Shieh (1977)	A. RECEPTOR-ORIENTED Eliassen and Saltbones (1975) Eliassen (1978) Ottar (1979) Szepesi (1978) Olson et al. (1979)
B. PSEUDOSPECTRAL METHOD Fox and Orszay (1973) Christensen and Prahm (1976) Prahm and Christensen (1977) Berkowicz and Prahm (1978)	Mills and Hirata (1978) Venkatrom et al. (1980) Fay and Rosenzweig (1980)	B. SOURCE-ORIENTED Wendell et al. (1976) Johnson et al. (1978) Mancuso et al. (1979) Bhumralkar et al. (1980) Heffter (1980) Powell et al. (1979) Maul (1977)
C. PIC METHOD Sklarew et al. (1971) Lange (1978)		C. HYBRID Draxler (1977, 1979) Meyers et al. (1979)

Eulerian Grid Models

     This type of model breaks up the geographical area or volume
of interest into a two- or three- dimensional array of grid cells.
The advection, diffusion, transformation, and removal (deposition)
of pollutant emissions in each grid cell is simulated by a set of
mathematical expressions, generally involving the K-theory assumption (assumption that the flux of a scalar is proportional to its
gradient). Some type of finite-differencing technique is usually
employed in the numerical solution of these equations.

     The major advantages and disadvantages of the Eulerian grid
approach are summarized below:

     Advantages:

          • Capable of sophisticated 3-D physical
            treatments

          • Can handle nonlinear chemistry

          • Data input is simplified on Eulerian grid.

Disadvantages:

- Require large amounts of computer time and storage

- Require large amounts of input data

- Cannot determine contribution from individual sources

- Artificial (computational) dispersion can be significant.

Although in principle this type of model is capable of incorporating more physical realism than some of the other model types, this advantage is largely offset in many applications by the high computational costs involved in its use.

A further problem is the artificial diffusion effect inherent in conventional finite-differencing techniques. This has led to the development of various modeling schemes to minimize this effect, such as the moment method, the pseudospectral method, and the particle-in-cell (PIC) method, as indicated in Table 1.

Statistical Trajectory Models

Although there are many different varieties of statistical trajectory models, these models have one or more of the following characteristic features that distinguish them as a type:

- Using climatological wind data, large numbers of air trajectories are calculated, either forward in time from source areas or backward in time from receptor areas, and the results are statistically analyzed to give average pollutant contributions, horizontal diffusion, etc.

- Meteorological variables are frequently averaged over long periods of time before application to calculate such parameters as concentrations and depositions.

Such models have the following advantages and disadvantages:

Advantages:

- Computational requirements are modest

- Well suited for repeated runs using alternative emissions scenarios

- No computational dispersion

- Can determine contributions from individual
  sources

- Pollutant mass balances can be determined.

Disadvantages:

- Most types not adaptable to short averaging
  times

- Dispersion and other processes are
  usually highly parameterized

- Dependence between meteorological variables
  (e.g., wind and precipitation) ignored in
  some types.

Statistical trajectory models enjoy a major advantage in having low
computational costs, but this is sometimes obtained at the expense
of physical realism.

Lagrangian Trajectory Models

    A characteristic feature of these models is that calculations
of pollutant diffusion, transformation, and removal are performed
in a moving frame of reference tied to each of a number of air
"parcels" as they are transported around the geographical region
of interest in accordance with an observed or calculated wind field.

    As indicated in Table 1, Lagrangian trajectory models can be
either receptor-oriented, in which trajectories are calculated
backward in time from the arrival of an air parcel at a receptor of
interest, or source-oriented, in which trajectories are calculated
forward in time from the release of a pollutant-containing air parcel from an emissions source. There are also a few hybrid approaches,
in which a Lagrangian trajectory technique is used to simulate horizontal transport and diffusion, and an Eulerian grid technique is
used to simulate vertical diffusion.

    Most source-oriented Lagrangian trajectory models simulate
continuous pollutant emissions by means of discrete increments or
"puffs" of emission occurring at set time intervals, usually between
1 and 24 hrs, depending upon the designed averaging time of the
model outputs. Such models simulate the movement and behavior of a
pollutant plume from a continuous source (as shown in the upper
left portion of Figure 1) by one of the three approaches illustrated
in the other portions of the figure.

    The "segmented plume" approach involves computational diffi-
culties when variable wind fields cause convoluted plume geometries.
The "puff superposition" approach avoids these problems, but can

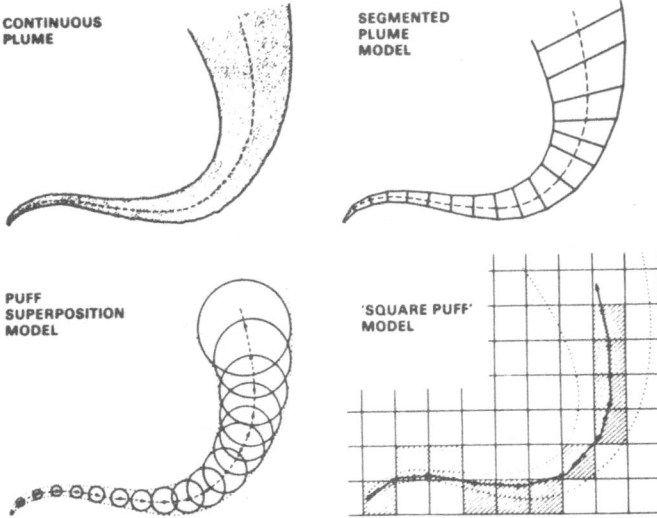

Fig. 1.   Various trajectory modeling approaches (from Bass, 1980).

lead to errors when the product of wind speed and computational time-step duration is sufficiently high to cause puffs to travel long distances (relative to their size) between time steps. Under these circumstances, which are more likely to occur during the initial portion of its travel, a puff can skip over and make no (calculated) contribution of pollutant concentration or deposition to certain geographical areas, as illustrated in Figure 1. The "square puff" approach simply keeps track of the receptor cell containing the puff center at each time step, and assigns the entire contribution of the puff to this cell. This approach tends to ignore the effects of puff spread caused by horizontal diffusion.

Some of the advantages and disadvantages of Lagrangian trajectory models are listed below:

Advantages:

- Can determine contributions from individual sources

- Relatively inexpensive to run on a computer

- Easy to keep track of pollutant mass balances

- Realistic treatment of wet deposition

- No computational dispersion

- Individual sources or receptors can be run separately.

Disadvantages:

- Extention to 3-D not straightforward
- Nonlinear chemistry difficult to incorporate
- Horizontal and vertical diffusion highly parameterized
- Errors can be introduced in interpolating results onto Eulerian grid.

The two most important features of this type of model are its capability for calculating detailed source-receptor contributions, and its computational efficiency. To achieve the latter, most models of this type are more highly parameterized, and thus potentially less physically realistic, than some Eulerian grid approaches.

A DETAILED LOOK AT ONE TYPE OF INTERREGIONAL MODEL--
THE EURMAP/ENAMAP APPROACH

To provide a fuller understanding of the design and application of interregional models, this section will examine one particular model--the EURMAP/ENAMAP approach--on the basis that it can be considered as a typical example of such models. The basic structure of this model will be discussed, and examples of various results (outputs) from its use will be presented for illustrative purposes.

EURMAP (European Regional Model of Air Pollution)

There are two versions of EURMAP--a long-term model (EURMAP-1), which calculates monthly, seasonal, and annual values, and a short-term model (EURMAP-2), which calculates 24-hourly values. The development and application of these models has been described in several publications: Johnson et al. (1978), Mancuso et al. (1978), Bhumralkar et al. (1979), Bhumralkar et al. (1981b).

EURMAP-1 is a practical air pollution model designed to have minimum computational requirements for use in making long-term calculations economically, while at the same time offering acceptable realism in simulating the most important processes involved in the transboundary air-pollution problem. The EURMAP-1 model can be used to calculate monthly, seasonal, and annual SO_2 and $SO_4^=$ air concentrations; SO_2 and $SO_4^=$ dry and wet deposition patterns; and interregional exchanges resulting from the SO_2 and $SO_4^=$ emissions over central and western Europe. The model uses long sequences of historical meteorological data as input, retaining all the original temporal and spatial detail inherent in the data.

The basic structure of the EURMAP-1 model is illustrated in Figure 2. Discrete puffs of SO_2 and $SO_4^=$ are assumed to be emitted at equal time increments from cells of an emission grid. This type

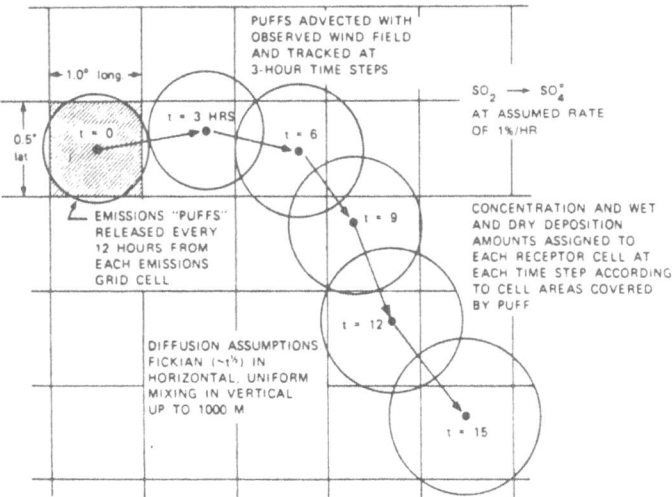

Fig. 2.   Emissions puff advection and diffusion scheme used in EURMAP-1.

of treatment provides a realistic representation of area sources. For a point source, it assumes that the pollutant expands initially to fill uniformly the volume of the cell from the point within the cell where the source is actually located. For each of the emission cells, the average annual or seasonal emissions are divided into discrete emission puffs released at 12-hour intervals and tracked at 3-hour time steps, until either they move outside the region of analysis or their concentrations drop to an insignificant level. The individual puffs are transported according to a transport wind field that is derived objectively from the available upper-air wind observations. Meteorological data in the form of wind and precipitation values from some 45 upper-air and 535 surface stations are input at 6-hourly intervals for use in the calculations of puff transport and wet deposition.

Since diffusion on the regional scale is not as important as the transport and removal processes, very simple treatments of vertical and horizontal diffusion have been used. Upon release, each puff is assumed to undergo instantaneous vertical diffusion to give uniform concentration in the layer between the surface and the top of the mixing height. Horizontal diffusion is treated by allowing the area of the puff to increase linearly with time on the basis of Fickian diffusion, assuming a horizontal eddy diffusivity of 36 km^2h^{-1}. During the transport of the puff, the model assumes that the pollutant concentration within a puff is always uniform.

1 Austria
2 Belgium/Luxembourg
3 Czechoslovakia
4 Denmark
5 East Germany (GDR)
6 France
7 Hungary
8 Holland
9 N. Italy
10 N. Spain
11 Poland
12 Romania
13 S. Sweden
14 S. Norway
15 Switzerland
16 U. K. and Ireland
17 West USSR
18 West Germany (FRG)
19 Yugoslavia

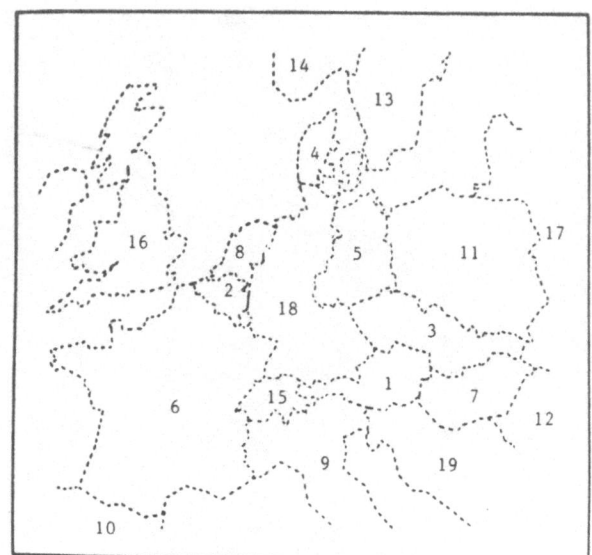

Fig. 3.   Geographical domain to which the EURMAP models have been
          applied.

        As illustrated in Figure 3, the model covers all or portions
of 19 countries in central and western Europe, a geographical area
2100 km by 2250 km in size. Available emissions data is partitioned
into a 32 x 36 array of grid cells covering this area.

        Figures 4 and 5 show how the puff advection/deposition scheme
operates. At each 3-hour time step a new puff position is computed
and a fraction of the SO_2 is converted to $SO_4^=$ at the specified
transformation rate. The amounts of SO_2 and $SO_4^=$ that are removed
from a puff during each time step are dependent on the specified
dry and wet deposition coefficients that are used; these amounts
are deposited within the appropriate cells of the receptor grid.
A wet deposition coefficient is used that depends upon precipitation
rate.

        The short-term model, EURMAP-2, uses the same general design
as EURMAP-1, but has a number of important differences, as summarized
in Tables 2 and 3. For comparison purposes, corresponding information for the NILU (Norwegian Institute for Air Research) trajectory
model (Eliassen et al., 1976) is also included in the tables. Most
of the differences between EURMAP-2 and EURMAP-1 are necessary in
order to incorporate more detail into the emissions and meteorological simulations to be consistent with the much shorter (24-hour)
averaging time of EURMAP-2. In particular, atmospheric boundary-
layer processes have been treated in a more realistic manner than
in EURMAP-1.

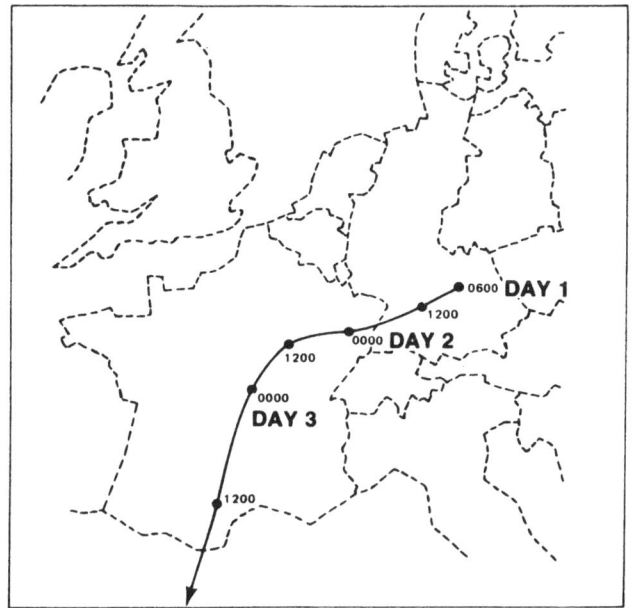

Fig. 4.   Example of puff advection with time.

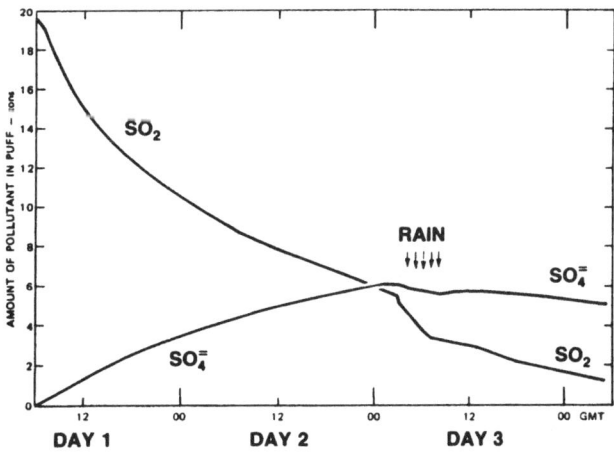

Fig. 5.   Daily variation in the amount of pollutant in a puff.

Table 2.   Basic Differences Between the EURMAP-1, EURMAP-2,
           and NILU Models

Model	Emission Puff Frequency (hours)	Puff-Tracking Frequency (hours)	Minimum Averaging Time for Concentrations and Deposition
EURMAP-1	12	3	1 month
EURMAP-2	6	1	1 day
NILU	6	2	1 day

EURMAP-1 and EURMAP-2 have both been extensively evaluated
(through comparison with available measurements) and applied to the
assessment of short-term and long-term pollutant concentrations,
depositions, and exchanges in the European domain illustrated in
Figure 3. The results from both models take the following forms:

- Graphical displays of the distribution of SO_2
  and $SO_4^=$ concentrations

- Graphical displays of the distributions of SO_2
  and $SO_4^=$ wet and dry depositions

- Tabulated results showing the interregional
  exchanges of sulfur pollution between individual
  source and receptor regions.

Examples are presented in Figures 6 and 7 and Table 4, respectively, of each of the above types of products resulting from the
application of EURMAP-2 to a selected two-day pollution episode
(25-26 August 1974). The meteorological data associated with this
episode and used as input to the model calculations are depicted
in Figure 8.

ENAMAP (Eastern North American Model of Air Pollution)

ENAMAP is a closely related version of EURMAP-1 that has been
adapted for application to the geographical region covering the
eastern United States and southeastern Canada, as illustrated in
Figure 9. This work has been conducted and described by
Bhumralkar et al. (1980a, 1980b, 1981a). Table 5 lists the
values of model elements used in ENAMAP; comparison with Table 3
reveals only a few minor differences.

Examples of ENAMAP monthly and annual results corresponding to
the same types of model outputs presented earlier for EURMAP-2 are
included as Figures 10 and 11 and Table 6. Some results from the
Canadian LRTAP model (Olson et al., 1979) prepared by Olson (1980)

Table 3. Differences in the Treatment of Some Elements Between the
EURMAP-1, EURMAP-2, and NILU Models

Element	Model		
	EURMAP-1	EURMAP-2	NILU
Emission data	For 13 countries	For 19 countries; includes emissions from all countries within grid	For polar stereographic grid covering Europe (emission from southeast section excluded)
	SO_2 amount constant and based on annual total	SO_2 amount consists of constant and variable components	SO_2 amount constant and based on annual total
		Variable component is based on degree days	
Mixing height (h)	Constant at 1 km	Varies diurnally	Constant at 1 km
Transport wind (v_T)	$v_T = 0.75\ v_{850}$ $\theta = \theta_{850} - 15°$	Based on surface and 850-mb winds and a power law of the form $$\frac{v_2}{v_1} = \left(\frac{z_2}{z_1}\right)^p$$	$v_T = v_{850}$ $\theta = \theta_{850}$
Horizontal diffusion	Fickian diffusion	Based on horizontal deformation of wind field	None stated
Vertical diffusion	Immediate uniform mixing in vertical Limited up to mixing height	Not instantaneous Puff is allowed to penetrate beyond mixing height	Immediate uniform mixing in vertical
Decay rates (% per hour)	Dry deposition 2.9 for SO_2 0.7 for $SO_4^=$ Wet deposition[*] 21.6 · R for SO_2 7.0 · R for $SO_4^=$ Transformation 1.0 for $SO_2 \to SO_4^=$	Dry deposition[†] $2.9/Z_p$ for SO_2 $0.7/Z_p$ for $SO_4^=$ Wet deposition[‡] 21.6 · R for SO_2 7.0 · R for $SO_4^=$ Transformation[‡] 1.0 for $SO_2 \to SO_4^=$	Decay rate 5.4 for SO_2 1.44 for $SO_4^=$ Transformation 0.72 for $SO_2 \to SO_4^=$

[*]R is rainfall rate in mm/hr

[†]Z_p is the height of the top of the puff in km

[‡]Same as EURMAP-1

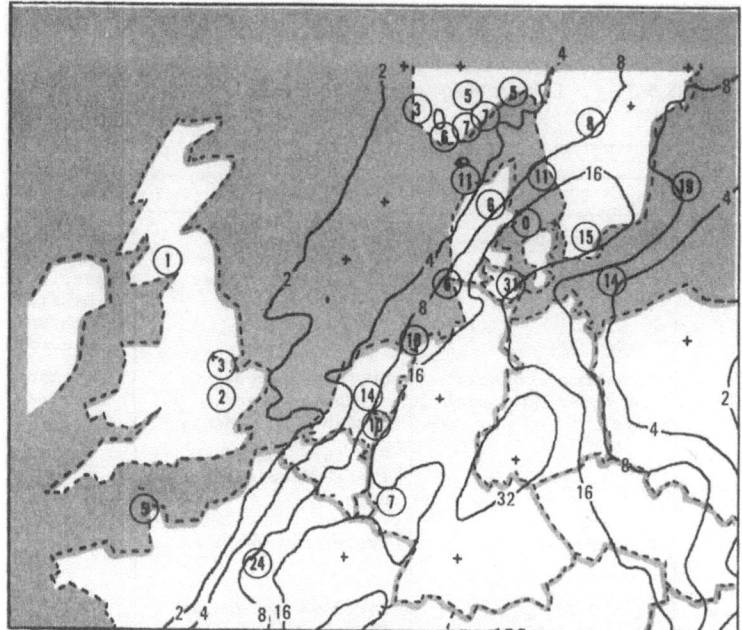

(c) SO$_4^=$ CONCENTRATION (μg/m^3), 25 AUG. 1974

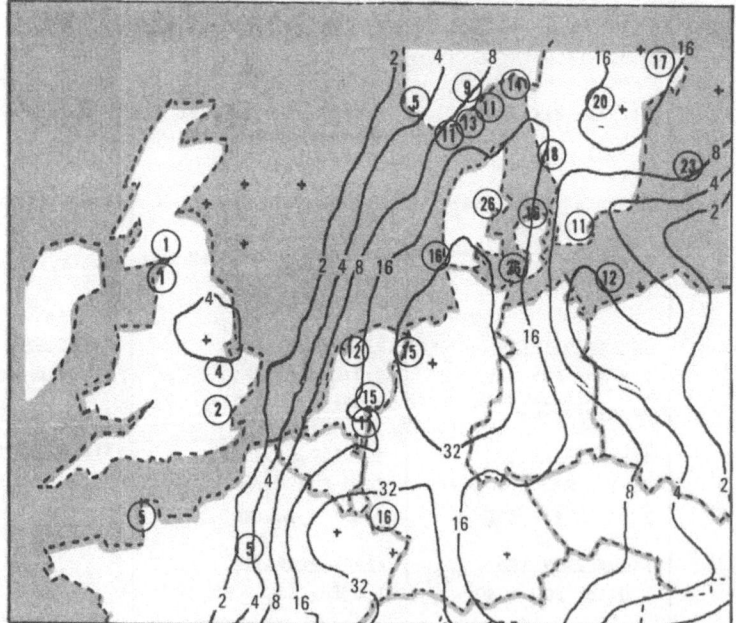

(d) SO$_4^=$ CONCENTRATION (μg/m^3), 26 AUG. 1974

Fig. 6.   Calculated and measured concentrations of SO$_4^=$ (μg/m^3)
           for 25 August (top) and 26 August (bottom) 1974.
           (Circles indicate measured values.)

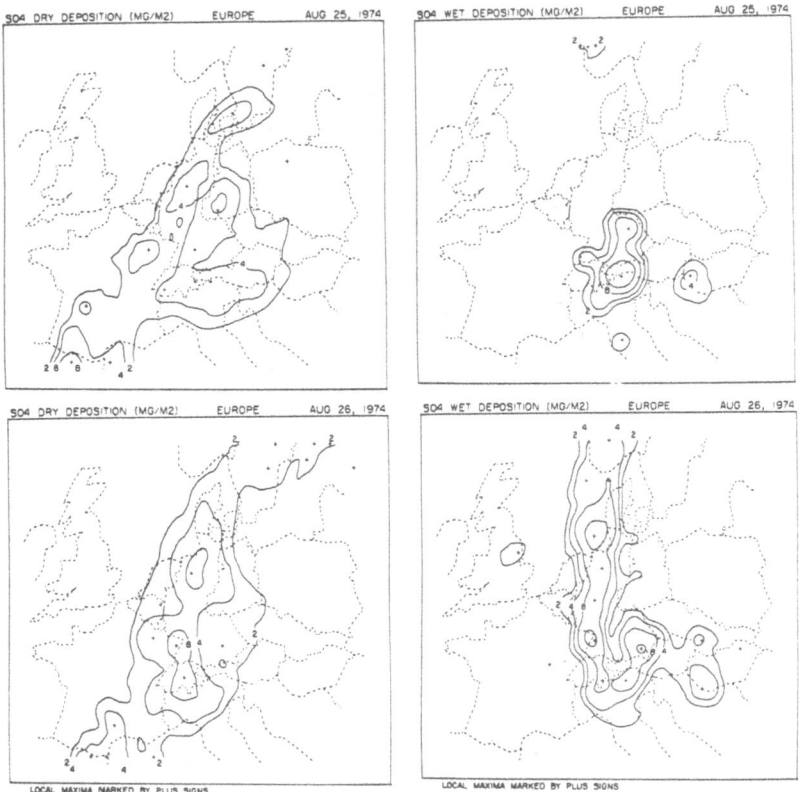

Fig. 7.   Calculated $SO_4^=$ dry and wet deposition from emissions in central and western Europe for 25 and 26 August 1974.

are also included in Figure 10 for comparison purposes. In general, the agreement between the two models is relatively good. However, the measurements indicate that the ENAMAP model apparently under-predicts $SO_4^=$ concentrations in the eastern Great Lakes area.

PROBLEMS AND PITFALLS IN INTERREGIONAL MODELING

This section will discuss some of the more significant diffi-culties which model developers face in their work. It is important that persons who <u>apply</u> models also understand these difficulties, since many of them result in model limitations or constraints which in turn affect the operational use of models.

Initialization

Before model results for a given time period can be considered valid, model calculations must be performed for a sufficient length

Table 4.   Contributions to Total Sulfur Depositions (Metric Tons) on 25 and 26 August 1974, Based on Emissions from Each of the 19 Countries Used in the EURMAP-2 Model Calculations

TOTAL CONTRIBUTIONS TO S DEPOSITIONS WITHIN RECEPTOR COUNTRIES

AUG 25, 1974

EMITTOR COUNTRY	1	2	3	4	5	6	7	8	9	10	11	12	13	14	15	16	17	18	19
1 FRG	2238	104	447	1	3	44	38	25	0	6	0	21	59	0	7	0	66	0	0
2 GDR	1000	1857	132	36	113	2	0	0	0	0	0	22	70	0	6	0	12	0	0
3 FRANCE	44	0	1455	0	0	1	30	115	0	0	0	0	12	0	0	0	121	0	0
4 POLAND	140	268	14	1083	521	0	0	0	0	0	0	302	29	22	198	23	1	0	0
5 CZECHOSLOVAKIA	647	133	16	35	864	0	0	0	0	59	0	531	117	104	445	73	1	0	0
6 DENMARK	104	6	41	40	0	62	0	0	0	4	0	0	0	0	0	0	0	28	0
7 HOLLAND	220	0	114	0	0	41	130	0	0	0	0	0	0	0	0	0	18	0	0
8 BELGIUM	47	0	169	0	0	9	135	211	0	0	0	0	0	0	0	0	35	7	0
9 U.K. + IRELAND	0	0	0	0	0	206	72	7	1020	328	303	0	0	0	0	0	33	0	0
10 S. SWEDEN	0	0	0	0	0	0	0	0	0	7	15	0	0	0	0	0	0	0	0
11 S. NORWAY	28	0	0	7	0	0	0	0	0	98	0	0	0	0	0	18	0	27	0
12 AUSTRIA	1	0	14	0	0	0	0	0	0	0	0	185	11	3	167	0	0	0	4
13 SWITZERLAND	0	0	0	0	20	0	0	0	0	0	0	1	32	0	0	0	0	0	0
14 HUNGARY	0	0	56	0	0	0	0	0	0	0	0	106	15	594	111	241	0	0	4
15 N. ITALY	0	0	0	0	0	0	0	0	0	0	0	2	0	40	667	802	0	0	6
16 YUGOSLAVIA	0	0	1	0	0	0	0	0	0	0	0	2	0	0	135	0	0	0	7
17 N. SPAIN	0	2	0	32	25	0	0	0	0	0	0	0	0	68	10	4	20	0	0
18 W. RUSSIA	0	0	0	0	0	0	0	0	0	0	0	12	0	137	2	132	0	151	0
19 ROMANIA	0	0	0	0	0	0	0	0	0	0	0	6	0	0	0	0	0	0	63
TOTAL (TONS S)	4470	2373	2467	1234	1547	366	405	364	1020	502	318	1189	345	967	1752	1292	307	223	100

AUG 26, 1974

EMITTOR COUNTRY	1	2	3	4	5	6	7	8	9	10	11	12	13	14	15	16	17	18	19
1 FRG	2533	47	372	0	4	260	400	69	0	72	62	8	27	0	0	0	16	0	0
2 GDR	1606	1325	177	13	9	351	3	0	0	28	11	0	26	0	1	0	6	0	0
3 FRANCE	171	0	1342	0	399	5	138	207	0	0	3	100	11	7	2	0	122	0	0
4 POLAND	596	524	20	1087	75	0	0	0	0	12	1	100	112	71	114	1	0	0	0
5 CZECHOSLOVAKIA	871	392	66	62	1158	0	0	0	0	68	0	334	168	0	240	0	1	0	0
6 DENMARK	0	1	0	13	0	65	135	7	0	37	28	0	0	0	0	0	0	0	0
7 HOLLAND	37	0	21	0	0	38	135	175	0	6	36	0	0	0	0	0	7	0	0
8 BELGIUM	126	0	58	0	0	31	19	22	0	0	14	0	0	0	0	0	4	0	0
9 U.K. + IRELAND	7	0	102	0	0	67	0	0	2341	207	428	0	0	0	0	0	24	0	0
10 S. SWEDEN	0	0	0	5	0	0	0	0	0	6	7	0	0	0	0	0	1	1	0
11 S. NORWAY	172	0	0	0	0	0	0	0	0	107	7	0	37	3	88	10	0	0	0
12 AUSTRIA	7	0	11	0	15	0	0	0	0	0	0	310	51	0	4	0	0	0	0
13 SWITZERLAND	10	0	18	0	0	0	0	0	0	0	0	3	0	797	267	65	0	2	0
14 HUNGARY	0	0	4	0	129	0	0	0	0	0	0	273	47	0	1063	978	0	0	11
15 N. ITALY	0	0	108	0	0	0	0	0	0	0	0	10	0	124	200	0	26	0	0
16 YUGOSLAVIA	0	0	0	0	0	0	0	0	0	0	0	0	0	0	16	0	0	164	14
17 N. SPAIN	8	2	2	63	94	0	-1	0	0	0	0	43	0	74	61	107	0	1	14
18 W. RUSSIA	0	0	0	0	0	0	0	0	0	0	0	12	0	168	0	0	0	164	0
19 ROMANIA	0	0	0	0	0	0	0	0	0	0	0	0	0	0	0	0	0	0	176
TOTAL (TONS S)	6143	2292	2304	1244	1814	913	971	481	2341	538	604	1098	481	1244	2078	1182	207	212	210

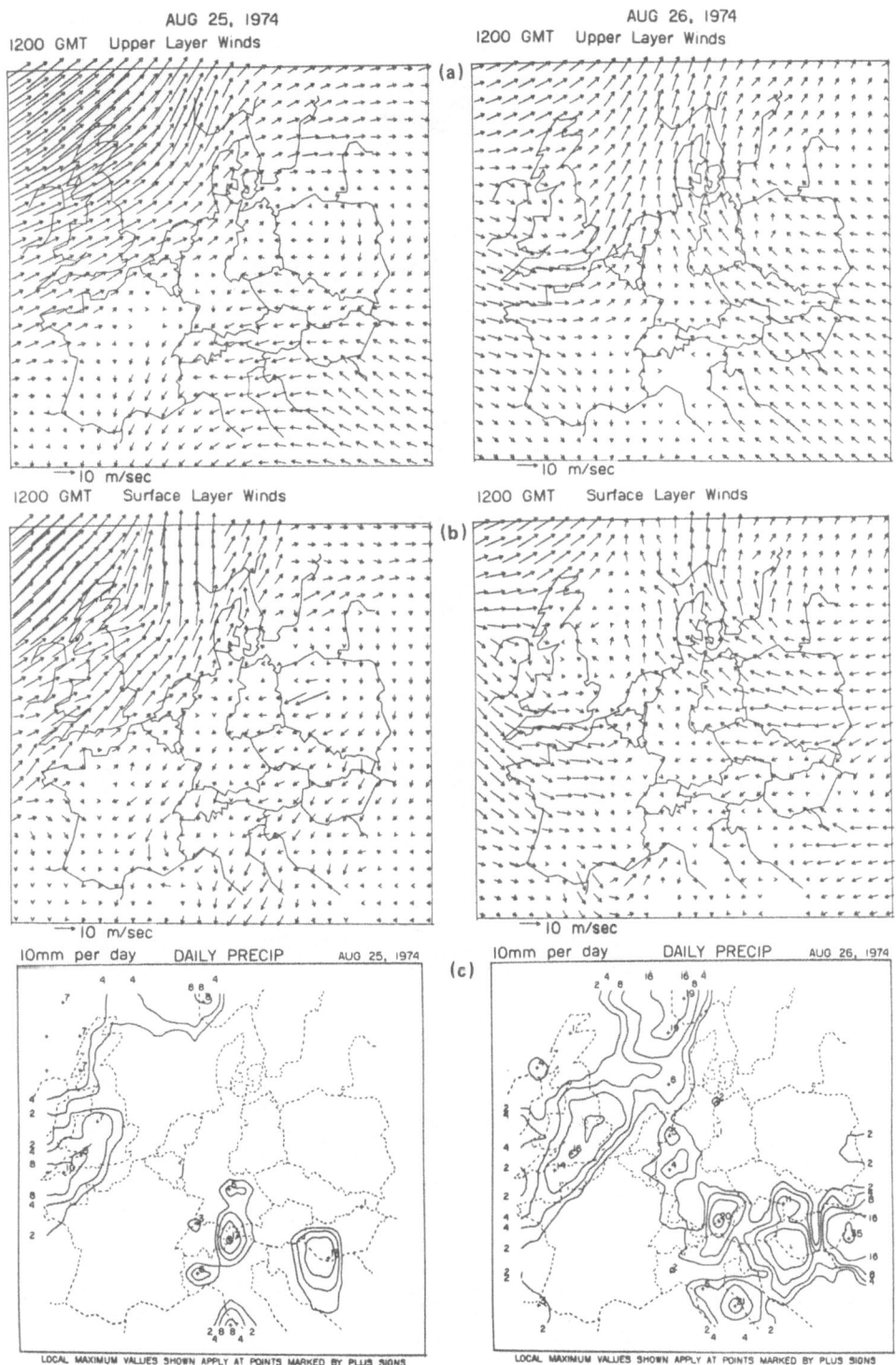

Fig. 8.   Sample weather data for 25 and 26 August 1974.

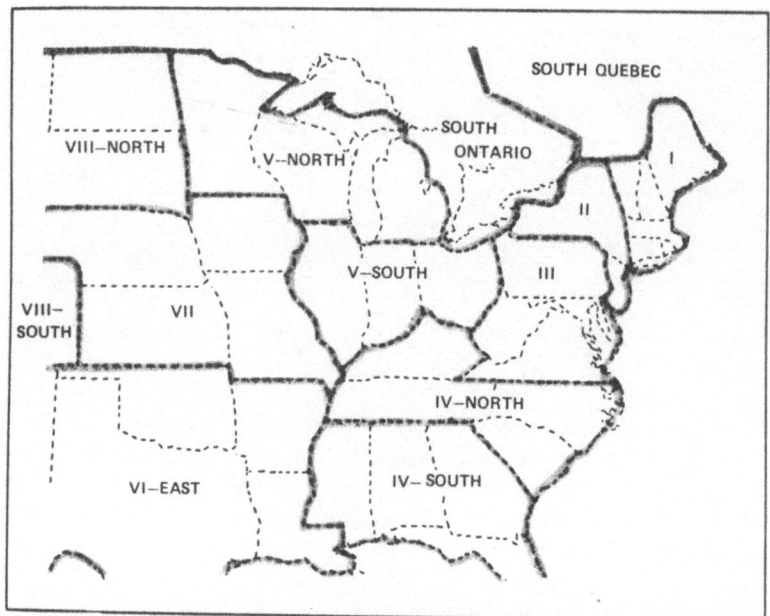

(a) EPA REGIONS USED IN THIS STUDY

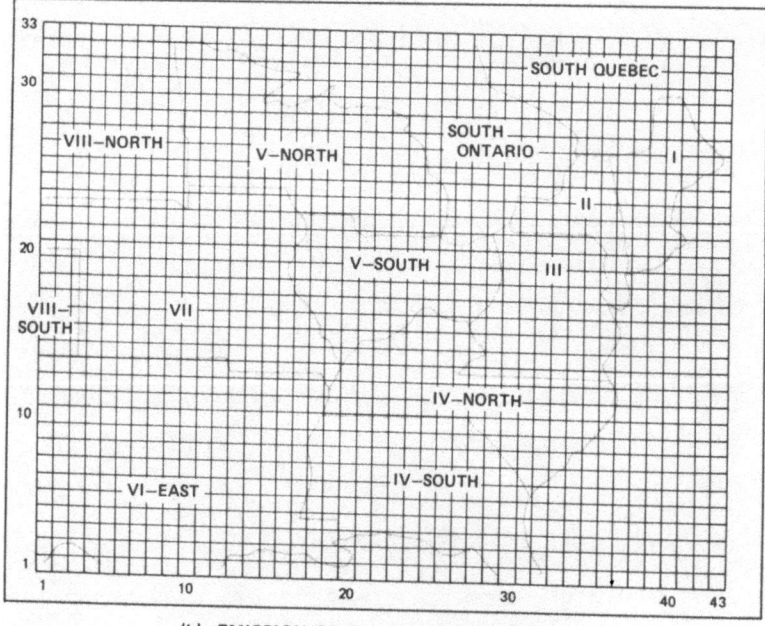

(b) EMISSION GRID AND MODEL DOMAIN

Fig. 9.   Eastern North American domain and EPA regions used in
          the ENAMAP modeling study.

Table 5. Element Values Used in the ENAMAP-1 Application to
          Eastern North America

Element	Values
Emission rate	Data provided by season
Transport windspeed (V) (ms^{-1}) and direction (θ)	Derived by integrating winds over boundary layer
Mixing height (km) $h = h_o + \zeta \Lambda^*$	$h_o = 1.3$ $\zeta = -0.15$
SO_2 deposition rates (hr^{-1}) Dry Wet	0.037 $0.28R^\dagger$
$SO_4^=$ deposition rates (hr^{-1}) Dry Wet	0.007 $0.07R^\dagger$
$SO_2/SO_4^=$ transformation rate (hr^{-1})	0.01

$^*\Lambda = +1$ in winter, -1 in summer, and 0 in spring and fall.

†R is the precipitation rate in mm/hr^{-1}.

of time--prior to the start of the period of interest--so that the
model has reached a steady-state condition by this time. As indicated in Table 7, achieving this condition for models covering large
geographical areas (EURMAP-2 in this case) requires an initialization period of approximately five days. The computational costs
for the initialization period can be a significant portion of the
total computational costs for applying an interregional model, if
the application is to simulate pollutant episodes of short duration.

Boundary Conditions

It is clear that conditions outside the geographical modeling
domain of interest can have an effect inside this domain. The air
that enters a modeling domain has already been polluted to some
undetermined degree by emissions outside the modeling grid. This
will result in model-calculated values of concentrations and depositions that are found to be too low when compared to measurements.
An example of this can be seen in some of the model results presented

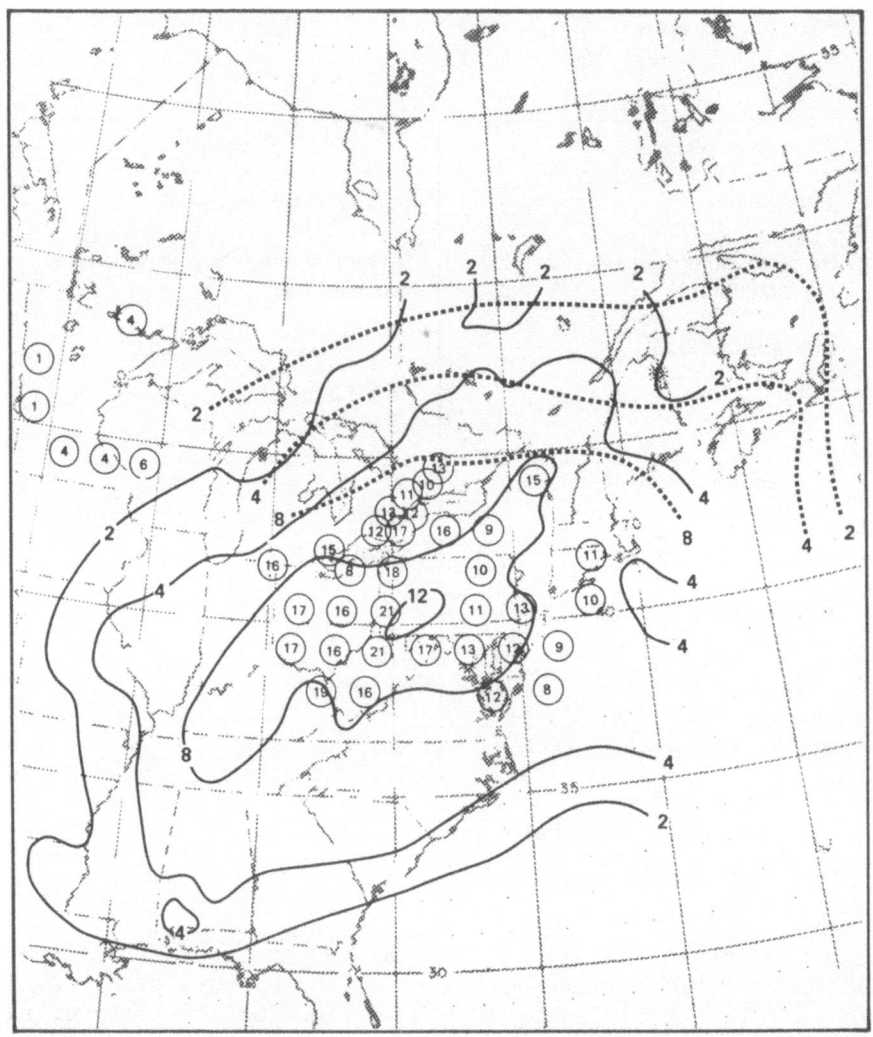

Fig. 10.   Calculated and measured $SO_4^=$ air concentrations for July
           1978.

           Solid lines: SRI ENAMAP-1 model results
                          (Bhumralkar et al., 1980)
           Dashed lines: Canadian LRTAP project model results
                          (Olson, 1980)
           Circled values: Available measurements
           (All values are in units of 10^{-6} g/m^3)

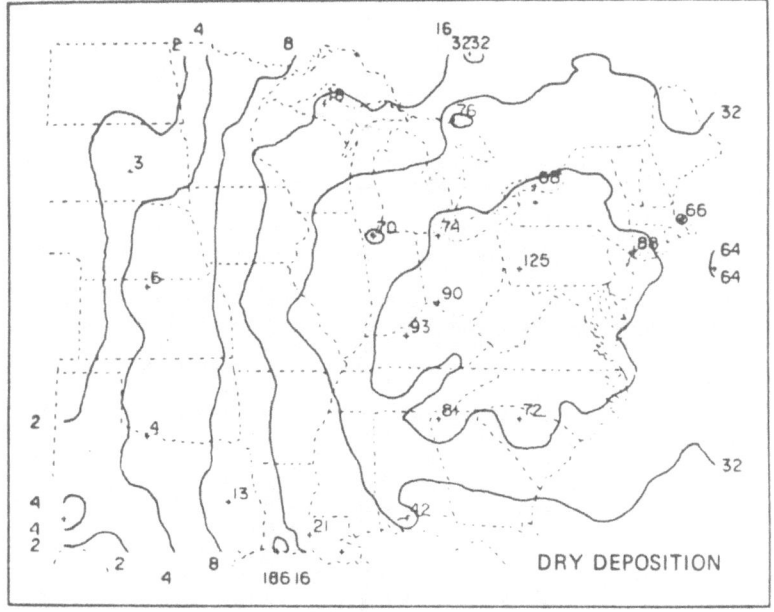

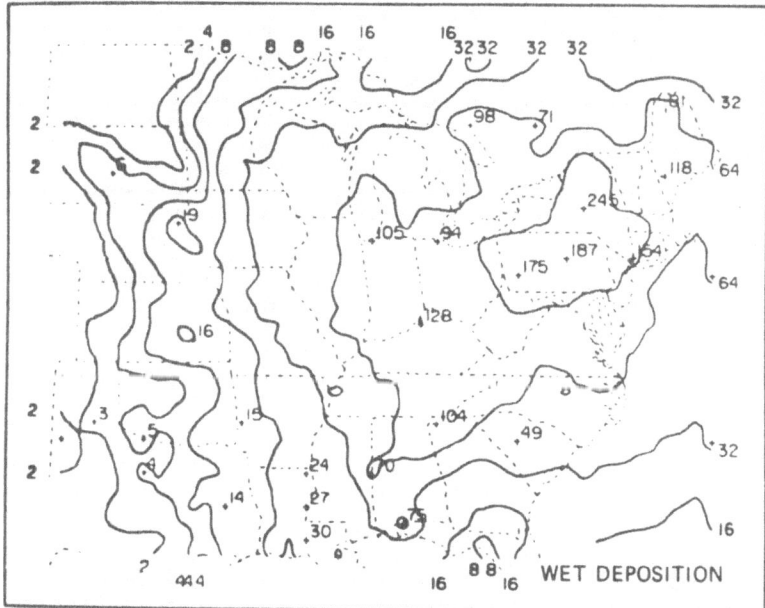

LOCAL MAXIMUM VALUES SHOWN APPLY AT POINTS MARKED BY PLUS SIGNS

Fig. 11. Calculated annual dry and wet depositions of $SO_4^=$
          (10 mg/m^2) for 1977.

Table 6.　Annual Interregional Exchanges of Sulfur Deposition for 1977 as Calculated by the ENAMAP-1 Model

TOTAL CONTRIBUTIONS TO S DEPOSITIONS WITHIN RECEPTOR REGIONS (kilotons)

EMITTER REGION	1	2	3	4	5	6	7	8	9	10	11	12	13
1 VIII-NORTH	10.	1.	0.	2.	0.	0.	0.	0.	0.	0.	0.	0.	0.
2 V-NORTH	3.	655.	290.	46.	0.	3.	229.	6.	24.	78.	50.	18.	23.
3 S. ONTARIO	0.	66.	820.	2.	0.	1.	49.	2.	7.	74.	87.	40.	87.
4 VII	1.	43.	10.	367.	0.	26.	137.	22.	41.	12.	3.	2.	2.
5 VIII-SOUTH	0.	0.	0.	0.	1.	0.	0.	0.	0.	0.	0.	0.	0.
6 VI-EAST	1.	4.	1.	40.	0.	401.	7.	35.	6.	1.	0.	0.	0.
7 V-SOUTH	2.	186.	145.	135.	0.	14.	1566.	59.	425.	520.	92.	30.	26.
8 IV-SOUTH	0.	8.	7.	16.	0.	44.	31.	949.	279.	25.	2.	1.	2.
9 IV-NORTH	0.	19.	24.	11.	0.	13.	221.	108.	929.	159.	15.	7.	6.
10 III	0.	11.	57.	3.	0.	0.	178.	14.	141.	1363.	179.	56.	21.
11 II	0.	1.	53.	0.	0.	0.	1.	1.	4.	37.	204.	65.	14.
12 I	0.	0.	1.	0.	0.	0.	1.	0.	2.	9.	91.	207.	22.
13 S. QUEBEC	0.	2.	105.	0.	0.	0.	1.	0.	0.	2.	8.	41.	204.
TOTAL (KTON S)	18.	997.	1514.	621.	1.	503.	2422.	1197.	1856.	2280.	732.	467.	407.

PERCENT CONTRIBUTIONS TO S DEPOSITIONS WITHIN RECEPTOR REGIONS

EMITTER REGION	1	2	3	4	5	6	7	8	9	10	11	12	13
1 VIII-NORTH	55.	0.	0.	0.	6.	0.	0.	0.	0.	0.	0.	0.	0.
2 V-NORTH	19.	66.	19.	7.	0.	1.	9.	0.	1.	3.	7.	4.	6.
3 S. ONTARIO	3.	7.	54.	0.	0.	0.	2.	0.	0.	3.	12.	9.	21.
4 VII	3.	4.	0.	59.	0.	5.	6.	2.	2.	0.	0.	0.	0.
5 VIII-SOUTH	0.	0.	0.	0.	92.	0.	0.	0.	0.	0.	0.	0.	0.
6 VI-EAST	7.	0.	0.	6.	0.	80.	0.	3.	0.	0.	0.	0.	0.
7 V-SOUTH	9.	19.	10.	22.	1.	3.	65.	5.	23.	23.	13.	6.	6.
8 IV-SOUTH	0.	1.	0.	3.	0.	9.	1.	79.	15.	1.	0.	0.	1.
9 IV-NORTH	2.	2.	2.	2.	0.	3.	9.	9.	50.	7.	2.	1.	1.
10 III	1.	1.	4.	1.	0.	0.	7.	1.	0.	60.	24.	12.	5.
11 II	0.	0.	4.	0.	0.	0.	0.	0.	0.	2.	28.	14.	3.
12 I	0.	0.	0.	0.	0.	0.	0.	0.	0.	0.	1.	44.	5.
13 S. QUEBEC	0.	0.	7.	0.	0.	0.	0.	0.	0.	0.	1.	9.	50.

Table 7. Amount of Material in Puffs from GDR at
          End of Each Day During the Initialization
          Period

	August				
	20	21	22	23	24
SO_2	0.8	2.5	5.0	7.8	8.5
$SO_4^=$	2.7	6.2	7.8	8.3	8.4

earlier. In Figure 6, the calculated sulfate concentrations for
26 August 1974 are significantly lower than measured values in the
region surrounding the Baltic Sea. This discrepancy between calculated and measured values can be attributed to the nonavailability
of emissions data from large areas of the USSR. This is particularly
important because the transport wind during this episode has a significant component coming from the east, as shown by the surface-layer
winds on August 26 (Figure 8).

One way to solve this problem is to establish a computational
grid area that is substantially larger than the area within which
the calculated values are taken to be valid. This, of course, can
significantly increase computational costs. A simpler, but less
accurate, method is to assign "background" concentrations to the
air coming from different directions. This assignment can be made
in accordance with available measurements, or by some other more
arbitrary method.

Data-Sparse Regions

As can be judged from Figure 12 (and Figure 8), interregional
models require copious amounts of meteorological and emissions data
for model input, and large amounts of air quality data for model
evaluation. Unfortunately, many modeling domains of interest, such
as that illustrated in Figure 13, cover geographical regions containing large areas where conventional data of all sorts are missing.
These data-sparse areas usually are large expanses of water, such
as the South China Sea as in Figure 13 or the North Sea as in
Figure 3.

Anthropogenic emissions from such areas can generally be considered to be nil, which obviates part of the problem, but wind
and precipitation data are still required. These data can be
estimated in some cases on the basis of information furnished by
meteorological satellites. Cloud-motion vectors can be used to
derive wind fields at one or more levels (Smith, 1975; Wolf et al.,

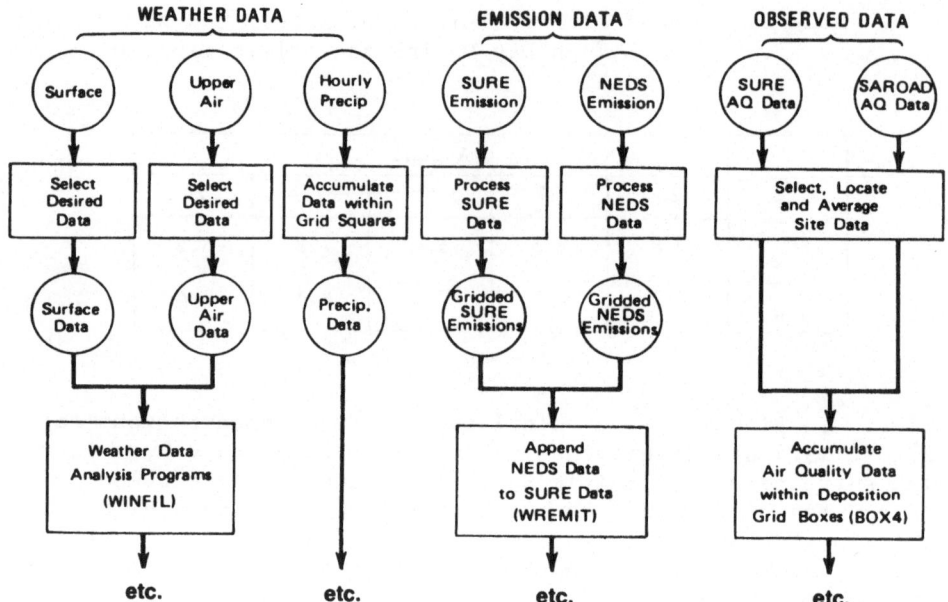

Fig. 12. Data flow for ENAMAP-1 calculations of monthly average
concentrations, depositions, and exchanges.

1977; Endlich and Wolf, 1981). In addition, rainfall patterns and
amounts can be estimated by means of empirical relationships that
have been developed (Woodley et al., 1980) linking satellite-derived
cloud heights (as represented by "brightness" temperatures) to rainfall rates. Such wind and precipitation estimates have been prepared from information obtained by the Japanese Geosynchronous
Meteorological Satellite (GMS) for use in applying ENAMAP to the
area shown in Figure 13 (Ruff et al., 1981).

Transformation and Deposition Coefficients

As pointed out by a number of reviewers, such as MacCracken
(1979), Eliassen (1980), and Hosker (1980), the values of transformation and deposition coefficients used in various LRT/interregional
models vary widely, sometimes by more than a factor of 10. This
partly is caused by the different model formulations, but also
reflects in a major way our basic lack of knowledge in this area.

Measurements of transformation and deposition coefficients for
many chemical species of interest are non-existent. Those measurements that are available show a major degree of variability even
when stratified (see, for example, Figure 14), indicating that the
values of the coefficients are influenced by a number of factors.

Some of the factors that are known to have significant effects on
dry and wet deposition are listed below.

Dry deposition:

- Atmospheric properties

  -- Solar radiation
  -- Wind speed
  -- Atmospheric stability
  -- Surface roughness
  -- Humidity

- Pollutant properties

  -- Form (and size distribution if particulate)
  -- Concentration vertical profile
  -- Solubility and reactivity

- Vegetation properties

  -- Type, size, leaf area, density
  -- Stomatal condition
  -- Growth stage
  -- Stress condition
  -- Wetness.

Wet deposition:

- Atmospheric properties

  -- Precipitation rate and type
  -- Cloud type and size
  -- Storm intensity
  -- Temperature and humidity

- Pollutant properties

  -- Form (and size distribution if particulate)
  -- Solubility and reactivity
  -- Concentration vertical profile
  -- Location relative to clouds

It is apparent that current models account for wet and dry
deposition with highly parameterized treatments that do not
explicitly include many of the factors in the above lists. Some
of the effects of these variables can be considered to be "averaged
out" over the long travel distances and large spatial averaging
areas involved in interregional-scale modeling. By comparing model-
calculated concentrations and depositions to available measured
values, useful information can be obtained to help select suitable
values for such "integrated" values of transformation and deposition
coefficients. In general, however, much additional fundamental
knowledge about the transformation and deposition processes is
needed to facilitate further progress in interregional modeling.

Model Evaluations and Intercomparisons

Models cannot be applied with any confidence until they have
been thoroughly evaluated by comparing calculated values against
measurements. However, for interregional-scale models in particular,
there is a major difficulty in this procedure: the spatial scales
represented by the model calculations and the measurements are
usually vastly different. The predictions from interregional models
apply as averages over individual grid cells covering some 3,000
to 10,000 km^2, while measurements are usually taken at a single
point.

Figure 15 (and Figures 6 and 10 presented earlier) give
examples of comparisons between modeled and measured values in
which one solution to this disparity in scale has been applied.
(As an aside, note that this type of geographical evaluation is
more revealing than the usual simple scatter plot and regression
of calculated versus measured values.) The spatial compatibility
problem was addressed in the examples in Figures 6 and 10 by
averaging all measured values falling within each grid cell, and
in Figure 15 by averaging the measurements from stations situated

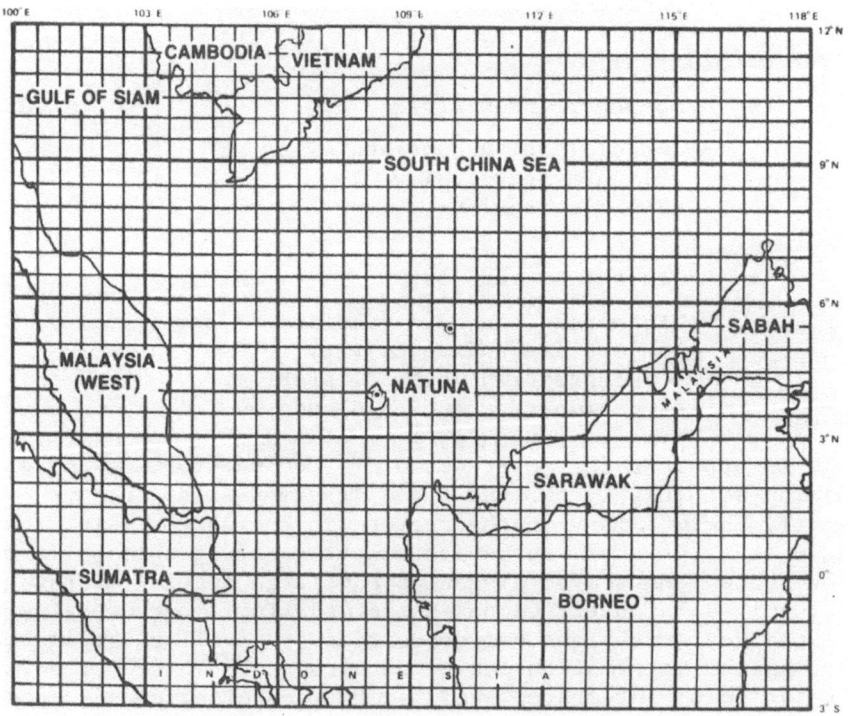

Fig. 13. Grid layout for application of ENAMAP models to a data-
          sparse region in the vicinity of Indonesia.

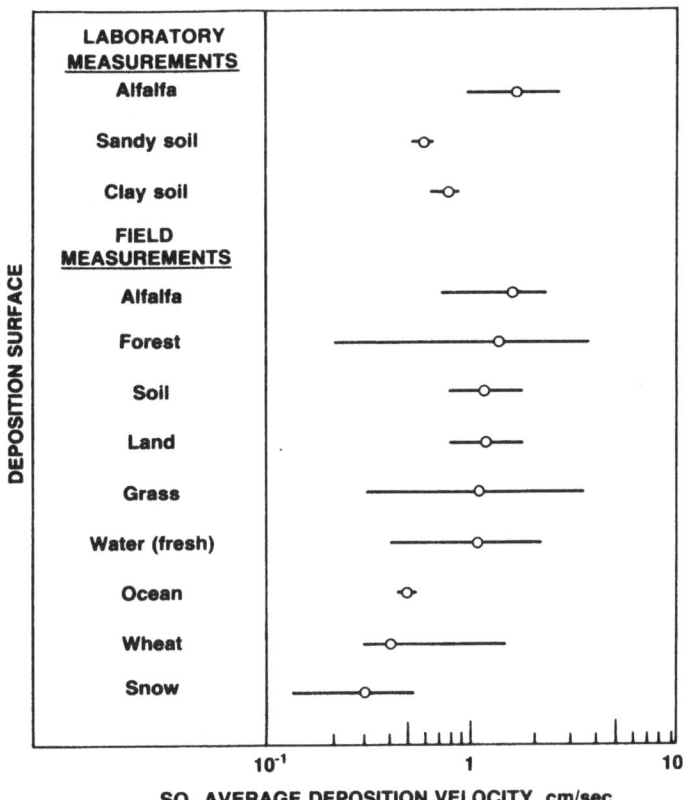

Fig. 14.   Measured deposition velocities for SO_2 for various
surfaces (from Demerjian, 1980).

up to about ±2° away from 10°E longitude (Eliassen, 1980). Clearly,
model evaluations of this sort require data from a relatively high
density of measurement stations to facilitate the comparison between
measured and modeled values.

Figure 15 illustrates an intercomparison of four different
models, while a model intercomparison in a somewhat different format is exemplified in Table 8. The latter example reveals the
difficulty of comparing the results from different models when the
basic elements of the model runs (e.g., emissions and meteorological
input data, geographical area covered) are not the same. The point
here is that model intercomparisons are likely to be most useful and
effective when model runs are designed, coordinated, and conducted
especially for this purpose.

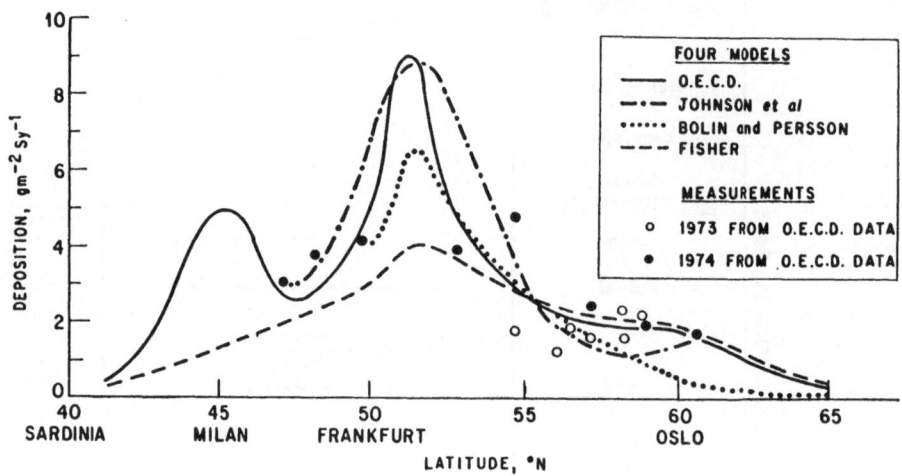

Fig. 15. Model estimates of annual sulfur deposition (wet plus
          dry) over Europe along longitude 10°E (from Eliassen,
          1980).

Table 8. A Model Intercomparison: Annual U.S./Canada Sulfur
          Exchanges as Estimated by the ENAMAP-1 (Bhumralkar et al.,
          1980) and ASTRAP (Shannon, 1979) LRT Models (All Values
          are in Units of 10^{12}g S/yr)

Source Region	Model	Year of Met. Data	Total Emissions	Sulfur Deposited in Receptor Regions			
				SE Canada		Eastern U.S.	
				Wet	Dry	Wet	Dry
SE Canada	ENAMAP-1	1977	1.9[a]	0.45	0.77	0.12	0.26
	ASTRAP	1975	2.1 [b]	0.58	0.36	0.14	0.10
Eastern U.S.	ENAMAP-1	1977	14.1[c]	0.33	0.38	4.01	6.70
	ASTRAP	1975	11.9 [d]	0.95	0.28	3.20	2.20

(a) ENAMAP-1 model area in Canada: Ontario and Quebec
     south of 50°N latitude.
(b) ASTRAP model area in Canada: Ontario and Quebec south
     of 55°N latitude; Maritime Provinces, Western
     Newfoundland.
(c) ENAMAP-1 model area in U.S.: East of 105°W longitude.
(d) ASTRAP model area in U.S.: East of 95°W longitude.

AN EXTENDED APPLICATION OF INTERREGIONAL MODELING: VISMAP

Because sulfur emissions have attracted the most attention to date in relation to the acid deposition problems in Scandinavia and eastern North America, most current interregional models have been designed to treat sulfur compounds. However, efforts are now underway to adapt some of these models to the treatment of other chemical species, and to apply the resulting models to other types of air pollution problems.

In this regard, a short-term version of ENAMAP-1 has been adapted by Dabberdt and Eigsti (1981) to the calculation of visual range on the regional scale; the modified model is called VISMAP (Visibility Model of Air Pollution). Figure 16 illustrates the nature of this simple approach. ENAMAP calculations of three-hourly averages of $SO_4^=$ concentrations are combined with measured values of relative humidity to calculate visual range by means of the empirical relationship given below (Trijonis, 1978):

$$b_{ext}(km^{-1}) \approx b_{scat}(km^{-1}) = 0.024 + \frac{0.004 \cdot (SO_4^=)}{(1-0.01 \cdot RH)}$$

where

b_{ext} = extinction coefficient, $\approx b_{scat}$ = scattering coefficient

$SO_4^=$ = sulfate concentration ($\mu g/m^3$)

RH = average relative humidity for the past three hours

(0.024 is a parameter representing average light scattering from remaining particles in the atmosphere).

Examples of model outputs are presented in Figures 17 and 18 (from Dabberdt and Eigsti, 1981). As indicated, in this case the ENAMAP $SO_4^=$ calculations and the resulting VISMAP calculations of visual range both agree reasonably well with available measurements.

CONCLUDING REMARKS

Evidence continues to accumulate, showing that polluted air can have adverse impacts at very long distances from its sources, perhaps even after travel across the Atlantic Ocean (Nyberg, 1976). This indicates that LRT/interregional models will continue to fill a strong need for techniques to estimate the contributions from individual source areas and to assess the effects of prospective changes in emission patterns, perhaps on even larger scales.

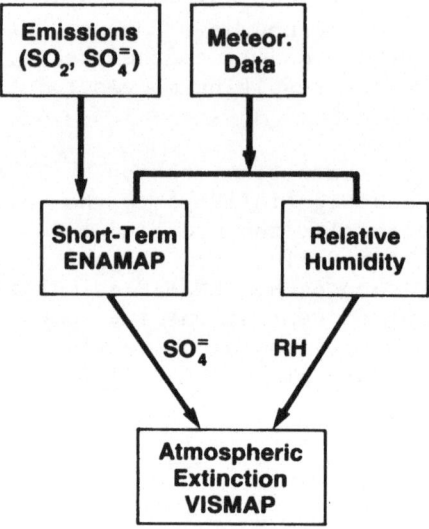

Fig. 16. Elements of VISMAP (Visibility Model of Air Pollution).

In the meantime, it is important that current efforts to improve existing models continue in such areas as the following:

- Incorporating the effects of complex terrain on transport wind and dry deposition

- Incorporating multiple layers in the vertical

- Treating boundary-layer processes and vertical diffusion more realistically

- Improving the treatment of dry and wet deposition rates

- Including satellite-derived wind and rainfall data over data-sparse regions.

With such improvements, LRT/interregional models can be expected to be even more useful in the future.

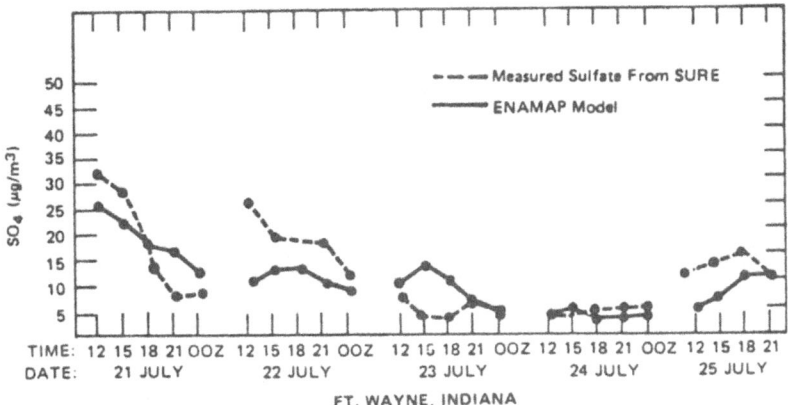

Fig. 17. Three-hourly ENAMAP calculations of $SO_4^=$ versus measured concentrations.

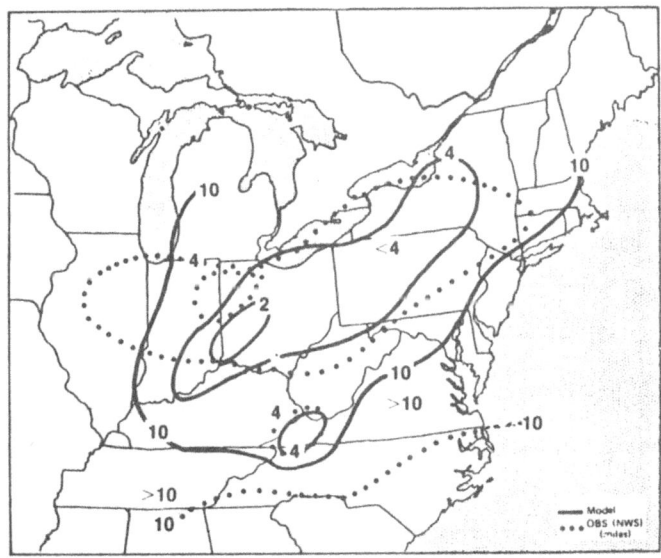

Fig. 18. VISMAP calculations versus measured visual range (miles) for 1500 Z, 21 July 1978.

ACKNOWLEDGMENTS

Contractual support for the development and application of the EURMAP model, examples of which were shown in this paper, was provided to SRI International by the Federal Environmental Agency (Umweltbundesamt) of the Federal Republic of Germany, under the direction of Dr. D. Jost and Dr. J. Pankrath. Similar contractual support for the ENAMAP model was provided by the U.S. Environmental Protection Agency, under the direction of Dr. K. Demerjian and Mr. T. Clark. The assistance of the following individuals at SRI International is also gratefully acknowledged: Dr. C. Bhumralkar, who provided technical assistance and carried out several of the model runs presented as examples; Mrs. D. Sevela and Miss S. Webster, who typed the manuscript; Mrs. V. Ramsay, who compiled the references; and Miss J. Kealoha, who prepared the illustrations.

REFERENCES

Bass, A., 1980: Modelling long-range transport and diffusion. In Proc. Second Joint Conference on Applications of Air Pollution Meteorology, New Orleans, LA, USA, 24–27 March 1980, 193–215, American Meteorological Society, Boston, MA, USA.

Berkowicz, R., and L.P. Prahm, 1978: Pseudospectral simulation of dry deposition from a point source. Atmos. Environ., 12, 379–387.

Bhumralkar, C.M., W.B. Johnson, R.L. Mancuso, and D.E. Wolf, 1979: Regional patterns and transfrontier exchanges of airborne sulfur pollution in Europe. Final Report, SRI Project 4797, SRI International, Menlo Park, CA, USA, prepared for the Federal Environmental Agency (Umweltbundesamt) of the Fed. Rep. of Germany, Berlin.

Bhumralkar, C.M., R.L. Mancuso, D.E. Wolf, R.H. Thuillier and W.B. Johnson, 1980b: Adaptation and application of the EURMAP-1 model to eastern North America. Final Report, Project 7760, SRI International, Menlo Park, CA, USA, prepared for the U.S. Environmental Protection Agency, Research Triangle Park, NC, USA, Contract No. EPA/68-02-2959.

Bhumralkar, C.M., R.L. Mancuso, D.E. Wolf, K.C. Nitz, and W.B. Johnson, 1980b: Adaptation and application of ENAMAP-1 model to eastern North America--Phase II. Final Report, Project 7760, SRI International, Menlo Park, CA, USA, prepared for the U.S. Environmental Protection Agency, Research Triangle Park, NC, USA, Contract No. EPA/68-02-2959.

Bhumralkar, C.M., R.L. Mancuso, R.H. Thuillier, K.C. Nitz, W.B. Johnson, and D.E. Wolf, 1981a: Transboundary exchanges of airborne sulfur pollution and deposition in eastern North America. Manuscript submitted to Atmos. Environ.

Bhumralkar, C.M., R.L. Mancuso, D.E. Wolf, and W.B. Johnson, 1981b: Regional air pollution model for calculating short-term (daily) patterns and transfrontier exchanges of airborne sulfur in Europe. Tellus, Feb. 1981 (in press).

Bolin, B. and C. Persson, 1975: Regional dispersion and deposition of atmospheric pollutants with particular application to sulfur pollution over Western Europe. Tellus, 3, 281-310.

Carmichael, G.R., and L.K. Peters, 1979: Numerical simulation of the regional transport of SO_2 and sulfate in the eastern United States. In Proc. Fourth Symposium on Turbulence, Diffusion, and Air Pollution, Reno, NV, USA, 15-18 January 1979, 337-344, American Meteorological Society, Boston, MA, USA.

Christensen, O, and L.P. Prahm, 1976: A pseudospectral model for dispersion of atmospheric pollutants. J. Appl. Meteor., 15, 1284-1294.

Dabberdt, W.F. and S.L. Eigsti, 1981: Regional visibility modeling for the eastern United States. Atmos. Environ. (in press).

Demerjian, K.L., 1980: Atmospheric transport, transformation and deposition. Chap. 6 in Air Quality Criteria for Sulfur Oxides and Particulate Matter, draft document for public comment (Dec. 1980), U.S. Environmental Protection Agency, Research Triangle Park, NC, USA.

Draxler, R.R., 1977: A mesoscale transport and diffusion model. Tech. Memo ERL-ARL-64, Air Resources Laboratories, National Oceanic and Atmospheric Administration, Silver Spring, MD, USA.

Draxler, R.R., 1979: Modeling the results of two recent mesoscale dispersion experiments. Atmos. Environ., 13, 1523-1533.

Egan, B.A., K.S. Rao, and A. Bass, 1976: A three-dimensional advective-diffusive model for long range sulfate transport and transformation. In Proc. Seventh International Technical Meeting on Air Pollution Modeling and Its Application, Airlie, VA, USA, 7-10 Sept. 1976, 697-714, a report of the Air Pollution Pilot Study, NATO Committee on the Challenges to Modern Society.

Eliassen, A., 1980: A review of long-range transport modelling. J. Appl. Meteor., 19, 231-240.

Eliassen, A., O. Jensen, and J. Saltbones, 1976: Model comparison using selected test results with NILU (trajectory) model. Report EMP 3/76 (November), Norwegian Institute for Air Research, Lillestrom, Norway.

Eliassen, A. and J. Saltbones, 1975: Decay and transformation rates of SO_2 as estimated from emission data, trajectories and measured air concentrations. Atmos. Environ., 9, 425-429.

Endlich, R.M. and D.E. Wolf, 1981: Automatic cloud tracking applied to GOES and METEOSAT observations. J. Appl. Meteor. (in press).

Fay, J.A. and J.J. Rosenzweig, 1980: An analytical diffusion model for long distance transport of air pollutants. Atmos. Environ., 14, 355-365.

Fisher, B.E.A., 1975: The long range transport of sulfur dioxide. Atmos. Environ., 9, 1063-1070.

Fisher, B.E.A., 1978: The calculation of long term sulphur deposition in Europe. Atmos. Envir., 12, 489-501.

Fox, D.G. and S.A. Orszag, 1973: Pseudospectral approximation to two-dimensional turbulence. J. Comput. Phys., 11, 612-619.

Heffter, J.L., 1980: Air Resources Laboratories Atmospheric Transport and Dispersion model (ARL-ATAD). Tech. Memo ERL-ARL-81, Air Resources Laboratories, National Oceanic and Atmospheric Administration, Silver Spring, MD, USA.

Hosker, R.P., Jr., 1980: Practical application of air pollutant deposition models--current status, data requirements, and research needs. In Proc. International Conference on Air Pollutants and Their Effects on the Terrestrial Ecosystem, Banff, Alberta, Canada, 10-17 May 1980, S.V. Krupa and A.H. Legge (editors), John Wiley and Sons, New York, NY, USA.

Johnson, W.B., D.E. Wolf, and R.L. Mancuso, 1978: Long term regional patterns and transfrontier exchanges of airborne sulfur pollution in Europe. Atmos. Environ., 12, 511-527.

Lange, R., 1978: ADPIC--A three-dimensional particle in cell model for the dispersal of atmospheric pollutants and its comparison to regional tracer studies. J. Appl. Meteor., 17, 320-329.

Lavery, T.F., R.L. Baskett, J.W. Thrasher, N.J. Lordi, A.C. Lloyd, and G.M. Hidy, 1980: Development and validation of a regional model to simulate atmospheric concentrations of SO_2 and sulfate. In Proc. AMS/APCA Second Joint Conference on Applications of Air Pollution Meteorology, New Orleans, LA, USA, 24-27 March

Air Pollution Meteorology, New Orleans, LA, USA, 24-27 March 1980, 236-247, American Meteorological Society, Boston, MA, USA.

MacCracken, M.C., 1979: Simulation of regional precipitation chemistry. In Proceedings: Advisory Workshop to Identify Research Needs on the Formation of Acid Precipitation, D.H. Pack (editor), 2-75 to 2-92, EPRI Report EA-1074, Electric Power Research Institute, Palo Alto, CA, USA.

Mancuso, R.L., C.M. Bhumralkar, D.E. Wolf, and W.B. Johnson, 1978: Evaluation and sensitivity analysis of the European Regional Model of Air Pollution (EURMAP-1). Progress Report, SRI Project 4797, SRI International, Menlo Park, CA, USA, prepared for the Federal Environmental Agency (Umweltbundesamt) of the Fed. Rep. of Germany, Berlin.

Mancuso, R.L., C.M. Bhumralkar, D.E. Wolf, and W.B. Johnson, 1979: The exchange of sulfur pollution between the various countries of Europe based on the EURMAP model. In Proc. Fourth Symposium on Turbulence, Diffusion and Air Pollution, Reno, NV, USA, 15-18 January 1979, 345-354, American Meteorological Society, Boston, MA, USA.

Maul, P.R., 1977: The mathematical model of the mesoscale transport of gaseous pollutants. Atmos. Environ., 11, 1191-1195.

McMahon, T.A., P.J. Denison, and R. Fleming, 1976: A long-distance air pollution transportation model incorporating washout and dry deposition components. Atmos. Environ., 10, 751-761.

Meyers, R.E., R.T. Cederwall, J.A. Storch, and L.I. Kleinman, 1979: Modeling sulfur oxide concentrations in the eastern United States--model sensitivity, verification, and application. In Proc. Fourth Symposium on Turbulence, Diffusion, and Air Pollution, Reno, NV, USA, 15-18 January 1979, 673-676, American Meteorological Society, Boston, MA, USA.

Mills, M.F., and A.A. Hirata, 1978: A multi-scale transport and dispersion model for local and regional scale sulfur dioxide/ sulfate concentrations--formulation and initial evaluation. In Proc. Ninth International Technical Meeting on Air Pollution Modeling and Its Application, Toronto, Ontario, Canada, 28-31 August 1978, a report of the Air Pollution Pilot Study, NATO Committee on the Challenges to Modern Society.

Niemann, B.L., A.A. Hirata, B.R. Hall, M.T. Mills, P.M. Mayerhofer, and L.F. Smith, 1980: Initial evaluation of regional transport and subregional dispersion models for sulfur dioxide and fine particulates. In Proc. Second Joint Conference on Applications

of Air Pollution Meteorology, New Orleans, LA, USA, 24–27 March
1980, 216–224, American Meteorological Society, Boston, MA, USA.

Nordø, J., 1974: Quantitative estimates of long range transport of
sulphur pollutants in Europe. Ann. Meteor., 9, 71–77

Nordø, J., 1976: Long-range transport of air pollutants in Europe
and acid precipitation in Norway. Water, Air Soil Pollut., 6,
199–217.

Nyberg, A., 1976: On transport of sulphur over the North Atlantic.
SMHI Report No. RMK 6 (1976), Swedish Meteorological and Hydro-
logical Institute, Norrköping, Sweden.

OECD, 1977: The OECD program on long-range transport of air pollu-
tants: Measurements and findings. Organization for Economic
Cooperation and Development (OECD), Dir. of Inf. OECD, Paris,
France.

Olson, M.P., 1980: Results from LRTAP model runs for July 1978.
Personal communication, Environment Canada, Downsview, Ontario,
Canada.

Olson, M.P., E.C. Voldner, K.K. Oikawa, and A.W. MacAfee, 1979: A
concentration/deposition model applied to the Canadian Long-
Range Transport of Air Pollutants Project: a technical description. Report No. LRTAP 79-5, Environment Canada, Downsview,
Ontario, Canada.

Ottar, B., 1978: The OECD study on long range transport of air
pollutants (LRTAP). Atmos. Environ., 12, 445–454.

Pedersen, L.B., and L.P. Prahm, 1974: A method for numerical solution of the advection equation. Tellus, 26, 594–602.

Perhac, R.M., 1978: Sulfate Regional Experiment in Northeastern
United States: The SURE program. Atmos. Environ., 12,
641–647.

Powell, D.C., D.J. McNaughton, L.L. Wendell, and R.L. Drake, 1979:
A variable trajectory model for regional assessments of air
pollution from sulfur compounds. Report No. PNL-2734,
Battelle Pacific Northwest Laboratory, Richland, WA, USA.

Prahm, L.P., and O. Christensen, 1977: Long-range transmission of
pollutants simulated by the 2-D pseudospectral dispersion
model. J. Appl. Meteor., 16, 896–910.

Rodhe, H., 1972: A study of the sulfur budget for the atmosphere
over northern Europe. Tellus, 24, 128–138.

Rodhe, H., 1974: Some aspects of the use of air trajectories for
    the computation of large-scale dispersion and fallout patterns.
    Advances in Geophysics, 18B, 95-109, Academic Press.

Ruff, R.E., R.L. Mancuso, and W.B. Johnson, 1981: Air quality
    assessment for the proposed Natuna LNG plant in the South China
    Sea. Final Report, SRI Project 1866, SRI International, Menlo
    Park, CA, USA, prepared for the Exxon Corporation (in
    preparation).

Scriven, R.A., and B.E.A. Fisher, 1975: The long range transport
    of airborne material and its removal by deposition and washout,
    I and II. Atmos. Environ., 9, 49-68.

Shannon, J.D., 1979: The advanced statistical trajectory regional
    air pollution model. In Proc. Fourth Symposium on Turbulence,
    Diffusion, and Air Pollution, Reno, NV, USA, 15-18 January 1979,
    376-380, American Meteorological Society, Boston, MA, USA.

Shieh, C.M., 1977: Application of a statistical trajectory model
    to the simulation of sulfur pollution over the northeastern
    United States. Atmos. Environ., 11, 173-178.

Sklarew, R.C., A.J. Fabrik, and J.E. Prager, 1971: A particle-in-
    cell method for numerical solution of the atmospheric diffusion
    equation, and application to air pollution problems. Report
    No. 3SR-844, Systems, Science and Software, La Jolla, CA, USA.

Smith, E., 1975: Man-computer interactive data access system.
    IEEE Trans. Geosci. Electron., GE-13, 123-126.

Smith, F.B., and R.D. Hunt, 1979: The dispersion of sulfur pollu-
    tants over western Europe. Phil. Trans. Roy. Soc. London,
    A290, 523-542.

Szepesi, D.J., 1978: Transmission of sulfur dioxide on local,
    regional and continental scales. Atmos. Environ., 12,
    529-535.

Trijonis, J., 1978: Visibility in the southwest—an exploration
    of the historical data base. EPA Report No. 600/3-78-039,
    U.S. Environmental Protection Agency, Research Triangle Park,
    NC, USA.

Venkatram, A., 1980: Role of models in long-range transport.
    Personal communication, Ontario Ministry of the Environment,
    Toronto, Ontario, Canada.

Venkatram, A., B.E. Ley, and S.Y. Wong, 1980: A statistical model
    to estimate long-term concentrations of pollutants associated

with long-range transport. Internal Report, Ontario Ministry
of the Environment, Toronto, Ontario, Canada.

Wendell L.L., C.D. Powell, and R.L. Drake, 1976: A regional scale
    model to computing deposition and ground level air concentra-
    tion of SO_2 and sulfates from elevated and ground sources. In
    Proc. of the Third Symposium on Atmospheric Turbulence, Diffu-
    sion and Air Quality, Oct. 19-22, 1976, Raleigh, NC, USA, pub-
    lished by the AMS, Boston, MA, USA. Also in Proceedings of
    the 7th Int. Tech. Meeting on Air Pollution Modeling and Its
    Application, 7-10 Sept. 1976, Airlie House, VA, USA, a report
    of the Air Pollution Pilot Study, NATO Committee on the
    Challenges to Modern Society.

Wolf, D.E., D.J. Hall, and R.M. Endlich, 1977: Experiments in auto-
    matic cloud tracking using SMS-GOES data. J. Appl. Meteor.,
    16, 1219-1230.

Woodley, W., C.G. Griffith, J.S. Griffin, and S.C. Stromatt, 1980:
    The inference of GATE convective rainfall from SMS-1 imagery.
    J. Appl. Meteor., 19, 388-408.

DISCUSSION

L. PRAHM   The computational effort with
              these models might not be much less than the effort
              with Eulerian grid point models if one takes the long-
              lived species such as SO_4 into account, because the
              trajectories become very long, in some cases longer than
              98 hours. Could you please give some comments on this.

W.B. JOHNSON   When the concentration level
              of a puff in a forward trajectory model becomes rela-
              tively low, then the puff is not accounted for any more.
              When large geographical areas are considered, the spa-
              tial resolution is cedreased to avoid an excessive com-
              putation effort.

E. RUNCA   One of the main problems of
              LRTAP is the formation of acid deposition. Most of
              this process occurs in clouds. To my knowledge this
              process is not properly taken into account in the mo-
              del you have mentioned. I would like you to comment
              on this point.

W.B. JOHNSON   This problem is handled by
              using two transformation rates in the model: one for

in-cloud processes and the other for clear air processes. The in-cloud transformation is about 10 times
greater.

M.L. WILLIAMS   Your paper had so many interesting features in it and we could discuss them for
a long time. One point of interest to modellers however concerns the shape of the power plant plume in
the stable layer before fumigation, which as visual
observations often suggest was much wider than it was
deep and was very far from being Gaussian. Presumably
this is due to the horizontal wind meander dominating
the turbulent processes - Could you elaborate on this ?

W.B. JOHNSON   The picture showed a snap shot
of the plume and one does not compare it with a Gaussian
plume. However, many pictures show similar characteristics and the behaviour of this plume in stable conditions could be studied in more detail, including the
effect of vertical shear of horizontal wind.

TRAJECTORIES AS TWO-DIMENSIONAL PROBABILITY FIELDS

Perry J. Samson and Jennie L. Moody

Department of Atmospheric and Oceanic Science
The University of Michigan
Ann Arbor, Michigan 48109

INTRODUCTION

The diagnoses of air pollution episodes involving regional-scale transport has been aided by the use of trajectory analysis. This analysis involves the use of interpolation from observed or computed winds to estimate either the path of air parcels leaving a source region or the path traversed by air sampled at a receptor location. This analysis has provided valuable qualitative insight into the nature of regional-scale transport.

As we strive to quantify the relationships between source regions and receptor locations it behooves us to understand clearly with what confidence we can specify individual trajectories and with what limitations we are constrained using different types of trajectory analyses. Pack et al. (1978) have summarized the potential errors inherent in trajectory analysis and described the efforts to validate various downstream models. It would be expected that upstream trajectories would suffer from identical sources of error. Such errors would, presumably, be random in time and could be filtered by examination of large data sets of trajectories. This "ensemble trajectory analysis" has been attempted by Samson (1980) for sulfate concentrations measured in the northeastern United States. Similar analyses have been performed for sites in Europe by Rodhe (1974), Eliassen and Saltbones (1975), and Smith and Hunt (1978).

Several techniques have been employed for computing trajectories. Hall et al. (1973) used an interpolation scheme proposed by Saucier (1955) to trace the movement of regions of reported haze in the United States based on observed 850 mb winds measured

43

every twelve hours. This wind field was also used by Nordo et
al. (1974) using the interpolation scheme suggested by Petterssen
(1956).

Prahm et al. (1976) used computed geostrophic winds over the
ocean to estimate the transport of air to islands in the North
Atlantic. Other investigators have assumed empirical approximations to describe the differences between calculated surface
geostrophic winds and the actual transporting winds. Chung (1978)
has used this approach, modifying the surface geostrophic winds
by a frictional term which is a function of stability, to estimate the upstream path of sampled air in Ontario, Canada. Smith
and Jeffery (1975) estimated winds by hand from surface pressure
charts with the deviations from geostrophic derived by empirical
comparisons. The use of surface geostrophic winds have the advantage that data is available from a much finer grid and at time
intervals of one hour as opposed to the upper-air winds which
are usually sampled at twelve hour intervals. However, such an
approach is limited by the accuracy and generality of the relationship used to describe the differences between the surface
winds and the actual winds in the layer transporting the pollutant.

Predicted winds have also been used to estimate potential
source regions for measured pollutants. Samson et al. (1975)
and Husar et al. (1976) have used the predicted trajectories
available from an operational model developed by Reap (1972).
This model is based on the predicted wind fields generated by the
National Meteorological Center six-layer primitive equation model.

A model used widely for estimation of upstream or downstream
path of a layer of the atmosphere is the ATAD model developed by
Heffter (1980). The model computes trajectories for the mixed-
layer of the atmosphere or for a layer prescribed by the user
using observed wind and temperature data from the first 3000 m
of the atmosphere.° The height of the mixed-layer is defined from
an estimation of the potential temperature profile as being the
height of the base of a layer whose gradient exceeds 0.5° C/ 100m.
The base of the layer is defined as the surface unless strong
wind velocity shears exist near the surface. In this situation
the height at which the shear drops below a critical value is
chosen as the base of the mixed layer.

° Earlier versions of the model either ignored the temperature
  structure altogether (Heffter et al., 1975) and required prescribed limits for the layer or used a predefined critical lapse
  rate to define the top of the mixed-layer (Heffter and Ferber,
  1977).

The model has been compared with other models by Hoecker (1977) against the recovery locations of tetroons over scales of 300 to 1300 km. The model predicted the recovery location more consistently than the other models tested, though the sample size was not large.

PROBABILITY FIELDS

When trajectories are computed for a layer of the atmosphere it must be recognized that the resulting path represents the path of maximum probability for air parcels arriving at or departing from a given trajectory origin. That is, the combination of vertical mixing and vertical wind velocity shear will act to disperse pollutants over a wide lateral extent so that a measured concentration, for example, could, potentially, have been influenced by sources over a wide area. Figure 1 shows the paths computed using the Heffter (1980) model upstream from a sampling point in the Adirondack region of upstate New York. These trajectories represent the movement of four sublayers of the atmosphere each 400 m thick extending from the surface to 1600 m above the surface. Note that the trajectories envelope a wide area even by 24 hours upstream. If the atmosphere had been well-mixed over this period then each of these trajectories could have contributed equally to the observed concentration.

Earlier work by Samson (1979) calculated upstream trajectories for four sublayers as demonstrated in Figure 1 and analyzed the divergence of the paths as a function of time upstream. The perturbations of the four sublayers from the location of the mixed-layer trajectory were computed and their distributions found to be normal for a sample of 66 summertime trajectories. The growth of the lateral distribution of endpoints was found to be proportional to time. Values of the along-trajectory perturbations were also calculated and found to grow at roughly the same rate. Thus, the probability of emissions from a grid cell centered at (x,y) of size $\delta x \delta y$ can be calculated as

$$p(x,y;t) = \frac{1}{2\pi\sigma_x(t)\,\sigma_y(t)}\ \exp\left\{-\ \frac{1}{2}\left(\frac{x'^2}{\sigma_x^2(t)} + \frac{y'^2}{\sigma_y^2(t)}\right)\right\}\delta x \delta y$$

(1)

where x' and y' are the distance along and across the axis of the trajectory, respectively, from the trajectory location at time t. This probability could, for example, represent the chance that emissions from (x,y) would contribute to an observed concentration if emitted t hours upstream of the sampling point.

Given a fixed grid of emissions cells, the probability of emissions from a grid cell influencing the resultant concentration can be obtained by integrating over the length of the trajectory, τ, as

$$p_{i,j} = \int_0^\tau p(x,y;t) \, dt \Big/ \int_0^\tau dt \qquad\qquad (2)$$

where i,j are the row and column indexes for the fixed grid.

This is, of course, a simplified estimate because the effects of precipitation, dry deposition and chemical conversion processes have not been included. The net effect of the first two will be a reduction in the probability field in an inhomogeneous manner reflecting the distribution of precipitation within the domain and the spatial and temporal inhomogeneity of dry deposition rates.

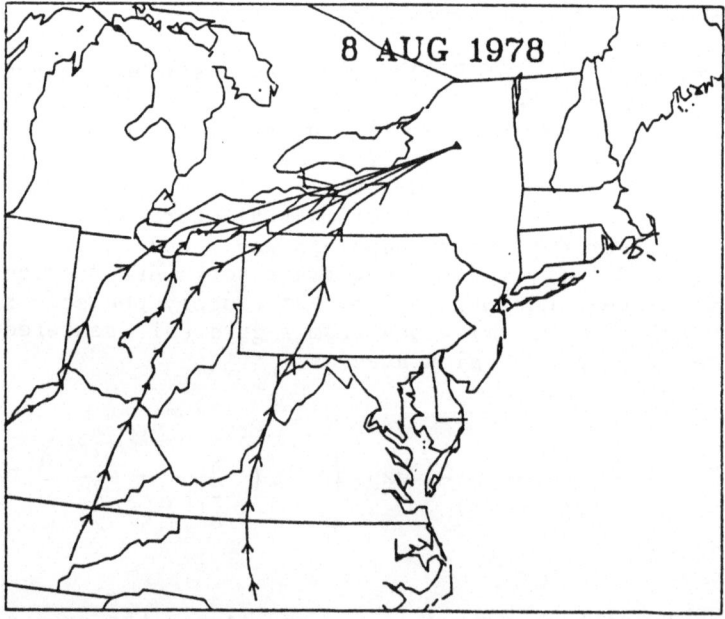

Fig. 1. Four constant layer trajectories representing flow, clockwise, for the sfc - 400 m, 400 m - 800 m, 800 m - 1200 m, and 1200 m - 1600 m layers.

DISPERSION PARAMETERIZATION

The original analysis by Samson (1979) has been reevaluated to ascertain whether the distribution of trajectory endpoints is isentropic about the mean location of the mixed – layer endpoints. Figure 2 shows the mean relative locations of the four sublayer endpoints with respect to the mean location of the mixed-layer endpoint at 24 hours upstream of the trajectory origin. A marked orientation of the endpoints exists with respect to the mixed-layer. The lowest layer systematically arrives from the left (looking upstream) of the mixed-layer with less average speed while the upper layers approach from the right with higher mean speed.

The mean locations of the sublayers form an angle of roughly 30° clockwise from the along-trajectory axis. By rotating the axis of probability through this angle, estimates were made of cross-ellipse, v, and along-ellipse, u, standard deviations. The cross ellipse standard deviations, σ_v, were found to increase linearly with time as

$$\sigma_v = 4.6 \ t \qquad\qquad\qquad (3)$$

and the along-ellipse standard deviations, σ_u, at a rate

$$\sigma_u = 7.5 \ t \qquad\qquad\qquad (4)$$

where the standard deviations are in kilometers and the time, t, is in hours.

APPLICATIONS

Assessment of probable Contribution to Measured Concentrations

Assuming a normally distributed probability field broadening with time as described above, individual mixed-layer trajectories can be integrated to derive a probability field of contribution to the concentration observed at the time of origin of the trajectory. This integration yields a probability of contribution field such as the one illustrated in Figure 3. This Figure shows the widening of the probability field upstream for the same time of arrival as was shown in Figure 1 earlier. This probability field assumes that the atmosphere was well-mixed, with inert pollutants and no removal en route.

The potential contribution of a particular source region could be calculated from this plot by integrating over the area of the source region and multiplying by the total emissions in the region during the period of the trajectory. This has been done

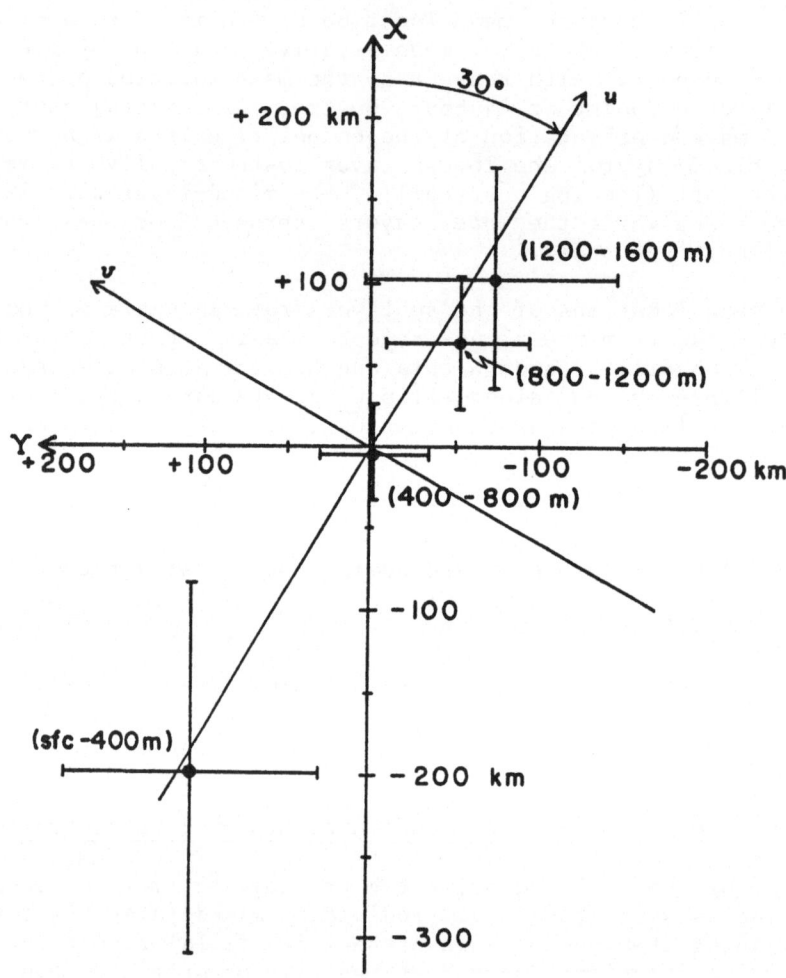

Fig. 2. The mean location of trajectory endpoints for
        four sublayers from the surface to 1600 m rela-
        tive to the mean locations of the mixed-layer
        trajectory. The bars represent the along, x,
        and across-trajectory, y, standard deviations.
        The axis of the mean sublayer locations is
        roughly 30° from the axis of the mixed-layer
        trajectory.

by Samson (1980) for a large grid of emissions cells to determine
the total potential loading of sulfur dioxide upstream of a
sampling point for comparison with observed sulfate concentrations.

Since atmospheric samples are often considered over some
averaging time of finite length it is also necessary to consider
the integration of the probability field over more than one trajectory (which are available at six-hour intervals using the
model of Heffter, 1980). The net effect of this will be a further
broadening of the probability field due to the fluctuations of
synoptic-scale wind patterns in time.

To analyze the results the probability fields produced for
each trajectory are weighted according to the concentration of
pollutant observed. The ensemble of individual trajectory probability fields for the weighted case is then compared with the
ensemble field of probability for the unweighted case. If the
two fields are statistically similar, then we can conclude that
there is no predominant pathway associated with high concentrations of pollutants.

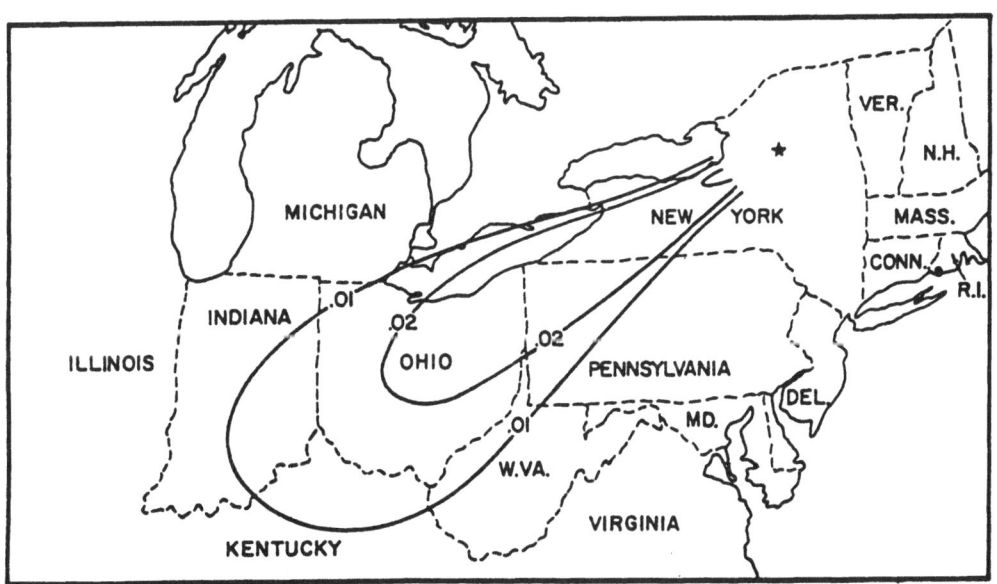

Fig. 3. The probability of contribution field for a
        sample collected in the Adirondack region of
        upstate New York based solely on atmospheric
        dispersion. Units are 10^{-4} km^{-2}.

Weighted fields estimating the probability of contribution
to samples of precipitation can also be calculated if care is
taken to isolate the hours of precipitation and the precipitation is collected on event or sub-event basis. The weighting
would be proportional to the concentration of pollutant in the
precipitation times the amount of precipitation which fell during
the event.

The net effect of precipitation along the course of the trajectory is to reduce the probability of contribution for those
regions in which precipitation is occuring and those regions upstream of the regions of precipitation. By summing precipitation
amounts over time interval commensurate with the time steps of
the trajectory model and introducing a first-order loss rate
which is proportional to the amount of precipitation we can then
recalculate the probability of contribution field. A sample
field where precipitation has occured in the hatched region shows
in Figure 4 the net effect on the probability field originally
calculated and shown in Figure 3.

This approach for diagnosing source-receptor relationships in
quantitative terms thus includes the effects of dispersion due
to vertical wind velocity shear and losses due to precipitation
along the route of the trajectory.

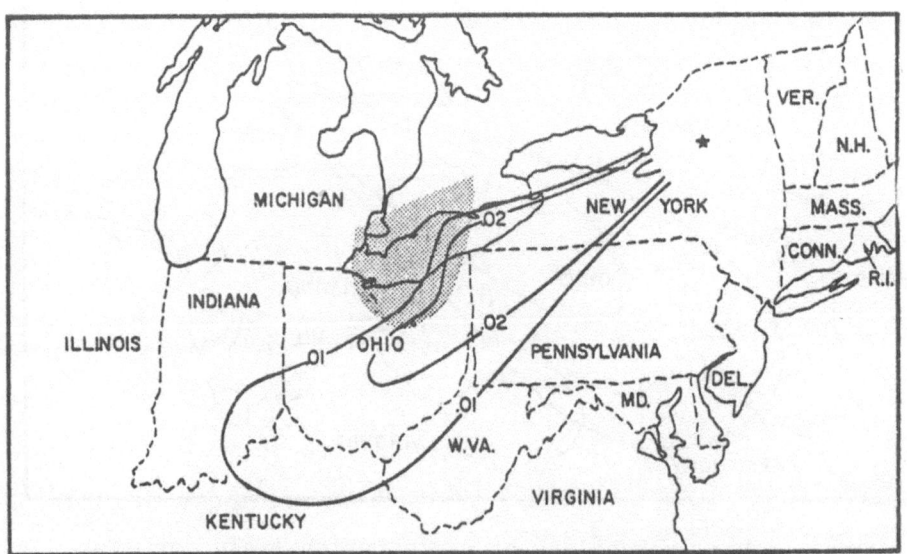

Fig. 4. Same as Fig. 3. but reflecting the effect on
        the probability field due to a precipitation
        event in the hatched region.

Assessment of Probable Impact of Individual Source Regions

The same techniques which are used for estimating upstream contributions to a measured concentration of pollutant can be turned around to estimate the regions of potential impact from a source or source region. Assuming the same dispersion parameters are applicable to the downstream transport of pollutants, areas of probable impact can be delineated.

In an attempt to identify the regions which would most probably be affected by ozone generation from the emissions of Detroit, Michigan, probability fields were calculated for the location of the plume six hours downstream of Detroit. It is thought that this time should be representative of time of maximum impact of ozone for emissions released during the morning hours. Trajectories started at 0700 EST from Detroit were used to define the probability of impact at six hours downstream. An ensemble of trajectories were chosen from those days when measured ozone levels in the Detroit area exceeded 120 ppb. These days were thought to exhibit conditions favorable for the generation of ozone from precursors. Using the ensemble of days collected from 1976 through 1979, the resultant probability of impact field is shown in Figure 5.

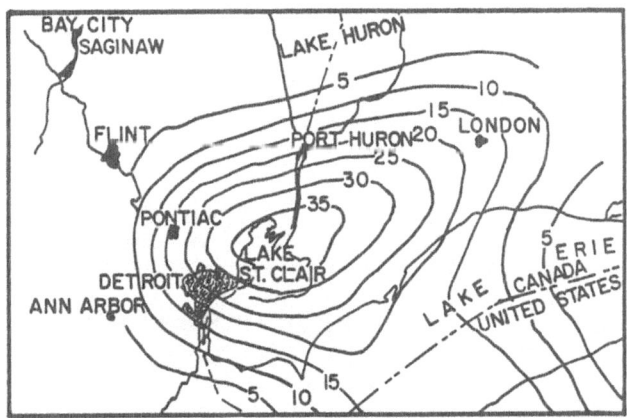

Fig. 5. The spatial probability distribution (10^{-4} km^{-2}) describing the likelihood of being affected by air leaving downtown Detroit at 0700 EST on those days during 1976-1979 when ozone concentrations exceed 120 ppb.

The figure indicates that the regions most likely to see an ozone increase due to Detroit emissions, if indeed one exists, would be located northeast of the downtown area. It further suggests that the area of potential effect covers an area of southwestern Ontario at least as large as the area of potential effect in Michigan.

SUMMARY

   While trajectory analysis has provided useful qualitative in-formation about the relationships between source regions and re-ceptor areas and the intervening meteorology, there is a need to provide more quantitative investigation of these relationships. This investigation, must, however, realize the inherent limita-tions of the trajectory model which is used.

   The methodology which has been discussed in this article in-corporates the dispersive action of the mixed-layer as deduced from an examination of the divergence of trajectories within the layer. Additionally, this methodology takes into account the precipitation which occurs within the area of potential contri-bution and reduces upstream potentials accordingly. The tech-nique can be used with large data sets of pollutant concentrations to attempt to explain the variance in concentrations. It can also be employed to estimate the climatological impact of a par-ticular source region given the climatology of trajectories and corresponding precipitation.

REFERENCES

Chung,Y.-S., 1977, Sources and sinks of photochemical ozone in the boundary layer, in : "Proc. 4th Inter. Clean Air Congress," Tokyo, Japan, 138.

Eliassen,A. and Saltbones, J., 1975, Decay and transformation rates of SO_2, as estimated from emission data, trajectories and measured air concentrations, Atmospheric Environment, 9: 425.

Hall, F. P., Duchon, C. D., Lee, L. G., and Hagan, R. R., 1973, Long-range transport of air pollution : a case study, August 1970, Mon. Wea. Rev., 101 : 404.

Heffter, J. L., 1980, "Air Resources Laboratories Atmospheric Trans-port and Dispersion Model (ARL-ATAD)," NOAA Tech. Memo, ERL-ARL-81, Silver Springs, MD.

Heffter, J. L., Taylor, A. D. and Ferber, G. J., 1975, "A Regio-nal-Continental Scale Transport, Diffusion, and Deposition Model," NOAA Tech. Memo. ERL-ARL-50, Silver Springs, MD.

Heffter, J. L., and Ferber, G. J., 1977, Development and verifi-cation of the ARL regional-continental transport and disper-sion model, in :"Proc. Joint Conf. on Appl. of Air Pollution Meteorology," Amer. Meteor. Soc., Boston, Mass.

Hoecker,W. H., 1977, Accuracy of various techniques for estimating boundary-layer trajectories, J. Appl. Meteor., 16 : 374.

Husar,R. B., Gillani,N. V., Husar,J. D., Paley, C. C. and Turcu, P. N., 1976, Long-range transport of pollutants observed through visibility contour maps, weather maps, and trajectory analysis, in : "Proc. Third Symp. on Atmos. Turb., Diff. and Air Quality", Amer. Meteor. Soc., Boston, Mass.

Nordo,J., Eliassen,A., and Saltbones, J., 1974, Large-scale transport of air pollutants, in : " Adv. in Geophysics," Vol. 13B, Academic Press, New York, NY.

Pack,D. H., Ferber, G. J., Heffter, J. L., Telegadas, K.; Angell J. K., Hoecker, W. H., and Machta, L., 1978, Meteorology of long-range transport, Atmospheric Environment, 12 : 425.

Petterssen, S., 1956, "Weather Analysis and Forecasting", 2nd Ed., McGraw-Hill Book Company, New York, NY.

Prahm, L. P., Torp, U., and Stern, R. M., 1976, Deposition and transformation rates of sulfur oxides during atmospheric transport over the Atlantic, Tellus, 23 : 355.

Reap, R. M., 1972, An operational three-dimensional trajectory model, J. Appl. Meteor., 11 : 1193.

Rodhe, H., 1974, Some aspects of the use of air trajectories for the computation of large-scale dispersion and fallout patterns, in : " Adv. in Geophysics", Vol. 18 B, Academic Press, New York, NY.

Samson, P. J., 1979, "Ensemble Trajectory Analysis of Summertime Sulfate Aerosol Concentrations in the Northeastern United States," Ph. D. Thesis, Univ. of Wisconsin-Madison, University Microfilms, Ann Arbor, MI.

Samson, P. J., 1980, Trajectory analysis of summertime sulfate concentrations in the northeastern United States, in : J. Appl. Meteor., 19 : (in press).

Samson, P. J., Neighmond, G., and Yencha, A. J., 1975, The transport of suspended particles as a function of wind direction and atmospheric conditions, J. Air Poll Control Assoc., 25 :1232.

Saucier,W. J., 1955, "Principles of Meteorological Analysis," Univ. of Chicago Press, Chicago, IL.

Smith, F. B. and Hunt, R. D., 1978, Meteorological aspects of the transport of pollution over long distances, Atmospheric Environment, 12 : 461.

Smith, F. B. and Jeffrey, G. H., 1975, Airborne transport of sulfur dioxide from the U. K., Atmospheric Environment, 9 : 634.

DISCUSSION

F. B. FISHER   Does the model show that
                    it is not possible to track trajectories
                    accurately beyond a certain distance, the dis-
                    tance depending on meteorological conditions,

thus putting a limit on any model for describing an episode of high concentration ?

P. SAMSON   Yes, for individual trajectories the uncertanties produced by horizon-
tal diffusion and random errors broaden the
probability of contribution field considerably.
For times greater than 48 hours upstream of
the sampling point the identification of the
relative contribution of different source
regions will have little confidence.
It is only through the examination of a large
ensemble of trajectories that we can hope to
identify consistens source-receptor relationships for large travel times.

THE USE OF A REGIONAL-SCALE NUMERICAL MODEL IN ADDRESSING

CERTAIN KEY AIR QUALITY ISSUES ANTICIPATED IN THE 1980s

Mei-Kao Liu
Philip M. Roth

Systems Applications, Inc.
950 Northgate Drive
San Rafael, California 94903

INTRODUCTION

Ever since air pollution is considered as a serious matter for public concern, emphasis has almost always been on problems occurred in the immediate vicinity of the emission sources. In the early 1970s, concern over carbon monoxide, as a consequence of vehicular emissions, is primarily limited to locations in the proximity of major roadways. Sulfur dioxide or particulate problems related to the operation of industrial point sources are also restricted to an area immediately downwind of the emission stacks. Even the concern over the photochemical smog, as a result of the mixing and reaction of reactive hydrocarbons and oxides of nitrogen, has been confined in the past to the urban area where the emission sources are located. However, over the last few years, there has been a marked shift in the emphasis of air pollution problems. Instead of the typically short or episodic problems of a local nature, significant interest has been placed recently on the degradation of air quality or related problems on regional scales caused by the transport of a variety of air pollutants from a large agglomerate of sources over long distances. Listed in Table 1 are the general characteristics of the local versus regional air pollution problems. Of particular importance is the fact that the latter can often influence regions that are larger in size than many states or even countries. Thus, with pollutants being carried across political boundaries from one jurisdiction to another, a host of issues can be raised that are difficult to disentangle. Furthermore, effects caused by regional air pollutant problems are generally of a long-term nature. With minute increments in pollution levels accumulated over long times, it is therefore quite difficult to ascertain their long-term trends

or to assess their impacts within a relatively short time period.   The
study of acid precipitation is a typical example.   These regional
problems appear to be the key air pollution issues to be resolved in
the 1980s.   In the present paper, the use of a regional-scale
numerical model to address these issues is discussed.

Table 1.   General Characteristics of Local Versus Regional
Air Pollution Problems

Characteristics	Local Air Pollution Problems	Regional Air Pollution Problems
Pollutant Species Primary/Secondary	CO, NO, SO_2, hydro-carbons particulates/ NO_2, O_3	SO_2, NO, hydro-carbons, fine parti-culates/O_3, sulfates, nitrates
Half Life	Less than 24 hours	Several days
Spatial Scales Affected	Less than 100 km	100 km to 1000 km
Important Physical Processes	Horizontal transport, vertical diffusion, chemical reaction	Horizontal transport, horizontal diffusion, chemical reactions, deposition
Effects	Acute health effects. Visibility degrada-tion.	Chronic health effects. Visibility degradation. Acid rain. Ecological impacts.

A REGIONAL-SCALE AIR QUALITY MODEL

A regional-scale air quality model capable of addressing issues
related to the long-distance pollutant transport must have the
following technical attributes:

> Ability to simulate pollutant concentrations at
relatively low levels at locations from 100 to
1000 kilometers away from the emissions sources.
> Ability to simulate the diurnal formation of
secondary pollutants such as sulfate and ozone
as a result of chemical reactions.
> Ability to simulate physical processes, such as
dry deposition, which are only important on
large time scales.

A regional-scale air quality model was first developed at Systems Applications, Inc. in the late 1970s that includes some of these attributes.[1] More recently, a complete regional-scale model possessing all desirable attributes listed above has been assembled. Basically, the model equation is based on the time-dependent multiple-species atmospheric diffusion equation in two dimensions, which can be expressed as follows:

$$\frac{\partial c_i^j}{\partial t} + u\frac{\partial c_i^j}{\partial x} + v\frac{\partial c_i^j}{\partial y} = \frac{\partial}{\partial x}\left(K_x\frac{\partial c_i^j}{\partial x}\right) + \frac{\partial}{\partial y}\left(K_y\frac{\partial c_i^j}{\partial y}\right) + R_i^j - S_i^j \quad ,(1)$$

$$i = 1, 2, \ldots, N \quad ,$$

wher c_i^j is the vertically averaged concentration of species i in

$$j = \begin{cases} 1 \text{ the mixed layer} \\ 2 \text{ the inversion layer} \end{cases}$$

u, v, K_x, and K_y denote wind speeds and the turbulent diffusivities in the x and y direction, respectively. The last two terms on the right-hand side of Eq. (1), R_i and S_i, represent the chemical reaction and pollutant sink terms.

There are two unique features in this regional-scale air quality model. The model has incorporated a photochemical kinetic module capable of handling nighttime chemistry and muliple-day model simulations, it also has included a sophisticated dry and wet deposition mechanism suitable for modeling pollutant transport over long distances. Only a brief description of these two features will be given here. A discussion of the detailed formulation can be found elsewhere.[1,2,3]

Treatment of Photochemistry

For the purpose of long-distance transport modeling or multiple-day simulations of photochemical pollutants, a chemistry package for nighttime will be apparently required. Simulation of the chemical reactions occurring during nighttime can cause computational problems in a conventional kinetic mechanism for urban smog. As the sun sets, the rates of production of many radicals approach zero. At night, as the production rates are virtually zero, the steady-state approximations for the radicals become indeterminate. Moreover, all of the reactions that produce nitric oxide are photolytic, therefore, the production rate of NO, in the presence of O_3, tends to approach zero rapidly via the following reaction:

$$NO + O_3 \rightarrow NO_2 \tag{2}$$

Numerical solutions become inefficient when the solutions approaching a zero asymptote. The time step required for avoiding negative
concentrations becomes extremely small. In addition, because the
destruction rates of some radicals (e.g., the radicals produced by
ozone-olefin reactions) are principally dependent upon the concentrations of NO, the concentrations of these radicals, as calculated
using the steady-state approximation, become large. The high
radical concentrations will further accelerate the rate the NO concentration to approach zero.

To resolve this problem, a different set of chemical reactions
appropriate for the nighttime case has been chosen by truncating
the full photochemical reaction set, the Carbon-Bond Mechanism,
proposed by Whitten et al.[4] The nighttime chemistry module can
account for the following chemistry known to occur during nighttime:

> The scavenging reaction between O_3 and NO
> The self-oxidation of NO at high concentrations
> The oxidation of olefins by ozone
> The decay of PAN in the presence of NO.

Treatment of Deposition Processes

Because of the long residence time, dry deposition becomes an
important process in the overall mass balance in a regional-scale
air quality model. A unique characteristic of the surface layer
is its diurnal variations in temperature, which is a result of daytime heating and nighttime cooling of the surface. This variation
affects the vertical pollutant distribution through atmospheric
stabilities and, consequently, the rate of surface uptake of
pollutants. An elaborate dry deposition parameterization was included in the original model developed by Durran et al.[1] to consider
this effect. The removal processes consist of the diffusion of
the pollutants through the surface layer to the ground, followed by
absorption or adsorption at the atmospheric-ground interfaces.

More recently, because of concern with acid rain problems, a
wet deposition scheme has also been built into the model. The
parameterization of pollutant scavenging by precipitation is
accomplished by the following equation:

$$S_i = \chi\ \rho_w\ R/H \tag{3}$$

where ρ_w is the density of water, R the rainfall rate, and h the
mixing depth. The ratios between the liquid-phase concentration
and the gas-phase concentration, χ, for SO_2 and sulfate must be
prescribed.

For SO_2, assuming that the falling hydrometeor is in equilibrium with the gas-phase SO_2 in the atmosphere, a formula proposed by Hales and Sutter[5] to relate the concentration of SO_2 in rain to the concentration of SO_2 in the layer of air beneath the clouds can be used for X. For the wet deposition of sulfate aerosols, it is assumed that the principle sulfate scavenging mechanism is the nucleation of cloud droplets around sulfate aerosols followed by particle growth through accretion of other sulfate-containing cloud droplets. A secondary removal mechanism is the collection of sulfate aerosol greater than 1 μm below the cloud base. Under these assumptions, the ratio between the sulfate concentration in rain to the sulfate concentration in the layer of air beneath the clouds has been derived by Scott.[6]

APPLICATION OF THE REGIONAL-SCALE MODEL

The regional-scale air quality model described above has been used to address a number of key air pollution problems. These applications are summarized in the following sections.

Analysis of SO_2/Sulfate/Visibility Problems

The first test of this model is its application to the Northern Great Plains region in the United States to assess the air quality impact of emissions from existing and proposed energy developments in that area. The model was exercised for all combinations of two emissions inventories--for the year of 1976 and 1986--and three meteorological scenarios--a strong-wind winter case, a stagnation spring case, and a moderate-wind summer case. The predicted SO_2 and sulfate concentrations were found to be generally greatest in the stagnant spring case, intermediate in the strong-wind winter case, and lowest in the moderate-wind summer case.[1] More recently, the analysis has been extended to the estimation of regional visibility problems via the following well known relationship,[7]

$$\text{Visual Range} = \frac{3.91}{b_{scat}} \tag{4}$$

where $b_{scat} = 7.6 \ m^2g^{-1} \ [\text{Sulfate}] + 2.4 \ m^2g^{-1} \ [\text{Nonsulfate Aerosol}]$ Examples of the computed SO_2 and visual-range distribution in the Northern Great Plains areas are shown in Figures 1 and 2. In this calculation, a first-order linear reaction between SO_2 and sulfate is used, and the non-sulfate aerosol concentrations are assumed to be proportional to the sulfate aerosol concentrations.[8]

Analysis of Rural Ozone Problem

As a result of the long-range transport of ozone and its precursors, elevated oxidant concentrations have been observed in many rural areas in the continental United States.

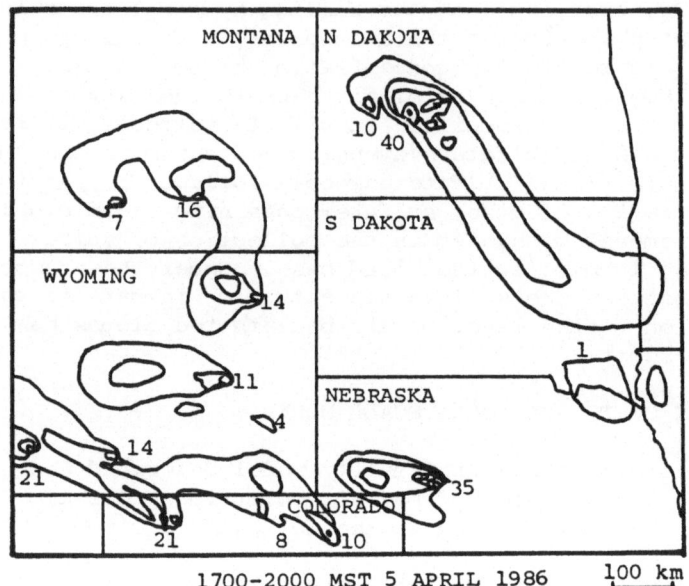

Fig. 1. Predicted Sulfur Dioxide Concentration
         Distributions in the Northern Great
         Plains (Concentrations in $\mu g/m^3$).

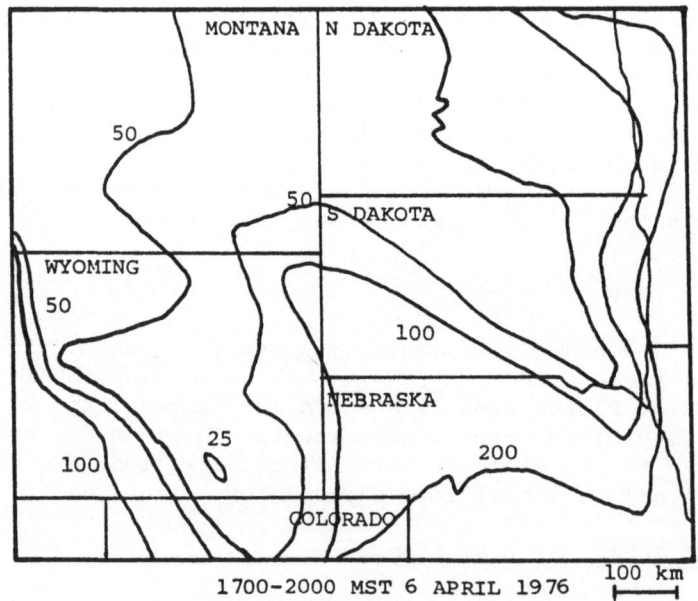

Fig. 2. Predicted Visual Range in Kilometers
         in the Northern Great Plains.

The two-layer mesoscale air quality model has been used to
examine this problem in the northeastern United States. The model
has been exercised to simulate the long-range transport and diffusion of photochemical pollutants under different meteorological
conditions over regions of 1000 km x 1000 km in size. Interaction
of emissions from both point and area sources, the dry deposition
and both the daytime and nighttime photochemistry have been included
in this version of the model to simulate the evolution of the NO_x/
HC/O_3/SO_2 concentration distributions. The input data for the model
has been derived from the 1975 Northeast Oxidant Transport Study.
The predicted results for the concentrations of ozone, nitrogen
oxides and hydrocarbons are generally in good agreement with the
data. For example, as shown in Figures 3 and 4, the computed ozone
concentrations for two consecutive hours on the third day range
from 0.05 ppm to 0.12 ppm within an 150 km radius of New York that is
comparable with the observed ozone distributions.[9]

Analysis of Acid Rain Problems

Increases in anthropogenic emissions have been implicated as
possible candidates for contributing to the increasing acidification
of rain, streams, and lake waters in Scandinavia and certain regions
of the United States. Photochemical or catalytic oxidation of
primary NO and SO_2 emissions form nitrates and sulfates which may
be subsequently introduced directly or indirectly into the recipient
water body via either dry or wet processes. A recent application of
the regional-scale model is an analysis to quantitatively assess
the relative contribution on both local and regional scales of dry
and wet depositions on an episodic basis. Parameterizations of SO_2
and sulfate wet processes described above are invoked in the one-
layer version of the long-range transport model.

The model has been exercised for a realistic rain episode in
the Great Plains region of the United States. The geographic area
studied in this analysis is identical to the one shown in Figures 1
and 2. Preliminary results show that, on a region-wide scale, wet
deposition of sulfate is 10-15 times more efficient than dry removal,
while SO_2 removal by either dry or wet processes is comparable.
Figures 5 and 6 show the overall atmospheric budgets for SO_2.
Locally, sulfate removal rates can exceed 60 percent per hour, and
be more efficient by order of magnitude than dry removal processes.
Details of this modeling analysis that provides a regional climatology of sulfur deposition are discussed in a separate paper.[10]

SUMMARY

A number of key issues have been raised recently concerning
the air quality impact as a result of the transport of atmospheric
pollutants over long distances. A regional-scale air quality model
based on the numerical solutions of the diffusion equation in

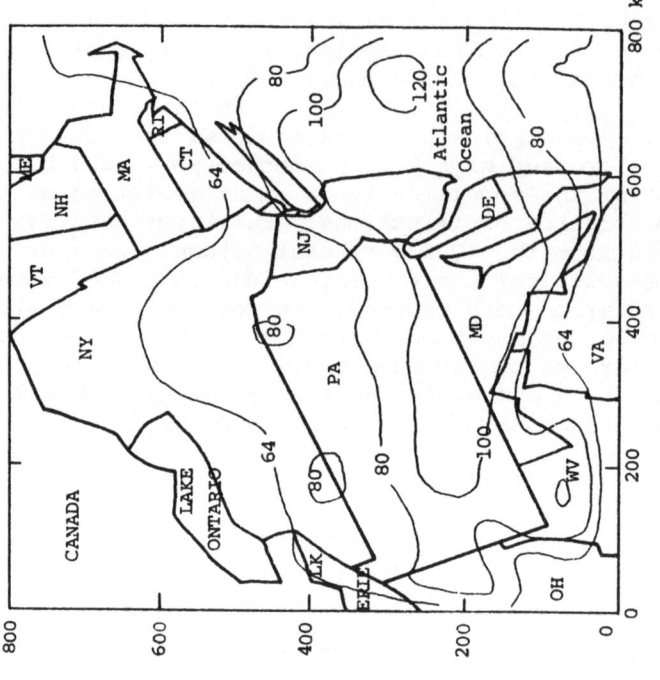

Fig. 4.   Predicted Ozone Distributions in the
          Northeast U.S. (1500–1600 30 July 1975;
          concentrations in ppb).

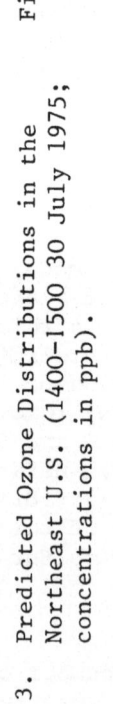

Fig. 3.   Predicted Ozone Distributions in the
          Northeast U.S. (1400–1500 30 July 1975;
          concentrations in ppb).

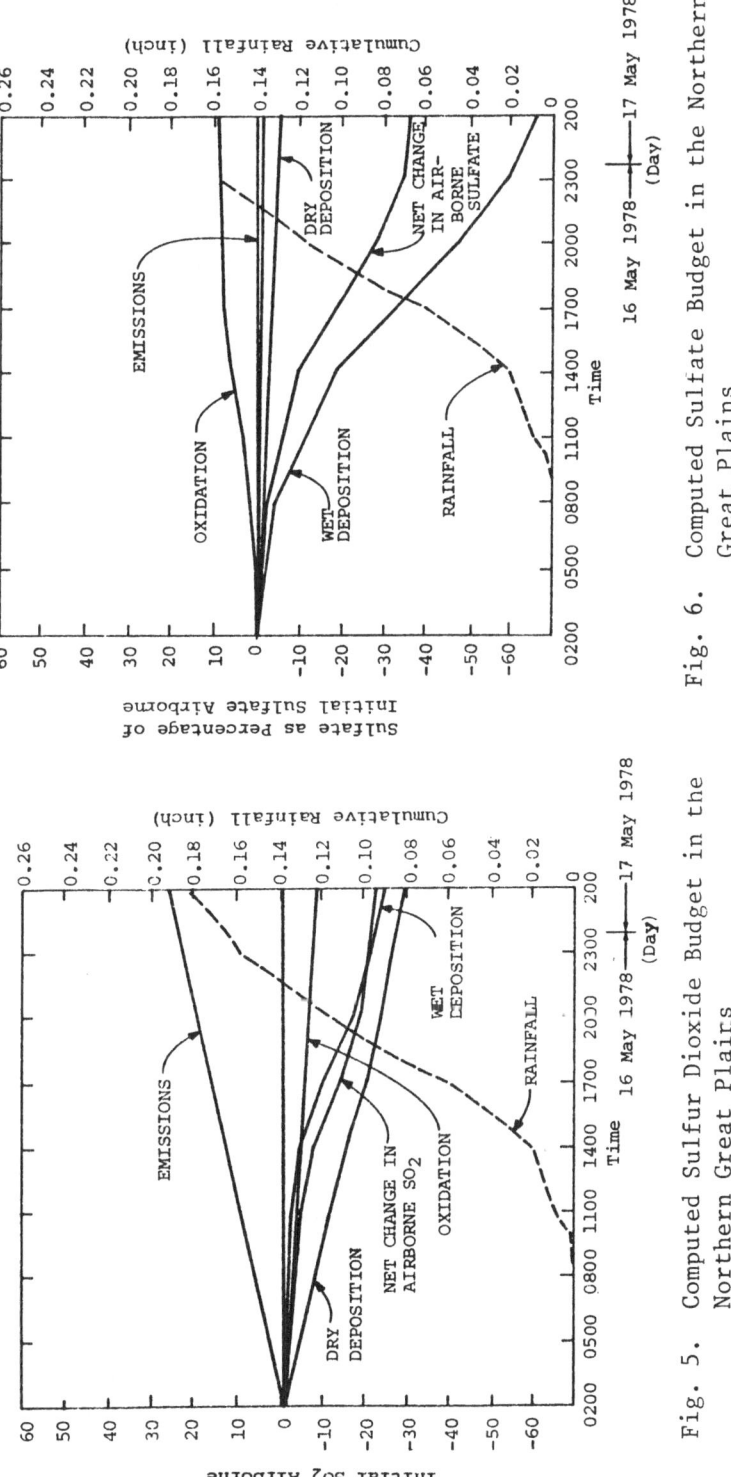

Fig. 6. Computed Sulfate Budget in the Northern
         Great Plains

Fig. 5. Computed Sulfur Dioxide Budget in the
         Northern Great Plains

Eulerian coordinates is used to address these issues. Examples of
the application of this model to regional SO_2/sulfate/visibility
problems, regional ozone problems, and acid precipitation problems
are presented in this paper.

ACKNOWLEDGEMENT

        In this paper, we have freely drawn upon the results from a
number of related studies carried out by our colleagues. We would
like to thank in particular Dale Durran, Mike Wojcik, Upender Kaul,
Jim Killus, and Doug Stewart in this regard. We would also like to
thank Don Henderson of the National Park Service for his foresight
on the regional air quality problems and his constant support and
encouragement during the course of the model development.

REFERENCES

1. D. Durran, M. J. Meldgin, M. K. Liu, T. Thoem, and D. Henderson,
      A Study of Long Range Air Pollution Problems Related to Coal
      Development in the Northern Great Plains, Atmos. Environ.,
      13:1021-1037 (1979).
2. R. W. Bergstrom, C. S. Burton, D. R. Souten, and M. K. Liu,
      Development and Evaluation of a Mesoscale Photochemical Air
      Quality Simulation Model, Quarterly Progress Report No. 6/7,
      Report EI79-51, Systems Applications, Inc., San Rafael,
      California (1979).
3. D. A. Stewart, M. K. Liu, and D. Henderson, An Assessment of
      Relative Contributions of Wet and Dry Deposition via a Long-
      Range Transport Model, Fifth Symposium on Turbulence,
      Diffusion, and Air Pollution, American Meteorological
      Society, Boston, Massachusetts (1981).
4. G. Z. Whitten, H. Hogo, and J. P. Killus, The Carbon Bond
      Mechanism: A Condensed Kinetic Mechanism for Photochemical
      Smog, Environ. Sci. Technol., 14:690-700 (1980).
5. J. M. Hales, S. L. Sutter, Solubility of Sulfur Dioxide in
      Water at Low Concentrations, Atmos. Environ., 7:997-1105
      (1973).
6. B. C. Scott, Parameterization of Sulfate Removal by Precipita-
      tion, J. Appl. Meteorol., 17:1375-1389 (1978).
7. W. E. K. Middleton, Vision Through the Atmosphere, University
      of Toronto Press, Toronto, Canada (1952).
8. C. S. Burton and M. K. Liu, Application of a Mesoscale Air
      Quality Model to Long Range Air Pollution Problems, Studies
      in Environ. Sci.,1:187-192 (1978).
9. M. Wojcik, T. C. Myers, J. P. Killus, O. Serang, and M. K. Liu,
      Development and Evaluation of a Mesoscale Photochemical Air
      Quality Simulation Model, Quarterly Progress Report No. 4,
      Report EI78-118, Systems Applications, Inc., San Rafael,
      California (1978).

10. M. K. Liu, D. A. Stewart, and D. Henderson, Numerical Simulation
    of Transport and Deposition of SO$_2$ and Sulfate by a Regional
    Model (in preparation).

DISCUSSION

P. J. SAMSON   What are the boundary con-
            ditions of your model ? I believe the episode you
            present is one where ozone concentration in excess
            of 100 ppb existed in the Midwest and were advected
            to the East Coast.

M. W. CHAN   The simulation considers con-
            centrations from the day before in the boundary
            conditions.

PROBABILITY PREDICTION OF THE WET DEPOSITION OF AIRBORNE POLLUTION

F. B. Smith

Meteorological Office
Bracknell
England

SUMMARY

This paper reviews various aspects of the problem of incorporating wet deposition of airborne pollution in long-range trans·port models. The probability of rain affecting an element of polluted air is studied in terms of wind speed and direction, altitude, and wet and dry synoptic regions. The accuracy with which the resulting wet deposition can be assessed from averaged rainfall data at nearby meteorological observing stations is studied by analysis of spatial and temporal rainfall intensities measured by the UK network of weather radar, and it is shown that current practice in operational models used in the EMEP programme may be satisfactory provided the station network is not too sparse.

Wet deposition

Many pollutants emitted into the atmosphere are subject to removal by precipitation either by a rain-out process, where the pollutant gets absorbed into the growing rain-drops in the clouds, or by a wash-out process, where the pollutant is captured by the drops as they fall through the polluted sub-cloud layer, or by both processes. (Note that for simplicity we will use the terms "rain" and "rainfall" in this paper to represent more general forms of precipitation). The rate of removal varies in a rather complex way with the rate of rainfall and with the pH of the rain-water as determined by the overall ensemble of pollutants within the drops. Other factors are also important, for example the chemical form of the pollutant ; for one of the commonest industrial pollutants, sulphur, for example, the rate of uptake depends

on whether the sulphur is as sulphur dioxide or whether this has
been oxidised to form sulphate ions.

Whether or not variations in removal rate for a particular
pollutant arising from these complex chemical dependencies are
really important in the real problems of long-range transport of
air pollution has never been firmly demonstrated. Whether effective removal of a pollutant like sulphur by a rain system on one
occasion takes place during 50 km travel or requires 250 km travel because of differences in air chemistry is probably not very
relevant to the long-term deposition pattern when the location,
movement and intensity of rain are so highly variable. One is
tempted to postulate that the only important factors are the
average removal rates for given pollutant concentrations in the
air and for given rainfall rates.

Wet deposition is naturally very variable in space and time,
in contrast with dry deposition which is almost a continuous and
rather constant loss process. In major source areas dry deposition is, over a long period of time, more important than wet deposition, but in more remote areas, and especially in mountainous
regions where rainfall tends to be heavy, the reverse is true.
Figure 1 shows the ratio of wet to dry deposition of sulphur over
Europe for the year 1974 as implied by daily measurements of sulphur dioxide in the air and sulphate in collected rainwater,
(Eliassen, 1978). Spain, Ireland, Wales, the highlands of Scotland, all of Scandinavia and the Balkans, all appear to have a
greater contribution from wet deposition than from dry deposition
(except possibly in a few localised urban areas). The ratio in
ecologically sensitive areas of Norway, for example, varies from
2 to 5. It is therefore clear that it is very important to treat
wet deposition processes adequatly in our modelling of long-range
transport and to gain some understanding of the likely errors
that will inevitably arise.

There are several difficulties that are immediately apparent

(i) in certain rain-systems, and in particular in warm-front
    situations, the pollution absorbed into the cloud drop-
    lets may be of significantly different origin than that
    captured in the sub-cloud boundary layer by wash-out
    processes.

(ii) At present rainfall long a trajectory has to be asses-
    sed from observations of rain made at a rather sparse
    net-work of meteorological stations. The situation is
    particularly acute over the sea where observations are
    few in number and quantitatively rather uncertain.
    Even over land the locations of the stations are not
    always without bias, the majority are in open-level

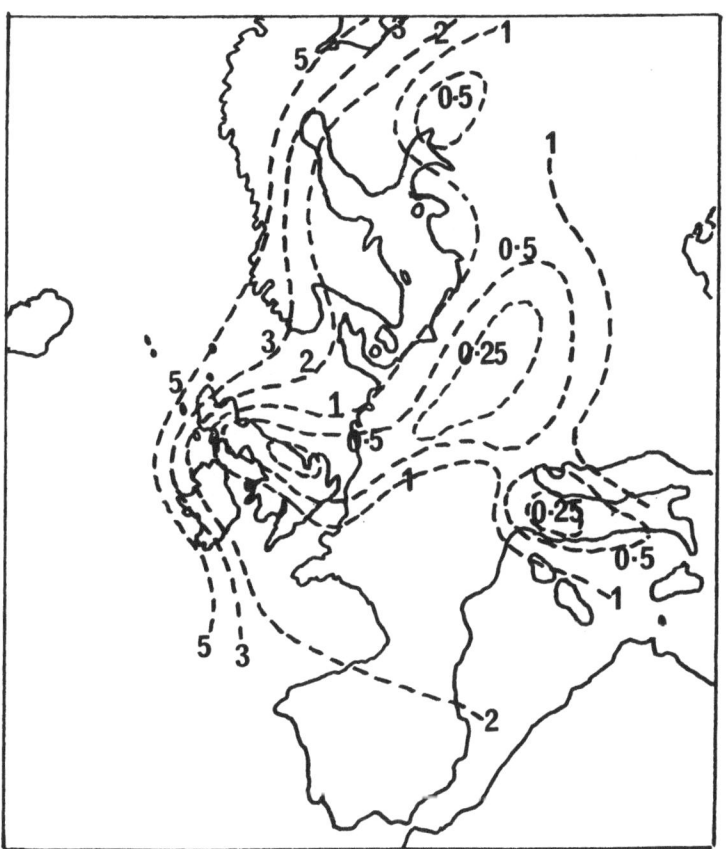

Figure 1. The ratio of wet deposition of
          sulphur to dry deposition of sul-
          phur in 1974. (OECD)

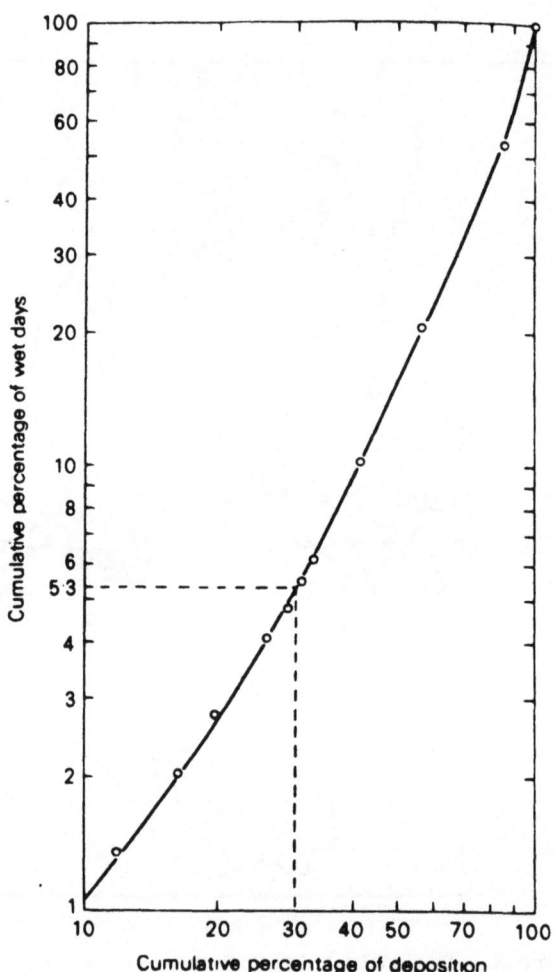

Figure 2a.   The cumulative percentage of deposition plotted against the cumulative percentage of days, plotted on logarithmic scales, at Cottered.

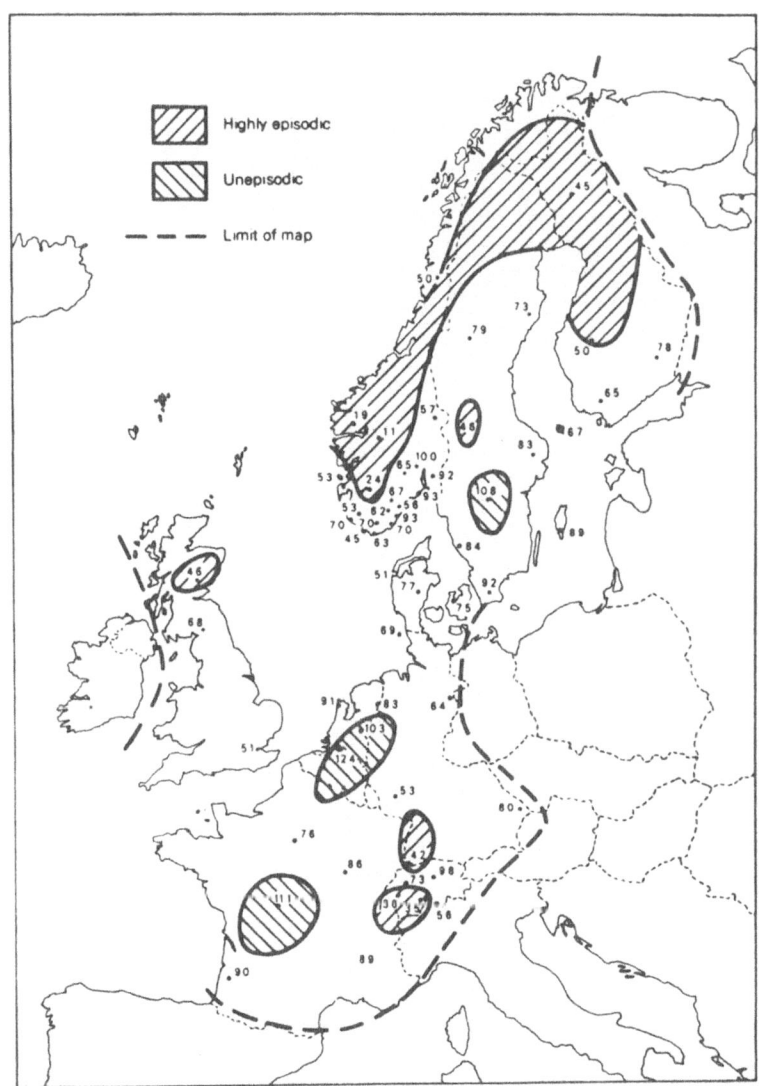

Figure 2b. The map of western Europe showing the distribution of episodicity for 1974.

terrain, in valleys or at coastal sites. Elevated
mountainous sites, where the rainfall is probably
greater than elsewhere, are usually unrepresentatively
few and far between. The rainfall affecting the trajectory has to be deduced either from spot observations made every one or three hours at these stations
or from accumulated rainfall observations made every
three or six hours.
Ultimately this situation may improve as more and more
of Europe and the surrounding sea areas are covered by
weather-radar networks which can make and record quantitative measurements of rainfall intensity with good
spatial and temporal resolution. It is clear such
data are potentially invaluable and provided the extra
cost of incorporating such information into the models
can be justified, one can foresee significant improvements in this particular respect.

(iii) The heaviest rainfall events are sometimes associated
with rather complicated baroclinic flow fields and
simple trajectory-tracking techniques may prove to be
rather inadequate in these important situations. This
consideration leads us to remind ourselves of the importance of so-called episodes which were a subject of
limited study in the OECD study (Smith and Hunt,1978).

Episodes of wet deposition

In the OECD study it became apparent that over most of Europe a quite large and important part of the total annual wet
deposition of sulphate occurred on relatively few days of the
year. Episode definitions are largely arbitrary. Smith and Hunt
defined 'episode-days' at a particular place as those days with
the highest wet depositions which, when summed, make up 30 % of
the annual wet deposition total. Figure 2 illustrates the definition for a site in S.E. England where 5.3 % of the days when
rain occurred in 1974 contributed 30 % of the total wet deposition. They further defined the term 'episodicity' as the ratio,
expressed as a percentage, of the number of episode-days to the
annual number of wet days, and then described an area as 'highly
episodic' if the episodicity was less than 5 %. As the number
of wet days in an area in a year is normally between 80 and 200,
this means the number of episode-days in a highly episodic area
is typically between 4 and 10. An area was defined as unepisodic
if its episodicity was greater than 10 %. Figure 3 shows the
highly episodic areas in western Europe which seem on the whole
to be remote from the chief industrial areas. Trajectories crossing these areas will generally be rather unpolluted, but occasionally they will have arrived from distant industrial areas and
if drawn into slow moving or very active rain systems, heavy

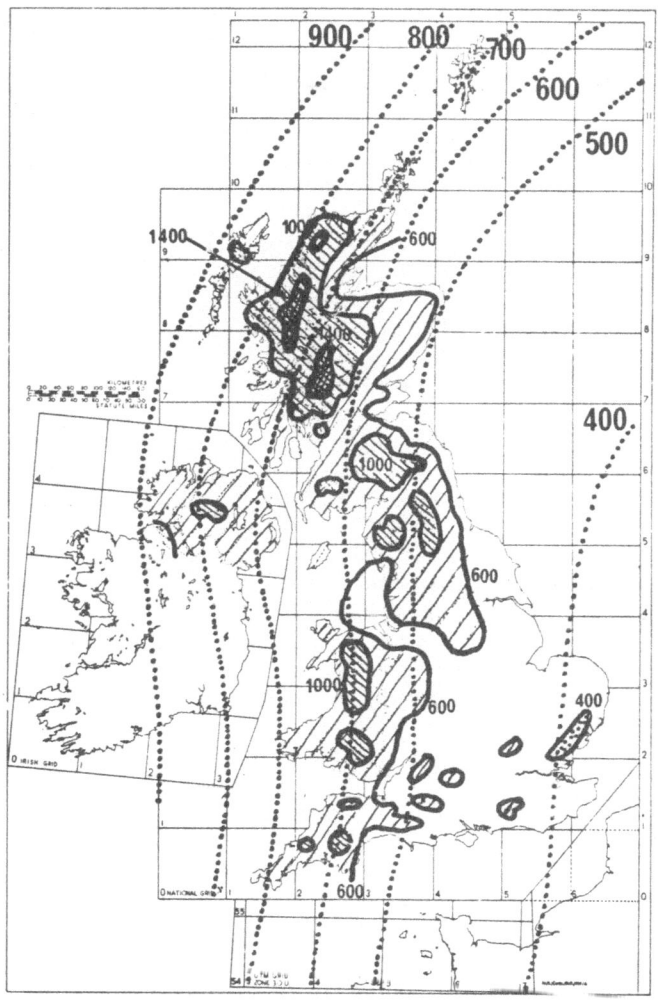

Figure 3. Contours of measured total duration
of rain,D, in units of hours per year, within
the U.K. The dotted lines are estimated sea-
level equivalent durations,S, interpolated
from coastal values. The effect of ground-
level altitude is very evident if this map is
compared with a relief map of the same area.
If h is this altitude in metres, then D ≈ S +0.9h;
and D calculated this way correlates with observed D with a correlation coefficient of
0.96.

depositions can result. Thunderstorms are sometimes associated with these episodes, and when they are it seems that the low-level convergence feeding these storms can give rise to sulphate depositions perhaps twice or three times that to be expected from simple air-trajectory modelling.

In parts of the Alps when thunderstorms are not uncommon and when they tend to be slow moving or 'locked-on' to particular mountains or other topographical features, the wet depositions could be very episodic. This could be very important not only for common pollutants, like sulphate, but also in the rare event of an accidental release of a toxic or radioactive substance from some rather distant installation if the plume were to be drawn into such a meteorological situation. Potentially dangerous depositions could then occur well away from the near-source zone normally considered.

It should be a matter of further study as to whether or not these episodes which although few in number are of relative importance in the annual budget of sulphur and other common pollutants and which apparently are often related to rather baroclinic situations, in which the rain systems may move slowly and not with the geostrophic wind, and are often of fairly small horizontal scale are well treated by the 'simple' statistical-type models in current use. If they are not then clearly model-predictions of annual wet deposition must be suspect.

The Statistical Variation of Rain

So far in this paper we have only referred to operational models of long range transport in which a continuous sequence of trajectories are followed through actually observed meteorological fields and implied depositions are calculated. A great number of occasions, stretching over many years, need to be investigated to build up a reasonable "climatology", for use in assessing the possible consequences of new technology or installations. A quicker, less costly approach may be to use a more statistical type of model in which known long-term wind-roses and knowledge of the statistics of rainfall can be used. Such models are clearly somewhat less reliable because there may be subtle inter-relationships between some of the meteorological factors which are not readily discovered and included. However allowing for the fact that even the operational day-by-day models are far from perfect, we may attempt to use the simpler statistical models to obtain a better qualitative understanding of the fate of pollution, remembering that actual point-estimates of deposition are bound to be subject to a fairly sizeable degree of uncertainty. Smith (1980b) has described a short study of rainfall character over the UK. The first part was to study the effect of ground-level altitude on rainfall amounts and duration. We are fortunate

in the UK in having a network of autographic raingauges from
which both these quantities can be deducted. Figure 3 shows a
map of the UK with contours of"mean hours per year when precipitation is falling", or "duration of rain". Even the quickest of
inspections reveals
     (i) duration increases with altitude
     (ii) there is an overall decrease in duration from west to
          east.
Closer inspection of coastal and lowland values indicates broad
contours of equivalent sea-level duration S can be drawn. These
are shown on the Figure and vary from 900 hours in the west to
400 in the east. This variation comes about largely because depressions tend to run SW to NE along the western side of the UK
with associated fronts affecting the western areas more than the
eastern areas.
Topography obviously has some part to play in determining this
common track of depressions.

    The difference between actual durations D and sea-level
durations S then correlate extremely well with broad-scale altitude $D-S = 0.9$ h, where h is ground-level altitude in metres, and
D and S are in hours. The correlation between D calculated on
this basis and D observed is 0.96.

    Figure 4 shows the annual rainfall R, as observed, and esti-
mated sea-level equivalents. Immediately it is clear there is a
good correlation between D and R. Figure 5 shows the overall relationship between the two and indicates that upland stations not
only have more hours of rain but also a higher average rainfall
rate (up to 2mm h^{-1} compared with 1.3 mm h^{-1} in lowland areas).
This is largely due to the fact that in frontal conditions rainfall in the mountains are sometimes four or more times heavier
than in equivalent lowland areas because the lifting of the airmass over the mountains encourages the development of so-called
low-level "feeder" clouds within the boundary layer.

    The following approximate empirical relation appear to hold :

$$R = D(1.25 - 1.15D' + 36.5D'^{2})$$

where $D^{1} = D \times 10^{-4}$

Whether or not this relationship holds in Europe generally remains
to be investigated. If it does then maps of D (and hence the probability of rain) could be readily deduced from existing R-maps.

    The probability of rain is also a function of wind speed
and direction. Figures 6 and 7 illustrate this for typical lowland UK stations. Figure 6 shows the probability of experiencing
rain in any one hour at a fixed point as a function of the magnitude

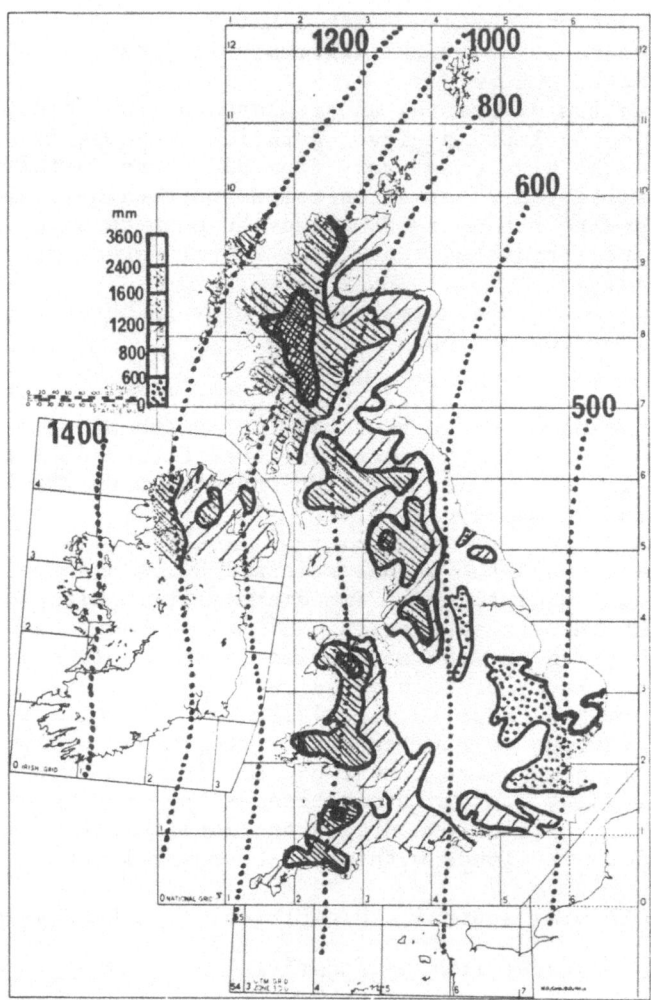

Figure 4. Contours of measured amounts of rain R in mm per year. Again a good correlation may be noted with altitude and with duration D. The dotted lines are estimated sea-level equivalent amounts.

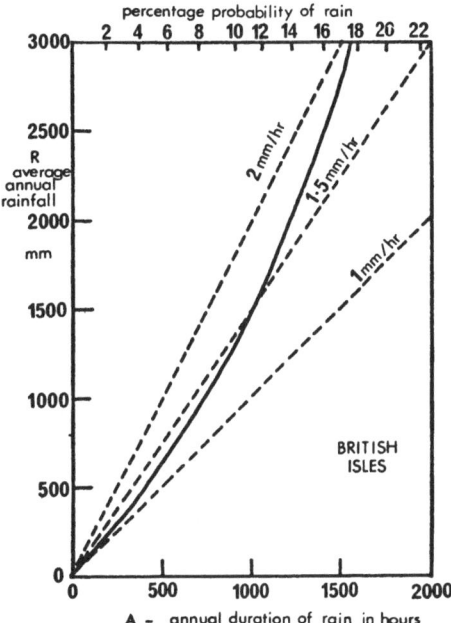

Figure 5. Comparison of the average annual rainfall
(mm y^{-1}) with annual duration of rain (hours y^{-1})
for many sites in the U.K. The curve shows that
rainfall rate tends to increase with R and D, and
hence with altitude. Lowland rates are typically
about 1.3 mm h^{-1} whereas upland rates approach
2 mm h^{-1} in the U.K.

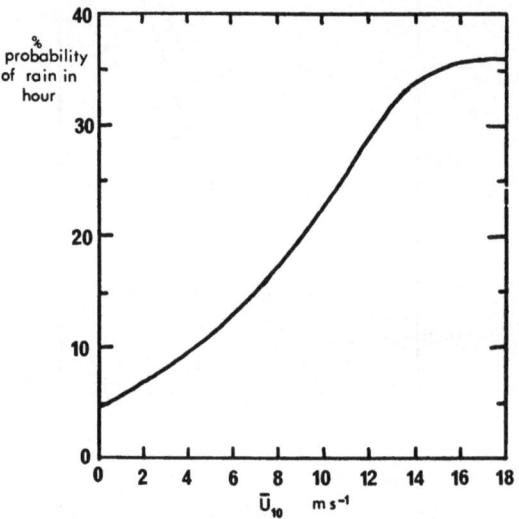

Figure 6. The probability that rain will occur in
any one hour increases with wind speed, at least up
to about 18 m s^{-1}. The curve represents an average
of data from six widely-separated U.K. lowland stations : Plymouth, Heathrow, Elmdon, Renfrew, Turn-
house, and Aldergrove.

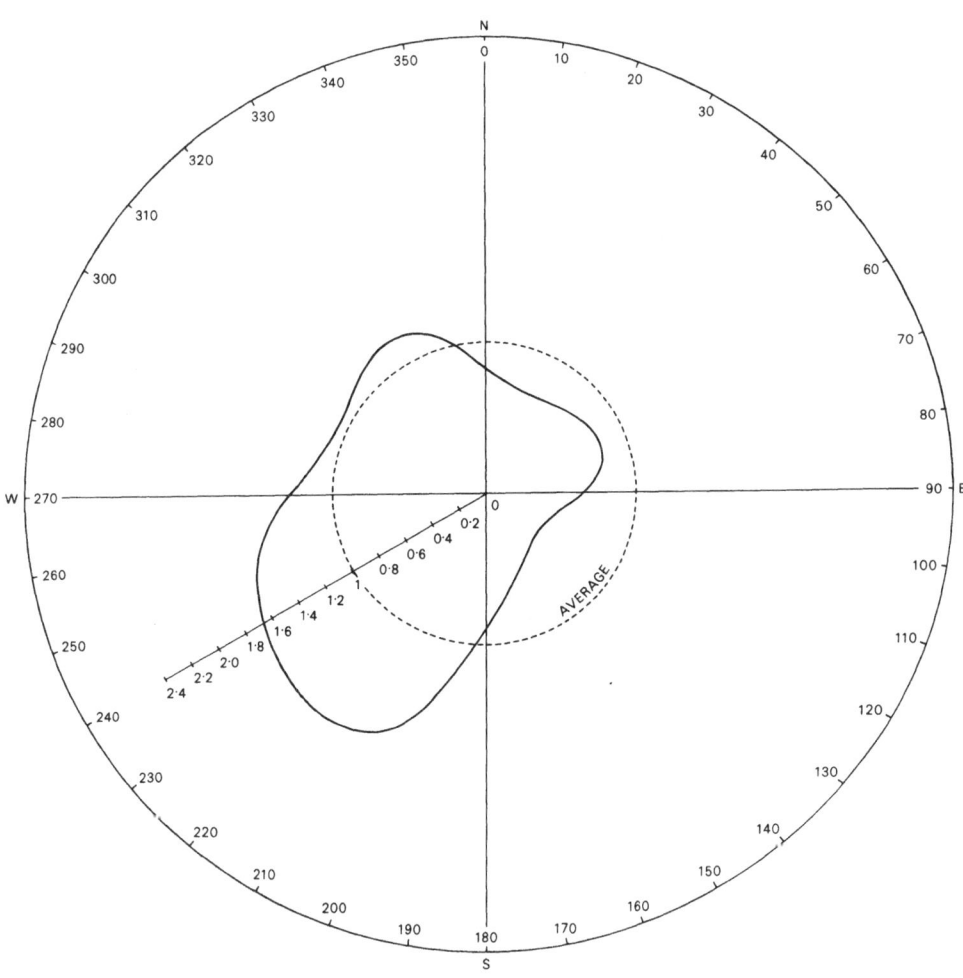

TYPICAL RAINFALL – DIRECTION ROSE

Figure 7.

of the 10-metre wind speed. Part of this variation is no more
than a reflection of the rate at which a pattern of rain cells of
specified spatial distribution can be advected over a fixed point.
Figure 7 is an estimate (based on limited statistics of surface
and geostrophic winds) of the variation of the probability of
rain with the mean boundary-layer wind direction in central
England.

    The results of this section are clearly only a tentative
and incomplete beginning to the true understanding of the statistics of rainfall in the context of the long-range transport
of pollution.

Wet and Dry Synoptic Regions

    Rodhe and Grandell (1972) in a study of the lifetime of
atmospheric aerosols were the first to consider the effect of
moving wet regions (like frontal zones and unstable polar outbreaks) and dry regions (like summer anticyclones) on the average
transport and deposition of airborne material. Fisher (1978,
and in several more recent unpublished notes) has followed a similar approach in studying the lifetime of sulphur in the atmosphere.
More recently Smith (1980) and Venkatram(1980) have studied more
general equations than those of the above authors, and the former
will be examined briefly in section 6. These various studies
show that Rodhe and Grandell's recognition of the importance of
allowing for the different properties of wet and dry regions was
a significant step forward, and that the various model results
give different average deposition rates to models which fail to
make this distinction.

    It should be emphasised that the use of the terms "dry" and
"wet" should only be taken to imply something about the precipitation characteristics and not immediately about the respective
boundary layer relative humidities (altough it may follow that
these are lower in dry regions than in wet).

    Since the synoptic patterns are in constant motion the wet
and dry regions will also move and evolve in time. Consequently
an observer at a fixed point or a moving element of boundary
layer air carrying pollution will experience these regions as wet
and dry periods, and the terms "regions" and "periods" may be
used equivalently according to which is best in the context. The
characteristics of the "experience" will naturally depend on
whether the receptor is the fixed observer (Eulerian experience)
or the air-element (Lagrangian experience), and the problem of
relating the two will be discussed below.

    Pollution released into the boundary layer will usually
remain in the boundary layer or be deposited onto the underlying

surface. The part that remains airborne will be advected by the
mean boundary layer wind until subject to some major "break-down"
of the boundary layer at, for example, a frontal surface. Since
such areas of breakdown are often associated with significant
rainfall, a pollutant which is efficiently removed by rain will
be largely wetdeposited in a relatively short period of time,
and so very little will escape to higher levels.

The boundary layer wind vectors are usually rather different
from those of the upper winds which carry the rain systems, and
so the pollution will tend to move from relatively dry regions
into relatively wet regions and vice versa. A trajectory might
be defined as having entered a wet region when one or more of
those meteorological observing stations within a specified number of kilometres (say 100 km) are reporting rain. Such a definition is clearly somewhat arbitrary and questions of whether
rainfall reported in the last hour or so at a station no longer
experiencing rain, qualifies or not for the trajectory being in
a wet region, require facing and answering.

ApSimon and Wrigley (1980) have made a limited study of
the durations of wet and dry regions as encountered by trajectories evaluated using their MESOS model. The rain data used in
the model over Europe has a time resolution as large as 3 hours,
so that the duration of rain is limited to multiples of 3 hours.
Furthermore the spatial resolution of rain is also limited; in
any 3 hour period the average rainfall in any 100 x 100 km gridcell is deduced either by averaging observations from stations
lying within the cell, or by interpolation from data lying outside the cell.

The end-point of each trajectory was followed from the
source until it advected out of the study-area, and the frequency
and duration of rain along each path was compared with the frequency and duration which would have been deduced if simultaneous
data at the source had been used throughout instead of local
data along the trajectory, i.e. Lagrangian experience of rain
was compared with Eulerian experience. Figure 8 shows their results which indicate no significant difference between the two
sets of statistics. The mean duration of wet periods is about
7 hours and that of wet periods about 18 hours. Both these estimates are subject to uncertainty. On the one hand the fact that
the duration of rain is limited to multiples of 3 hours tends to
overestimate the mean duration of rain and underestimate the
mean dry-duration. On the other hand the finite area covered by
the analysis (Pyrennees to 65°N, Ireland to the Soviet western
borders) underestimates both durations (but principally the mean
duration of dry periods) since once the trajectory leaves the area
the duration is terminated. Further discussion of the relative
Lagrangian and Eulerian probabilities appear in the next section.

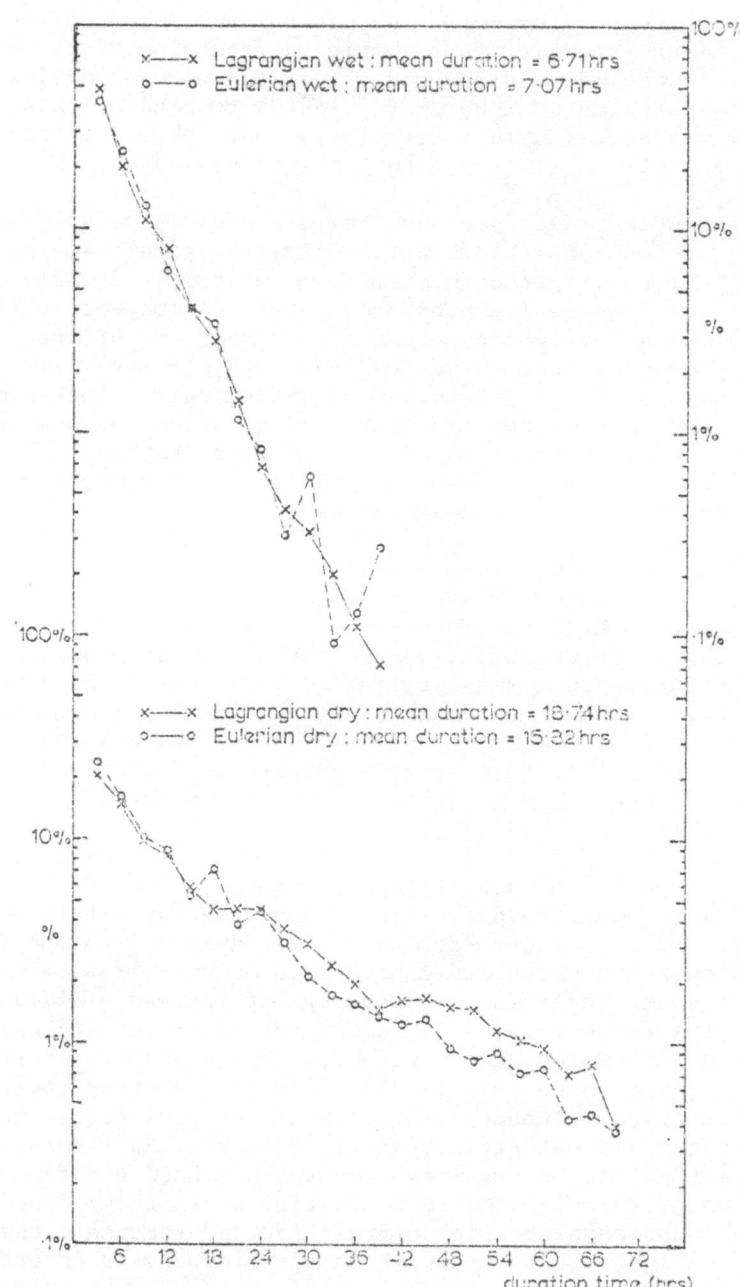

Figure 8. Comparison of statistics for dry and wet periods
during puff trajectories.

The Use of Weather-radar Data

A joint Meteorological Office/Royal Signal and Radar Establishment Group at Malvern have four coordinated weather radar under their control capable of monitoring quantitatively the rainfall distribution over most of England, Wales, the Irish Sea and the English Channel, and displaying the output in real time, or recording the data on magnetic tape or as paper print-outs for future analysis. Figure 9 gives a part of one such print-out for the 1st May, 1980, at 02Z. The resolution of the system is such that the average intensity of the rain can be determined in 5 x 5 km squares (corresponding to British Ordnance Survey Map squares), although it can be tuned to even smaller 2 x 2 km squares. The situation is recorded and displayed every 15 minutes, and so a very complete and detailed picture of the rain-field and its evolution can be obtained. Typically the pattern displays a lot of coherence from one picture to the next, especially on the larger scale. On the morning of May 1st all four radars were operational when a series of active rain-cells spread across central southern England and South Wales from the eastsoutheast between midnight and 07Z. Individual cells, typically 100-150 km across, could be followed across the area for most of this period and smaller "hot spots" of heavy rain associated with individual cloud-cells imbedded in the larger rain-cells persisted for around 30-45 minutes, being picked up on two or three consecutive displays.

During this night of April 30th-May 1st 1980 an anticyclone to the north of Scotland was drifting very slowly northeastwards keeping a northeasterly surface flow over all parts of the British Isles (see Figure 10). A baroclinic trough across southern areas was drawing in warmer air northwestwards and this led to the thundery outbreaks of rain observed by the radar. The surface synoptic pattern changed only marginally over the period 00Z to 07Z maintaining almost uniform steady boundary layer winds of about 50 km h^{-1} from the ENE, whilst the rain-cells were moving at about 40 km h^{-1} from the ESE (veered about 30° from the mean boundary layer wind).
The boundary layer winds were obtained by summing vectorally $\frac{3}{4}$ x the surface geostrophic wind and $\frac{1}{4}$ x the average observed surface wind. Relative to the rain system therefore the boundary layer air was moving from the north at approximately 25 km h^{-1}
The horizontal shape of the individual rain areas naturally varied to some extent, but on the whole there appeared to be a very rough 2:1 ratio between the E-W chord length and the N-S chord length. The consequence of this is that to a first approximation the boundary layer air typically took just about the same time to pass through an individual rain-cell as the cell took to cross a fixed ground station in its path.

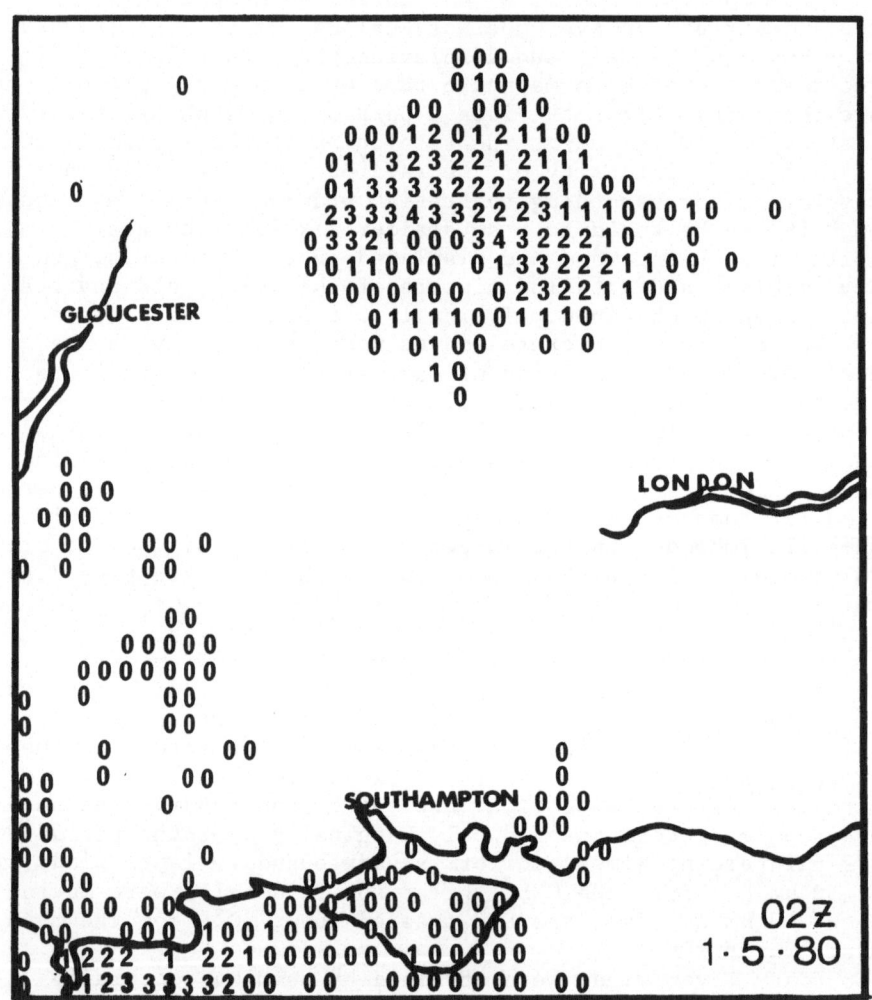

Figure 9. Part of the weather-radar display at 02ƶ on
May 1st, 1980. For purposes of display rainfall intensities are classified as follows   0 : 0 - 1 mm h^{-1}
1 : 1 - 2 mm h^{-1}, 2 : 2 - 4 mm h^{-1}, 3 : 4 - 8 mm h^{-1},
4 : 8 - 16 mm h^{-1}, 5 : 16 - 32 mm h^{-1}.

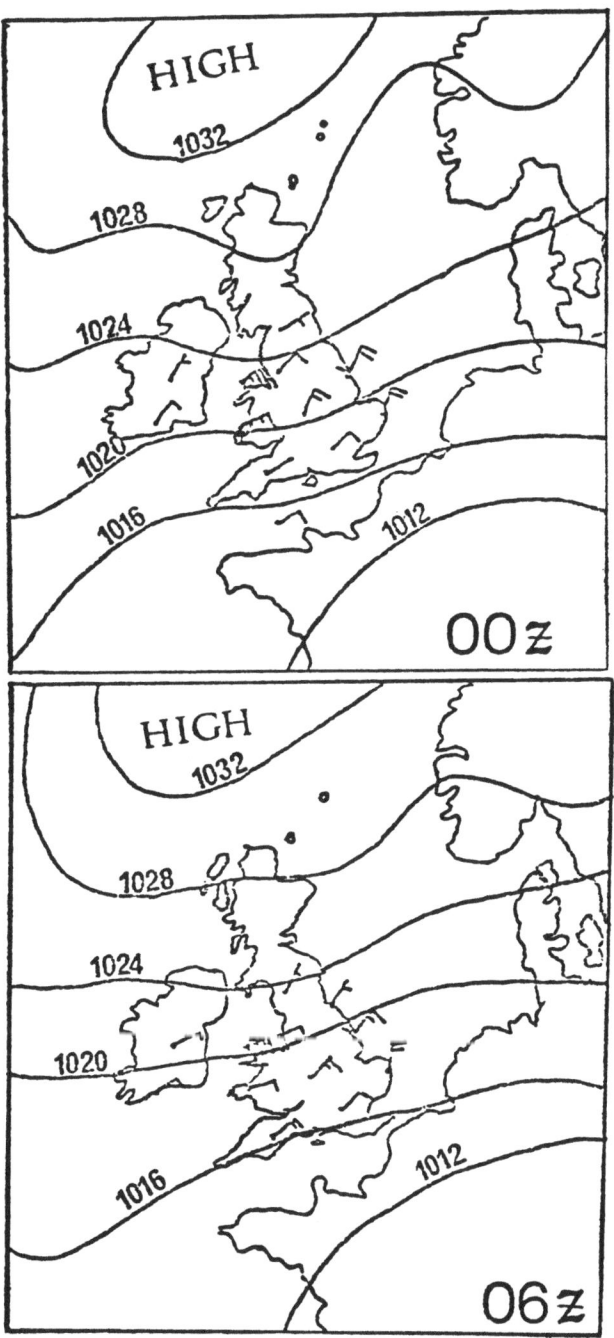

Figure 10. The rather static synoptic situation over
the U. K. on May 1st, 1980.

On this occasion wet regions were defined as all areas within 35 km of rain areas consisting of 3 or more 5 x 5 km squares on each of the recorded radar scans. Both Eulerian rainfalls at individual points (including meteorological observing stations) and Lagrangian rainfalls following boundary layer trajectories were evaluated from the radar-observed rainfall intensities, which were assumed to persist for the 15-min time intervals between scans. Both the Eulerian and Lagrangian records could be subdivided into wet periods and dry periods according as to whether or not the point at a particular time was in a dry or a wet region.

Considering the observed occurrences of rain within wet periods the probabilities of rain and no rain could be determined both following the trajectories and at fixed points : see Table 1.

		Lagrangian	Eulerian
	(rain	0.34	0.28
	(no rain.	0.66	0.72
The probability of	(rain at t+15 given rain at t	0.72	0.66
	(rain at t+15 given no rain at t	0.17	0.14
	(no rain at t+15 given rain at t	0.28	0.34
	(no rain at t.	0.83	0.86
The mean duration of rain when it rained for at least 2 scans		65 mins	65 mins

Table 1 : Statistics of rain and no-rain in wet periods experienced between 00Ζ and 07Ζ on May 1st. 1980.
Roughly an equal number of cases occurred with only 1 period of rain as with 2 or more periods.

The results displayed in Table 1 are of course particular to this one rain occasion, but are at least useful in indicating the likely ratios between Lagrangian and Eulerian statistics. Since, as we have seen, the geometry of the situation suggests that the rain-cells were likely to be sampled with much the same duration by the trajectories and by surface stations, it is not altogether surprising to see a strong agreement between the Langrangian and Eulerian figures in the Table. This confirmation must imply that the evolutionary changes in the rain-cells are of secondary

importance in determining the probabilities. If however the rain-
cells had moved twice as fast relative to the surface stations
as relative to the trajectories, one might reasonably expect the
mean Lagrangian duration to be twice that of the Eulerian and
the probabilities of change from dry to wet and from wet to dry
to be twice as great in the Eulerian frame as in the Lagrangian
frame.

Other useful information can be gained from this study.
For example it is profitable to consider how accurate the current
procedure is as used in the operational long-range transport models to assess wet deposition along the trajectories. In the
Norwegian model (Eliassen 1980), rainfall data, from standard
meteorological observing stations transmitted between countries
on the normal met data communication channels, are used. These
data are rainfall amounts over consecutive 6-hourly periods. The
spacing of the stations varies from about 60 km in lowland areas
of western and central Europe to perhaps 100 km or more in mountainous areas, to very large spacing in sea areas. Stations in
mountainous areas may not always be representatively positioned
and may be biased towards valley sites where the rainfall is likely to be somewhat less than average. Rainfall is interpolated to
the moving trajectory end-point using these 6-hourly amounts and
the wet depositions deduced :

$$\text{Wet deposition rate } \frac{dD}{dt} = \lambda Q = -\frac{dQ}{dt}$$

where Q is the amount of the pollutant above unit surface
area and λ is believed to vary with rainfall rate R in a
way that can be approximated by a linear law at small R or
by a parabolic law at larger R (Kallend, 1980). In our
study we will examine two dependencies appropriate for airborne sulphur :

(i) $\lambda = 10^{-4} R^{\frac{1}{2}}$ (R in mm h^{-1})

(ii) $\lambda = 0.944 \times 10^{-4} R$

These two forms give identical values for λ in lowland
England where the average rainfall rate is 1.3 mm h^{-1}

Our study consists of comparing rainfall and wet depositions
experienced along each of a series of boundary layer trajectories
intercepting the rain area with what would be impled by interpolating between rainfall experienced at nearby actual meteorological stations. The effects of cutting the number of stations and
thereby increasing the average spacing is examined, as is the
effect of averaging the 15-minute radar-rainfalls over 6 hours
so that they are comparable to the rainfalls used in the operational models.

Table 2 shows the results. $Q_{observed}$ is the fraction of
the original S still remaining at the end of the path through
the rain belt using the appropriate radar-rainfalls along the
trajectories whereas $Q_{interpollated}$ refer to the equivalent a-
mount remaining using rainfalls interpolated from the station
network. The σ's refer to the standard deviations of the Q's
about their means; r is the correlation between the observed
and interpolated Q's and n is the number of separate trajectories
considered.

R_{obs} and R_{int} are the corresponding rainfalls experienced
or implied either along the whole trajectories or at individual
points along the trajectories.

The following conclusions may be drawn from the data in
the Table :

(i)   increasing the station spacing decreased the implied
      wet deposition on this occasion and decreased the
      correlations using the 15-min rain data. It also
      decreased the variability in wet deposition from one
      trajectory to another.

Whole trajectory

$$\lambda = 10^{-4} R^{\frac{1}{2}}$$

	15 min data		6 hr data	
	Small Spacing	Wide Spacing	Small Spacing	Wide Spacing
$\overline{Q}_{int}$	.43	.70	.37	.40
Standard deviation σi	.20	.12	.07	.06
$\overline{Q}_{obs}$	.52	.52	.53	.53
σ_o	.27	.27	.29	.29
r (correl)	.76	.31	.84	.85
Number of trajectories	12	12	8	8

$$\lambda = .944 \times 10^{-4} R$$

	15 min data		6 hr data	
	Small Spacing	Wide Spacing	Small Spacing	Wide Spacing
$\overline{Q}_{int}$	.43	.77	.53	.65
Standard deviation σ_i	.27	.17	.12	.09
$\overline{Q}_{obs}$	.44	.44	.47	.47
σ_o	.35	.35	.38	.38
r(correl)	.69	.35	.83	.78
Number of trajectories	12	12	8	8

Whole trajectory and individual points

	Whole traj. 15 min data		Individ. points 15 min data	
	Small Spacing	Wide Spacing	Small Spacing	Wide Spacing
$\overline{R}_{int}$	12.30	3.90	12.30	3.90
σ	7.73	2.67	–	
$\overline{R}_{obs}$	19.21	19.21	19.21	19.21
σ	20.56	20.56	–	
r	0.49	0.30	0.26	−0.17
n	12	12	300	300

Average spacing : small spacing $\sim$ 65 km
wide spacing $\sim$ 145 km.

Table 2 : Results of the analysis of the situation of May
1st, 1980, over southern England and Wales. Q represents
the fraction of sulphur remaining in the atmosphere after
having passed through the rain belt. R is the rainfall
experienced along a trajectory. Other notation is explained in the text.

(ii) the use of the 6 hour rain data rather than the 15-
      minute data significantly reduced the σ_{int} but, perhaps
      surprisingly, improved the correlation coefficients
      to quite respectable levels. The estimates of $\overline{Q}$ using
      the small-spacing network also turned out to be about
      as good as using the 15-minute data. The $\overline{Q}$ estimates
      for the wide-spacing network were equally poor in both
      cases.

(iii) Rainfall predictions are much less reliable than the
      wet deposition estimates and the correlations using
      the 15-minute data are rather low and decrease with
      increasing station-separation. The correlation be-
      tween observed and interpolated rainfall intensities
      at individual points is so low as to be not significant.

(iv) There is no significant difference in the statistics
      between the two forms of λ .

Clearly not too much in the way of general conclusions should be
drawn from just one single case study. We hope to repeat the
analysis on several other situations in the near future to see
whether or not station-spacing appears to be the critical factor
it seem to have been here. If it is, it implies that wet deposition estimates over the sea, and perhaps also over the mountains,
will always be grossly unreliable until an adequate weather radar
network is established to cover these areas.

Statistical long-term deposition from a continuous steady source

    For planning purposes it is often necessary to estimate the
long-term depositions that are likely to occur as a result of
emitting a pollutant from a proposed installation. As we have
seen, whereas the dry deposition may be fairly continuous and
quite readily assessed, the wet deposition depends on several
factors which makes it's estimation somewhat harder. For example
rain depends on wind speed and direction, rain intensity and
frequency depend on topography, and the probability of rain changes
dramatically as the pollution is advected from wet to dry regions
or vice versa. Smith (1980) and Venkatram (1980) have recently
investigated this problem using the basic ideas of Rodhe and
Grandell (1972). We will now briefly outline the method and

results. Because wet and dry regions have such different characteristics, it is necessary, taking sulphur as an example, to define the relative fractions of SO_2 and sulphate that exist in a
statistical sense after a given time of travel t in dry regions
and in wet regions. If q refers to SO_2, Q to sulphate, suffix
D to dry regions and W to wet regions, then four Langrangian rate
of change equations are required to describe the statistical fate
of the original emission :

$$\frac{d}{dt} q_D = S_W q_W - S_D q_D - \frac{v_g}{h} q_D - \alpha_D q_D$$

$$\frac{d}{dt} q_W = S_D q_D - S_W q_W - \frac{v_g}{h} q_W - \alpha_W q_W - A_{q_W}$$

$$\frac{d}{dt} Q_D = S_W Q_W - S_D Q_D + \alpha_D q_D$$

$$\frac{d}{dt} Q_W = S_D Q_D - S_W Q_W + \alpha_W q_W - A Q_W$$

S_D is proportional to the probability that the individual element
of polluted air that is being followed will, at any instant of
time, flow from a dry region into a wet region. S_W is likewise
proportional to the probability of flowing from a wet region into
a dry region. v_g is the deposition velocity of SO_2 to the ground
and h is the depth of the polluted layer usually assumed constant.
The terms including v_g and h therefore represent, albeit
somewhat crudely, the loss by dry deposition. Similar terms can
also appear in the last two equations for the sulphate, as Venkatram has done, but v_g for sulphate is normally very small
and we chose to ignore them. α represents the rate of oxidation
of SO_2 to sulphate and since α depends to some extent on relative
humidity we allow α_W to be larger than α_D on the assumption that
the humidity is likely to be greater in wet regions than in dry.
A represents the removal rate by precipitation and we assume this
is only significant in wet regions. A can be made to depend on
wind direction and on the known local average rainfall (which
must reflect the effect of altitude and thereby allow varying
probabilities of rainfall intensity along the trajectory).

These linear first-order simultaneous differential equations are readily solved by conventional means.

As an example Smith considered the case of 12 continuous
unit sources, emitting SO_2 into the atmosphere, uniformly spaced
around Frankfurt in West Germany at a range of 900 km. Expected
wet depositions were evaluated for Frankfurt for the 12 different
wind directions carrying pollution from each source and experien-

cing very different amounts of rainfall (in a statistical sense)
en route. The following parameters were assumed : a uniform wind
speed of 10 m s^{-1},   v$_g$ = 1 cm s^{-1}, h = 1 km, an average duration of dry periods of 40 hours and of wet 8 hours (although
other values were also tried). α_D was taken to be $\frac{1}{2}$ % h^{-1} and
α_W 5% h^{-1}.
The linear form of λ on R quoted earlier was assumed, where R was
taken to be a function of wind direction and the actual annual
average rainfall at each point.

     The equations were solved numerically and led to the follow-
ing three conclusions, given in Smith's paper :

   (i)   The wet depositions resulting at Frankfurt are im-
          plied to be significantly different, using the model
          equations described above from those obtained using
          the rather too-simple model which fails to differen-
          tiate between mobile wet and dry synoptic regions.

   (ii)   The results are rather insensitive to the selected
          magnitudes of the parameters (like S$_D$ and S$_W$), pro-
          vided they are reasonably selected, since these are
          rather imprecisely known.

   (iii) The wet depositions at Frankfurt are not as sensitive
          to the upwind average rainfall probability distri-
          butions as one might pre-suppose, and show only an
          implied 1.7 : 1 variation with wind direction com-
          pared with a 5 : 1 ratio in maximum to minimum
          average rainfall. (See Figure 11.)

Short period releases and synoptic swinging

     Another aspect of forward planning required when contem-
plating new industrial plant which might in rare accidental situations release large amounts of toxic or radioactive material
into the atmosphere. The probability of certain air dosages and
ground depositions need to be assessed to evaluate the potential
risks to surrounding populations, and what level of emergency
services should be available to deal with such an accident.

     Most releases of this kind would persist for hours, rather
than seconds, since often the cause is a ruptured pipe or lining
which limits the flow rate but which nevertheless takes time to
repair or shut off. During this time the character of the atmosphere is likely to be changing and in particular the wind direction will not remain constant so that different downwind areas
are likely to be affected as the release continues. Small-scale
turbulence has a limited effect on this, but larger synoptic or
meso scale "turbulence" can be very important. This is equally
true for both air dosages and ground depositions (including wet

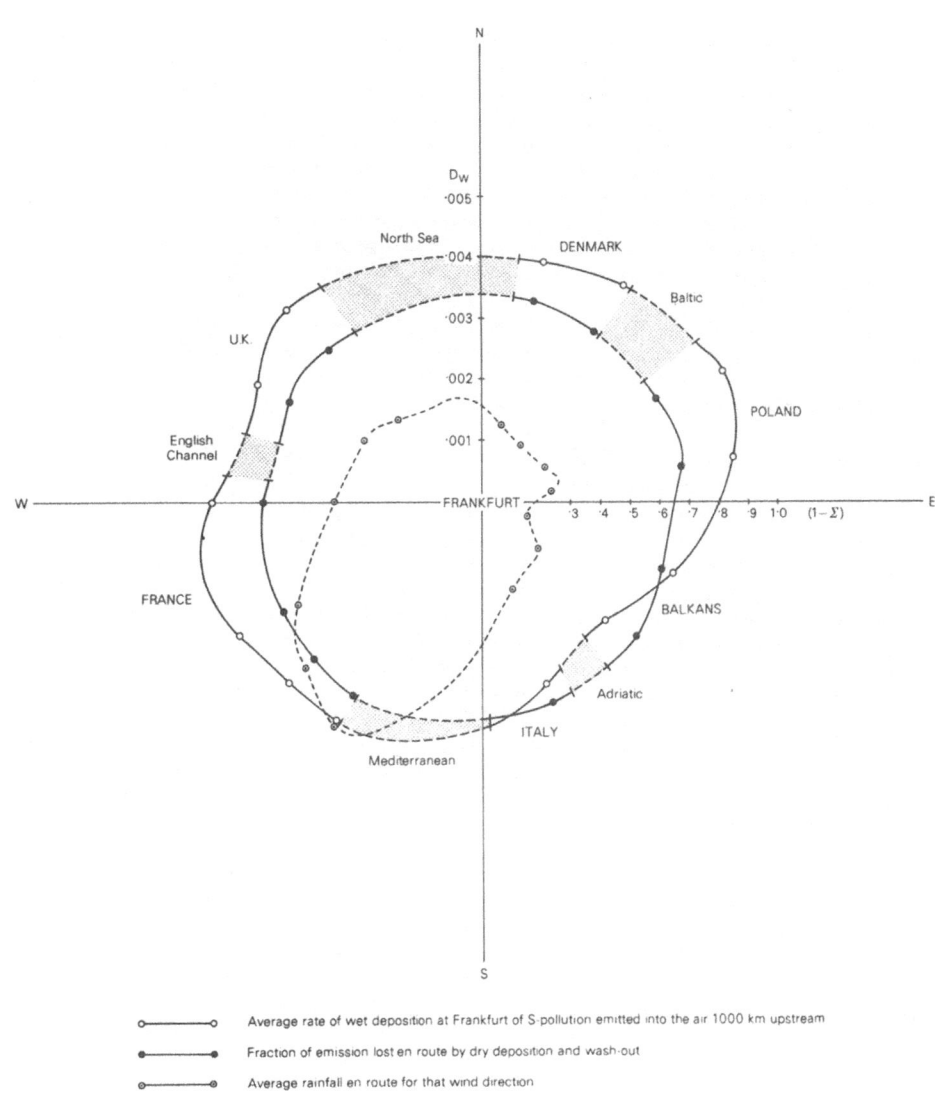

Figure 11.

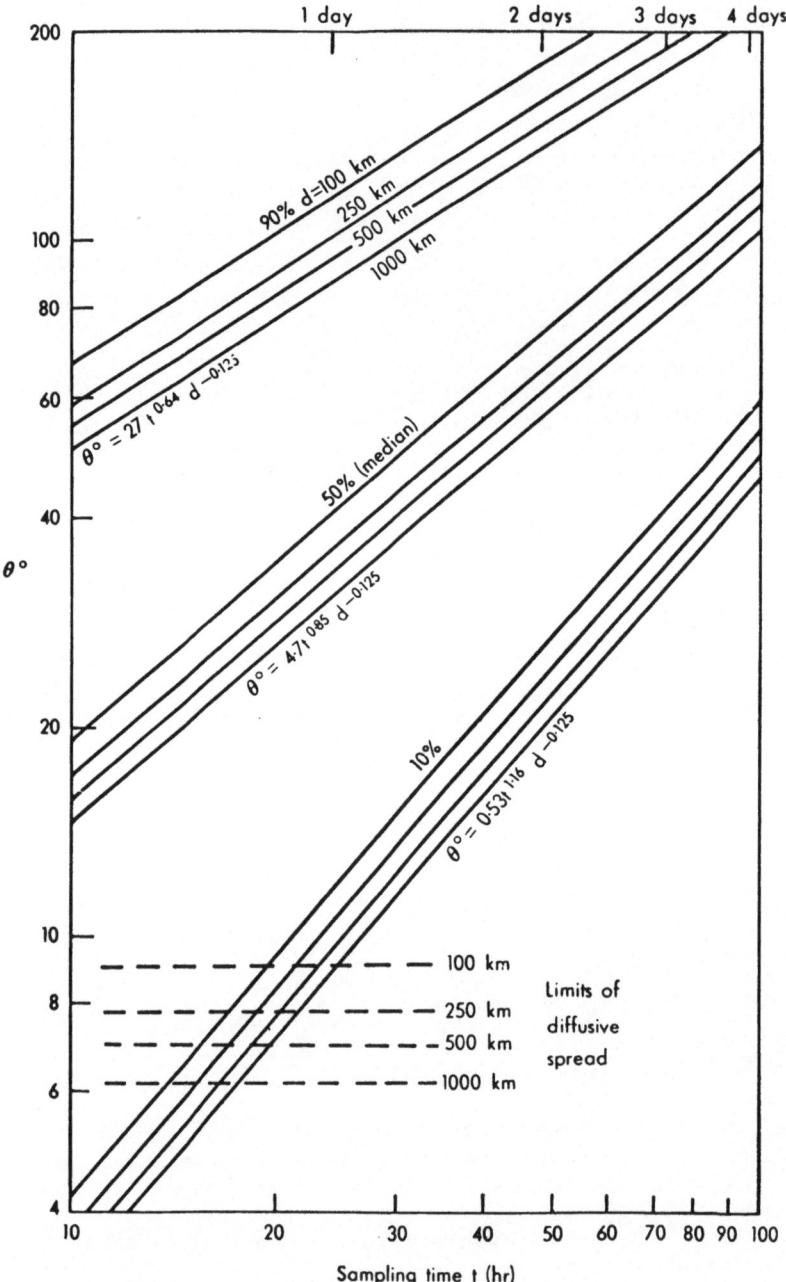

Figure 12. Angular spread θ ° as a function of distance
d from source and sampling time t due to synoptic swinging.

depositions). We call this movement in the trajectories "synoptic swinging" for obvious reasons.

One rather successful way of approaching this problem has been suggested by Smith (1980b). This is to consider, from the analysis of a long sequence of trajectories stretching out from the source for 2000 km or more, the frequencies with which a plume at a given range will swing through specified angles in specified periods of time. Figure 12 shows the kind of result that can be obtained from an analysis of over 2000 sequential surface-geostrophic wind trajectories spaced 12 hours apart. As an example the Figure would tell us that for a release persisting over 20 hours and at a range of 100 km from the source, on half the number of occasions the plume would swing through less than about 35° whereas on 10 % of occasions it could be as little as 9° and on another 10 % it could be more than 100°.

This sort of information could be vital in assessing the probability that wet deposition at a given location could exceed a certain threshold limit in a certain kind of accident.

REFERENCES

ApSimon, H. M., and Wrigley, J., 1980, Notes on the statistics of the frequency and duration of rainfall, Eulerian v Lagrangian. Unpublished Report SAF-TN-6 Env. Safety Group., Imperial College, London.

Eliassen, A., 1978, Atmospheric Env., 12, pp 479-487. The OECD Study of long-range transport of air pollutants : long-range transport modelling.

Eliassen, A., 1980, Cooperative programme for monitoring and evaluation of the long-range transmission of air pollutants in Europe (EMEP). Summary Report of the Western Meteorological Synthesizing Centre. Norw. Met. Instit., Oslo.

Fisher,B. E. A., 1978, The calculation of long-term sulphur deposition in Europe. Atmos. Environ., 12, pp 489-501.

Rodhe, H., and Grandell, J., 1972, On the removal time of aerosol particles from the atmosphere by precipitation scavenging. Tellus XXIV, 5, pp 442-454.

Smith, F. B., 1980, The significance of wet and dry synoptic regions on long-range transport of pollution and its deposition. Atmos. Environ. (in press).

Smith, F. B., 1980(b), The influence of meteorological factors on radioactive dosages and depositions following an accidental release. Proc. of CEC Seminar on "Radioactive releases and

their dispersion in the atmosphere following a hypothetical
reactor accident". Riso, Denmark.

Smith, F. B., and Hunt, R. D., 1978, Meteorological aspects of
the transport of pollution over long distances. Atmos.
Environ., 12, pp 461-477.

Venkatram, A., 1980, A scheme to incorporate scavenging processes
in statistical long-range transport models. Unpublished
note.

Kallend, A. S., Billingsley, J., and Marsh, A. R. W., 1976,
Washout of sulphur dioxide by rain, Central Electricity
Generating Board Report RD/L/N85/76.

DISCUSSION

H. VAN DOP   Lagrangian and Eulerian
coefficients are approximately equal ; you would
expect larger Langrangian correlation coefficients.
The fact however, that they are approximately
equal might explain why the rainfall is bounded to
fixed places (for instance orographic) in the analysis region. Was this the case in your analysis ?

F. B. SMITH   I believe you refer to the
results in Table 1 where the probabilities of
rain, or no rain, given the state during the previous interval, are approximately equal when
viewed either in the Lagrangian frame or in the
Eulerian frame. The explanation for this is that
the trajectories were made to move with the boundary layer winds, which on this occasion were
such that on average the trajectory end-points
passed through individual rain systems in about
the same time that these systems took to cross
fixed points on the ground. There did not appear
to be any marked tendency on this night for the
rain systems to lock-on to any orographical features although some local intensification of rainfall may have occurred. In other cases where the
boundary layer winds might be much closer to the
upper winds, the implied Lagrangian correlations
would presumably be greater than the Eulerian ones.

W. KLUG   The author spoke about trajectories of the air masses. I assume that these
trajectories are boundary layer trajectories which
do not coincide with trajectories in the layers
where precipitation starts and originates. Is
this interpretation correct ?

F. B. SMITH   This is correct. There was roughly a 30° wind-direction shear between the low-level mean boundary-layer wind carrying the pollution and the upper winds advecting the rain systems.

A. VENKATRAM   Does the length of the wet period limit the spatial resolution of the deposition field ?

F. B. SMITH   Assuming a sufficient ensemble of wet and dry periods is taken then I think the answer is no. The spatial resolution of the wet deposition field (or equivalently the optimum spacing of monitoring stations) should be determined by the spatial variability of the precipitation field in a climatological sense (including the effects of topography) and by the relative positions of major pollution sources and the most common tracks of depressions. By the latter I mean for example if depressions tend to cross western parts of the U. K. and run up into Scandinavia, then pollution originating from the industrialised central parts of Europe which gets drawn into these depressions might tend to get washed out differentially in some intermediate band on the southern or south-eastern edge of the depression-belt. This would be an area requiring many monitors.

W. JOHNSON   Do you have any suggestions regarding techniques for obtaining needed precipitation data over data-sparse areas such as the North Sea ? Are there any weather radar that could be used for this purpose ?

F. B. SMITH   Perhaps more effort could be put into equipping oil rigs with precipitation monitors. However reliable measurements are not easily obtained in these environments and perhaps weather radar are the best remedy. These are costly units of course and would require greater justification than can be found in long-range transport studies. At the present time none of the U. K. radar cover more than a small part of the North Sea.

R. STEENKIST   What is known about air movement in a shower ? Is it possible that the air is recycled ?

F. B. SMITH   Some clouds do exhibit rather
organised circulations and these are described in
modern cloud-physics text-books. As an example of
the consequences of this I believe observations
have been made near St. Louis of the rain-out of
a tracer from a 'family' of storms lying across-
wind in circumstances where the tracer must have
first entered only one storm but was in part passed on systematically down the line from one storm
to the next, at each stage being further depleted
by rain-out. Clearly the capacity of any one cloud
to remove the inflowing pollution is limited.
Nevertheless convective-type cloud do need to draw
into themselves a continuous supply of boundary-
layer air to maintain the cloud against loss of
water through precipitation, and this air will
often carry pollution.

K. GRONSKI   Have you considered the correlation between wind direction and wet deposition
in areas other than the Frankfurt area ?

F. B. SMITH   No I haven't. Frankfurt was
chosen as an example since it lay central to Europe
in an area where the average annual rainfall decreases rather rapidly from south to north, and
therefore seemed an ideal choice for testing the
importance of the factors involved in the theory.
I agree that it would be useful to check the conclusions against other sites.

THE STATISTICS OF PRECIPITATION SCAVENGING DURING LONG RANGE TRANSPORT

B.E.A. Fisher

Central Electricity Research Laboratories,
Leatherhead, Surrey KT22 7SE
England

ABSTRACT

The statistical properties of rainfall duration along air trajectories are derived on the basis that the onset of wet and dry periods occur at random. These properties are used to formulate a statistical model of pollution transport over long distances which includes the effect of precipitation scavenging. This enables an earlier model of the annual average pattern of air concentration and deposition around source regions to be extended to include the frequency with which given concentration or deposition levels are exceeded. Some support for the assumptions of the model is given by an analysis of rainfall duration over a 30 year period at three sites in England.

INTRODUCTION

An adequate treatment of removal processes involving precipitation is one of the major problems when treating the long range transport of airborne material. Besides the issues of whether the removal processes can be described by a single rate constant and the value that this parameter should take, the distribution of rainfall in time and place is not easy to specify. The network of ground stations measuring precipitation is not dense enough over land to obtain the necessary resolution and is absent over the sea. Even if the necessary network were available, the direct treatment of precipitation scavenging by considering regions where a parcel of air crossed wet and dry zones would present a considerable data-handling problem.

Some progress has been made using the direct approach. In the OECD project[1] to study sulphur transport over Europe,objectively analysed six-hour precipitation fields were used to assess the removal of sulphur oxides from the air.An empirical relationship was then used to relate sulphur in the air to the sulphate concentration in precipitation.Therefore the estimated wet deposition is not exactly equal to the sulphur removed from the air during transport along a trajectory. An improved method has been tried in the current EMEP exercise in which modelled sulphur concentrations over Europe have been compared with measured concentrations from the EMEP network[2]. The method is based on an objective assessment of the rate of removal of water vapour from the boundary layer over an interval of six hours and the use of the same value of the rate constant to describe the removal of sulphur from the air. This approach has a firmer physical basis, but has not brought improved agreement between model predictions and measurements.This is presumably because the empirical relationship fixes a satisfactory "average" ratio between sulphur in the air and sulphur in precipitation.

An alternative method to the direct approach,used successfully by Rodhe and Grandell[3],is to assume that the incidence of precipitation along the trajectory of an air parcel can be treated as a random Markov process.One is then able to calculate the main parameters needed to estimate the effect of precipitation on the long term average concentrations of airborne material.These parameters are $\bar{f}_p$,the mean fraction of material which survives a period of travel of length T hours and $\bar{f}_w$ the mean fraction of material which survives, given that it is raining at the end of the period of travel. These two parameters are all one needs to include wet deposition in a calculation of the long term average distribution of airborne material in which removal by precipitation has been neglected[4].

In some situations it is not the long term average concentration that is required,but the probability of the concentration occuring in a given range. For example for assessing risks from accidental releases, one may wish to know, given the release of a given quantity of airborne material, what the probability is that the wet deposition at a site at a long distance from the source,is exceeded. The purpose of this paper is to show how the statistical treatment of precipitation developed by Rodhe and Grandell[3] can be extended to provide such probability distributions.In order to do this it is clearly necessary to be able to specify the probability $p(t,T)$ that the sum of all wet periods during an interval T is t. An equation for $p(t,T)$ will be derived.

Before presenting the derivation it is worth stating some simple approximate ways of specifying the proportion of time it has been raining along the trajectory of a plume. One assumption that is sometimes made is that conditions at the time of emission persist throughout the plume lifetime. Thus there is a probability,

p_w, that material is subject to precipitation throughout the whole of its travel. p_w is equal to the fraction of the time rain falls. There is also probability $p_D = 1 - p_w$ that the material is not subject to any precipitation throughout its travel. p_D is the fraction of the time that conditions remain dry. This approximation is reasonable near to the source, but incorrectly implies no wet deposition at long distances from the source.

The other assumption that is sometimes made is that material is subject to continuous removal by wet deposition, but at a reduced rate $p_w \Lambda$, where Λ is the scavenging coefficient (equal to the fraction of material removed per unit time in precipitation). This method treats wet deposition in a similar manner to dry deposition. It is clearly in error at long distances since it implies that a fixed fraction of material equal to $\exp - p_w \Lambda T$ always survives after a travel time T, whereas there is a finite probability that all the material has survived, given by the proportion of occasions on which dry weather has persisted throughout the period of travel.

DERIVATION OF EXPRESSION FOR p(t,T)

In this section an expression for p(t,T), the probability that the sum of all wet periods during an interval T lies between t and t + dt, is derived. There are four different situations each of which will be treated in turn.

(a) It is assumed that it is dry at the start of period (t = 0) and at the end of the period (t = T). The probability of a change to a wet period during an arbitrary interval, dt, in a dry period is dt/t_D. Similarly the probability of a change to a dry period during an arbitrary interval dt in a wet period is dt/t_w. The probability of n changes during the total time interval, t, of wet periods, is given by the Poisson distribution.

$$\frac{1}{n!} (t/t_w)^n e^{-t/tw} \tag{1}$$

Similarly the probability of m changes during the total period T - t of dry conditions is

$$\frac{1}{m!} \left(\frac{T - t}{t_D} \right)^m e^{-(T - t)/t_D} \tag{2}$$

Hence, if the complete period of lenght T starts dry and finishes dry, the probability that the total length of wet weather lies between t and t + dt, is

$$P_{DD} = P_D \sum_{n=0}^{\infty} \frac{1}{n!} \left(\frac{t}{t_w} \right)^n e^{-t/t_w} \frac{1}{n+1!} \left(\frac{T-t}{t_D} \right)^{n+1} e^{-(T-t)/t_D} \frac{dt}{t_w} \quad (3)$$

where $p_D = \dfrac{t_D}{t_D + t_w}$ is the probability of the period starting in dry conditions.

The probability of no rain throughout the period, which is not included in this sum, is $p_D e^{-T/t_D}$.

(b) If it is assumed that the period starts dry (at $t = 0$) and finishes in wet conditions (at $t = T$), then the probability of wet weather of total length between t and $t + dt$ is

$$P_{DW} = P_D \sum_{n=0}^{\infty} \frac{1}{n!} \left(\frac{T-t}{t_D} \right)^n e^{-(T-t)/t_D} \frac{1}{n!} \left(\frac{t}{t_w} \right)^n e^{-t/t_w} \frac{dt}{t_D} \quad (4)$$

(c) If it is assumed that the period starts wet (at $t = 0$) and finishes in wet conditions (at $t = T$), then the probability of wet weather of total length between t and $t + dt$ is

$$P_{WW} = P_W \sum_{n=0}^{\infty} \frac{1}{n!} \left(\frac{T-t}{t_D} \right)^n e^{-(T-t)/t_D} \frac{1}{n+1!} \left(\frac{t}{t_w} \right)^{n+1} e^{-t/t_w} \frac{dt}{t_D} \quad (5)$$

where $p_w = 1 - p_D$ is the probability of the complete period starting in wet conditions. The probability of rain throughout the period, not included in this sum, is $p_w e^{-T/t_w}$.

(d) If it is assumed that the period starts wet (at $t = 0$) and finishes in dry conditions (at $t = T$), then the probability of wet weather of total length between t and $t + dt$ is

$$P_{WD} = P_W \sum_{n=0}^{\infty} \frac{1}{n!} \left(\frac{t}{t_w} \right)^n e^{-t/t_w} \frac{1}{n!} \left(\frac{T-t}{t_D} \right)^n e^{-(T-t)/t_D} \frac{dt}{t_w} \quad (6)$$

The probability density that the total length of wet periods lies in the range $(t, t + dt)$, regardless of starting or finishing conditions, is thus

$$p(t,T) = P_{DD}(t,T) + P_{DW}(t,T) + P_{WD}(t,T) + P_{WW}(t,T) \quad (7)$$

The probability density that the total length of wet periods lies in the range $(t, t + dt)$ subject to the condition that it is raining at the end of the period is

$$p_1(t,T) = p_{DW}(t,T) + p_{WW}(t,T) \qquad (8)$$

These expressions for p_{DD}, p_{DW} etc. look complicated but are easily evaluated numerically. The expressions may be derived from first principles by summing and integrating all possible arrangements of wet and dry periods, provided the total length of wet periods is t^5. A number of important measures of the effect of precipitation scavenging on long range transport can be derived from $p(t,T)$ and $p_1(t,T)$.

FRACTION OF MATERIAL REMAINING AIRBORNE

The probability density distribution that a fraction f_p of the material remains airborne in the plume at time T is $p(t_*,T)$, where $t_* = - (\log_e f_p) \Lambda$ or equivalently $f_p = \exp(-\Lambda t_*)$. Thus the probability that the fraction of the plume surviving is greater than f_p is given by

$$\text{Prob}(f > f_p) = \int_0^{t_*} p(t,T) \, dt \qquad (9)$$

Similarly the mean fraction of material remaining airborne at time T is

$$\bar{f}_p = \int_0^T e^{-\Lambda t} p(t,T) \, dt \qquad (10)$$

The mean value of the air concentration, including the effects of precipitation scavenging, is obtained from the mean air concentration calculated without including precipitation scavenging, by multiplying by this factor.

The maximum rate of wet deposition D_{max} at any receptor is obtained by multiplying the integrated vertical concentration in the absence of precipitation by the scavenging coefficent. This is of course a very pessimistic assumption as it assumes no precipitation has occurred between the source and the receptor. The wet deposition, D, is likely to occur at a fraction f_w of the maximum rate of deposition with a probability $p_1(t_*,T)$, where

$$t_* = -(\log f_w) / \Lambda \text{ or equivalently } f_w = \exp -\Lambda t_*.$$

Thus the probability that the rate of wet deposition is greater than
a fraction f_w of the maximum possible rate of wet deposition is
given by

$$\text{Prob}(D/D_{max} > f_w) = \int_0^{t*} p_1(t,T) \, dt \qquad (11)$$

Similarly the mean rate of wet deposition $\bar{D}$ at time T is obtained
by multiplying the integrated vertical concentration in the absence
of precipitation by the factor $\Lambda \bar{f}_w$, where

$$\bar{f}_w = \int_0^T e^{-\Lambda t} p_1(t,T) \, dt \qquad (12)$$

Knowledge of the probability distributions $p(t,T)$ and
$p_1(t,T)$ is very sparse. Until a time when better information on
rainfall statistics are available, the only practical alternative
is to make use of equations (3) to (6) of the previous section.
Examples of the shape of the probability distributions given by
equations (9) and (11) are shown in Figs. 1 and 2.

From equations (7), (8), (10) and (12) explicit expres-
sions for $\bar{f}_p$ and $\bar{f}_w$ can be derived. The derivation is given in
Fisher[5] and involves evaluation of the expressions

$$G_{DD}(T) = \int_0^T p_{DD}(t,T) e^{-\Lambda t} dt, \quad G_{DW}(t) = \int_0^T p_{DW}(t,T) e^{-\Lambda t} dt \quad \text{etc.}$$

It is more convenient to consider the Laplace transforms of these
expressions

$$G_{DD}(p) = \int_0^\infty e^{-pT} \int_0^T p_{DD}(t,T) e^{-\Lambda t} dt \, dT \quad \text{etc.}$$

Use of the convolution theorem enables the series in equa-
tions (3) to (6) to be summed and $G_{DD}(p), G_{DW}(p)$ etc. to be evaluated.
The alternative derivation, which is the one originally used by
Rodhe and Grandell[3], is to consider the equations satisfied by the
functions $G_{DD}(T)$, $G_{DW}(T)$ etc. For example $G_{DD}(T)$ is the probability
of a particle of material released in dry conditions surviving until dry conditions at time T. Thus $G_{DD}(T)$ satisfies the differential equation

$$\frac{\delta G_{DD}}{\delta T} = -\frac{1}{t_D} G_{DD} + \frac{1}{t_w} G_{DW}$$

subject to $G_{DD}(T) = p_D$ at $T = 0$. G_{DW}, G_{WW} and G_{WD} satisfy similar types of equation which on taking the Laplace transform lead to identical expressions to those derived by the direct evaluation of equations (7) and (8). $\bar{f}_p$ and $\bar{f}_w$ are given by

$$\bar{f}_p = G_{DD}(T) + G_{WD}(T) + G_{DW}(T) + G_{WW}(T)$$

$$= \frac{1}{E_1 - E_0}\left[\left(E_1 + \frac{\Lambda t_w}{t_w + t_D}\right) e^{E_0 t} - \left(E_0 + \frac{\Lambda t_w}{t_w + t_D}\right) e^{E_1 t}\right]$$

while   (13)

$$\bar{f}_w = G_{WD}(T) + G_{WW}(T)$$

$$= \frac{p_w}{(E_1 - E_0)}\left((E_1 + \Lambda) e^{E_0 t} - (E_0 + \Lambda) e^{E_1 t}\right)$$ (14)

where $2E_{0,1} = -(\Lambda + t_w^{-1} + t_D^{-1}) \pm \sqrt{(\Lambda + t_w^{-1} + t_D^{-1})^2 - 4\Lambda t_D^{-1}}$

(15)

As a check these expressions have been evaluated for the examples shown in Figs. 1 and 2. They clearly give the correct mean values of the fractions surviving.

SIMPLE LIMITS

Equations (3) to (8), though not difficult to evaluate, are difficult to interpret at a glance. In practice the scavenging coefficient is not known to great accuracy nor are the statistics of the mean length of wet and dry periods well known. Hence the consideration of simple limits of these formulae would be helpful.

If the mean length of wet periods t_w is longer than the inverse of the scavenging coefficient Λ^{-1}, then on average little material will survive one or more wet periods. One need only consider the lowest order terms. So assuming $t_w \gg \Lambda^{-1}$ and $T \gg t_w$, and allowing just one wet period during the total period of length T, the probability of more than a fraction $1/e$ of the particles surviving, is from equations (3), (4), (5), (6) and (7) given by

$$\text{Prob}(f > 1/e) = P_D \int_0^{\Lambda^{-1}} e^{-(\Lambda + 1/t_w)t_*} \left(\frac{T - t_*}{t_w}\right) e^{-(T - t_*)/t_D} \frac{dt_*}{t_D} \quad (16)$$

if the wet periods lies in the middle of the interval,

$$\text{Prob}(f > 1/e) = P_D \int_0^{\Lambda^{-1}} e^{-(\Lambda + 1/t_w)t_*} e^{-(T - t_*)/t_D} \frac{dt_*}{t_D} \quad (17)$$

if the wet period lies at the end of the interval,

$$\text{Prob}(f > 1/e) = p_w e^{-T/t_w} \quad (18)$$

if it rains throughout the period, but since $T \gg t_w$, this probability is virtually zero, and

$$\text{Prob}(f > 1/e) = p_w \int_0^{\Lambda^{-1}} e^{-(\Lambda + 1/t_w)t_*} e^{-(T - t_*)/t_D} \frac{dt_*}{t_w} \quad (19)$$

if the wet period lies at the beginning of the interval.
Since $\Lambda^{-1} \ll t_w \ll T$, the sum of these contributions is approximately

$$\left(p_D(T/t_D) + 2p_w\right) e^{-T/t_D} \left(\frac{1 - e^{-(1 + 1/\Lambda t_w)}}{1 + \Lambda t_w}\right) \quad (20)$$

In addition there is the probability of no rain and therefore no loss of material by scavenging during the time T, and this occurs with the probability $p_D e^{-T/t_D}$. Hence the overall probability of more than $1/e$ of the particles surviving an arbitrary period of length T reduces to

$$\text{Prob}(f > 1/e) = P_D e^{-T/t_D} \left(1 + (T/t_D + 2t_w/t_D) \frac{(1 - 1/e)}{\Lambda t_w}\right) \quad (21)$$

This approximate formula for the probability that more than $1/e$ of the airborne material survives has been evaluated using the values needed to plot Fig. 1 and shows good agreement with the exact expression.

The worst situation from the point of view of the airborne concentration of particles is when all the material survives and this occurs with probability given by

$$\text{Prob}(f = 1) = p_D e^{-T/t_D} \quad (22)$$

which is an exact limit of the full expression.

The worst situation from the point of view of the wet deposition of material is the situation when material is released in dry conditions, but is subject to rain at the end of the period. In the limit $\Lambda^{-1} \ll t_w \ll T$, the probability that the fraction of wet deposition is close to its maximum is given by

$$\text{Prob}(D/D_{max} > f_w) = P_D \int_0^{t_*} e^{-(\Lambda + 1/t_w)t} e^{-(T - t)/t_D} \frac{dt}{t_D} \qquad (23)$$

assuming that a single wet period occurs at the end of the period, f_w is close to one, and making use of equation (17). t_* in equation (23) is equal to $-(\log f_w)/\Lambda$. On integration, one obtains from equation (23), the relationship :

$$\text{Prob } (D/D_{max} \quad f_w) = P_D \frac{e^{-T/t_D}}{\Lambda t_D} (1 - f_w) \qquad (24)$$

which applies when f_w is close to one. This simple approximate relationship is plotted on Fig. 2 and is in good agreement with the exact expression.

CONCLUSIONS

It has been shown that on the assumption that the onset of a wet period or dry period is a random process all the statistical functions required for including the occurrence of rainfall into models of long range transport can be included. This approach has already been used for calculating long term average concentrations, but an extension of the method has been demonstrated which enables one to determine quantities such as the frequency with which a given concentration is exceeded. In certain limits, which are likely to occur in many pratical situations, the main results, i.e. expressions for the probability distribution of the number of particles surviving a given period either subject or not subject to rain at the end of the period, have been reduced to very simple algebraic expressions. The method is recommended as an alternative to the direct approach, which involves the difficult and time consuming extraction from meteorological data of information on the occurrence of rainfall along the path of an air mass, and is especially useful in areas, such as over the sea, where meteorological data on rainfall is very sketchy.

The problem remains of choosing appropriate values of the parameters t_D and t_w, the mean lengths of wet and dry periods. Most sets of data of rainfall statistics refer to a single sampling station rather than follow the incidence of rainfall along the trajectory of an airmass. It is hoped thought that results are relatively insensitive to values of t_D and t_w, provided the ratio is kept constant. This assumption is also supported by some recent work of Smith[6].

Some support for the assumptions about the incidence of wet and dry periods is given by a study of rainfall duration at three sites in the United Kingdom and this data is briefly outlined in an Appendix.

ACKNOWLEDGEMENTS

The author is grateful to the British Meteorological Office for undertaking the analysis of rainfall described in the Appendix. This work is published with the permission of the Central Electricity Generating Board.

REFERENCES

1.   A. Eliassen, "The OECD study of long range transport of air pollutants : Long range transport modelling", Atmospheric Environment, 12, 479-487, 1978.

2.   Western Meteorological Synthesizing Centre, "Some experiments with the long range transport model of MSC-W on data from a winter month", EMEP/MSC-W Report 1/80, 1980.

3.   H. Rodhe and J. Grandell, "On the removal time of aerosol particles from the atmosphere by precipitation scavenging" Tellus, 24, 442-454, 1972.

4.   B.E.A. Fisher, "The calculation of long term sulphur deposition in Europe", Atmospheric Environment, 12, 489-501, 1978.

5.   B.E.A. Fisher, "The statistics of precipitation scavenging during long range transport", CEGB Report RD/L/N 42/79, 1979.

6.   F.B. Smith, "The significance of wet and dry synoptic regions on long range transport of pollution and its deposition", Atmospheric Environment, to be published, 1980.

DISCUSSION

A. BERGER   Have you tried to use a 4th order Markov process, as used in some applications at our institute, to fit the data on rainfall duration ?

B.E.A. FISHER I have prefered to use a simple first order process, but it may be worthwhile to investigate a higher order process.

APPENDIX : DURATION OF SEQUENCES OF WET AND DRY WEATHER

The persistance of rainfall at three sites in the U.K. (London, Birmingham and Manchester airports) was investigated. The data base consists of hourly readings, with rain defined as precipitation rate greater than 0.1 mm h^{-1} and covers the period 1950 to 1978 inclusive. The data from three sites exhibit almost identical distributions of the frequency of duration of wet and dry weather.

The fraction of time that wet periods occur is 0.12 at Manchester Airport, 0.10 at Birmingham Airport and 0.09 at London Airport. The number of dry periods of duration t hours should be proportional to t_D^{-1} exp $-t/t_D$, if the assumption that sequential periods are independent of one another (the Markov assumption) is true. A negative exponential probability density function of this form fits the data for durations of dry weather greater than 40 hours, in which case the parameter t_D takes the value 67 hours.

The negative exponential probability density function does not fit the data for periods of dry weather of duration less than 40 hours. The onset of wet conditions occurs more frequently for the shorter dry periods. One reason for this discrepancy may lie with the Eulerian nature of the statistics. Material which is emitted and moves largely within a dry region is unlikely to be affected by rain. Thus Eulerian statistics of periods of duration less than 40 hours may reflect the movement of the airmass over the rain collector. If a Lagrangian basis were available, the use of $t_D \simeq 67$ h might be justified for the whole range of durations of dry weather.

The data on the frequency of occurrence of durations of wet weather are also very similar at the three sites. The value of t_w in a negative exponential distribution fitted to the data is 3 h. Statistics on rainfall duration appropriate to moving air masses might imply a larger value of t_w.

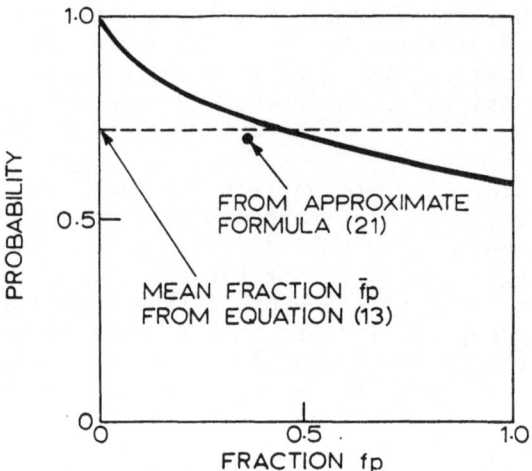

Fig. 1.   Probability more than given fraction survives PROB(f>fp)

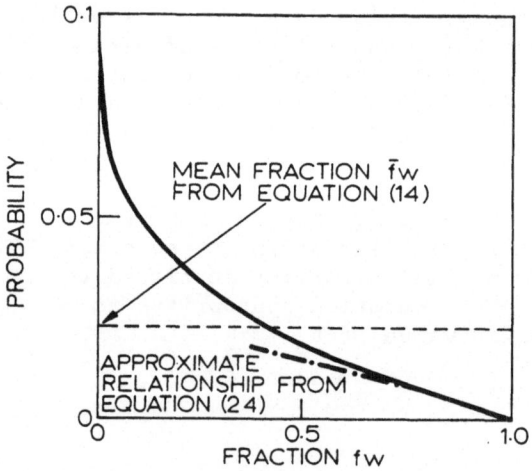

Fig. 2.   Probability that rate of wet deposition is more than given
          fraction of maximum possible deposition PROB(D/D$_{max}$>fw)
          Parameter values   T = 30h, t$_D$ = 70h,
                              t$_w$ = 7h, Λ = .3h^{-1}

AIR POLLUTION TRANSPORT OVER WESTERN EUROPE; EXCHANGES BETWEEN GERMANY AND THE NETHERLANDS

N.D.van Egmond and H.Kesseboom

National Institute of Public Health

P.O.Box 1, Bilthoven - The Netherlands

INTRODUCTION

Quasi-persistent high pressure areas over central-Europe in many cases result in air pollution episodes in the northwestern part of the European continent. Primary and secondary pollutants are accumulated in stable air which is transported over hundreds of kilometers in northwesterly directions. Situated downwind of this circulation the air pollution levels in the Netherlands are dominated by long range and mesoscale transported pollutants; the small dimensions of the country (200 x 300 km) imply an important role for the import from neighbouring countries. In figure 1 the SO_2-emissions for the Netherlands and surrounding countries are given (TNO-1979, LRTAP-1974, Bundesministerium der Innern, Bonn). Dutch sources totally amount 60 tons/h SO_2 and foreign sources in the map-area represent a total of about 400 tons/h. With respect to source strength and geographical position the German Ruhr-area is one of the most important sources.

In 1974 a bilateral project was initiated between Germany and the Netherlands to study the transport of SO_2 between the two countries. Measurements, data-analyses and modelling efforts were mainly directed to the estimation of SO_2- and SO_4-fluxes and decay during westerly and easterly transports, downwind of respectively the Ruhr-area and the Rotterdam-Rijnmond industrial area. In this paper the results of that project will be summarized; furthermore recent investigations, with respect to the mesoscale transports in the map area given by figure 1 will be described, involving also parts of Belgium and France.

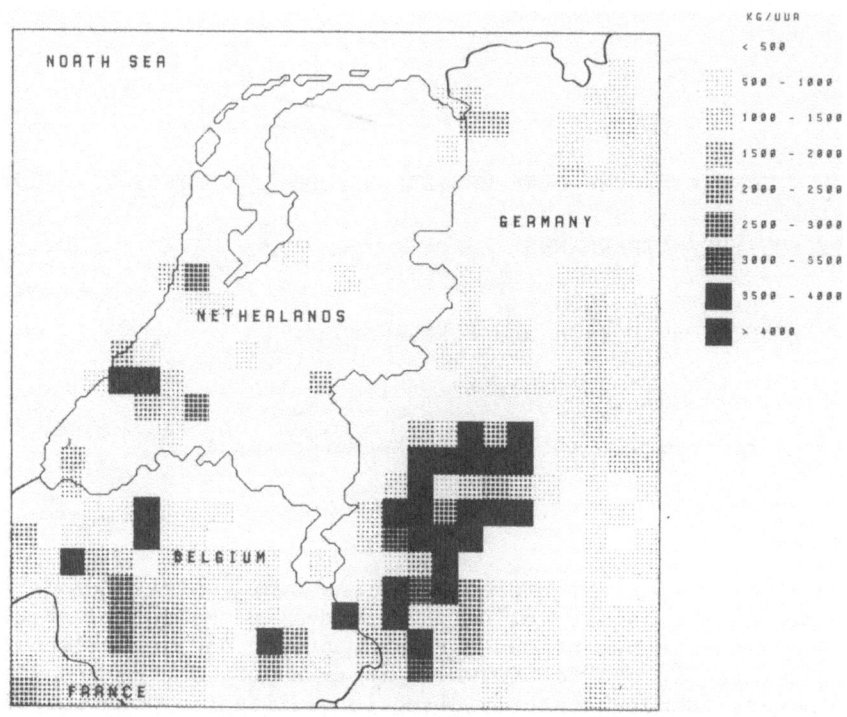

Figure 1 - SO_2-emissions in kg/h 400x400 km^2: Ruhr and Rijnmond are
           the major source areas in resp. Germany and the Nether-
           lands.

LONG-TERM AVERAGE SO_2-CONCENTRATIONS; ESTIMATION OF TRANSFRONTIER
CONTRIBUTIONS

        Estimations of the transfrontier SO_2-exchanges in Europe are
given by the LRTAP-model (OECD, 1979) and by the EURMAP-I model
(Johnson et al., 1978). Both (lagrangian) models are directed to
transport over large distances as they describe the horizontal dispersion in a completely mixed boundary layer of constant height.
As further the spatial resolutions of both models are rather poor
(LRTAP; 127 km and EURMAP; 50 km grid size) the estimations of contributions from emittor-countries to the Netherlands should be interpreted carefully.
The contributions to dry deposition, based on the emissions for 1974
(LRTAP) and 1973 (EURMAP) are given in Table I.

To arrive at a more accurate estimation of the contributions of surrounding countries a statistical-analytical (Gaussian) model was
applicated to the long term average concentrations measured in the
Netherlands (Van Egmond et al., 1979). This type of statistical analytical approach was earlier followed by Heimbach and Sasaki (1975),
Gustavson and Kortanek (1977) and is comparable with the statistical (Lagrangian) regional source quantification approaches as described by Eliassen and Saltbones (1975) and recently by Prahm et
al. (1980).

Table 1   Percent contributions to SO_2-dry deposition
            in the Netherlands by LRTAP and EURMAP models

emittor country	LRTAP	EURMAP
Netherlands	60%	51%
Germany	10	18
Belgium	7	15
U.K.	10	10
others	13	6

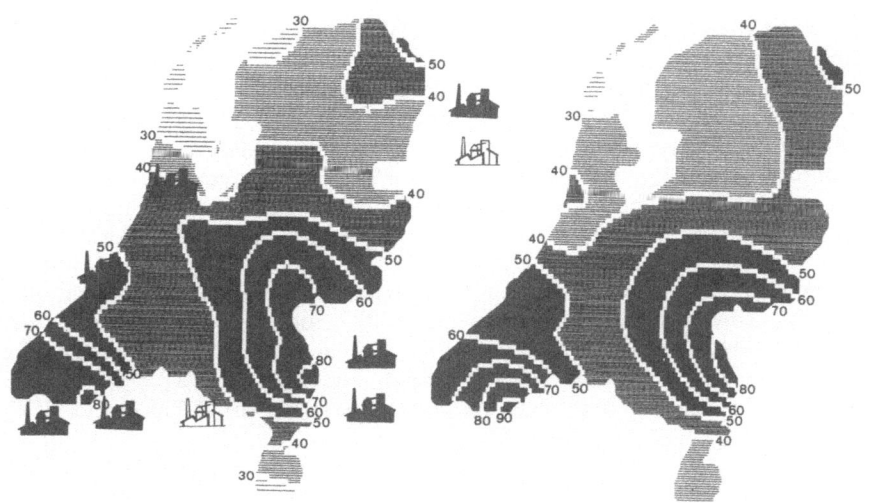

Figure 2. Measured (left) and modelled (right) longterm average SO_2-
          concentration field for south-easterly winds, 3-5 m/s
          (1977).

Individual hourly average or long term average concentrations
per wind direction are assumed to be given by the product of the
unknown source strength Q_i and transfer coefficient a_i (x,y), summed over all possible sources i.
The Gaussion type transfer coefficient is given by

$$a(x,y) = \frac{\exp{(-\frac{1}{2}(y/\sigma_y)^2)} \exp{(-v_g x/uL)}}{\sqrt{2 \pi} \ \sigma_y \ uL}$$

in which all parameters have their usual meaning and $\sigma_y = p \ x^{0.89}$.
For given v_g and source receptor configuration $a(x,y)$ is fully
determined by p and uL, which are choosen such that the least
squares discrepancy between measured and estimated concentrations
$C(x,y)$ is minimal. Rather than interpreting the relative source
strenghts Q, the contributions to $C(x,y) = \Sigma \ Q_i \ a_i$ (x,y) are determined.

An example of the method for the field of average concentrations is given in figure 2. At parameter values v_g = 0.01 m/s, u = 4
m/s, L (mixing height) = 350 m and σ_y = 400 $x^{0.89}$(x in km) effective SO_2-source strengths are determined as 54 tons/h (+ 6) for the
Ruhr-area, 33 tons/h (+ 7) for source areas in northern Germany
24 tons/h (+ 2) for areas in Belgium and 4 tons/h (+ 1) for both
IJmond (Amsterdam) and Rijnmond areas. Sources at larger upwind
distances cannot be estimated individually as they do not result
in "recognizable" spatial patterns (plumes). Together they contribute about 10 $\mu g/m^3$ SO_2 over the area, whereas Dutch sources (assumed to be homogeneously distributed) account for about 13 $\mu g/m^3$.
The modelled concentration field (figure 2) explains 96% of the
spatial variance of the measured field. The effective source
strenghts, although irrelevant to the estimation of concentration
contributions, are rather low, in agreement with the findings that
in the most cases a substantial part of the SO_2-flux is transported
above the ground based layer, as will be described below. In cases
of individual hourly concentration fields at extremely strong inversion conditions, the estimated effective source strenghts approximate the figures obtained from emission inventarisations
(figure 1).

By adding all estimates for 12 winddirection and 4 windspeed
categories, the contributions to the 1977 average concentration
fields were estimated. The model finally accounts for 85% of the
measured spatial variance. The percent contributions, which are
comparable to the dry deposition contributions of LRTAP and EURMAP
(table I) are given in table II.

The large estimate in the category of non separable foreign
(distant) sources is rather inaccurate as this allocation is based
on the large scale concentration gradient. However, investigation
of this category as a function of winddirection indicated that the
major part of this contribution originates from south-westerly
directions. Recent transmission (Cospec) measurements have shown

Table II.   Percent contribution of emittor countries to the 1977
average SO_2-contributions in the Netherlands
(statistical-derterministic model)

Netherlands	30%
Germany	20
Belgium	15
Non separable foreign sources	35
(Germany, Belgium, France)	

that source areas in the northern part of France (figure 1) might
explain this contribution to a large extent. It should be noted
that the contributions of Germany and Belgium given by the EURMAP
model (table I) correspond to the nearby SO_2-sources, which were
determined with high accuracy in the statistical model (table 2).
It consequently is concluded that the total contribution of these
emittor countries must be higher than the LRTAP and EURMAP estimates.

An intermediate between long term average concentration patterns and individual hourly concentration fields is given by eigenvector analysis of hourly average concentrations, observed over a
longer period. The individual hourly fields (location x, time t)
are described as

$$C(x,t) = \sum_k f(k,t) e(x,k)$$

where $e(x,k)$ the k-th eigenvector of the N x N concentration or
covariance matrix for N monitoring stations and $f(k,t)$ the eigen
coefficient for vector k at time t. So the sequence of concentration fields is described by a set of time-independent spatial patterns e and space-independent temporal factor f. In the framework of
the Dutch-German bilateral project, Dietzer (1978) based a statistical model on eigenvector analysis by relating the factors f to meteorological variables such as windspeed, winddirection, temperature, dewpoint and thermal stratification by means of multiple regression analysis. The model was applied to the monitoring network
between the Ruhr area and the central part of the Netherlands.
The concentration patterns, including the mesoscale transport
plumes could be described very well. With 22 stations, 98.8% of
the spatial variability could be explained by the first 7 eigenvector patterns.
An example of a modelled concentration field is given in figure 3. The modelled field is a reconstruction on the basis of five
eigenvectors and explains 86% of the measured spatial variability.
This good result was obtained by including thermal stratification
as independent variable in the regression of eigencoefficients.

A same tendency is found for concentration fields over the
Netherlands. An analysis for 60 monitoring stations indicates that
92% of the total variance, with respect to the overall space-time
mean, can be explained by 12 eigenvectors. The patterns suggest
that the long term concentrations in the southern part of the
Netherlands are strongly affected by source areas in easterly and
southerly directions.

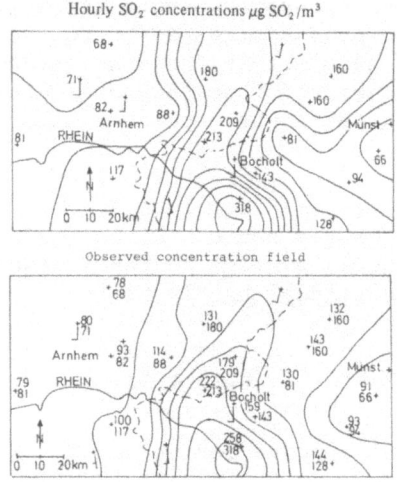

Figure 3. Measured (above) and statistically modelled (below) SO_2-
    concentrationfield (5-eigenvectors) for 23 febr. 1975,
    11.00 h. (Dietzer, 1978).

Although such pure statistical models are attractive tools to
analyse and describe the spatial and temporal variability of concentrations they only partly give insight into the mechanisms of
pollutant transport. This is mainly due to the fact that the eigenvectors are mutually orthogonal, which does not imply that they
have fysical meaning; especially in cases of transient phenomena
the patterns might be arbitrary in character. As pointed out by
Klug (1978) the combination of eigenvector analysis and deterministic modelling is promising as the statistical model only recon-
structs the significant spatial variability and eliminates local
disturbances of measured concentration fields, ensuring good comparability with modelled fields.

SHORT TERM SO_2-CONCENTRATION LEVELS; MESOSCALE TRANSPORT DURING EPISODES

In the framework of the bilateral Dutch-German project Klug and Hermann (1978) developed a numerical model to describe the SO_2-transport downwind of the Ruhr-area. This Time Dependent Multi Box-model covered the 200 x 160 km^2 border area between the two countries by a 10 km grid. Windfields were obtained from both measurements at 10 m, 200 m and from synoptic charts (850 mb topography). K-profiles were determined from temperature gradients at different heights. Comparison of modelled and measured concentrations led to the following conclusions (Klug, 1978).
1. The detailed structure of the concentration field is not reproduced by the model, using the available emission and meteorological data; a comparison between measured and modelled concentration fields with correlation coefficient 0.73 is given in figure 4.

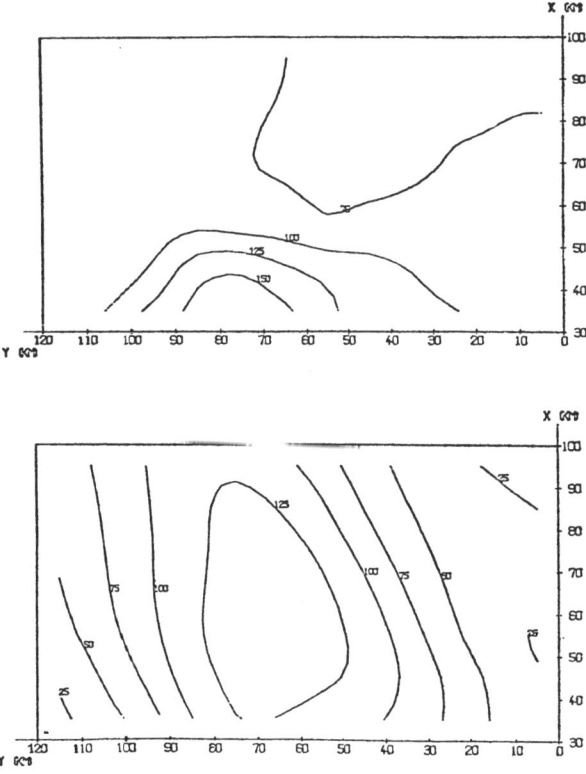

Figure 4. Measured (above) and modelled SO_2-concentration field, downwind of Ruhr area (Klug and Hermann, 1978).

2. The influence of the magnitude of K_z is small as a consequence of the complementary configuration of ground level and elevated sources.
3. Plume rise effects and horizontal diffusion are of minor importance.
4. Simulation of sedimentation could improve the results substantially.

The discrepancy between measured and modelled concentration fields might be explained from the size of the modelled area, relative to the surrounding source areas which contribute to high background levels. The emission in the model-area amounts only 70 tons/h while the emission for the Ruhr-area as a whole, extending to south of Cologne is estimated as 200 tons/h (Bundesministerium des Innern).

This estimate corresponds to the SO_2-fluxes of 200-300 tons/h SO_2 over the Ruhr-plume as computed from Cospec-measurement results (Ván Egmond et al., 1978). From comparison of measured SO_2-fluxes and ground level concentrations during daytime, when enhanced vertical mixing in the ground based layer can be assumed, it was concluded that in most cases a substantial part of the SO_2-transport takes place above this layer. This was confirmed by both Dutch and German aircraft measurements. An example of a vertical profile obtained about 100 km downwind of the Ruhr-area at a temperature inversion at 300 m is given in figure 5, in combination with ground level concentration field and SO_2 gasburden profile.

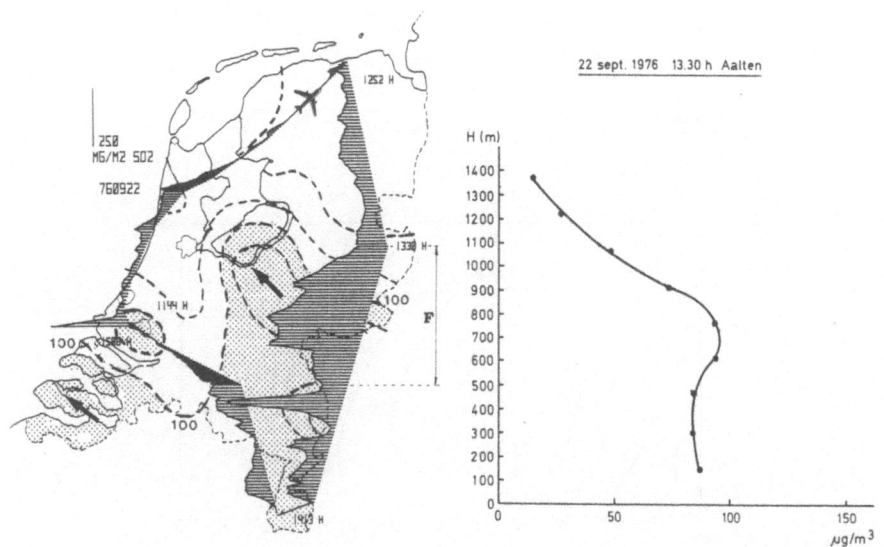

Figure 5. SO_2; vertical profile, ground level concentration field ·($\mu g/m^3$) and gasburden profile (22 september 1976).

From the concentration maximum at about 700 m it is concluded that high sources contribute substantially to the total flux, but have a relatively small impact on the measured ground concentrations. The concentration profile below the inversion at 300 m is rather uniform suggesting a well mixed ground based layer, which can be considered as a box from which SO_2 mainly is removed by dry deposition. From a large number of such daytime measurements, deposition velocities were estimated by computation of SO_2-removal over the central part of the Netherlands, where local SO_2-sources are extremely scarce; from estimated mixing heights deposition velocities of about 1 cm/s were found.

The SO_2 vertical structure recently was studied by van Dop et al. (1980). Average diurnal concentration patterns at 3 m, 100 m and 200 m were derived from measurements at a 213 m meteorological tower; the patterns are given in figure 6.

At 100 and 200 m during the nighttime hours higher levels are measured, while in daytime conditions smaller gradients between the three levels are observed, corresponding to the diurnal cycle in the atmospheric stability. Apparently the concentrations between 0 and 200 m primarily are affected by horizontal transport and dry deposition at the surface. For stable conditions the deposition velocity (at a reference height of 1 m) was estimated from concentration gradients between 3 and 100 m, and from temperature and wind profiles which allowed determination of friction velocity (u_*) and Obukhov length (L). From a data set of 51 cases for which L $\gtrless$ 50 m, a deposition velocity v_g = 1.7 cm/s was obtained.

An example of the profiles of SO_2 and potential temperature is given in figure 7. The corresponding concentration fields are given in figure 8. During the morning hours of 20 december the vertical stability decreases due to the increased heat flux from the surface, resulting in large scale fumigation. SO_2 from elevated layers is transported rapidly to ground level over the entire advection zone, in which the concentrations increase from 75 to 175 $\mu g/m^3$ between 11.00 and 14.00 hours (10.00 and 13.00 GMT) as shown in figure 8.

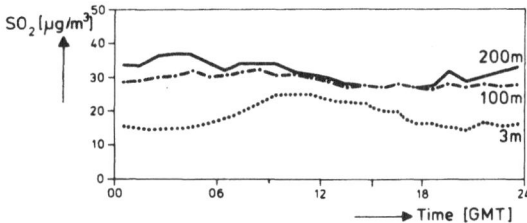

Figure 6. Average diurnal SO_2-concentration patterns at 3, 100 and 200 m (van Dop et al., 1980).

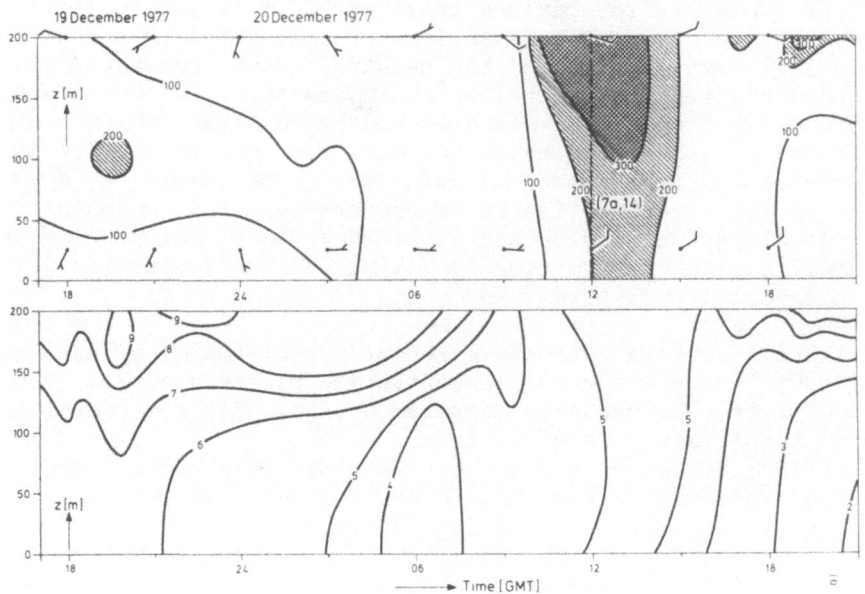

Figure 7. Profiles of SO_2 ($\mu g/m^3$) and potential temperature for 19 and 20 december 1977 (van Dop et al., 1980).

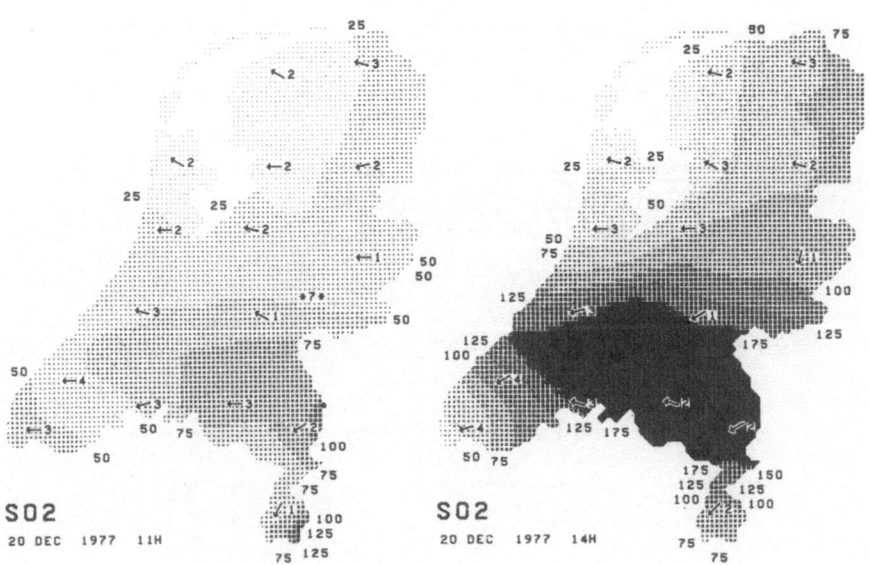

Figure 8. SO_2-concentration fields for december 20, 1977 at 10.00 and 13.00 GMT ($\mu g/m^3$).

The value of v_g = 1,7 cm/s is rather high compared to the commonly accepted value of 0,8 cm/s (Garland, 1978). An even higher deposition velocity was obtained by Bingemer and Georgii (1978). From the decrease of SO_2 at increasing transport time they derived an overall decay constant of 3,1 x 10^{-5} s^{-1}. At pollution heights

$$H = \int_o^\infty C(z)dz/C(o)$$

in the order of about 800 m they found an overall deposition velocity of 2,6 cm/s. As suggested by Fisher (1978) this might be due to the unrealistic assumption of intensive vertical mixing (box-model) combined with ground level concentrations higher than the box-average concentration. However, the decay of 3,1 x 10^{-5} s^{-1} (11-14% h^{-1}) agrees with values obtained elsewhere. In figure 9 the SO_2- and SO_4 fluxes are given as a function of travel time, starting from an initial SO_2-flux of 38 kg/s (137 tons/h) and an oxidation rate of 1% h^{-1}; 90% of the initial SO_2 mass is removed within 20 hours of travel, corresponding to a distance of 500 km at 7 m/s.
The SO_2-fluxes in the lee of the Ruhr-area ranged from 90 to 190 tons/h. Downwind of the Rijnmond and IJmond industrial areas in the western part of the Netherlands, fluxes of about 100 tons/h SO_2 were measured, in contrast with mobile gasburden flux estimates which ranged from 30-60 tons/h (van Egmond et al., 1978). The decay of SO_2 in these fluxes derived from the German aircraft measurements was estimated at 18-19% h^{-1}. This high figure is explained from strong convection during westerly circulations, transporting SO_2 into higher atmospheric levels.

In order to integrate the results of the above mentioned measurements and modelling efforts, a simple numerical mesoscale model

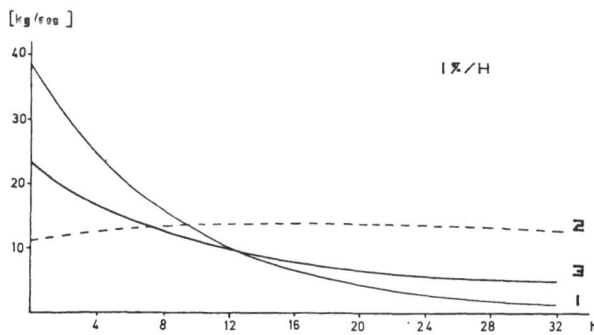

Figure 9. Calculated SO_2 (1), SO_4 (2) and total sulfur (3) for Ruhr-
fluxes as a function of travel time (Bingemer and
Georgii, 1978).

was developed which would account for the findings obtained so far
(van Egmond en Kesseboom, 1980):

1. Vertical stratification in three layers, accounting for mesoscale transport in an elevated reservoir layer and large scale
   fumigation during the morning hours when the inversion height,
   i.e. the depth of the mixed layer increases (figure 10).
2. Limited transport of SO_2 to the ground (deposition) during
   stable (nighttime) conditions. The concentration profile and
   the aerodynamic resistance in the surface layer are derived
   from the surface layer profiles of $u(z)$ and $K(z)$ (Businger,
   1973). Applying the empirical relation between Pasquill stability classes and the Obukhov Length L, as given by Golder (1972)
   the friction velocity u_* is computed from L and wind observations. From L and u_* the aerodynamical resistance r_a and the
   surface resistance r_s are derived. Together with the canopy
   stomatal resistance $r_c > 0.70$ s/cm (Wesely and Hicks, 1977), the
   deposition velocity at the top of the surface layer is given by
   $v_g = (r_a + r_s + r_c)^{-1}$. As such this deposition varies from about
   0.20 cm/s for nighttime conditions (Pasquill E) to 0.90 cm/s for
   unstable daytime conditions (Pasquill A). Oxidation to sulfate
   is assumed to be 1% h^{-1}. The concentration at the level of 4m
   (Dutch monitoring network) is obtained from the corresponding
   surface layer (resistance) profile.
3. From the 10 m winds, hourly measured in a network of 48 stations
   and from the windspeed and -direction measurements at 5 tv-towers
   at levels between 150 and 300 m, the wind fields in the middle
   of the reservoir layer and at the level of 50 m between surface
   and mixed layer are derived by in- and extrapolation. Vertical
   directional interpolations are made by fitting a (Taylor)spiral

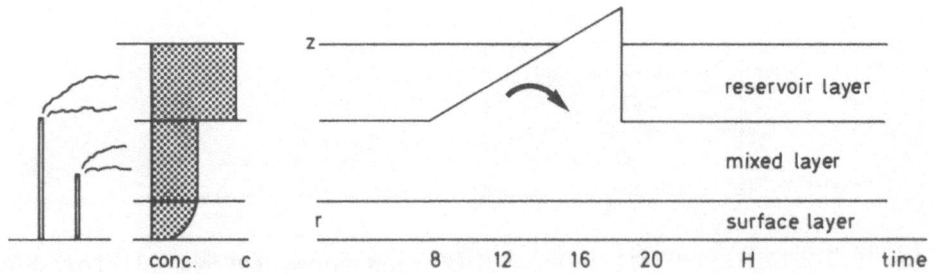

Figure 10. Schematic diagram of numerical mesoscale advection model.

profile to the observations, under the unrealistic assumption of vertically constant K_z. Wind speed is interpolated by fitting a power low wind profile to the observations. Geostrofic winds are not (yet) included in the model, which might result in underestimation of advection in the reservoir layer at strong inversions above 300 m.

4. The horizontal advection is based on the pseudo-spectral numerical scheme as given by Christensen and Prahm (1976). Spatial derivatives are determined by one-dimensional Fast-Fourier Transforms, as advised by Prahm (1979). To avoid aliasing errors the concentration fields were filtered (Lanczos-Sigma filter) every 3 hours of transport time. Periodic boundary conditions were realised by combination of exponential decay at the downwind side, and a cosine-taper over all edges of the grid. The time step was limited to an advection of 3 km at the grid size of 15 km. The 32x32 grid effectively covers all emissions over a 450x450 km^2 area. The results are presented over the area given in figure 1 (400x400 km^2), in which also the SO_2-source strengths are mapped.

5. Emissions are given for high and low sources, emitting in the reservoir or in the mixed layer depending on mixing height. Larger individual point sources were allocated to the grid according to a Gaussian plume model over the first 60 km downwind.

The SO_2 in the large afternoon mixing volume partly is brought into the nocturnal 'mixing' and into the reservoir layer; the remainder is assumed to leave the model area and is excluded from the continuing transport.

An example of the performance of the model is given by a simulation for 20 and 21 february 1980 under conditions representative for the frequently occurring anty-cyclonic south-easterly circulations (weathertype HM) with low mixing heights, moderate wind speed and moderate concentration levels (about 100 µg/m^3 SO_2). The diurnal variability of v_g (top of surface layer) and K_z, obtained from fitting a spiral wind direction profile, is given in figure 11. In future applications the stability information to derive v_g might be based on this empirically obtained K_z and on (measured) solar radiation flux. The acoustic sounding profile for 21 februari is given in figure 12. The mixed layer depth increased from about 175 m in the morning to about 300 m in the afternoon. As large scale fumigation is observed in the morning hours and windspeeds aloft are high in comparison to 10 m levels, a shallow (hypothetical) reservoir layer of 75 m is assumed (figure 10).

The resulting concentration field for 12.00 hours over the 400x400 km^2 area is given in figure 13. At the north-east side of the Ruhr transport zone concentrations are increased as a consequence of fumigation of SO_2 from the reservoir layer in which the transport takes place with a strong, veered wind in more northerly directions. This directional wind shear broadens the mesoscale

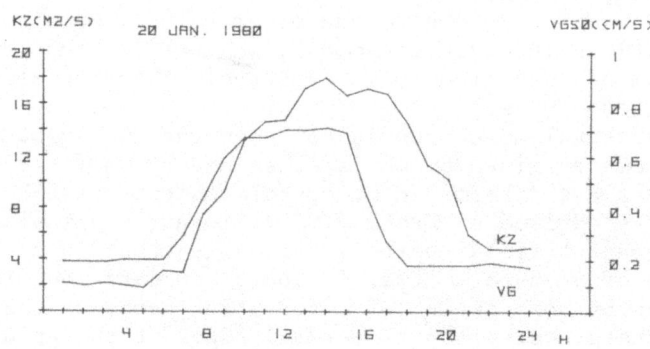

Figure 11'. Diurnal profiles for v_g and K_z applied in the numerial
      mesoscale model (20 febr. 1980).

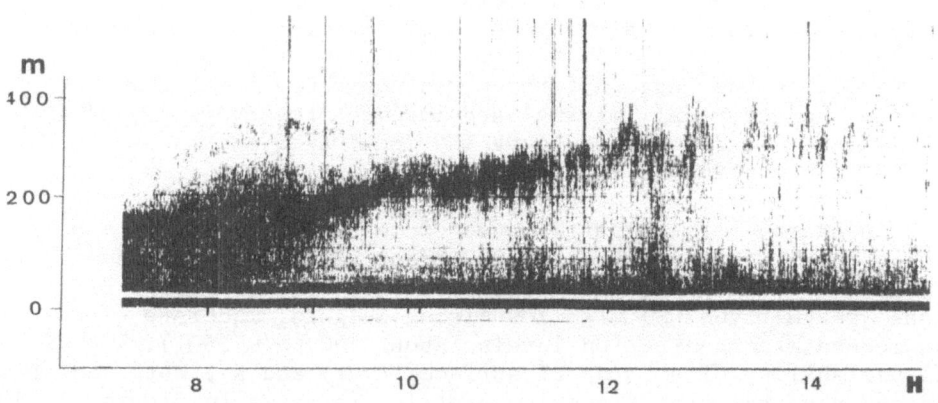

Figure 12. Acoustic sounding for 21 febr. 1980.

plume. The results of SO_2-network and mobile (Cospec) gasburden measurements are given in figure 14, in combination with the modelled
field over the measurement area. Modelled and measured field are
in good agreement. The gasburden profile shows two zones of higher
concentrations, extending over the entire cross-section of the
modelled plume. It is concluded that the pronounced directional
shear, demonstrated by the model causes the second zone of higher
gasburden levels, representing the transport of SO_2 from elevated
sources (in the reservoir layer) in more northerly directions.

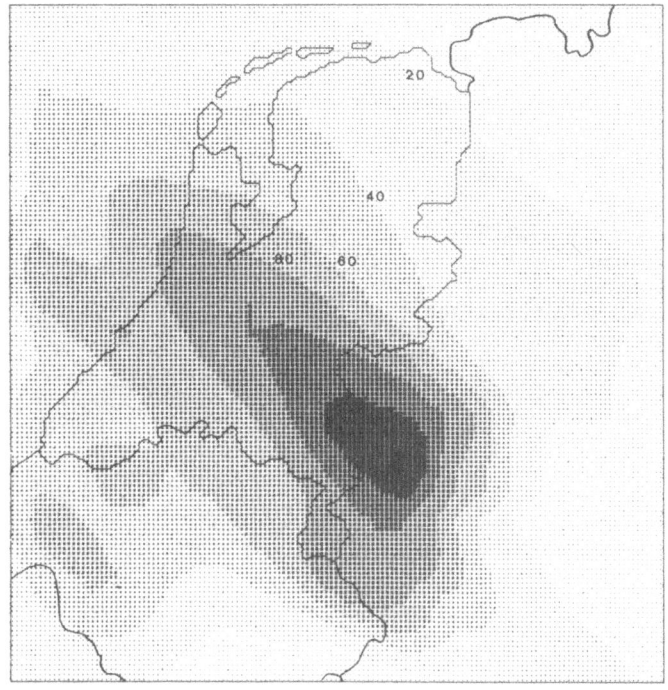

Figure 13. Modelled SO_2-concentration field ($\mu g/m^3$) for 21 febr. 1980, 12.00 hours (400×400 km^2).

The gasburden level overestimates the modelled SO_2-flux, but cannot be expected to give quantative results at the low UV-intensities at this time of the year in slow (ground level) mobile applications at 52° latitude. The deposition velocity of 0.8 cm/s as reciprocal resistance between the completely mixed layer and the SO_2-removal at the ground is active over the morning hours (figure 11) and results in realistic ground level gradients. At the mixed layer of about 250 m this value implies a decay constant K = 3.2 x10^{-5}, corresponding to the decay observed in the German aircraft measurements. Apparently the high deposition velocity of 2.6 cm/s is the result of an overestimation of the vertical scale (pollution height) over which the pollutants are effectively mixed. Accounting for this effect the results of the German aircraft measurements correspond to both the numerical model and the estimates of SO_2-removal from the effective mixing layer in daytime conditions, mentioned above. Supported by the results of a number of case studies, which

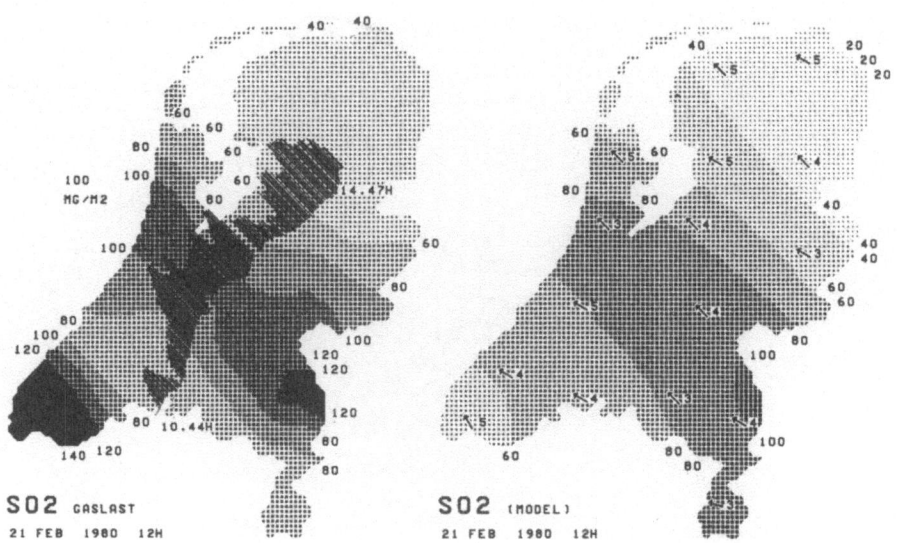

Figure 14. Measured (left) ground concentration field ($\mu g/m^3$), gas-
          burden profile and modelled (right) concentration field
          for 21 febr. 1980, 12.00 hours.

will be reported elsewhere, it is concluded that the relatively
simple 3-layer model gives a realistic description of the SO_2-con-
centrations in the Netherlands, in dependance of the emissions in
the 450x450 km^2 surroundings. The model seems sufficiently accurate
to enable interpretation of measured SO_2-concentrations in terms of
source contributions. For example the high residual concentrations
for south-westerly winds, as found in the above mentioned statistical model in which only the nearby foreign sources in the north of
Belgium were taken into account, are fully explained by the model;
source areas in the southern part of Belgium and in the French
Lille-Roubaix-Tourcoing area also substantially contribute to the
south-west background SO_2-concentrations.
        Nevertheless it is doubted that the detailed mechanisms of me-
soscale transport can be fully described by the 3-layer model. The
meteorologically more advanced multi-layer model developed by van
Dop and de Haan (1980, this meeting) will be more suited to this
purpose.

OTHER POLLUTANTS

Dutch aircraft measurements (Altena and Schneider, 1978) indicated that the transport processes described above also hold for other components than SO_2. Fluxes of NO_2 in the lee of the Ruhr area were found to range from 50 to 230 tons/h. Depending on O_3-background concentrations, NO-fluxes were found up to 35 tons/h. This is confirmed by the results of the Dutch monitoring network and ground level mobile NO_2-measurements. Modelling efforts for photo-chemical pollutants are undertaken by Builtjes et al. (1980, this meeting), who applied the SAI-photochemical model to the 200x200 km^2 area covering the central part of the Netherlands and nearby foreign source areas.

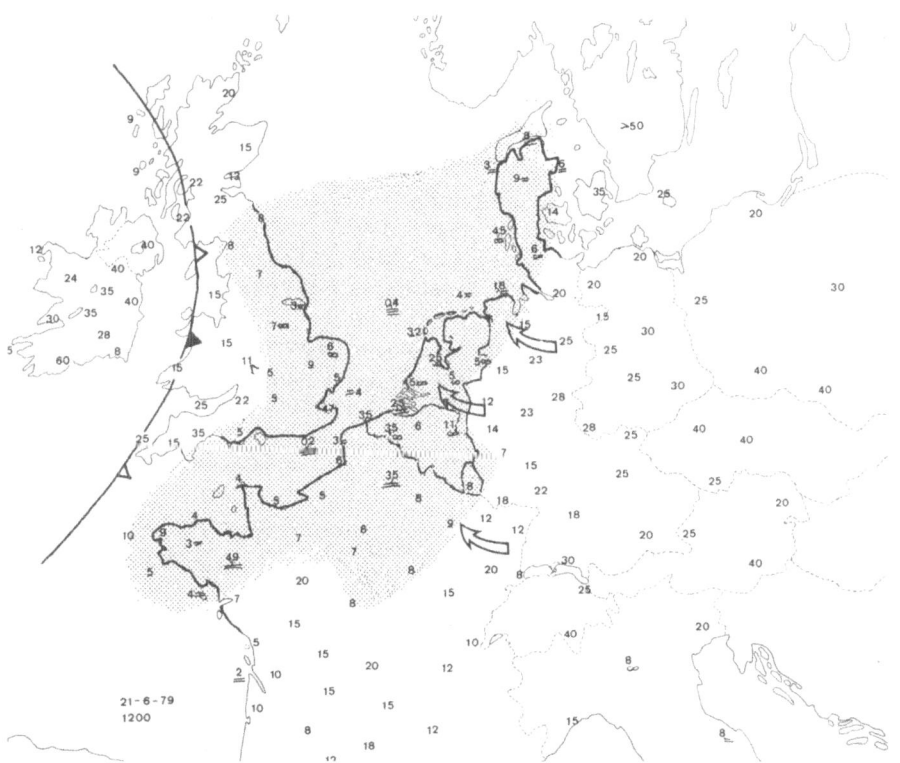

Figure 15. Visibility (in km) on june 21, 1979; visibility in the shaded area is less than 10 km (data supplied by Royal Dutch Meteor. Institute).

Apart from the large scale EURMAP and LRTAP models, sulphate
is not yet modelled for the study-area. The 3-layer model described above is extended for the computation of sulphate patterns,
but results cannot be presented yet. Aircraft measurements by
Bingemer and Georgii (1978) have shown only minor spatial variations
of the sulphate concentrations; there was no clear relation to an-
tropogenic emissions. Considering the low deposition velocity of
SO_4-aerosol, the sulphate which is formed in the study area is expected to be a small part of the SO_4-background. To illustrate this
the visibility over Europe on june 21, 1979 is presented in figure
15. As pointed out by several authors and confirmed by a long series
of measurements in the Netherlands, there is a very high correlation
between the (nephelometer) extinction coefficient b_{scat} and the
(submicron) sulfate-aerosol concentration.
On june 21, the daily average b_{scat} is 6×10^{-4}, corresponding to
a visibility of about 7 km, and the SO_4-concentration is 30 $\mu g/m^3$.
The noon visibility (rel.hum. = 58%) shows a large (shaded) area
of visibility less than 10 km, indicating that higher sulphate concentrations extend over a large part of Europe. The high levels
are explained from the accumulation of pollutants in the anticyclonic south-east circulation. This example illustrates that
modelling these types of pollutants on the mesoscale only partly
can explain the observed concentrations.

REFERENCES

1. Bundesministerium des Innern
   Raumliche erfassung der Emission ausgewahlter luftverunreini-
   gender Stoffe ans Industrie, Haushalt und Verkehr in der Bun-
   desrepubliek Deutschland 1960-1980.
2. Builtjes, P. H. J., Van den Hout, K. D., Veldt, C., Huldy. H. J.
   Hulshof, J., Basting, E. en Van Aalst, R. Application of a
   photochemical dispersion model to the Netherlands and its surroundings.
   Proc. 11th Int. Tech. Meeting on Air Poll. Mod. and its Appl.
   Amsterdam, 1980.
3. Businger, J. A. (1973)
   In: Workshop on Micrometeorology, Haugen, D. A. (Ed.),
   American Meteorological Society.
4. Christensen, O. and, Prahm, L. P. (1976).
   A pseudospectral model for dispersion of atmospheric pollutants.
   J. Appl. Meteor. 15, no 12 pp 1284-1294.
5. Bilat. Cooperation between the Netherlands and the Federal Re-
   public of Germany on air pollution problems; interregional
   transport of air pollutants, 1978.
   Umwelt Bundesamt, Bismarckplatz 1, Berlin (West) 33.
   - Altena, D. and, Schneider, T. Plume profiles and airpollu-
     tion transport.
   - Bingemer, H. and, Georgii, H. W.,

Mesoscale transport of airpollutants; aircraft measurements.
- Dietren, B., Development and application of a statistical model.
- Hermann. K., Development and application of a deterministic model.
- Klug, W., Summary of modelling experience and the climatological results from the project.
- Fischer, B., A short report from the Expert Meeting.
6. Van Dop, H., De Ridder, T. B., Den Tonkelaar, J. F. and, Van Egmond, N. D. (1980).
Sulphur dioxide measurements on the 213 metre tower at Cabauw, the Netherlands.
Atm. Env. 14, 933-945.
7. Van Dop, H., De Haan, B.J. and Cats, G.J. (1980)
Meteorological input for a three dimensional medium range air quality model. Proc. 11th Int. Meeting on Air Poll. Modell. and its Appl.
8. Van Egmond, N.D., Tissing, O., Onderlinden D. and Bartels, C. (1980).
Quantitative evaluation of mesoscale air pollution transport.
Atm. Env. 12, 2279-2287.
9. Van Egmond, N.D., Tissing, O. and Kesseboom, H. (1979).
Estimating contributions of source areas to the measured yearly average SO_2-concentration field in the Netherlands by dispersion-model parameter-optimization.
Atm. Env. 13, 1551-1557.
10. Van Egmond, N.D. and Kesseboom, H. (1980).
Numerical 3-layer mesoscale SO_2/SO_4 model (to be published).
11. Eliassen, A. and Saltbones, J. (1975).
Decay and transformation rates of SO_2, as estimated from emission data, trajectories and measured air concentrations.
Atm. Env. 9, 425-429.
12. Garland, J.A. (1978).
Dry and wet removal of sulphur from the atmosphere.
Atm. Env. 12, 349-362.
13. Golder, D. (1972).
Relations among stability parameters in the surface layer.
Boundary-layer Meteorolgy 3, 47-58.
14. Gustavson, S.A., Kortanek, K.O. and Sweigart, J.R. (1977).
Numerical optimization techniques in air quality modelling: objective interpolation formulas for the spatial distribution of pollutant concentration.
J. appl. Met. 16, 1243-1255.
15. Heimbach, J.A. and Sasaki, Y. (1975).
A variational technique for mesoscale objective analysis of air pollution.
J. appl. Met. 16, 1243-1255.
16. Johnson, W.B., Wolf, D.E. and Mancuso, R.L. (1978).
Seasonal and annual patterns and international exchanges of

SO_2-concentration and deposition in Europe, as simulated by
the Eurmap model.
Atm. Env. 12, 511-517.
17. OECD (1979).
OECD programme on long range transport of air pollutants;
Measurements and Findings.
Organisation for Economic Cooperation and Development, 2, rue
André Pascal, 75775 Paris Cedex 16, France.
18. Prahm, L.P., Conradsen, K. and Nielsen, L.B. (1980).
Regional source quantification model for sulfur oxides in
Europe.
Atm. Env. 14, 9 pp 1027-1054.
19. Prahm, L.P. (1979).
Personal communication.
20. TNO - Rekensysteem Luchtverontreiniging. II.
21. Weseley, M.L. and Hicks, B.B. (1977).
Some Factors that affect the deposition rates of sulphur
dioxide and similar gases on vegetation.
Journ. Air Poll. Contr. Ass., vol 27 no. 11.

DISCUSSION

J. KRETZSCHMAR   Why a Gaussian type model
            was applied in the statistical-deterministic
            modelling instead of a box-model of the Hanna-
            Gifford type.

N. D. VAN EGMOND   At the time this modelling
            was done, computer facilities and know-how were
            not available to apply box-model computations.
            Further the observed fields showed (mesoscale)
            plumes which reasonably could be fitted to Gaussian
            plumes profiles.

J. KRETZSCHMAR   How did you take into ac-
            count the dependence of the duration and/or fre-
            quence of rain as a function of winddirection,
            e.g. SW has a higher rain frequency than E-wind.

N. D. VAN EGMOND   The model was applied for
            every winddirection separately. So decay compo-
            nents, including the effect of wet deposition was
            accounted for at every winddirection-windspeed
            class. The results of 48 separate windclasses
            finally were added to the overall average concen-
            tration. Apart from this the presentation of dr.
            F. B. SMITH has pointed out that wet deposition
            in the study is not overly important.

M. WILLIAMS   How did you decide which sources emitted into the respective layers of the model ?

N. D. VAN EGMOND   From the emission-inventarisation (bilateral-project) of the Ruhr-areas, it was known that about 50 % of the SO_2 sources is higher than 120 m.
The effective source heights were arbitrarely set to 250 m for the higher sources and 35 m for lower sources. According to Gaussian plume modelling experiences, the emission is allocated to the reservoir layer when the mixing-height is smaller than 3/2 times the effective source height of 250 m. Otherwise the emission is brought into the mixed layer.

A. BERGER   Referring to the first slides (statistical-deterministic model) it was surprising to see how much of the total variance is explained by the model. However, as far as I understood, there are two degrees of freedom in the model :
1) the way pollutants are diffusing (the model) and
2) the sources. It is assumed that the model is Gaussian and the sources were fit to the observed concentrations. It is thus a good descriptive technique. But what about the cause-effect relationships and to what extend can we have confidence in the identified sources ?

N. D. VAN EGMOND   The relevant parameters are stability and mixing volume uL, which affects the decay in the profile (exp - v_g x/uL). The model was not very sensitive to these values. However the source strenghts are actually estimated as Q/uL, so a twofold mixing volume results in a twofold source-strength Q. The regression procedure gives estimates for the confidence-interval of estimated effective source strength Q.
For example, the Ruhr-area source strength was estimated as 60 tons/h with a 95 % confidence interval of 52-68 tons/h. This source strength has to be considered as the part of the true source strength which effectively affects the ground-level SO_2 concentrations.

K. GRONSKEI   When you compare observed and calculated concentration values, do you have to consider the vertical concentration gradient in the surface layer ? (numerical model)

N. D. VAN EGMOND   Yes, the concentration profile in the surface layer is modelled by assuming
that the concentration gradients are proportional
to the resistances over the surface layer. So
the ratio of concentration at 4m to the concentration at the top of the surface layer is found as
the fraction of the resistances below 4 m (incl.
surface and plant canopy resistances) to the total resistance over the surface-layer interface.

B. FISHER   Did you study the sensitivity for source height in the numerical mixed layer-
reservoir model ?

N. D. VAN EGMOND   No, the model is developed
recently and sensitivity analysis certainly still
has to be made. Up till now only several case studies havebeen performed.

W. KLUG   Comparisons also have been
made in Germany. We found that in some cases the
Ruhr area caused its own air pollution problems.
However with an easterly wind it appears that
pollution comes from East-Germany and Poland.

A CHARACTERIZATION OF INTERREGIONAL TRANSPORT OF

OZONE AND PRECURSORS INTO AN URBAN AREA

Michael W. Chan
Douglas W. Allard

Environmental Programs Division
AeroVironment Inc.
Pasadena, CA 91107

ABSTRACT

Evidence of the interregional transport of ozone and its precursors has been documented by many investigators. Monitoring air pollution upwind of urban areas in order to characterize the ozone and precursor transport into the city is not common, however, and understanding this phenomenon is essential to properly design control strategies to attain and maintain the ozone ambient air quality standard, and to develop models for assessing the impact of interregional transport of ozone and precursors.

To address the issue of transport, a study was performed during the summer of 1978 in the vicinity of Philadelphia, Pennsylvania. Data was collected at special surface transport sites upwind and downwind of Philadelphia at already existing surface monitoring sites. Data was also collected by an instrumented helicopter. No obvious transport of significant concentrations of ozone and precursors was observed along the surface into Philadelphia during the early morning hours; neither was transport of significant precursor concentrations aloft apparent; transport of ozone aloft, in excess of 0.08 ppm, was observed, however. There is no significant spatial variation in ozone concentrations in the air mass aloft along the upwind boundary of Philadelphia. On days where strong transport into the city was observed, the incoming mass generally approaches Philadelphia from the west in a broad, anticyclonically-curved path and does not routinely traverse any particular urban area upwind.

INTRODUCTION

The transport of ozone and its precursors has been reported by many investigators. Evidence of the transport of oxidants from urban to rural locations has been reported for the Washington, D.C. area,[1] New Jersey,[2] Staten Island and The Bronx,[3] and Santa Ynez Valley in California.[4] Long-range transport of ozone and its precursors by air masses is indicated by data obtained in the Los Angeles Basin,[5,6] the Washington-Baltimore area,[7] Southern Ontario,[8] Wisconsin,[9] and Connecticut.[10] The ozone-laden plume from St. Louis was tracked out to 240 km and was mapped in detail out to 160 km by White et al.[11] Cleveland et al.,[12] demonstrated by statistical analysis that ozone was transported over 300 km from New York to northeastern Massachusetts.

These and other studies have demonstrated that ozone and precursors can be transported over long distances. Since ozone and precursors transported into an urban area may have an impact on the maximum ozone concentration downwind of the urban area, control strategies designed for attainment of the ozone standard in individual urban areas need to take into consideration the impact of transported ozone and precursors. This impact must also be considered in the development of photochemical models.

A study was conducted during the summer of 1978 in Philadelphia and its vicinity to characterize the phenomenon of ozone and precursor transport into an urban area. Five surface air quality stations were established during the study to augment six other stations operated routinely by state agencies. Figure 1 shows the locations of these stations. In addition, vertical profiles and horizontal gradients of various pollutants were measured by an instrumented helicopter.

This paper analyzes the air quality and meteorological data collected during that study. A photochemical model is currently being applied to the Philadelphia area, taking into account the observations described here.

DATA ANALYSIS AND RESULTS

Considerable data regarding vertical and horizontal gradients of ozone and precursors upwind of and near Philadelphia have been analyzed to extract information on transported pollutant levels.

Overview of the Data

The maximum surface ozone concentration observed during the study period was 0.25 ppm at the Franklin Institute near downtown Philadelphia. Since Sites 1, 2, and 3 were upwind of Philadelphia during predominant flow, they are the most frequent indicators of surface transport into Philadelphia. Concentrations in excess of 0.12 ppm were observed at these stations on several occasions and concentrations in excess of 0.08 ppm were observed frequently. Somerville, New Jersey, and Site 5 were in the general downwind region. The highest ozone concentrations observed at these sites were 0.192 and 0.138 ppm, respectively.

Figure 2 gives some indication of the wind directions associated with the surface transport of ozone at a predominantly upwind site. This figure shows an ozone wind rose for Site 2 (Hockessin, Delaware). Missing wind direction data from Site 2 was filled by data collected at nearby Wilmington Airport. This ozone wind rose depicts, simultaneously, the frequency distribution of ozone concentration and surface wind direction. The wind directions associated with high ozone concentrations are easily identified from this figure. Surface transport of concentrations in excess of 0.08 ppm is most often associated with wind directions from the southwest to west-northwest. Ozone wind roses from the other predominantly upwind sites show a similar distribution. These directions put Sites 1, 2, and 3 upwind of Philadelphia and indicate significant surface transport of ozone. Wind directions from north to northeast are almost never associated with high ozone concentrations, probably because these wind directions are often associated with extensive cloudiness and a resultant inhibition of photochemical activity.

Vertical Profiles of Ozone and Precursors

Morning soundings of ozone, NO/NO_x, temperature, and wind speed/wind direction over upwind locations on days with strong transport reveal a consistent pattern of depleted ozone in the surface layer capped by much higher concentrations (0.05 ppm to 0.09 ppm) aloft, while NO_x concentration levels were low (less than 0.02 ppm) above the surface. Pibal data indicate that the general transport aloft was westerly. Figure 3 shows a sounding over Site 1 (Woodstown, New Jersey) during the morning hours of August 24, 1978. An ozone reservoir aloft with concentrations approaching 0.08 ppm is evident with advection from the southwest at approximately 8 meters per second. NO_x concentrations are generally less than 0.01 ppm aloft and 0.03 ppm in the surface layer.

Helicopter soundings performed over upwind locations in the late morning of transport days, usually after 1000 EDT, indicate some build-up in surface ozone, although higher concentrations continue to exist aloft. In the afternoon, a nearly uniform ozone profile to the top of the mixed layer was observed.

Soundings of ozone, NO_x, temperature and wind were also examined for two different cases: (1) days of significant photochemical activity, but little transport, and (2) days of little photochemical activity. These soundings were examined to further define the

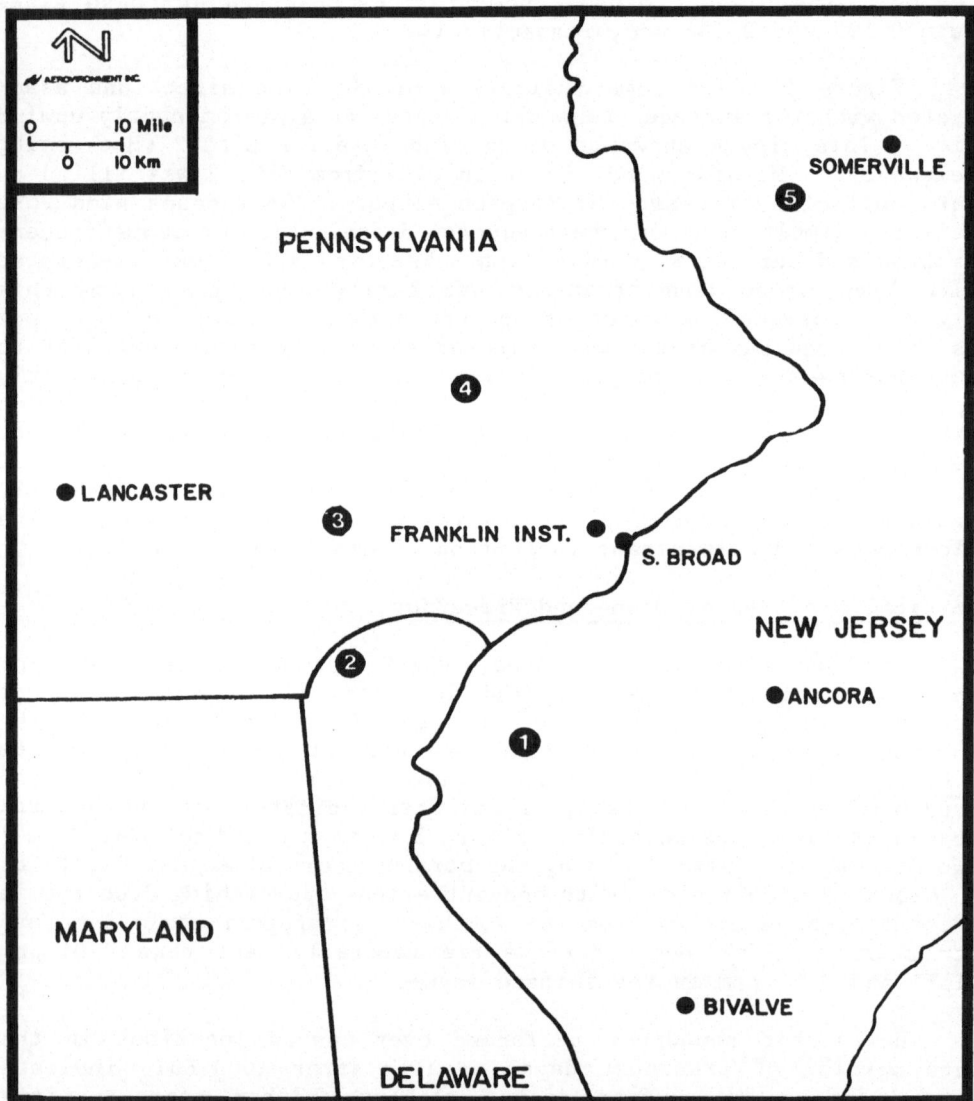

Fig. 1.   Locations of monitoring sites. Numbered sites are transport
          stations and dots representation supplemental stations.

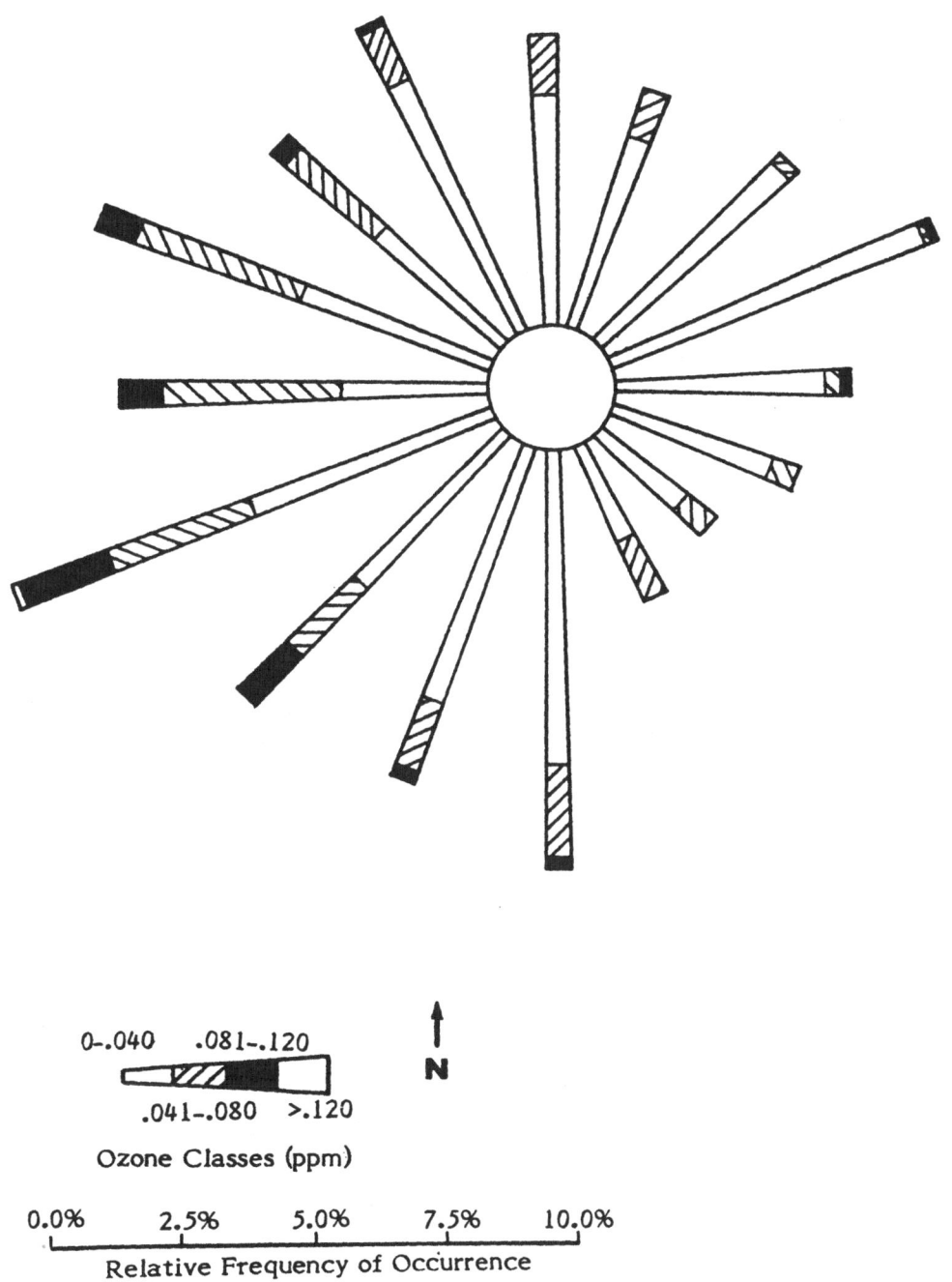

Fig. 2.   Ozone wind rose for Site 2.

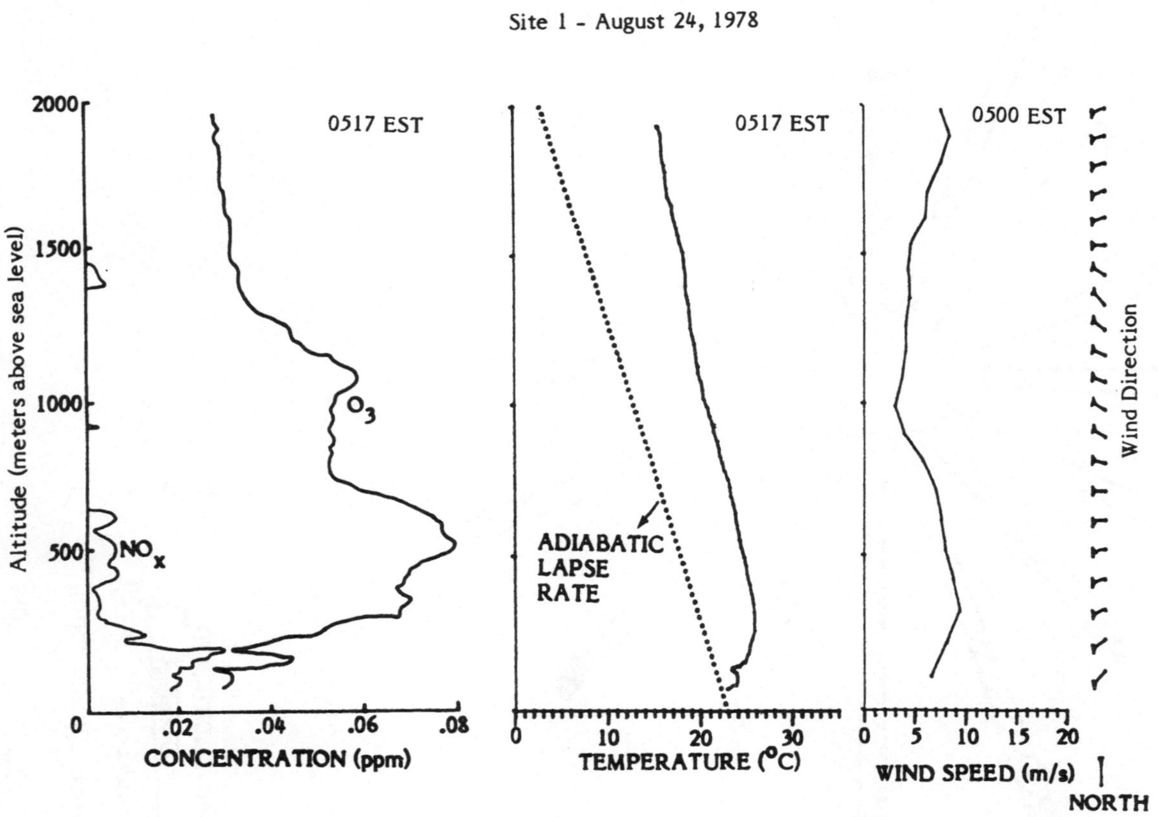

Fig. 3. Vertical profiles of ozone, oxides of nitrogen, and temperature obtained by the instrumented helicopter over Site 1 at 0517 EDT on August 24, 1978. Also shown is the corresponding vertical wind profile.

differences between these types of days and days of strong transport. One significant difference observed on September 30, 1978, a day of significant photochemical activity but little observed transport, was a persistent elevated inversion which separated the ozone reservoir aloft from the mixed layer. This inversion persisted throughout the day, effectively minimizing the ozone contribution from aloft.

Early morning profiles obtained on days of little photochemical activity indicate somewhat lower ozone concentrations aloft than do those obtained on days of significant photochemical activity. Surface depletion of ozone is still noticeable, however. It is interesting to note that concentrations near the surface in the morning hours are of the same magnitude as on transport days. Thus, until the time of mixing, surface ozone monitors contribute little or no information about ozone levels aloft.

Horizontal Gradients of Ozone and Precursors

The gradients of ozone and precursors along the upwind boundary of the study area (at the surface and aloft) and from upwind locations to primary source areas were investigated to determine the relative homogeneity of the transported air mass.

Ground level gradients were found to be generally substantial. Ozone gradients of 0.03 ppm or more are common, while precursors, although at low concentrations, also exhibit substantial differences. The differences in ozone and precursor concentrations observed at surface stations are probably due to differences in source strengths of ozone scavengers and precursors upwind and in the vicinity of the surface monitoring stations.

Gradients of ozone transported aloft are generally more minor. Helicopter transects from site to site along the study area upwind boundary were examined on five days of significant observed transport. A total of 21 transects was examined. The upwind boundary was determined by pibal-derived wind data collected every half-hour. A typical transect between sites takes approximately 20 minutes, allowing little time for increase in photochemical activity. The average absolute difference between concentrations at the beginning and end of a transect along the upwind boundary was 0.012 ppm ozone and 0.009 ppm NO_x. Occasionally, large differences were observed along the transect. These may be attributed to point sources upwind.

For example, the transect between Sites 1 and 3 on August 9, beginning at 0553 EDT, showed a difference in concentrations aloft over Sites 1 and 3 of 0.006 ppm O_3. However, a difference as large as 0.031 ppm was observed along the transect, but this was of relatively narrow horizontal extent, implying a plume crossing.

Helicopter transects from upwind areas to primary source areas near downtown Philadelphia were examined on these same five days of observed transport. Most of these were flown during the late morning hours. The data indicate that average ozone concentrations at the transect level over upwind areas were nearly identical to those over downtown locations, while NO_x levels were only slightly lower at upwind locations. The average absolute difference between concentrations observed at the transect beginning and end points (i.e., upwind boundary and downtown) was 0.011 ppm for ozone and 0.020 ppm for NO_x. As was the case for transects along the upwind boundary, somewhat higher differences were observed <u>along</u> the transect.

Case Study

August 22, 23, and 24 represent the only extended "transport episode" observed during the monitoring period for which data are available. This period illustrates the cumulative effect of a somewhat prolonged transport situation.

A high pressure cell was centered over the study area on August 22, moved to the south and was centered over North Carolina by August 24. A frontal passage on August 25 terminated the episode. Figure 4 shows the surface synoptic weather situation on August 23. The high was also observed over the Washington, D.C. area on several other strong transport days.

Figure 5 shows trajectories of air parcels arriving in Philadelphia at 1400 EDT on August 22, 23, and 24, respectively. As the high pressure cell moved over the study area and then to the south, flow changed from variable, mostly northerly, to southwesterly. The last two days show parcels approaching along a broad, anticyclonically curved arc. This trajectory is typical of several other strong transport cases which were examined.

Vertical profiles obtained upwind over Site 2 by the instrumented helicopter during the early morning hours show that average ozone concentrations aloft increase slightly from day to day. Late morning flights actually show a decrease in both peak and average concentrations aloft from August 22 to August 24. At the same time, surface concentrations were increasing from day to day as shown in Table 1. This table presents 1100 EDT and peak surface ozone concentrations, as well as average concentrations aloft at upwind sites during the early morning for the period under study. Concentrations aloft were determined by taking the average in the transport layer (above the layer of surface depletion). Table 1 shows that during the episode there was a slight day-to-day increase in the early morning transport of ozone aloft and a substantial day-to-day increase in the late morning surface ozone concentration.

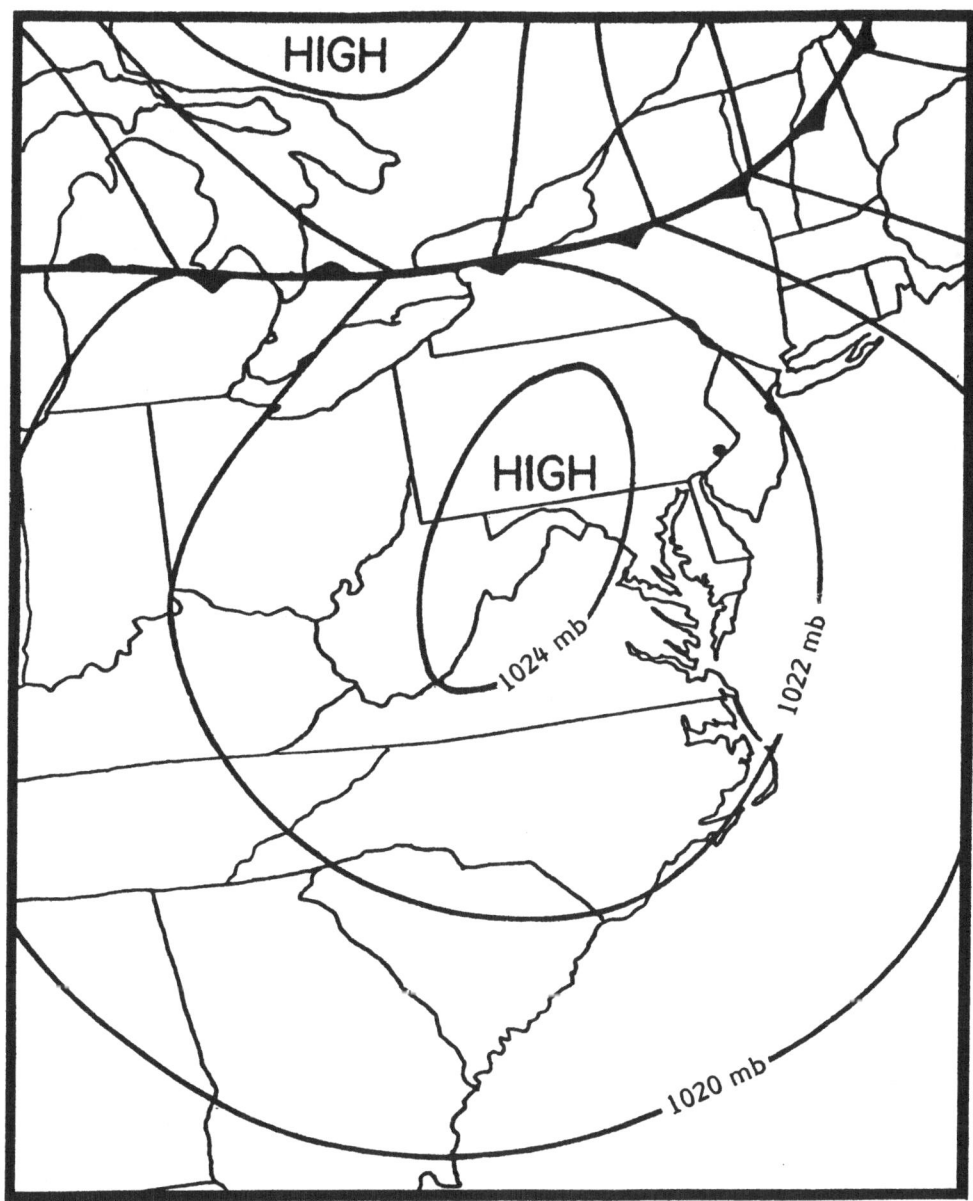

Fig. 4.   Synoptic situation, 0700 EDT, August 23, 1978.

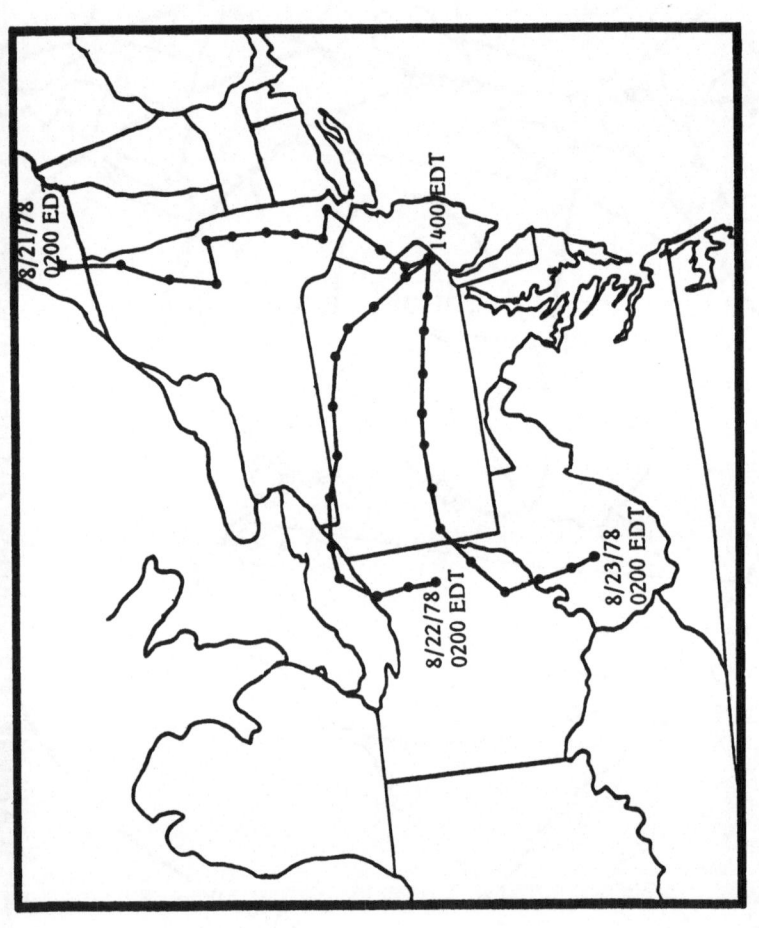

Fig. 5. Estimated 36-hour air parcel trajectories beginning 0200 EDT on August 21, 22, 23, and ending 1400 EDT on August 22, 23, and 24, respectively. ( ) indicates position every three hours.

Table 1. Ozone concentrations (in ppm) at upwind stations on August 22, 23, and 24, 1978.

Site	Surface						Aloft		
	1100 EDT			Maximum			Average of Transport Layer[a]		
	8/22	8/23	8/24	8/22	8/23	8/24	8/22	8/23	8/24
1	--	--	--	--	--	--	.070	.070	.072
2	.058	.063	.076	.079	.080	.121	.062	.068	.070
3	.069	.075	.103	.108	.108	.133	.052	--	--
4	.056	.074	.093	.098	.101	.126	--	--	.055

[a] obtained by instrumented helicopter at approximately 0500 EDT.

CONCLUSIONS

From the data collected during this study we obtained certain useful information on the transport of ozone and precursors into Philadelphia, including:

1. From the fixed sites located upwind for this study, only minimal transport of ozone, oxides of nitrogen, and non-methane hydrocarbons was observed along the surface into Philadelphia during the early morning hours.

2. Transport of oxides of nitrogen aloft was slight.

3. Transport of ozone aloft, in excess of 0.08 ppm, was some-times observed during transport days.

4. There is no significant difference in ozone concentrations laterally along the upwind boundary of Philadelphia in the air mass aloft. Differences have been observed at surface stations, probably due to variabilities in the strengths of ozone scavengers and in the strengths of sources of precursors in the vicinity of the monitoring stations.

5. Trajectories on photochemically active days with strong transport indicate that the incoming air mass generally approaches Philadelphia from the west in a broad, anti-cyclonically-curved path. While some urban areas lie upwind along the path, their particular contribution to the ozone problem cannot be determined by these trajec-tories.

Acknowledgements

This work was sponsored by the Office of Air Quality Planning and Standards of the U.S. Environmental Protection Agency (Contract No. 68-02-3027).

REFERENCES

1. R.C. Wanta, W.B. Moreland, and H.E. Heggestad, Tropospheric ozone: an air pollution problem arising in the Wash-ington, D.C. metropolitan area, Monthly Weather Review 89:289 (1961).
2. I.A. Leone, E. Brennan, and R.H. Daines, The relationship of wind parameters in determining oxidant concentrations in two New Jersey communities. Atmos. Environ. 2:25 (1968).
3. Scott Research Labs, Atmospheric reaction studies in the New York City area, Volume 1. Program design and methodology, data summary and discussion, Prepared for the Coordinating Research Council, Inc. and Air Pollution Control Office (1970).

4. L.B. Baboolal, M.I. Smith, D.W. Allard, L.G. Wayne, and J.W. Mortz, A climatological and air quality characterization and air quality impact assessment for various future growth alternatives in the Santa Ynez Valley, AeroVironment Inc., Pasadena, California (1975).

5. J.D. Edinger, M.H. McCutchan, P.R. Miller, B.C. Ryan, M.J. Schroeder, and J.V. Behar, Penetration and duration of oxidant air pollution in the South Coast Air Basin of California, J. Air Pollu. Con. Assoc. 22:882 (1972).

6. E.R. Stephens, Chemistry and meteorology in an air pollution episode. J. Air Pollu. Cont. Assoc. 25:521 (1977).

7. U.S. Environmental Protection Agency, Investigation of high ozone concentration in the vicinity of Garrett County, Maryland and Preston County, West Virginia, Publication No. R4-73-019, Office of Research and Monitoring (1973).

8. D. Yap, and Y.S. Chung, Relationship of ozone to meteorological conditions in southern Ontario, Paper presented at the 70th annual meeting of the Air Pollution Control Association, Toronto, Ontario, Canada, 20-24 June (1977).

9. W.A. Lyons, and H.S. Cole, Photochemical oxidant transport: mesoscale lake breeze and synoptic-scale aspects, J. App. Met. 15:773 (1976).

10. G.T. Wolff, P.J. Lioy, G.D. Wight, R.E. Meyers, and R.T. Cederwall, Transport of ozone associated with air mass, Paper presented at the 70th annual meeting of the Air Pollution Control Association, Toronto, Ontario, Canada, 20-24 June (1977).

11. W.H. White, J.A. Anderson, D.L. Blumenthal, R.B. Husar, N.V. Gillani, J.D. Husar, and W.E. Wilson, Jr., Formation and transport of secondary air pollutants: ozone and aerosols in the St. Louis urban plume, Science 194:187 (1976).

12. W.S. Cleveland, B. Kleiner, J.E. McRae, and J.L. Warner, Photochemical air pollution: transport from the New York City area into Connecticut and Massachusetts, Science 191:179 (1976).

AIR QUALITY PROJECTIONS FOR THE OHIO RIVER BASIN

M.T. Mills, E. Y. Tong, A. Hirata and A. Van Horn
Teknekron Research, Inc.
Waltham, Massachusetts USA

L.F. Smith
Environmental Protection Agency, Washington, DC - U.S.A.

INTRODUCTION

The purpose of the OHIO River Basin Energy Study (ORBES) was to assess the potential environmental, social and economic impacts associated with the construction of additional power plants in the Ohio River Basin. The study area, as mandated by the U.S. Congress, included parts of the six states of Illinois, Indiana, Kentuckly, Ohio, West Virginia and Pennsylvania. The evaluation of air quality changes for different assumptions of electricity demand, regulatory constraints and power plant retirement schedules played an important role in this assessment. This paper presents estimates of future air quality in the ORBES region based upon dispersion model calculations and projected emissions. While SO_2 emissions are projected to generally decrease in the future years, along with regional background concentrations, there will be areas which will have problems meeting ambient air quality standards. For this reason, the paper also deals with the problems involved in the use of standard regulatory models for air quality planning in this high emission density area. The results presented here pertain only to the current and future emissions of sulfur dioxide (SO_2), which is the constraining pollutant in terms of the growth of new sources and the attainment of National Ambient Air Quality Standards (NAAQS). Emissions of SO_2 are also a factor in the formation of fine particulates, acid precipitation and regional visibility degradation.

CURRENT AND PROJECTED SULFUR DIOXIDE EMISSIONS

As can be seen from Table 1, over 40 percent of the U.S. total SO_2 emissions in 1976 were contributed by the six ORBES states. Over 75 percent of this six state total was due to the electric utilities. The locations of these fossil steam plants is shown in Figure 1. In this region the net generation of power should more than double during the next 20 years if one assumes a high growth in electricity demand of 3.7 percent per year. Under a low growth scenario, the net generation will increase by only 20 percent during this time period. The projected SO_2 emissions, however, are relatively insensitive to the growth in electricity demand. This is due to the fact that new coal fired power plants in the Ohio River Basin are required to meet much stricter SO_2 emission standards than the existing power plants. Those plants construced before 1971 are subject to State Implementation Plan (SIP) emission limitations, which were chosen to ensure the attainment and maintenance of National Ambient Air Quality Standards (NAAQS). The average SO_2 SIP emission limit for coal fired power plants in the ORBES region is 3.3.lbs of SO_2 per 10^6 BTU heat input. The emission limit for a particular plant is often based upon point source dispersion model calculations which are then compared with 3-hour and 24-hour NAAQS for SO_2. In practice it is the 1300 $\mu g/m^3$ 3-hour SO_2 standard which governs the individual emission limitation for some of the older tall stack power plants. Coal fired power plants built after 1971 but before 1979 are subject to a New Source Performance Standard (NSPS) of 1.2 lbs $SO_2/10^6$ BTU. After 1979, new plants are required to remove 90 percent of the SO_2 emissions unless it is possible to reach a 0.6 lbs $SO_2/10^6$ BTU emission limit with a lower percent SO_2 removal. In any case a minimum of 70 percent of the SO_2 must be removed. For the ORBES region this revised NSPS means that new plants will generally have an SO_2 emission limitation of 0.6 lbs $SO_2/10^6$ BTU. In effect, the SO_2 emissions from a new coal fired power plant will generally be a factor of 5 below those for plants operating before 1971.

Due to the much higher SO_2 emissions from the older plants, the projected emissions in the ORBES region depend critically upon assumptions regarding when and if the plants are brought into compliance with SIP emission limits and the retirement schedule of the older plants. The importance of these factors is shown in Figure 2 which gives the ORBES region total projected SO_2 emissions under the assumptions of SIP compliance, no SIP compliance and compliance with strict SIP emission limits, which currently only apply to urban areas. The emission projections were carried out by use of the Teknekron Utility Simulation Model[1] which considers projected power demands, pollution control costs, electricity prices and regulatory constraints. Under the non-compliance scenario, SO_2 emissions actually increase until the mid 1980's after which they decrease due to the retirement of the older power plants. Under the compliance

Table 1. 1976 SO_2 Emissions (in 10^6 tons) for the
Nation and the six ORBES states

	Sulfur Oxides	
Source Category	U.S.	ORBES
Total	32.56	13.39
Stationary Fuel Combustion	25.27	12.32
Electric Utilities	20.01	10.40
Industrial	3.80	1.57
Residential, Commercial, Institutional	1.45	0.36
Industrial Processes	6.44	0.88
Transportation	0.77	0.17
Solid Waste	0.08	0.01
Miscellaneous	0.00	0.00

Note : The six ORBES states are Illinois, Indiana, Kentucky,
       Ohio, Pennsylvania, and West Virginia

Source : Extracted from U.S. EPA, 1976 National Emission
         Report, EPA-450/4-79-019
         (Research Triangle Park, N.C., 8/79).

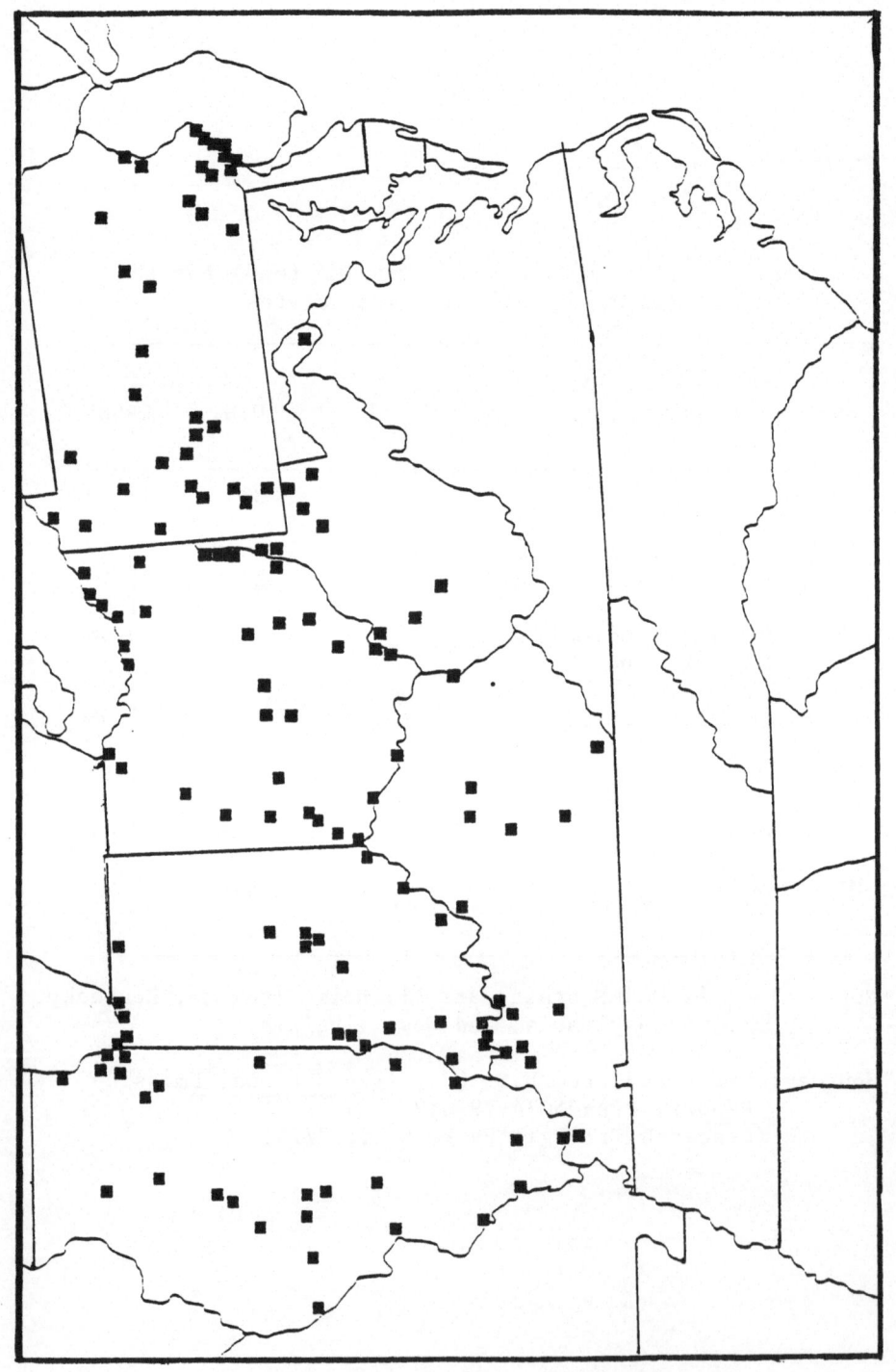

Figure 1. Fossil Steam Plants in the Ohio River Basin States as of 1976.

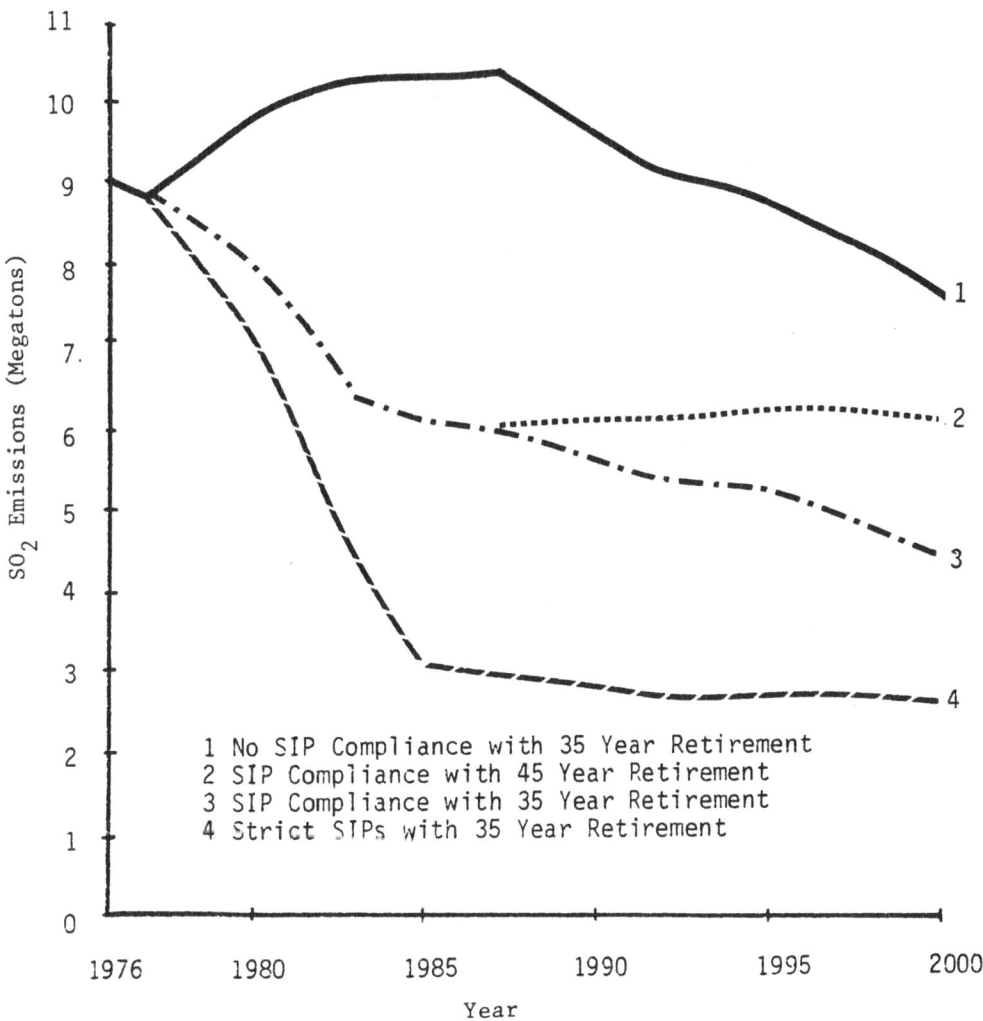

Figure 2. Projected ORBES Region Electric Utility SO_2 Emissions
Totals for Different Regulatory and Unit Lifetime
Assumptions.

scenario the emissions decrease sharply through 1985, the year in
which all sources are expected to be in compliance. After 1985 the
emissions decrease at the same rate as in the non-compliance case.
The assumption of the strict urban SIPs even more drastically reduce
the emissions by 1985, after which they decline only slightly. If
one assumes SIP compliance with a 45 year retirement schedule instead of 35 years, then there is virtually no change in emissions
from 1985 to 2000.

REGIONAL SO$_2$ AND SULFATE CONCENTRATIONS

          These SO$_2$ emission projections may be translated to
regional air quality impacts for both short and long concentration
averaging times. Calculated 24-hour average SO$_2$ concentrations for
the ORBES region are given in Figure 3(a) for the SIP compliance and
non-compliance scenarios based upon the meteorology of the 3 day period ending August 27, 1974, the day for which the calculated concentrations were reported. Projected 24-hour sulfate concentrations
based upon these conditions are shown in Figure 3 (b). While there
is some shift in the SO$_2$ and sulfate concentration patterns for different projection years and emission control scenarios, the primary
change is in the overall magnitude of the calculated concentrations.
These concentrations were calculated by use of the two dimensional
pseudospectral grid model[2] and input windfields derived from upper
air measurements in the eastern United States. This period was
chosen for the simulation because it represented the clearest example
of potential additive source impacts under persistend wind flow.
During this period the agreement between the magnitude and pattern
of measured and calculated sulfate concentrations[3] was relatively
good. Evaluation of the SO$_2$ concentration calculations is more difficult since they represent regional background concentrations which
are rarely measured in the absence of individual source contributions. In addition to the persistent flow regional episode such as
the August 27, 1974 case, the ORBES region can experience high regional pollutant concentrations under multi-day stagnation conditions
which may occur several times during the summer months.

          Baseline and projected annual SO$_2$ and sulfate concen-
trations were estimated by use of an analytical solution to the two
dimensional diffusion equation with pollutant transformation and removal included[4]. The SO$_2$ transport from a source grid depends upon
the annual resultant wind speed at the 600 meter level and a regional
scale diffusivity which was chosen as 2×10^6 m^2/sec for this analysis.
In effect, the diffusivity exerts a dominant role in the calculation
so that the "upwind" contribution of a source is significant. In
addition to the regional diffusivity, the sulfate concentration is
governed by the assumed transformation rate, chosen as 1 percent per
hour on an annual average basis. The calculated annual SO$_2$ and sulfate concentrations are given in Figures 4 (a) and (b) respectively
for the compliance and non-compliance scenarios. As was the case for

Base Case 1976 Emissions

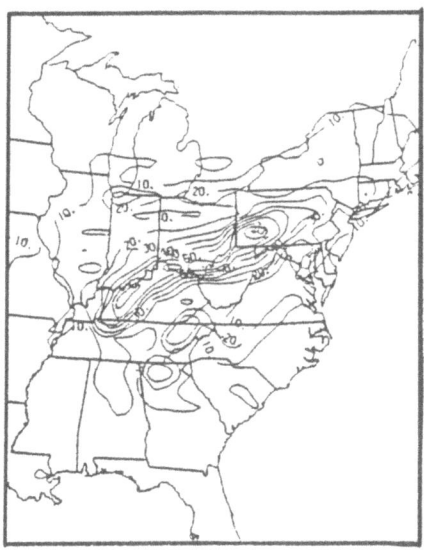

Year 1985- Non Compliance   Year 2000- Non Compliance

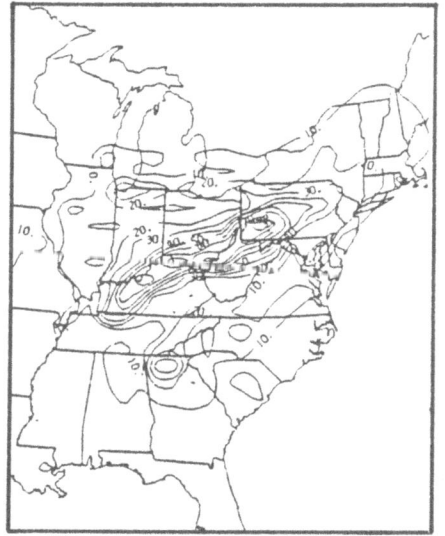 

Figure 3(a). Regional Baseline and Projected 24-Hour SO_2 Concentrations ($\mu g/m^3$) for SIP Compliance and Non-Compliance Scenarios. Based Upon Meteorology for the 3 Day Period Ending August 27, 1974. Utility Emissions Only.

Year 1985 - Compliance   Year 2000 - Compliance

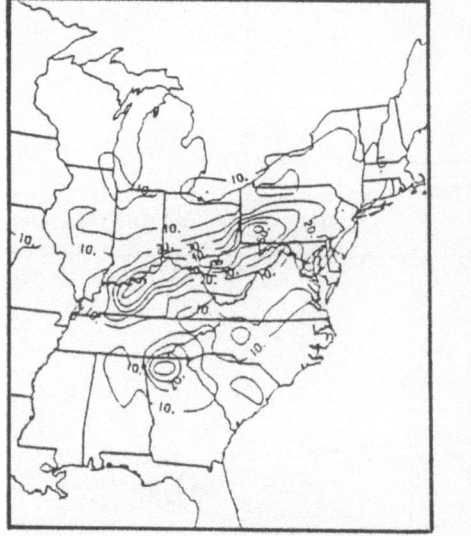

Figure 3(a). Continued

Base Case 1976 Emissions

Year 1985 - Non Compliance   Year 2000 - Non Compliance

Figure 3(b). Regional Baseline and Projected 24-Hour Sulfate Concentrations ($\mu g/m^3$) for SIP Compliance and Non-Compliance Scenarios. Based Upon Meteorology for the
3 Day Period Ending August 27, 1974. Utility Emissions Only.

Year 1985 - Compliance   Year 2000 - Compliance

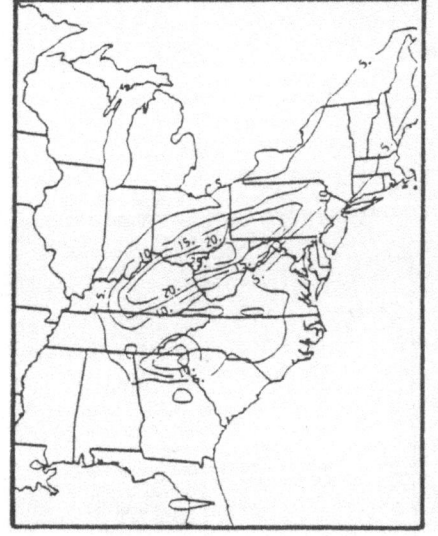

Figure 3(b). Continued.

Base Case 1976 Emission

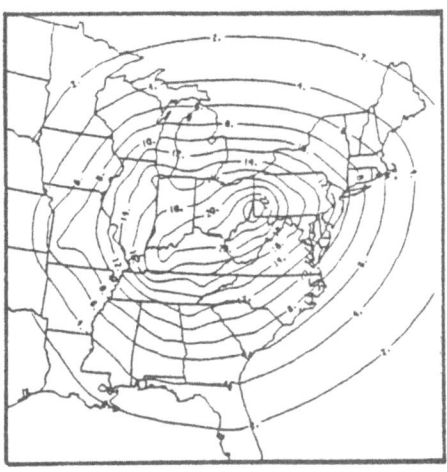

Year 1985 - Non Compliance

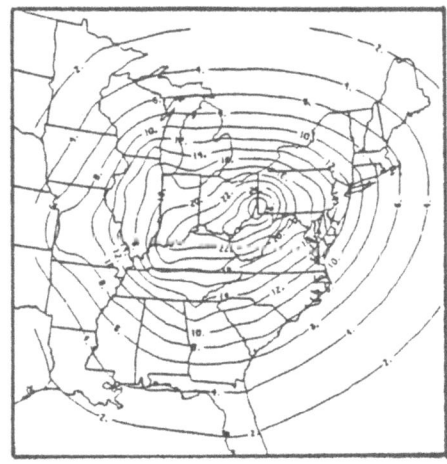

Year 2000 - Non Compliance

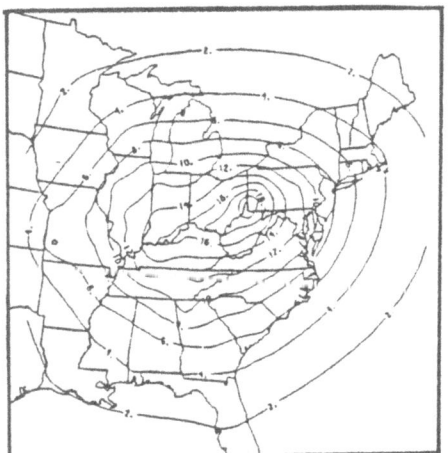

Figure 4(a). Regional Baseline and Projected Annual Average SO_2 Concentrations ($\mu g/m^3$) for SIP Compliance and Non-Compliance Scenarios. Utility Emissions Only.

Year 1985 - Compliance   Year 2000 - Compliance

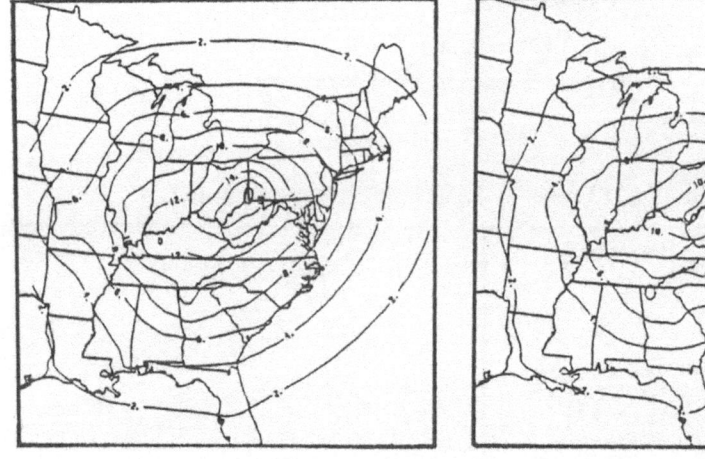

Figure 4(a). Continued.

Base Case 1976   Emissions

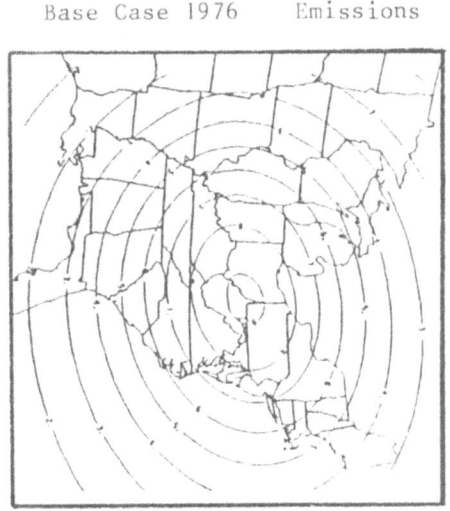

Year 1985 - Non Compliance   Year 2000 - Non Compliance

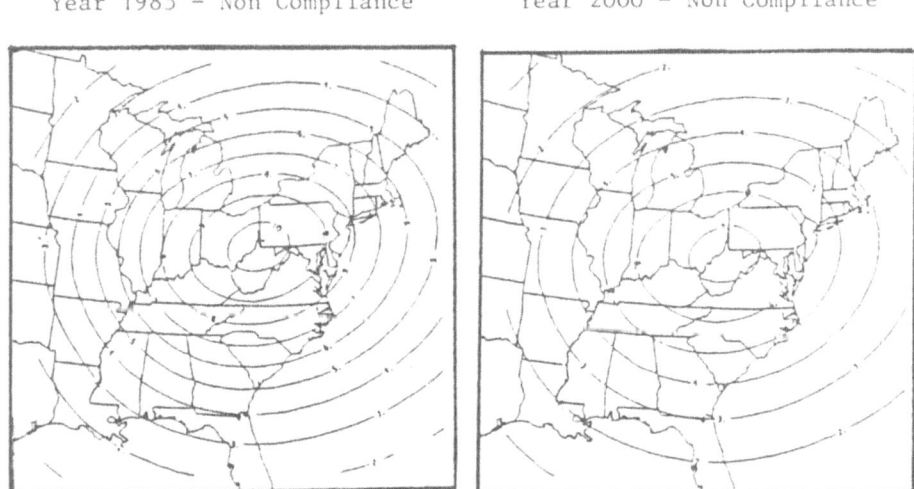

Figure 4(b). Regional Baseline and Projected Annual Average Sulfate Concentrations ($\mu g/m^3$) for SIP Compliance and Non-Compliance Scenarios. Utility Emissions Only.

Year 1985 - Compliance   Year 2000 - Compliance

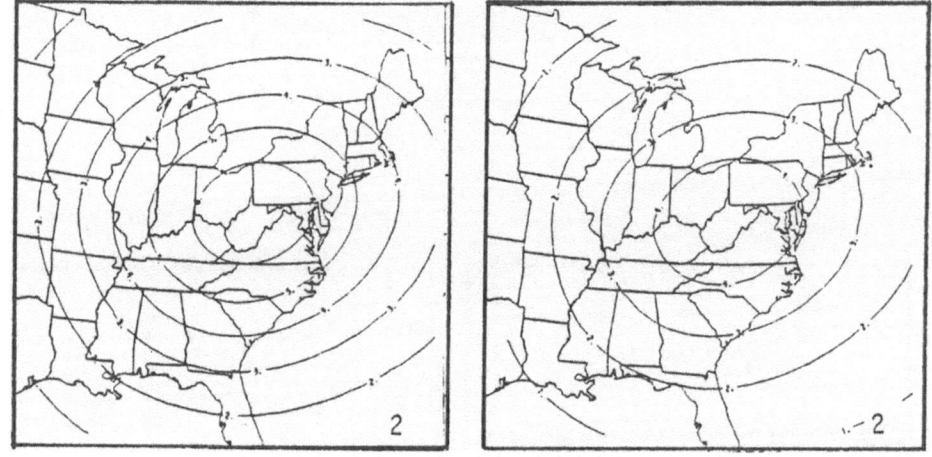

Figure 4(b). Continued.

the episode concentration calculations there is little change in the spatial pattern of concentrations for different projection years and emission control scenarios. Only the magnitude of the concentrations change and then generally in proportion to the overall emission changes given by curves 1 and 3 in Figure 2.

LOCAL AND SUBREGIONAL AIR QUALITY IMPACTS

While regional model calculations indicate that, even with the assumption of SIP non-compliance and a power plant lifetime as great as 45 years, background SO_2 and sulfate concentrations will decrease in the ORBES region by the year 2000, there is the possibility that the 24-hour SO_2 air quality standard of 365 $\mu g/m^3$ will be exceeded more than the allowable once per year. In some areas even the annual SO_2 standard of 80 $\mu g/m^3$ is exceeded, a situation which may continue even in the near future. This may be a significant problem in the upper Ohio River Basin where existing and proposed power plants are located along a direction of frequentlyoccurring persistent winds. In addition to the problem of non-attainment of the SO_2 standard, there is an additional constraint that concentration impacts for sources permitted after January 1975 not exceed allowable concentration increments under regulations for Prevention of Significant Deterioration (PSD). For most of the ORBES region, these allowable increments for SO_2 are 512 $\mu g/m^3$ and 91 $\mu g/m^3$ for the 3-hour and 24-hour averaging times respectively. As in the case with the SO_2 short term ambient standards, these PSD increments may only be exceeded once per year.

The decision as to whether emissions from a proposed new source will cause standards to be exceeded or exacerbate an existing non-attainment problem is made on the basis of dispersion model calculations. Obviously a new source can impact a non-attainment area to some degree no matter how distant. For SO_2, the U.S. Environmental Protection Agency has specified concentration levels below which new source impacts are considered to be insignificant. These concentration limits for SO_2 are 25, 5 and 1 $\mu g/m^3$ for the 3 hour, and 24-hour and annual averaging times. For a particular source, the calculation of this "radius of significance" is carried out by use of a steady state gaussian type model,[5,6]. These models calculate hourly concentrations at a large number of receptor locations based upon surface and upper air meteorological data obtained from the most representative sites. At each of the model receptor locations the highest 3-hour, 24-hour and annual SO_2 concentrations are compared with the significance level. The use of this type of model to calculate long range impacts is questionable considering that the models assume that a concentration profile is established out to an infinite distance for each hour. This assumption generally leads to an overestimate of maximum concentration impacts at these long distances. In light of this deficiency, the calculation of new source impacts is generally not required beyond distances of 50 km.

The same models are used to track the consumption of the PSD increment in the ORBES region.Even for the annual averaging times,the consumption of concentration increment by more than one source will not be additive since the maximum impact due to emissions from one source will generally not coincide with that of another source unless they are co-located with similar emission characteristics.In the case of 3-hour and 24-hour SO_2 concentrations, additive consumption of the increment will not be calculated to occur unless the new sources are sited exactly along a line. Even in this case more than 3 identical sources will have to be located along this line to give a joint maximum concentration impact which is greater than the maximum impact due to any one of the sources by itself.For those sources not located along a line,the gaussian models will underestimate additive impacts since they do not account for the SO_2 emissions form earlier hours or the variation of wind speed and direction with distance.As a practical matter,however,the allowable PSD concentration increments are not an important constraint to the siting of new coal fired generating capacity in the ORBES region since the new source emissions will be so strictly controlled.The presence of areas with concentrations near or above the ambient standard constitutes a more important constraint to the siting of new sources in the region.

It is this question of non-attainment which poses a real difficulty for the dispersion models currently used in regulatory applications.The problem with the models is not their accuracy in the prediction of short term concentrations, although some questions have been raised concerning the model predictions close to the source under very unstable conditions. Instead,it is the inability of the models to simulate pollutant transport in high emission density areas such as the upper Ohio River Basin.To account for the regional accumulation of polluant concentrations requires the use of a numerical grid or puff trajectory[7] model or a combination of both. The puff trajectory type model should be used for the large point source emissions to avoid the initial dilution of emissions associated with the use of a grid type model. The 24-hour average calculated SO_2 concentrations during a selected period due to emissions from 6 existing coal fired power plants along the upper Ohio River are shown in Figure 5. One set of SO_2 concentrations is calculated by a puff trajectory model while the other set is based upon the application of a steady state gaussian model used routinely in regulatory applications.The period used in the puff model simulation, April 12-15,was characterized by a large high pressure system moving slowly southeastward from the midwest. The concentrations calculated by the gaussian model were much lower due to neglect of the carryover of concentrations from one hour to the next and the fact that the surface data,which was required input for this model,dit not show the same persistence in wind direction as the 600 meter wind data used in the puff model simulation of elevated plume transport.

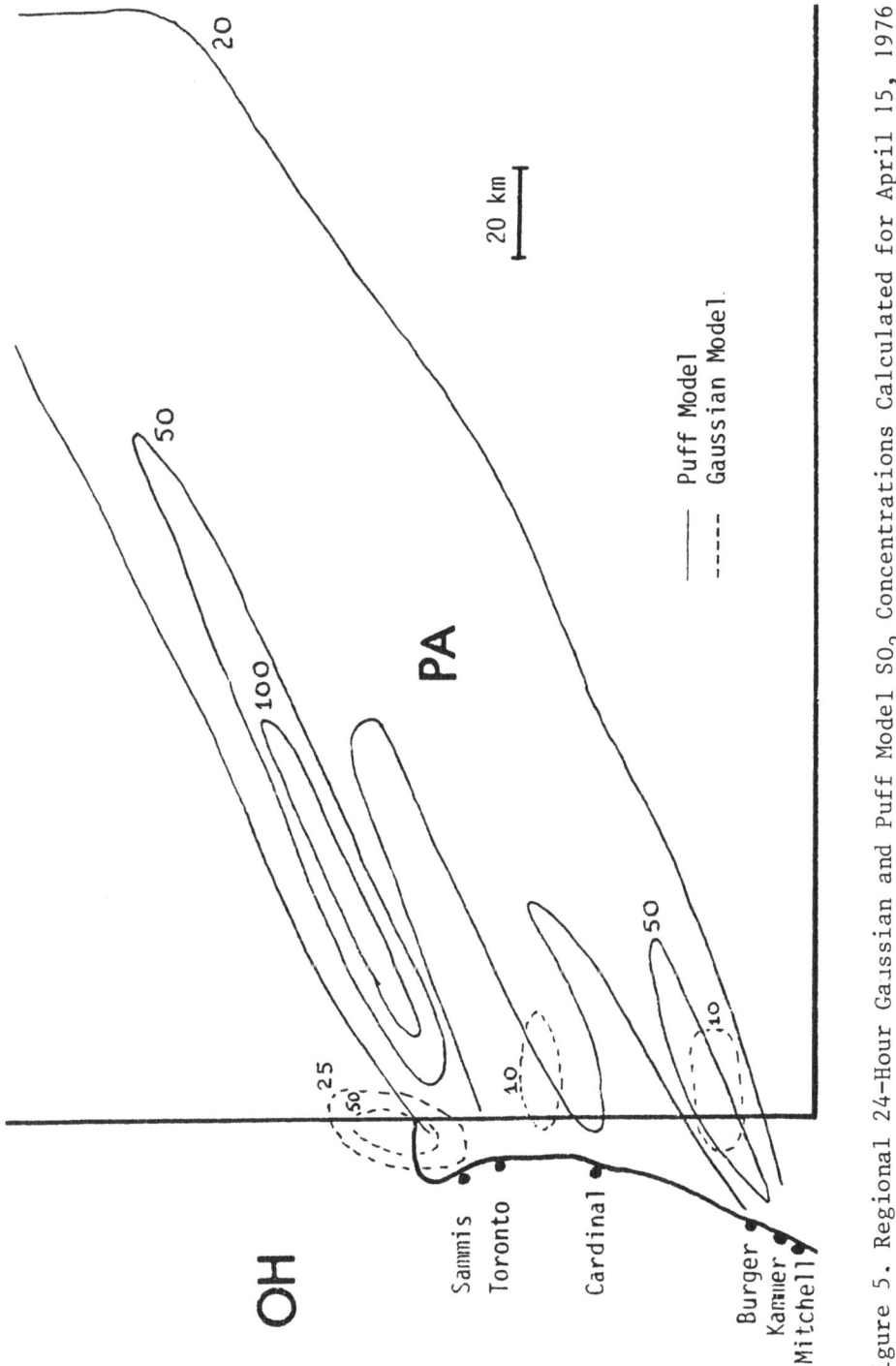

Figure 5. Regional 24-Hour Gaussian and Puff Model SO$_2$ Concentrations Calculated for April 15, 1976 for the Upper Ohio Valley Based Upon Emissions from 6 Coal Fired Power Plants. Meteorology for the Period April 12-15, 1976 was used in the Puff Model Calculation.

Modeling studies such as these, in conjunction with extensive analyses of monitoring data, have shown the need for a multi-scale modeling approach for the development of emission control measures for the attainment and maintenance of SO_2 standards in an area such as the Upper Ohio Valley. Within 10 km of the source, concentrations would still be calculated by use of the steady state gaussian models since the high puff release rate for this short distance would be prohibitive in terms of computation time. Beyond this distance the puff trajectory model would be used for a selected number of the largest point sources under consideration. The remaining emissions would be input to a grid model which would use the same windfield as the puff trajectory calculation. This methodology would provide a more realistic simulation of background SO_2 concentrations, which on occasion may be a significant fraction of the 24-hour standard or even exceed the standard.

CONCLUSIONS

For the ORBES region as a whole, there should be a reduction in SO_2 emissions from coal fired power plants by the year 2000. The timing and amount of this reduction depends upon the degree to which the existing plants achieve compliance with their established emission limitations. A second important factor is the assumed lifetime of the existing power plants. The projected SO_2 emissions depend only slightly upon the assumed growth rate in electrical demand since the power plants which will fill this demand have much stricter emission control requirements than the existing facilities. Based upon these emission reductions, there is a projected decrease in both regional SO_2 and sulfate concentrations in the ORBES region.

While SO_2 and sulfate concentrations are projected to generally decrease, there will remain areas in which 24-hour SO_2 concentrations exceed the ambient standard or where there is only a small margin left for the growth of new sources. In these areas, strategies for the attainment of the standard and the allocation of the remaining air quality resource requires the use of models which describe the transport of SO_2 on both local and regional scales.

REFERENCES

1. A. Van Horn, Selected Impacts of Electric Utility Operations in the Ohio River Basin (1976-2000) : Application of the Utility Simulation Model. Vol. 3, ORBES Air Quality and Related Impacts. Teknekron Report n° R-001-EPA-8.
2. L.P. Prahm and O. Christensen, Long Range Transmission of Pollutants Simulated by a Two-Dimensional Pseudospectral Dispersion Model, J. Appl. Meteor. 16 : 898 - 910 (1976)

3. M.T. Mills and A.A. Hirata, A Multi-Scale Transport and Dispersion Model for Local and Regional Scale Sulfur Dioxide/Sulfate Concentrations : Formulation and Initial Evaluation, Paper presented at the Ninth International Technical Meeting on Air Pollution Modeling and Its Application, NATO/CCMS Air Pollution Pilot Study Assessment, Methodology and Modeling, Toronto, Canada, August 28-31, 1978

4. J.A. Fay and J.T. Rosenzweig, An Analytical Diffusion Model of Long Distance Transport of Air Pollutants, Atmospheric Environment 14, pp. 355 - 365 (1980)

5. Monitoring and Data Analysis Division, User's Manual for Single-Source (CRSTER) Model, U.S. Environmental Protection Agency. Research Triangle Park, NC. EPA-450/2--7-013, 1977 (NTIS accession number PB 271-360).

6. D. Bruce Turner and Joan Hrenko Novak, "User's Guide for RAM", U.S. Environmental Protection Agency. Research Triangle Park, NC, EPA-600/8-78-016, November 1978.

7. A Bass, Modeling Long Range Transport and Diffusion, Second Joint Conference on Applications of Air Pollution Meteorology, American Meteorological Society, Boston, MA, March 1980.

2: METEOROLOGICAL PARAMETERS FOR USE IN ADVANCED AIR DIFFUSION MODELS

Chairmen: W. B. Johnson
L. Niemeyer
T. Turner
E. Runca

Rapporteurs: L. P. Prahm
F. Nieuwstadt
K. E. Grønskei
A. Venkatram

ESTIMATION OF TURBULENCE VELOCITY SCALES IN THE STABLE AND THE UNSTABLE BOUNDARY LAYER FOR DISPERSION APPLICATIONS

A. Venkatram

Air Resources Branch
Ontario Ministry of the Environment
880 Bay Str., 4th Floor, Toronto, Ont., M5S 1Z8

INTRODUCTION

Most of the information required to estimate the concentration of a pollutant released in the planetary boundary layer is contained in three variables namely the surface friction velocity u_*, the convective velocity scale w_* and the mixed layer height z_i. Although these variables are more directly related to the dispersive ability of the PBL than the various stability classification schemes their use has been largely ignored by the air pollution modeling community. This situation is a result of the relative lack of acceptable methods to relate these variables to routinely available meteorological measurements such as wind speed and solar radiation. In this note we present some techniques to do just this. We will also briefly describe the use of these variables in air pollution modeling.

AIR POLLUTION MODELS

A dispersion model is essentially a description of our understanding of the relevant physical system in the precise and concise language of mathematics. In most cases the physical system is so complex that we have to be satisfied with identifying the "controlling" variables. Furthermore, the concentration field is governed by turbulence; so we are forced to understand dispersion in terms of ensemble averages. On the other hand observed concentrations represent members of the population whose mean should correspond to the model prediction. Therefore, model predictions will deviate from observations and it is very difficult to "validate" a model on the basis of a few measurements. This suggests that we cannot choose between models only on the basis of a necessarily limited

comparison between model predictions and observations. We have to
insist that a good model incorporate our best understanding of the
physics of the situation. This is the only way of making sure that
the model can reproduce the general behaviour of the physical system
under study. This discussion of air pollution models highlights the
need for greater use of our knowledge of the planetary boundary layer
in interpreting dispersion of pollutants. As we will see in the
following sections the variables u_*, w_* and z_i are the most important descriptors of the dispersive ability of the PBL.

THE UNSTABLE BOUNDARY LAYER

The PBL is said to be unstable or convective when its turbulence
is generated by the upward heat flux originating from the ground
heated by solar radiation. Under these conditions the relevant
convective velocity scale w_* is given by:

$$w_* = \left(\frac{g}{T_o} Q_o z_i \right)^{1/3}$$ (1)

In (1), Q_o is the surface kinematic heat flux. z_i the height of the
mixed layer and g/T_o is the buoyancy parameter. Observations
(Kaimal et al., 1976) show that above $0.1\, z_i$, $\sigma_w \approx 0.6\, w_*$. Shear
generation of turbulence is confined primarily to heights below the
Monin-Obukhov length L which is given by

$$L \equiv -u_*^3\, T_o / gkQ_o$$ (2)

where u_* is the surface friction velocity. In this layer turbulent
velocities are proportional to u_*. The distance d beyond which u_*
is not important in controlling dispersion can be estimated by equat-
ing the height of a cloud σ_z released at surface level to $|L|$. Taking
$\sigma_z \approx u_* x/u$ we find

$$d = |L|\, u/u_* \approx |L|/C_D$$ (3)

where C_D is a drag coefficient. We can estimate $|L|$ through the
identity (Venkatram, 1980c)

$$|L| = z_i \left(\frac{u_*}{w_*} \right)^3 \frac{1}{k}$$ (4)

Taking u_* to be the relatively large value of $0.4 ms^{-1}$, and w_* to be
$1.5 ms^{-1}$ and $z_i \sim 1000m$ we find $|L| \approx 50m$. With a typical $C_D \approx 0.1$ we find
that u_* is not important beyond a kilometer from a surface release.
Note that vertical growth of the cloud is governed by w_* beyond
$\sigma_z \approx |L|$ (See Nieuwstadt, 1980).

It is found (McBean et al., 1979) that the motions of the large eddies extending through the depth of the convective PBL control the horizontal velocity fluctuations. Consequently we find that

$$\sigma_\alpha \approx 0.6 w_* \quad (\alpha = u, v); \quad |L| \le z \le z_i \qquad (5)$$

The time taken after release for pollutants to become well mixed through the depth of the boundary layer is approximately z_i/w_*. This introduces the non-dimensional distance X given by

$$X \equiv \frac{w_*}{z_i} \frac{x}{u} \qquad (6)$$

where u is the mixed layer wind and x is the downwind distance from the source. Observations (Deardorff and Willis, 1975; Venkatram, 1980a) show that pollutants are virtually uniform in the vertical beyond X=2 from the source. With the relationship $\sigma_y \alpha w_* x/u$ the expression for the ground-level concentration beyond this distance can be written as (Venkatram, 1980a)

$$C = \frac{0.9Q}{w_* z_i x} \qquad (7)$$

where Q is the emission rate. For distances less than X=2, the ground-level concentration is determined by the distance at which the plume is brought down to the ground by convective downdrafts whose velocities scale with w_*.

The foregoing discussion clearly shows the importance of w_* in controlling dispersion in the convective boundary layer. Note that w_* depends on the 1/3 power of Q_o and z_i. This suggests that an estimate of w_* is relatively insensitive to errors in Q_o and z_i. The surface heat flux can be related to the incoming solar radiation and surface albedo (Briggs, 1975). The mixed layer height z_i can be computed using simple models (See Tennekes, 1973). It can be shown (Venkatram, 1978) that w_* can be estimated from the following simple equation

$$w_*(t) = 1.5 \ Q_m^{\frac{1}{2}} \ (\frac{g}{T_o})^{1/3} \left[\frac{\tau}{\gamma(1-2f)} \right]^{1/6} \sin^{2/3}(\frac{\pi t}{4\tau}) \cos^{1/3}(\frac{\pi t}{4\tau}) \qquad (8)$$

In (8), Q_m is the maximum surface heat flux, γ is the potential temperature gradient of the stable layer capping the mixed layer, f is an entrainment constant ($\approx 1/7$) and τ is the half-period of the assumed sinusoidal surface heat flux. Note the weak dependence (of the 1/6 power) of w_* on τ, γ and f. This means that we can determine w_* as a function of time t from sunrise if we can estimate Q_m. The mixed layer height z_i can be estimated from

$$z_i = \left[\frac{8\tau Q_m}{\pi \gamma (1 - 2f)} \right]^{\frac{1}{2}} \sin(\frac{\pi t}{4\tau}) \tag{9}$$

It is clear that the calculation of w_* and z_i is more straightforward and objective than the specification of conventional stability classification schemes. Furthermore, these variables are more relevant to the physics of dispersion than the so-called stability classes. It is useful to point out that w_* can also be used to estimate final plume rise in convective conditions (See Briggs, 1975). We conclude this section by noting that the PBL is unstable more often than is normally believed (Venkatram, 1980c).

THE STABLE BOUNDARY LAYER

Our understanding of the stable boundary layer is far less complete than that of the convective boundary layer. This is because the stable boundary layer (SBL) does not often meet stationarity and horizontal homogeneity conditions which are extremely useful aids in the description of the boundary layer. The time scale of the evolution of the SBL does not usually allow us to assume quasi-steadiness except close to the ground. This means that although there is a definite boundary layer in which pollutants can be trapped (See Briggs, 1976) its height cannot be always related to surface parameters such as u_*. However, the observations of Caughey et al., (1979), indicate that the turbulence is in equilibrium with the mean temperature and velocity profiles in the SBL. So we can probably characterize the dispersive ability of the boundary layer with near surface meteorological observations.

The stable boundary layer is most important with reference to near surface releases. Recent work (Briggs, 1976; Van Ulden, 1978) indicates that the relevant variables for modeling vertical dispersion in this case are u_*, L and z_o. Van Ulden (1978) has proposed a simple technique to use these variables to compute the cross-wind integrated concentration $\overline{C}^y$. Briggs (1978) suggests the following simple form for the ground-level $\overline{C}^y$.

$$\overline{C}^y = Q/uh \tag{11a}$$

where

$$h/L = X/(1 + X^{\frac{1}{2}}) \tag{11b}$$

and

$$X \equiv u_* x/uL \tag{11c}$$

In (11), Briggs took u to be the wind speed measured at a height of 8m from the ground.

Venkatram (1980d) finds that the following simple empirical equation explains the behaviour of ground-level concentrations very well

$$\overline{C}^{\,y} = 0.38 Q u_*^{-1.34} x^{-0.83} \tag{12}$$

where u_* is in m/s and x is in m. Note that the use of (12) requires only knowledge of u_* which can be readily related to the wind speed using the method described in the following paragraph.

In order to use these formulations for the concentrations we need to know the surface heat flux Q in the stable boundary layer.

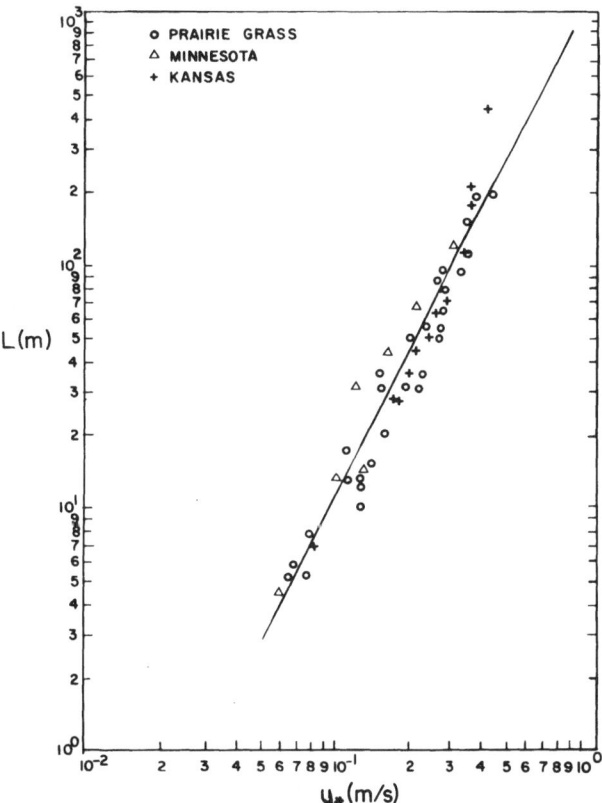

Fig. 1.   Variation of L with u_* for data collected during the Prairie Grass, Kansas and Minnesota experiments (stable conditions).

This information is not usually available on a routine basis. To overcome this problem we have proposed an empirical method to compute L. Figure 1 shows a plot of L against u_* for data collected in three separate field experiments. Although the variation of L for a given u_* is not small everywhere the correlation between these two variables appears to be useful for dispersion calculations. The best fit line in the figure corresponds to

$$L = 1.1 \times 10^3 u_*^2 \tag{13}$$

To understand the possible reason for (13) let us rewrite the definition of L in terms of $T_* \equiv -Q_o/u_*$

$$L \equiv T_o u_*^2 / g k T_* \tag{14}$$

Equations (13) and (14) tell us that the variation of T_* in the SBL is small. In other words, an increase in u_* results in an increase in the downward heat flux into the ground. This is clearly plausible. However, we suspect that the proportionality "constant" between Q_o and u_* will depend on the physical properties of the soil-air interface. It is useful to point out that Briggs(1978) noticed the strong (useable) correlation between L and the wind speed at a specific level. This corresponds to (13) at moderate wind speeds when u_*/u is not a strong function of Q_o.

Using (13) and the similarity velocity profile (See Businger, 1973) for the surface SBL we can write the following equation for u_*

$$u_* = C_{DN} u \left\{ \frac{1}{2} + \frac{1}{2} \left[1 - \left(\frac{2u_o}{C_{DN}^{\frac{1}{2}} u} \right)^2 \right]^{\frac{1}{2}} \right\} \tag{15}$$

where

$$C_{DN} = \frac{k}{\ln(z_m/z_o)} \tag{16a}$$

and

$$u_o^2 = \beta z_m / k A \tag{16b}$$

In (16) $\beta = 4.7$, $A = 1.1 \times 10^3$ (from (13)), z_o is the roughness height and z_m is the height at which u is measured. It is seen from (15) that with an estimate of z_o we can compute u_* from wind speed measured at one level. This result is very useful in view of the routine availability of wind speeds at most sites. Note that u_* tells us a great deal about the turbulence and hence dispersion in the SBL.

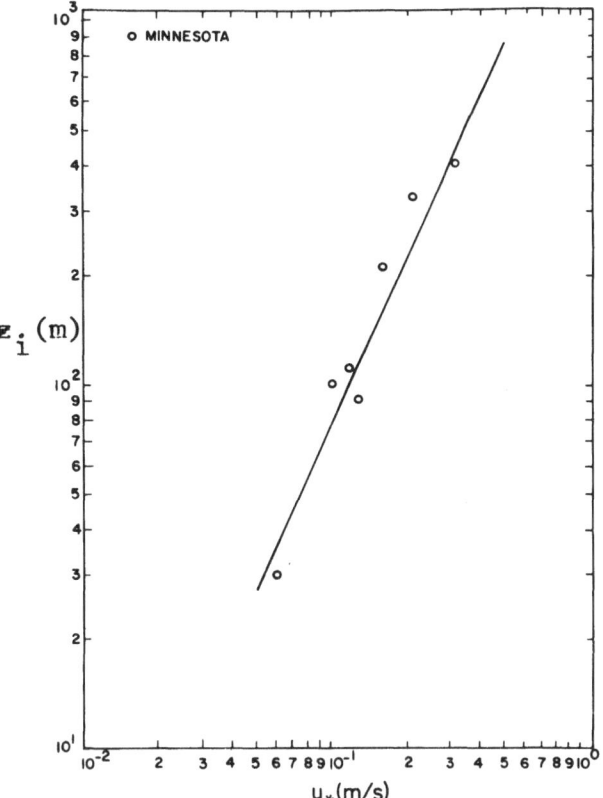

Fig. 2. Variation of z_i with u
during stable conditions in the
Minnesota experiment.

In the absence of a detailed boundary layer model we can probably estimate the height z_i of the SBL using the steady state expression suggested by Zilitinkevich (1972).

$$z_i = d \left(\frac{u_* L}{f} \right)^{1/2} \tag{17}$$

where f is the Coriolis parameter and d is a constant. The difficulty associated with measuring the steady state boundary layer
height has not allowed an unambiguous determination of the value of
d. The observations of Caughey et al. (1979) suggest that $d \approx 0.7$
while the modeling study of Wyngaard (1972) indicates that $d \approx 0.2$.
There is probably more general agreement on the form of (17). Note
that our empirical equation for L (Eq. 13) when combined with (17)
leads to

$$z_i \alpha u_*^{3/2} f^{-1/2} \tag{18}$$

Figure (2) shows a plot of data presented by Caughey et al. (1979). The best fit line of $z_i \approx 2400\ u_*^{3/2}$ suggests that (18) is an useful approximation which can be used to compute the maximum extent of vertical dispersion in the SBL.

The horizontal spread of a pollutant released in the SBL is controlled by mesoscale turbulence which leads to meandering of the plume. It is often impossible to estimate this spread which is tied to site-specific features. A conservative estimate of σ_v can be obtained from theories (Hunt and Weber, 1979) which relate horizontal spread to micrometeorological turbulence. In this connection we point out that there are methods (Binkowski, 1979) to estimate the standard deviations of turbulent velocity fluctuations σ_v and σ_w from Monin-Obukhov variables such as z/L and u_*.

THE NEUTRAL BOUNDARY LAYER

The ideal neutral boundary layer is one in which turbulence is entirely generated by shear. As the PBL is almost always associated with a surface heat flux neutral conditions are clearly very rare. However, when the wind speeds are high, the PBL is approximately neutral especially close to the ground. When this type of boundary layer is capped by a stable layer as it normally is the boundary height z_i is given by

$$z_i = \frac{au_*}{(fN)^{\frac{1}{2}}} \; ; \quad a=\text{constant} \tag{19}$$

where N is the Brunt-Vaisala frequency in the stable layer. Apart from its dimensional correctness there is little experimental evidence to indicate that (19) is appropriate. As mentioned earlier our lack of knowledge of the neutral boundary layer should not be of concern as it occurs so rarely.

SUMMARY

Stability classification schemes (See Pasquill, 1974) are methods of characterizing the dispersive ability of the PBL. Although they are quite useful they cannot be always related to turbulence levels which control dispersion. Recent work (Deardorff and Willis, 1975; Venkatram, 1980; Van Ulden, 1978) indicates that the micrometeorological variables u_*, w_*, L and z_i are much better descriptors of the dispersion process in the PBL. [1] These variables can also be used to estimate the statistics of atmospheric turbulence. In this paper we present simple methods to estimate w_*, u_*, L and z_i from routine meteorological observations. We also show that these methods are a practical yet superior alternative to commonly used stability classification schemes.

LIST OF REFERENCES

Binkowski, F. S., 1979, A simple semi-empirical theory for the
    turbulence in the atmospheric surface layer. Atmospheric
    Environment, 13 : 247-253.
Briggs, G. A., 1975, Plume rise predictions. Lectures on Air Pol-
    lution and Environmental Impact Analyses. American Meteo-
    rological Society, Boston, Mass., 59 : 105.
Briggs, G. A., 1976, Predictions of nocturnal mixing layer para-
    meters. ATDL Contribution File No. 76/22, available from
    author at Post Office Box E, Oak Ridge, Tennessee 37830.
Briggs, G.A. and McDonald, K. R., 1978, Prairie Grass revisited :
    optimum indicators of vertical spread. Proceedings of the
    Ninth International Technical Meeting on Air Pollution
    Modeling and its Application, NATO/CCMS report No. 103,
    209-220.
Businger, J. A., 1973, Turbulent transfer in the atmospheric sur-
    face layer. Workshop on Micrometeorology, D. A. Haugen
    (Editor), American Meteorological Society, Boston, Mass.,
    392 pg.
Caughey, S. J., Wyngaard, J. C. and Kaimal, J. C., 1979, Turbu-
    lence in the evolving stable boundary layer. J. Atmos. Sci.,
    36 : 1041-1052.
Deardorff, J. W. and Willis, G. E., 1975, A parameterization of
    diffusion into the mixed layer. J. Appl. Meteor., 14 :
    1451-1458.
Hunt, J. C. and Weber, A. H., 1979, A Lagrangian statistical ana-
    lysis of diffusion from a ground-level source in a turbu-
    lent boundary layer. Quart. J. R. Met. Soc., 105 : 423-443.
Kaimal, J. C., Wyngaard, J. C., Haugen, D. A., Cote, O. R., Izumi,
    Y., Caughey, S. J., and Readings, C. J., 1976, Turbulent
    structure in the convective boundary layer. J. Atmos. Sci.,
    33 : 2152-2169.
McBean, G. A. (Editor), 1979, The Planetary Boundary Layer, World
    Meteorological Organization, Technical Note No. 165, 201 pg.
Nieuwstadt, F. T. M., 1980, Application of mixed-layer similarity
    to the observed dispersion from a ground-level source.
    J. Appl. Meteorology, 19 : 157-162.
Pasquill, F., 1974, Atmospheric Diffusion, John Wiley and Sons,
    Chicester, 374 pg.
Tennekes, H. E., 1973, A model for the dynamics of the inversion
    above a convective boundary layer. J. Atmos. Sci., 30 :
    558-567.
Van Ulden, A. P., 1978, Simple estimates for vertical diffusion
    from sources near the ground. Atmospheric Environment.
    12 : 2125-2129.
Venkatram, A., 1978, Estimating the convective velocity scale
    for diffusion applications. Boundary-Layer Meteorology,
    15 : 447-452.
Venkatram, A., 1980a, Dispersion from an elevated source in a

convective boundary layer. Atmospheric Environment,14: 1-10.

Venkatram, A., 1980b Estimating the Monin-Obuhov length in the
    stable boundary layer for dispersion calculations.
    Boundary-Layer Meteorology, in press.

Venkatram, A., 1980c, The relationship between the convective boun-
    dary layer and dispersion from tall stacks. Atmospheric
    Environment, 14: 763-767.

Venkatram, A., 1980d, A semi-empirical method to estimate ground-
    level concentrations associated with surface releases in
    the stable boundary layer. Submitted for publication,
    manuscript available from author.

Wyngaard, J. C., 1975, Modeling the planetary boundary layer--
    extension to the stable case. Boundary-Layer Meteor.,
    9: 441-460.

Zilitinkevich, S. A., 1972, On the determination of the height of
    the Ekman boundary layer. Boundary-Layer Meteor.,
    3: 141-145.

DISCUSSION

G. SCHAYES   W. is certainly an important
physical parameter of the unstable PBL , but
how can you relate it to be the diffusivity K
used in Eulerian numerical models ?

A. VENKATRAM   You are assuming that the
concept of eddy diffusivity is valid. I am not
comfortable with the idea of relating W. to K.
If you want to use   a Eulerian model, higher
order closure schemes might be more appropriate.

L. P. PRAHM   Eddy diffusivities might be
related to W. through the standard deviation
of the velocity fluctuations σ w,h and a
length-scale. For σ w,u see BERKOWICZ and
PRAHM W,H
1980 ibid.

R. BERKOWICZ   How can the coriolis parameter enter the expression for the stable PBL
height at the equator ?

A. VENKATRAM   Another type of model must
be applied for such regions.

L. P. PRAHM   You mentioned that the PBL
height can be estimated by simple models.
Most comparisons between models giving PBL
height are compared with Wangara day 33. Is
there any evidence, that, these models can be

applied in practice for e.g. for European
weather conditions.

A. P. VAN ULDEN   TENNEKES and VAN ULDEN applied
the inversion rise model by TENNEKES to about
60 days in 1973 with more than 50 % sunshine.
They used a simple parameterization of the surface heat flux. The method uses standard
meteorological parameters. They obtained satisfactory results as is shown in their paper
presented at the AMS-conference on turbulence
and diffusion (Santa Barbara, 1974).
So these models can be used in practice indeed.

METEOROLOGICAL INPUT FOR A THREE DIMENSIONAL

MEDIUM RANGE AIR QUALITY MODEL

H. van Dop, B.J. de Haan and G.J. Cats[*]

Royal Netherlands Meteorological Institute, De Bilt
[*]European Centre for Medium Range Weather Forecasting,
Reading, U.K.

INTRODUCTION

In the early fifties the first attempts have been made to calculate the dispersion of air pollutants. Most, if not all calculations, were based on the so-called "Gaussian plume model". Due
to the restrictions of this model, dispersion calculations could
only be made over distances ranging from 1-50 km, while reliable
predictions of ground level concentrations could but be obtained
for average values over a period of at least two months. The conditions which limited the applicability of the Gaussian plume model
were the uniformity of the wind velocity and the homogeneity of the
turbulence.

The advent of larger computers allowed the introduction of
much more meteorological details into (numerical) air pollution
models, so that in principle the dispersion of individual releases
of pollutants could be described over larger distances. The two
main advantages of numerical models are that a realistic 3-dimen-
sional windfield can be introduced and that the turbulent diffusion
can be described according to our present knowledge of turbulence
in the atmospheric boundary layer.

This paper describes the lay-out of the meteorological input
for an air quality model, to be applied in a highly industrialized
area in Western Europe, i.e. the Northwestern part of France,
Belgium, Luxemburg, The Netherlands, and the Western part of the
Federal Republic of Germany (Fig. 1). According to Semb (1978) the
annual emission in 1973 amounted to 2.10^6 tonnes of sulphur in the
area considered, which covers about 500 x 400 km^2.

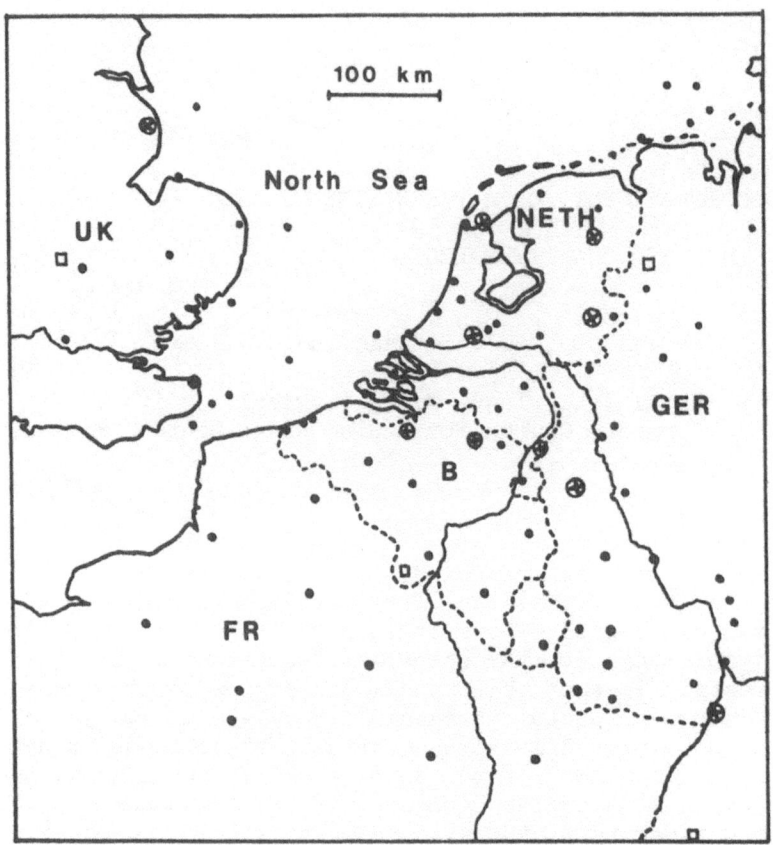

Fig. 1. Locations of synoptic stations (●), meteorological towers
(⊗) and 850 mbar gridpoints (□).

MODEL EQUATION

Advection and diffusion will be described by the semi-empirical equation of turbulent diffusion,

$$\partial c/\partial t + u\partial c/\partial x + v\partial c/\partial y = \partial/\partial z(K\partial c/\partial z) + S. \tag{1}$$

Here, u and v are the horizontal components of the average wind velocity, and K=K(z) is the eddy diffusivity. The last term on the righthand side, S, contains the sources and sinks.

The following additional assumptions have been made
- diffusion in the horizontal direction is neglected.
- the vertical component of the average wind velocity is neglected.

The accuracy of the description of the transport of air pollutants is affected by (i) the crudeness of the approximations used in Eq. (1), (ii) the accuracy of the numerical schemes used to solve Eq. (1), and (iii) the quality of the wind- and turbulence fields, and S.

The model is thought capable to describe the dispersion of pollutants during episodes which last one or two days. More details will be given in a future publication. In this paper we will focus our attention on the windfield, the eddy-diffusivity and the inversion height. This data base is deduced from mainly synoptic data. In the next section it will be indicated how from these data the input for the model in obtained. Finally some interpolation procedures will be discussed.

METEOROLOGICAL PARAMETERS

The major part of the contamination of the air takes place at the ground level or in its near vicinity. From there it is advected horizontally by the mean wind, and diffused vertically by turbulence.

The horizontal displacement considered here is of the order of a few hundred km. Therefore, the correspondent vertical scales do not exceed a height of 1500-2000 m, incidental case with strong vertical convection excepted. It should further be noted that the area is considered flat. Only in the Southeastern part some gently sloped hills exceed a height of somewhat more than 500 m.

THE WIND FIELD

There are two regions where, under some conditions, a rather detailed knowledge exists on the variation of wind speed with height. The first is near the surface where the wind profile obeys the similarity laws, expressed in Obukhov Length L, and the frictionvelocity u_* (Businger, 1973). The second region is in the "free atmosphere", where the windvelocity is in geostrophic balance. In the intermediate region the wind field is obtained by interpolation between the surface layer wind, $U_s(z)$[†], and the upper air geostrophic wind profile, $U_a(z)$, as will be indicated below. The former profile is given by

$$U_s = U_{10} \cdot [\ln(z/z_o) - \psi(z/L)]/[\ln(10/z_o) - \psi(10/L)], \qquad (2)$$

where z_o and L denote the roughness length and Obukhov length respectively, and ψ is a stability dependent function given by Businger (1973). The 10 m wind is denoted by U_{10}. It is obtained by interpolation of the synoptic wind observations. The interpolation

---

[†]In the following we will proceed for simplicity with the description of the U-component of the wind velocity only, the derivations for the V-component being similar.

procedures will be discussed later. Eq. (2) applies in the (stationnary) surface layer, which extends to a height of roughly $2|L|$, if not larger than say $0.2\,z_i$ (z_i = inversion height).

The geostrophic wind profile, U_g, is determined from the surface pressure gradient, U_g^o and the geostrophic wind at the 850 mbar level, U_g^1. The latter data are obtained from a numerical weatherforecast model, which is currently in use at the institute (Heijboer, 1977). Having no other data available we propose a linear geostrophic wind profile:

$$U_g = \{(z-z_1)U_g^o - (z-z_o)U_g^1\}/(z_o-z_1),\qquad(3)$$

where z_o and z_1 denote the levels where the geostrophic wind is given.

Eq. (2) is assumed to be valid for values of $z > z_i$. The (linear) combination of Eqs. (2 and 3) leads to a wind profile, U_o, given by

$$U_o = f_s U_s + f_g U_g,\qquad(4)$$

where f_s and f_g are weighting functions ($f_s + f_g = 1$), which are determined by the above considerations of ranges of validity of the surface layer wind and geostrophic wind respectively. Further we impose $U_o(z)$ to be continuous and differentiable.

$$f_s = 1 \qquad\qquad\qquad\qquad z < 0.2\,z_i \text{ and } z < 2|L| \qquad(5a)$$

$$f_s = 0 \qquad\qquad\qquad\qquad z \geq z_i \qquad(5b)$$

$$f_s = \{\cos[\pi/8\,(5z/z_i-1)]\}^2 \quad z_i \leq 10|L| \text{ and } 0.2\,z_i < z < z_i \quad(5c)$$

$$f_s = \{\cos[\pi/2(\frac{z-2|L|}{z_i-2|L|})]\}^2 \quad z_i > 10|L| \text{ and } 2|L| \leq z \leq z_i \quad(5d)$$

In Fig. 2 is indicated which of the expressions should be used in the respective regions.

In this way of construction, a maximum amount of synoptic observations is incorporated into the wind field (Eq. (4)). In the area considered, however, we are so fortunate as to have at our disposal also a relatively large amount of wind data along masts and towers (Fig. 2). They provide useful information on the wind field in the intermediate region between 50 and 300 m. These data can be introduced into the windfield by means of the following tentative interpolation formula,

$$U = U_o + \{\sum_{m=1}^{N} (U_m-U_o).\exp[-(d_m/1)^2]/d_m^2\}/\sum_{m} 1/d_m^2.\qquad(6)$$

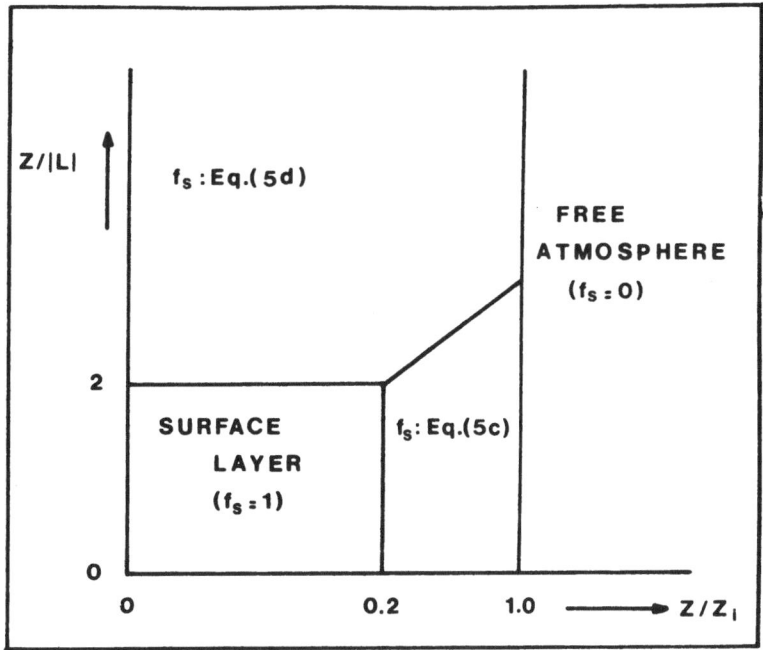

Fig. 2. Regions of application of the various weighting functions (Eqs. (5)). The Obukhov length and inversion-height are denoted by L and z_i respectively.

The collection of data obtained from masts and towers is denoted by U_m, and their location by the coordinates X_m, Y_m and Z_m. The parameter d_m is given by

$$d_m = \left[(X-X_m)^2 + (Y-Y_m)^2 + \alpha(Z-Z_m)^2 \right]^{\frac{1}{2}}.$$

With $\alpha = 1$, d_m would represent the distance between the location of observation, (X_m, Y_m, Z_m) and the location where wind data are desired. However, to account for the fact that vertical separation in general has a stronger influence on the variation of wind speed we will use α as a tuning parameter ($\alpha \gg 1$). The parameter 1, which appears in the exponential term in Eq. (6) can be considered then as a measure for the horizontal radius of influence, and can be chosen independently, according to, for instance, the homogeneity of the flow pattern. Preliminary experimental studies are necessary for the determination of α and 1.

THE EDDY-DIFFUSIVITY FIELD

In the model equations a first order closure scheme is applied. As a consequence mathematical expressions for the eddy diffusivity coefficient K_z, has to be known as a function of time and location.

Moreover, it should be possible to derive its value from synoptic observations. Here we will adopt an expression given by Brost and Wyngaard (1978) for the stable boundary layer.

$$K_Z = \frac{kzu_*.(1-z/z_i)^{1.5}}{\phi(Z/L)} \quad , \quad (z < z_i) \tag{7}$$

where k is Von Karmann's constant (k=0.35). The function, ϕ, is chosen such that in the surface layer, $(z/z_i \ll 1)$, K_Z approaches the eddy diffusivity of heat as given by Businger (1973). In this way a kind of consistency is achieved with the wind profile in the surface layer.

Eq. (7) satisfies also another important characteristic of the (stable and unstable) boundary layer, namely that K_Z vanishes near the inversion height. We apply Eq. (7) also in unstable conditions, where of course the appropriate expression for ϕ has to be used.

THE DETERMINATION OF AUXILIARY METEOROLOGICAL PARAMETERS

The expressions for the wind and eddy diffusivity profiles contain parameters which are not directly measured or available from synoptic data. In this section we shall indicate how we obtain the surface layer parameters u_* and L, the roughness length, z_o, and the inversion height, z_i. In table I a summary is given of the hierarchy of the deduction of all the necessary parameters.

-L and u_*

These parameters are determined from the relations

$$U = (u_*/k).\{ln(z/z_o) - \psi(z/L)\} \tag{8}$$

and

$$L = Tu_*^3/(kgQ/\rho c_p).$$

Here, ψ denotes again the stability function (Businger, 1973). The surface heat flux is given by Q. When at a given height within the surface layer the wind speed (U) is known, u_* and L can be obtained from Eqs. (8) by means of an iterative procedure, provided that an estimate of Q can be given. For this purpose we use a procedure, which is followed by Burridge and Gadd (1977). The relevant synoptic input parameters are the temperature, solar elevation, and cloud cover. Over a water surface L and u are determined by a method given by Nieuwstadt (1977).

Table I. Hierarchy of the deduction of meteorological parameters.

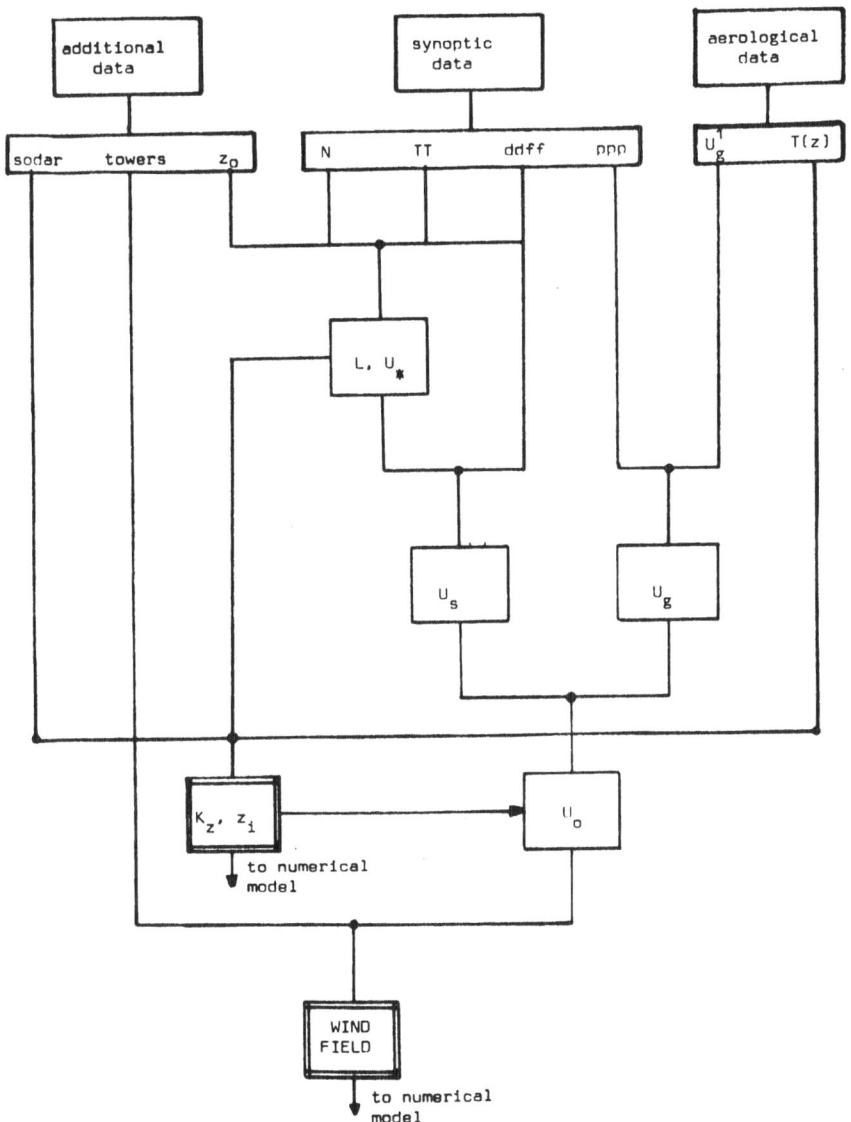

$^{-z}$o
The estimation of z_o requires a terrain description of the area.
This is realised by inspection of topographical maps. Squares of
1'x1' are analysed and classified according to their main topographical features. Next a value of z_o is alotted to each classification according to a scheme of Smith and Carson (1977).

$-z_i$

It is assumed that in the area considered only one single inversion will be present: during nighttime a ground based radiation inversion, and during day time an elevated inversion which caps the mixed-layer.

During stable conditions we use an expression which is derived from similarity prediction (Brost and Wyngaard, 1978), and which attains the proper limit in the neutral case (Nieuwstadt, 1980). Whence,

$$z_i = 0.3 \, u_*/f.(1 + 1.9 \, z_i/L)^{-1}. \tag{9}$$

For the Coriolis parameter, f, a constant value of $1.12 \ 10^{-4} \mathrm{s}^{-1}$ is taken, according to the average latitude of 50°.

During unstable conditions we shall use the model of Tennekes (1973).

We admit that the estimates of z_i are very crude and inaccurate ones, which must be considered as a first step only towards a more refined and realistic description of the atmospheric stability regime.

INTERPOLATION PROCEDURES

The synoptic wind, temperature and pressure field are horizontally interpolated by means of the optimum interpolation method (Lorenc et al, 1977). This well-known method in dynamical meteorology was adapted to ground-based observations by Cats (1980). The basic interpolation formula is given by

$$p_a = <p_a> + \sum_{i,j=1}^{n} \overset{n}{S} \ c_{ij}^{-1}.c_{ja}.(p_i-<p_i>). \tag{10}$$

Here, p is the parameter to be interpolated. The location where its value is derived is denoted by the subscript $a(X_a, Y_a, Z_a)$, whereas we have at our disposal n synoptic observations of p in the area, p_i (i = 1...n). (Note that Z_a is the height of the horizontal plane of interpolation). The climatic averages, denoted by the brackets, and the covariance matrix C, with elements c_{ij} (i, j=1...n) were determined beforehand. Mathematical models were obtained which expressed average values and covariance in terms of the horizontal coordinates (Cats, 1980). The data base used to construct these models consisted of 4 years of hourly (1200 GMT) observations at 19 selected synoptic stations within the Netherlands.

The optimum interpolation method has the advantage that horizontal gradients of p can be easily obtained by differentiation of Eq. (10) with respect to X_a or Y_a.

It should be noted that the values of the wind velocity are often affected by local terrain influences, which one would like to remove before interpolation. Therefore, the measured wind data are adjusted to a standard roughness of 0.03 m. After interpolation the wind velocity is adjusted again to the corresponding local roughness.

REFERENCES

Brost, R.A. and Wyngaard, J.C. (1978). A model study of the stably stratified planetary boundary layer. J. Atmos. Sci., 35, 1427-1440.

Burridge, D.M. and Gadd, A.J. (1977). The meteorological Office operational 10-level numerical weather prediction model (December 1975). Scientific paper no. 34, Met. Office, London, England, 39 pp.

Businger, J.A. (1973). In: Workshop on Micrometeorology. Haugen, D.A. (Ed.), American Meteorological Society.

Cats, G.J. (1980). Analysis of surface wind and its gradient in a mesoscale wind observation network. To be published in Mon. Wea. Rev.

Heijboer, L.C. (1977). Design of a baroclinic three-level quasi-geostrophic model with special emphasis on developing short frontal waves. Thesis, Royal Netherlands Meteorological Institute.

Lorenc, A., Rutherford, I. and Larsen, G. (1977). The ECMWF Analysis and Data Assimilation Scheme: Analysis of mass and wind fields. European Centre for Medium Range Weather Forecasts, Technical Report no. 6.

Nieuwstadt, F.T.M. (1977). The dispersion of pollutants over a water surface. Proceedings of the 8th International Technical Meeting on air pollution and its application, September 20-23, 1977.

Nieuwstadt, F.T.M. (1980). The steady-height and resistance laws of the nocturnal boundary layer: theory compared with Cabauw observations. To be published in Boundary Layer Met.

Semb, A., (1978). Sulphur emissions in Europe. Atmospheric Environment, 12, 455-460.

Smith, F.B. and Carson, D.J. (1977). Some thoughts on the specification of the boundary-layer relevant to numerical modelling. Boundary Layer Met., 12, 307-330.

Tennekes, H. (1973). A model for the dynamics of the inversion above a convective boundary layer. J. Atmos. Sci., 30, 558-567.

DISCUSSION

A. BERGER   You did not speak about the
interpolation method used in your model. Would
you comment briefly about them ?

H. VAN DOP   In the surface layer we
use the optimum interpolation method used in nume-
rical weather forecast models.
For the free atmosphere, simple interpolation
schemes as $\sim 1/r^2$ are used.

W.B. JOHNSON   Have you automatic procedures for interpretation of sodar measurements ?

H. VAN DOP   No, the measurements are
visually inspected.

INTEREST OF AN ATMOSPHERIC MESO-SCALE MODEL

FOR AIR POLLUTION TRANSPORT STUDIES OVER MEDIUM DISTANCES

Christian Blondin

Etablissement d'Etudes et de Recherches Météorologiques
73-77 Rue de Sèvres
92100   Boulogne-Billancourt (France)

INTRODUCTION

Air pollution transport over medium distances is perhaps one of the most complex question to study , but also a field in which a lot of problems is to be investigated .

One aspect of this complexity lies in the fact that the computation of pollutant concentrations needs to take into account both meteorological and physico-chemical parameters . The atmospheric conditions influence the trajectory patterns , mainly through orographic effects , diffusion processes and dry deposition due to the vertical static stability , wet deposition,transformation rate of a lot of pollutant species depending on moisture , cloudiness and precipitations .

In spite of this wide range of interferences between the meteorology and the physico-chemical behaviour of atmospheric pollutants , it is worth furthering researches in atmospheric meso-scale modelling , even if the pollutants themselves are roughly treated , as being inert for instance , and if this approach represents only one of the facets ot the problem .

As far as more or less simplified models are used for practical applications , the best adjustment of the parametrizations of complex phenomena needs a lot of experimental data , but also comparison with more sophisticated models which must be provided with the most convenient data sets . In that way , the development of complete atmospheric meso-scale models must allow to lead to realistic wind and temperature fields and information about turbulent parameters .

Table 1. Validity domain for different approximations

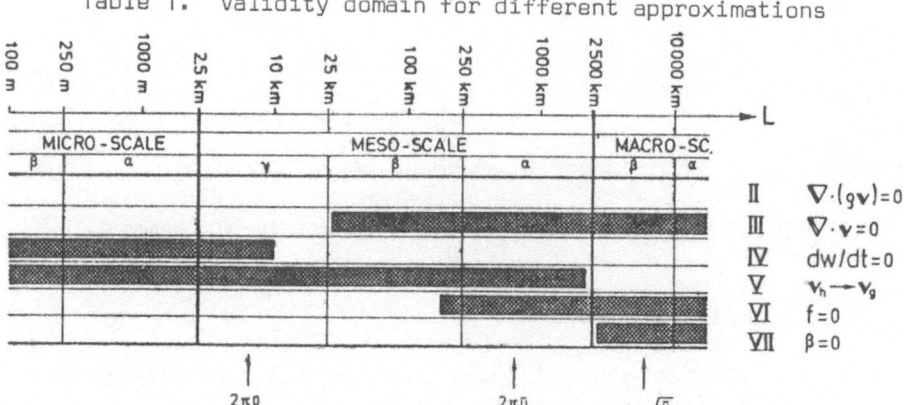

The horizontal axis represents the characteristic length of
the phenomena which are classified according to the
Orlanski's nomenclature . The roman numbers correspond to
II:anelastic assumption
III:incompressible assumption
IV:hydrostatic approximation
V:quasi-geostrophic approximation
VI:neglecting the Coriolis force
VII:beta-plane approximation
The dashed areas indicate the range where the approxima-
tions are not valid .   (Wippermann 1980)

     The aim of this paper is to show the framework of the work
done in this field in the French Meteorological Office and to
illustrate with a simple study case the usefulness of that kind
of model for air pollution transport studies .

DESCRIPTION OF THE MODEL

Assumptions - Variables - Equations

     For computer time or storage restrictions , or cost conside-
rations , it is often impossible to deal with the complete equations
of the meteorology , even regardless of the diabatic processes
parametrizations . The choice of the dimension number and of the
main assumptions ought to be the result of the best practical
compromise between computer facilities and problems to solve .

In this respect , a linear analysis of the equations system gives a first idea of the validity of the main assumptions used in meteorology . Roughly speaking , it can be seen that the study of the γ meso-scale is possible when using the incompressible and hydrostatic approximations provided that the mean flow in the tro-posphere has a stable stratification . A lot of aspects of the atmospheric circulation in the meso-scale can be described in that context , and , as winter stable synoptic conditions are the most propitious to severe pollution episodes , the use of an incompres-sible and hydrostatic model allows a wide range of interesting studies .

Under these assumptions , the equations of the model are the following ones :
the momentum equation

$$\frac{d\vec{V}_h}{dt} = - \theta . \vec{\nabla}_h \Pi - f . \vec{k} \times \vec{V}_h + \vec{\nabla} . (K^m \vec{\nabla} . \vec{V}_h)$$

the heat equation , using the potential temperature :

$$\frac{d\theta}{dt} = \vec{\nabla} . (K^\theta . \vec{\nabla} \theta)$$

the moisture conservation equation :

$$\frac{dq}{dt} = \vec{\nabla} . (K^q . \vec{\nabla} q)$$

the pressure equation , using the Exner variable :

$$\frac{\partial \Pi}{\partial z} = - \frac{g}{\theta}$$

the continuity equation :

$$\frac{\partial w}{\partial z} = - \vec{\nabla}_h . \vec{V}_h$$

the upper surface evolution equation :

$$\frac{\partial s}{\partial t} = - \int_{z_g}^{s} (\frac{\partial u}{\partial x} + \frac{\partial v}{\partial y}) \, dz$$

This last equation is obtained by a vertical integration of the continuity equation , with the assumption that the upper sur-face is a stream one .

In practice , these equations are written using the reduce variable :

$$z^* = \bar{s} \frac{z - z_g}{s - z_g}$$

where $\bar{s}$ is a reference height and z_g stands for the ground altitude . doing so , the integration domain remains unchanged .

As the physico-chemical behaviour of a lot of pollutant species are influenced by the water content of the atmosphere , two other variables can be used under certain circumstances , namely the liquid water potential temperature :

$$\theta_L = \theta - (\frac{\theta}{T} \frac{L}{C_p}) q_{liq}$$

and the total water mixing ratio :

$$q = \left\{ \begin{array}{ll} q_{vap} & \text{if } q_{liq} = 0 \\ q_{sat} + q_{liq} & \text{if } q_{liq} > 0 \end{array} \right.$$

The equations are slightly modified to take into account the new state equation :

$$P = \rho RT(1 + Eq_{vap}) \qquad E = 0.608$$

Parametrizations schemes

The equations system is closed by using diffusion coefficients We use the surface fluxes computation and the turbulence schenme proposed by J.F Louis (1979) . In this formulation , the height of the planetary boundary layer is not a relevant parameter . This is a practical advantage because it does not exist any convenient definition of this parameter for horizontally non homogeneous situations .

For sea-land breezes studies , we have adapted the radiation scheme developped by J.F Geleyn (1979) .

Numerical procedure

The set of equations is solved by the finite difference technique . The numerical scheme is centered both in time and space . For stability reasons , the external gravity wave propagation and the vertical diffusion are treated implicetly .

Starting from appropriate initial conditions , the equations
are integrated using the following boundary conditions :
- the values of the different variables remain unchanged at the
inflow points ,
- null gradient conditions are applied to the outflow points .

Characteristics of the experiments

The model has been applied to study the influence of the orographic effects on the synoptic flow structure in complex terrain
areas , namely the Rhin Valley and the Rhône Valley . Numerical
validation tests have been carried out using actual data collected
during intensive measurement campaigns , devoted to investigate
wintertime flow pattern structures on these regions , especially
under temperature inversion conditions .

Unfortunately , even during those campaigns , the information
concerning the surface parameters does not allow an objective analysis of the surface fields . In the numerical experiments , it
must have been assumed that the surface layer is neutral . Rewri-
ting the vertical advection terms under the form :

$$w^* \frac{\partial \alpha}{\partial z^*} = \frac{\partial (w^* \alpha)}{\partial z^*} - \alpha \frac{\partial w^*}{\partial z^*}$$

it can be seen that the surface values do not interfere in the computation . Doing so , the results only describe the pure dynamic
forcing of the orography on the meso-scale flow .

PRACTICAL EXAMPLE OF A NUMERICAL EXPERIMENT

Description of the problem : the Alsace Plain case

The Alsace Plain is situated in the eastern part of France and
corresponds to the middle Rhin Valley . This area is surrounded by
the Vosges , the Jura and the Black Forest massif .

One of the characteristics of its climate is the possibility
that stable cold air of a few hundred meters depth advecting
during anticyclonic synoptic situations can remain trapped under a
strong inversion layer during more than a week , even if the synoptic conditions have completely changed .

Pollution risks are deeply enhanced under such episodes , because of the atmospheric stability but also because of the weak
winds which might occur simultaneously .

The meteorological network in this area is not dense enough
to give a complete vision of the wind flow pattern str-cture and

the model has been applied to try to palliate this lack of information .

In order to show the interest of the knowledge of a realistic
wind flow pattern and the possible consequences on air pollution
dispersion and trajectories , a conservation equation of an inert
pollutant concentration has been added to the model equations set :

$$\frac{dr}{dt} = \vec{V}.(K^r.\vec{\nabla}r) + S$$

It is solved using the diffusion coefficients computed for
the heat exhanges . Any special numerical procedure is applied to
avoid the numerical dispersion . Here , S stands for the source term
and any deposition removals or transformation processes have   been
introduced .

Results for the 2nd February 1976

The data collected on this day shew the main features we spoke
about previously : the whole eastern part of France was covered by
a strong inversion layer (10 degrees amplitude) , the base of
which was about 800 meters above the sea (600 meters above the
Plain level) . Under this layer , the wind blew from the East as
indicated by the radio-sounding of Nancy and the data especially
collected in Eckwersheim (near Strasbourg) , with a 7 m/s average
speed . At Fessenheim , in the south-east part of the Alsace Plain ,
the wind blew southwards and its speed was lower than 3 m/s .

The figure 1 shows the model results for this day . On this
horizontal cross section (500 meters above sea level) , a kind
of rotating flow appears to take place in the Alsace Plain which
agrees with and explains the Fessenheim's measurements .

As there is a nuclear power plant at Fessenheim , an hypothetic
continous release of inert pollutant emitted above this point at
300 meters above the ground level has been simulated . S , in the
equation of r , took the value 4.10^{-8} s^{-1} .

The figures 2 , 3 , 4 exhibit the iso-lines -5 and -6 of the
variable $C = Log_{10}$ r,in the same horizontal plane that the one of
the figure 1 , respectively 3 , 6 , 9 hours after the beginning of
the emission .

In order to estimate the realism of these concentration fields
trajectory test studies have been carried out . They proved that ,
in spite of the numerical dispersion , the highest concentration

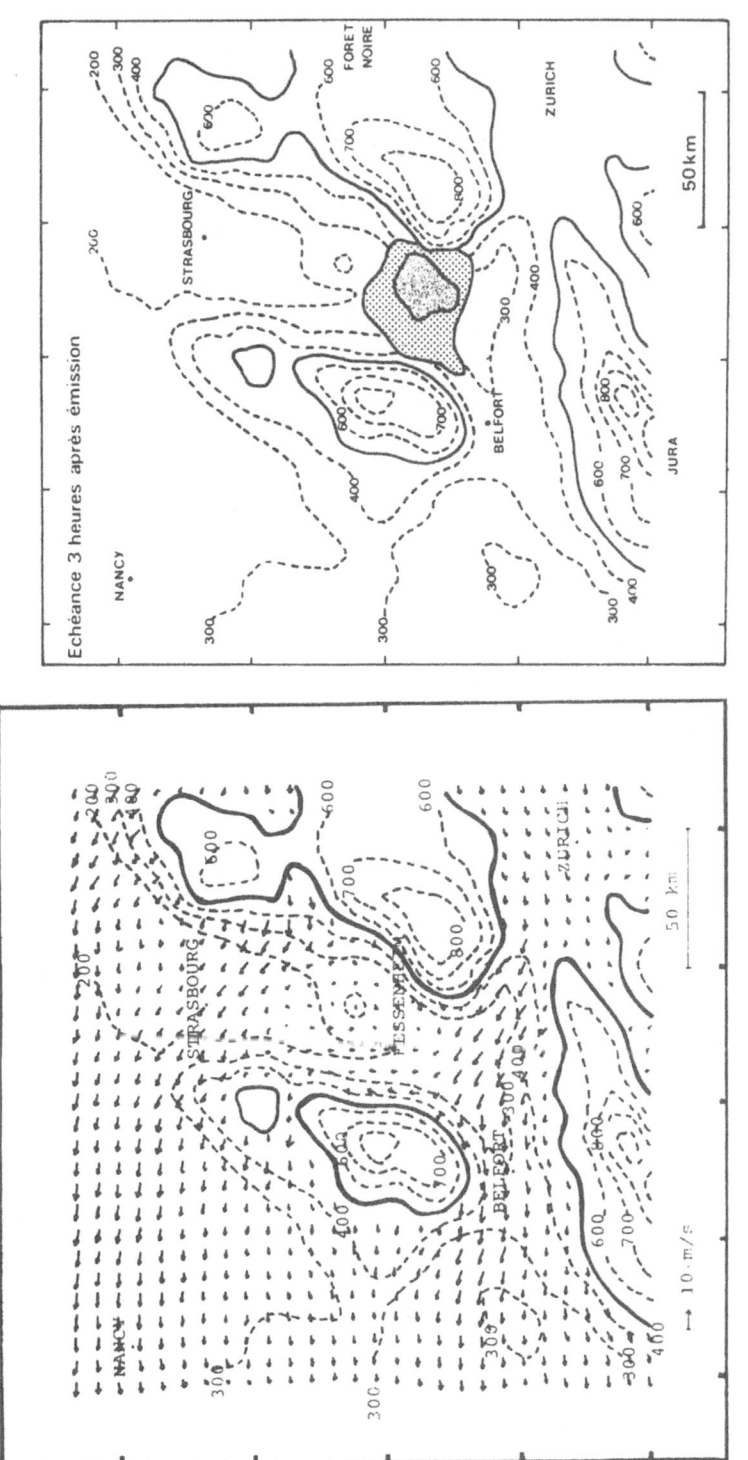

Figure 2

Figure 1

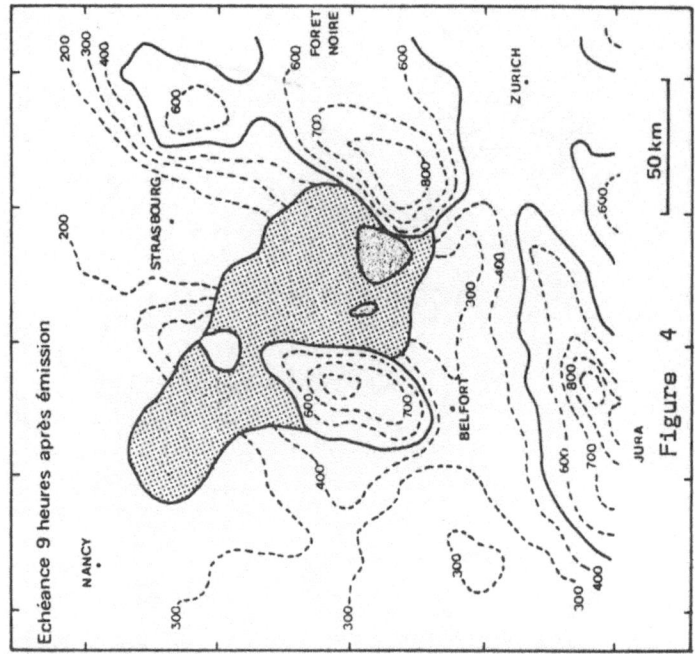

Figure 4

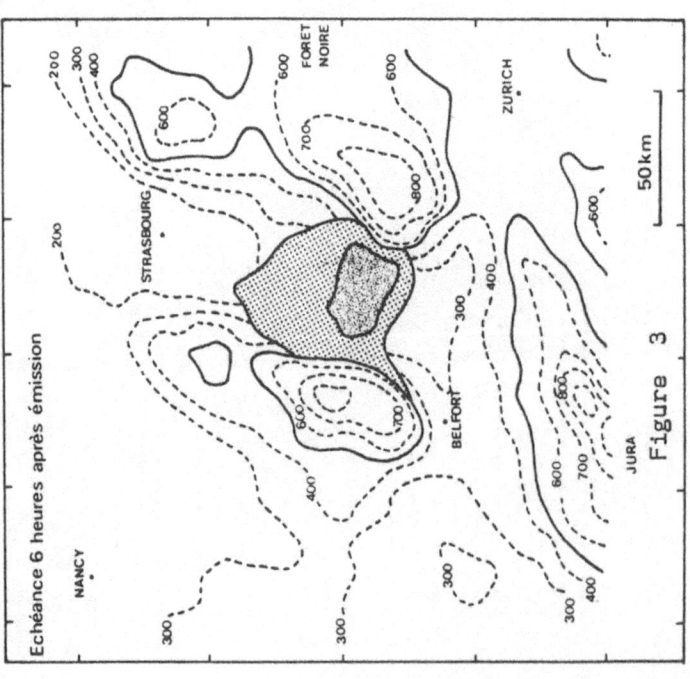

Figure 3

area was well located . It must be pointed out that the axis of this
"plum" is very different of the expected one if only the wind direction at the emission point has been considered .

CONCLUSION

   This simple experiment shows clearly that for a lot of practical applications ignoring meteorological features , such as mesoscale wind pattern , can produce quite erroneous results .

   Of course , the links between atmosphere and pollutant behaviour
are too complex to say that the add of a conservation equation for
an inert pollutant is enough to change an atmospheric meso-scale
model into an air pollution transport model over medium distances .

   Nevertheless , as far as the transport must essentially be studied as a lagrangian process , it is of a primary importance to get
realistic winds and turbulent parameters . In that respect , the
improvements in the atmospheric meso-scale modelling form fundamental
steps towards a better understanding of the air pollution problem .

REFERENCES

Anthes, R.A.,and Warner, T.T, 1978, Development of hydrodynamic
        models suitable for air pollution and other meso-meteo-
        rological studies, Mon.Wea.Rev., 106:1045-1078.
Blondin, C., 1980, Simulation numérique de l'atmosphère à méso
        échelle, in: Proc.Sym. on Physico-chemical behaviour of
        atmospheric pollutants, Ispra, Italy.
Geleyn, J.F., and Hollingworth, A., 1979, An economical analytical
        method for the computation of the interaction between
        scattering and line absorption of radiation, Con.Atm.Phy.
        ,52:1-10.
Louis, J.F., 1979, A parametric model of vertical eddy fluxes in
        the atmosphere, Boun.Layer.Met., 17:187-202.
Pielke, R., 1974, A three dimensional numerical model of the sea
        breezes over South Florida, Mon.Wea.Rev., 102:115-139.
Tapp, M.C., and White, P.W., 1976, A non-hydrostatic meso-scale
        model, Qua.Jou.Roy.Met.Soc., 102:277-296.
Wippermann, F., 1980, The applicability of several approximations
        in meso-scale modelling:a linear approach, in: Proc. of
        Meeting of the German Meteorological Society, Mannheim,
        Germany.

DISCUSSION

A. BERGER   I would like to underline
                    that one of the most important features for eva-

luation of models is the reliability of the models.
Moreover it comes clearly from the presentations
made by Blondin that, most of the time, sophisticated models have to be used to reproduce observations properly.

A. VENKATRAM   How do you account for the
effect of topography on the turbulence structure ?

C. BLONDIN   The effect of topography is
not explicitly included in the surface flux formulation. An adaption of the OBUKHOV similarity
theory is used (reference J.F. LOUIS ECMWF).

A MASS CONSISTENT WIND FIELD MODEL OVER THE MID-RHINE VALLEY

Patrick Racher[1], Robert Rosset[1] and Yves Caneill[2]

(1) LAMP, Univ. of Clermont II, France
(2) EDF, Etudes et Recherches, Chatou, France

INTRODUCTION

The transport and diffusion of atmospheric pollutants at all scales and particularly at the mesoscale is a major problem in environmental studies, both numerically and experimentally. Numerically, two general schematic types of modelling are to be considered. The first type refers to models with detailed physics which simulate the fields of wind, temperature, pollutant concentrations, water vapour and various microphysical parameters. These models are often physically satisfactory but, apart from the numerical difficulties encountered in some regimes, they are computationally expensive and they cannot be run on an operational basis. This draw back is not found in the second type of models, an example of which is given by Sherman,[1]. These latter models generally appear as three-dimensional objective analysis procedures applied to the data collected at a number of stations. We have retained here such a procedure, while improving the basic model. These improvements bear upon the adoption of a terrain-following vertical coordinate σ instead of the geometric height z, a better objective interpolation in the initial reduction of the observed surface winds to their gridded values and the adjustment of vertical profiles to the available soundings in the domain. We then proceed to the application of our model to some situations observed during an intensive field study performed in the mid-Rhine valley during october 1976.

VARIABLES AND EQUATIONS OF THE MODEL

The model is based on the variational objective analysis methods put forward by Sasaki,[2] and Sherman,[1].

201

After having specified the initial wind field $\vec{V}_o(u_o, v_o, w_o)$ as
a function of the three space coordinates, we minimize in the
least-squares sense the discrepancy between this field and the
final wind field $\vec{V}$ which must satisfy the incompressible continuity equation. For this purpose, we define in the domain of
study (Ω), the following functional :

$$J = \frac{1}{2} \int_{(\Omega)} (\vec{V} - \vec{V}_o)^T \ [W] \left(\vec{V} - \vec{V}_o \right) d\Omega \qquad (1)$$

with the constraint of incompressibility $\vec{\nabla}.\vec{V} = 0$ applying in
(Ω) and with $[W]$, a diagonal weighting matrix.

In fact, in view of the complex topography of the domain
with steep slopes, we introduce a new vertical coordinate σ,
defined as (fig. 1) :

$$\sigma = \frac{z - h(X,Y)}{H(X,Y) - h(X,Y)} \qquad (2)$$

where z is the height of a given σ-level above a reference level
(e.g. the sea-level), h(X,Y) is the ground height above this
same level and H(X,Y) the height of the top level in the domain.

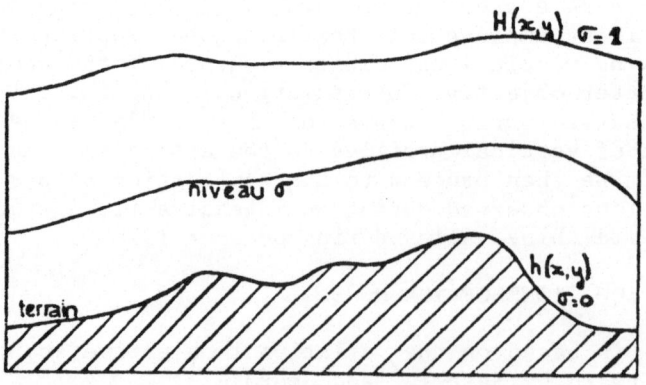

Fig. 1. The vertical coordinate σ

In the system of coordinates (X, Y, σ), the spatial derivatives express as :

$$\left. \frac{\delta}{\delta X} \right)_z = \left. \frac{\delta}{\delta X} \right)_\sigma - \left. \frac{\delta z}{\delta X} \right)_\sigma \cdot \frac{\delta}{\delta z}$$

$$\left. \frac{\delta}{\delta Y} \right)_z = \left. \frac{\delta}{\delta Y} \right)_\sigma - \left. \frac{\delta z}{\delta Y} \right)_\sigma \cdot \frac{\delta}{\delta z}$$

$$\frac{\delta}{\delta z} = \frac{1}{(H-h)} \cdot \frac{\delta}{\delta z} \qquad (3)$$

With these new coordinates, the divergence of the wind field is :

$$\vec{\nabla} \cdot \vec{V} = \frac{\delta u}{\delta X} + \frac{\delta v}{\delta Y} + \frac{1}{(H-h)} \left[\frac{\delta w}{\delta \sigma} - \frac{\delta z}{\delta X} \cdot \frac{\delta u}{\delta \sigma} - \frac{\delta z}{\delta Y} \cdot \frac{\delta v}{\delta \sigma} \right] \quad (4)$$

After introduction of the Lagrange multiplier λ, minimization of the functional J in (1) leads us to the following system :

$$u = u_o + \frac{\delta \lambda}{\delta X} - \frac{1}{(H-h)} \cdot \frac{\delta z}{\delta X} \cdot \frac{\delta \sigma}{} \qquad (5)$$

$$v = v_o + \frac{\delta \lambda}{\delta Y} - \frac{1}{(H-h)} \cdot \frac{\delta z}{\delta Y} \cdot \frac{\delta \lambda}{\delta \sigma} \qquad (6)$$

$$w = w_o + \frac{K}{(H-h)} \cdot \frac{\delta \lambda}{\delta \sigma} \qquad (7)$$

$$(H-h) \frac{\delta u}{\delta X} + \frac{\delta v}{\delta Y} + \frac{\delta w}{\delta \sigma} - \frac{\delta z}{\delta X} \cdot \frac{\delta u}{\delta \sigma} - \frac{\delta z}{\delta Y} \cdot \frac{\delta V}{\delta \sigma} = 0 \qquad (8)$$

with the boundary conditions :

$$\lambda = 0 \text{ on } (\Gamma_L) \qquad (9)$$
$$\text{and } \vec{n} \cdot \vec{V} = 0 \text{ on } (\Gamma_V) \qquad (10),$$

where (Γ_L) and (Γ_V) are respectively the lateral and vertical boundaries of the domain (Ω). K is an empirical parameter of similitude giving the ratio between the relative contributions in the domain of the horizontal and vertical air components. K is a complex function of the atmospheric static stability

(Haltiner,3).

Combining equations (5), (6), (7) and (8), we get a differential equation of the second order with respect to λ. With the boundary conditions (9) and (10), this equation is solved by a finite difference method on a grid of mesh lengths ΔX, ΔY and $\Delta \sigma$. Then the three wind components u,v and w are obtained at each grid point through equations (5) to (7).

In most of the applications performed until now, $\Delta X = \Delta Y = 10$ km and $\Delta \sigma = 1/10$ or $1/15$. Nevertheless, other tests have been made with a finer grid mesh embedded within the largest grid

($\Delta X = \Delta Y = 5$ km) which covers only part of the domain. This focusing has proven to be very useful in the zones of strong topographic gradients, as regards the secondary circulations induced.

SPECIFICATION OF THE INITIAL WIND FIELD $\vec{V}_o$

In the above methodology, the specification of the initial wind field appears as largely determining the final relaxed field computed. So, it is imperative to make use of the maximum meteorological information available in the domain of study. This information is in the form of wind measurements at the anemometer level for a number of stations (about 10 to 15 for the plain of Alsace) and possibly, radiosoundings at one or several stations (1 to 3 in the applications reported below).

For specifying the initial wind field $\vec{V}_o$, we first proceed to an objective analysis of the wind data at the anemometer level so as to interpolate the horizontal wind components at the nodes of a regular computational grid covering the domain.

Instead of using a classical inverse square distance weighting scheme,[4], for this interpolation, we have adopted a least square polynomial interpolation scheme,[5]. In this latter scheme, the domain is subdivided into a set of triangular sub-regions defined by the data points themselves. This scheme has the advantage of retaining much of the information contained in the original data, without too much smoothing.

Once the horizontal gridding at the surface has been realized, a definition of the vertical structure is in order. This problem is rendered difficult due to the loose spacing between the aerological stations. Three vertical procedures have been considered. The first one was based on a power law for the vertical dependence of the winds with height : needing no radiosounding at all, this law is very simple, through not satisfactory physically. The second law adopted was an Ekman law modified so as to include the thermal wind effects,[6]. Finally, to take a full account of all the data available in most of the applications, we have fitted a curve to the wind soundings : this is our third procedure of vertical interpolation. In the following applications, this latter procedure has been retained.

SOME RESULTS

The results reported here refer to an intensive field study performed by E.D.F. in the French part of the mid-Rhine valley in october 1976. The domain includes the Vosges massif, the Black Forest in Germany, the plain of Alsace and the northern part of the Jura mountains. This domain which is approximately 250 km x 250 km in horizontal extent is represented in figure 2, with the topographic isocontours. The surface wind data of 12 to 15 french and german stations have been collected for every three-hour period during the above field experiment. Further- more, the 6.00 or 12.00 Nancy radiosounding is used in our analysis for defining the vertical variations of the wind fields.

Several values of the empirical coefficient K (in equation (7)) have been tested. The horizontal velocities are not very sensitive to these changes whereas the vertical velocity field is more affected. In the following, we have adopted the value $K = 10^{-4}$.

Two contrasted meteorological regimes have been observed in october 1976 : during the first 10 days, the domain is under the influence of an area of high pressure and the winds are light. Then, during the rest of the month, a series of pertur- bations rapidly sweep the domain giving strong winds at times.

Schematically, according to a subjective meteorological analysis, four typical wind field patterns have been defined for october 1976, namely light and strong southwesterly wind regimes, northeasterly and southerly regimes.

In the following, we report some examples of our calcula- tions for two selected days in october 1976, namely october 6th (fig. 3) and october 12th (fig. 4), at 12.00 GMT.

Are given for each case the surface winds together with the winds on horizontal cross-sections at different heights (z = 300 m, 600 m and 1800 m) (the blank areas corresponding to the protruding topography above the section level).

In the two cases reported, we can see the diverting effect of the relief and the channelling effect of the Rhine valley.

In figure 3-d, we give a vertical cross-section at the ab- scissa x = 100 km : here again the topography appears as blank areas. Large differences in wind speeds are apparent with a general acceleration after flowing over the Vosges massif.

Similar results have been obtained for other days during the 1976 field experiment : we have just chosen here to report two typical situations.

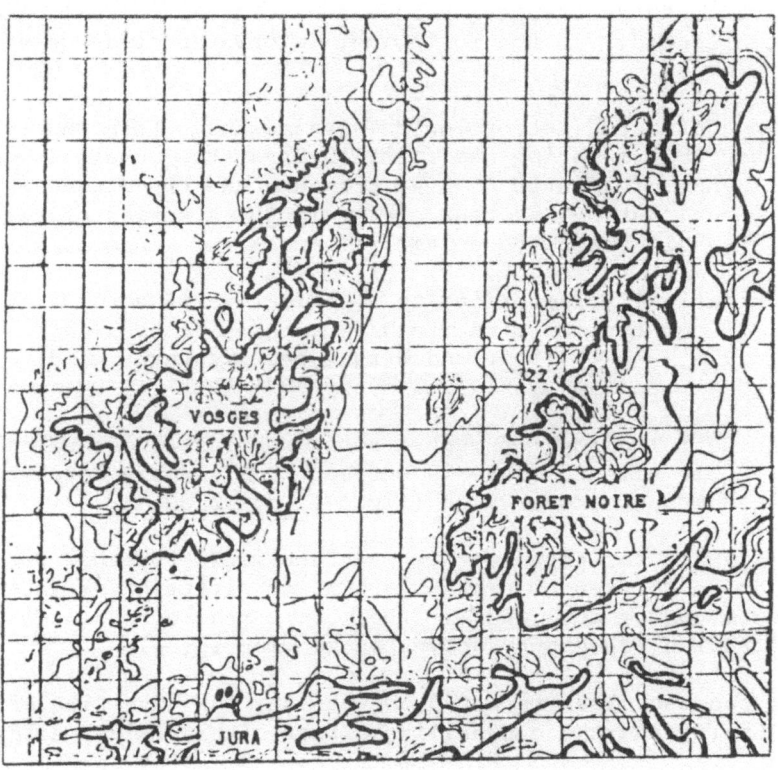

Fig. 2. Topography of the domain

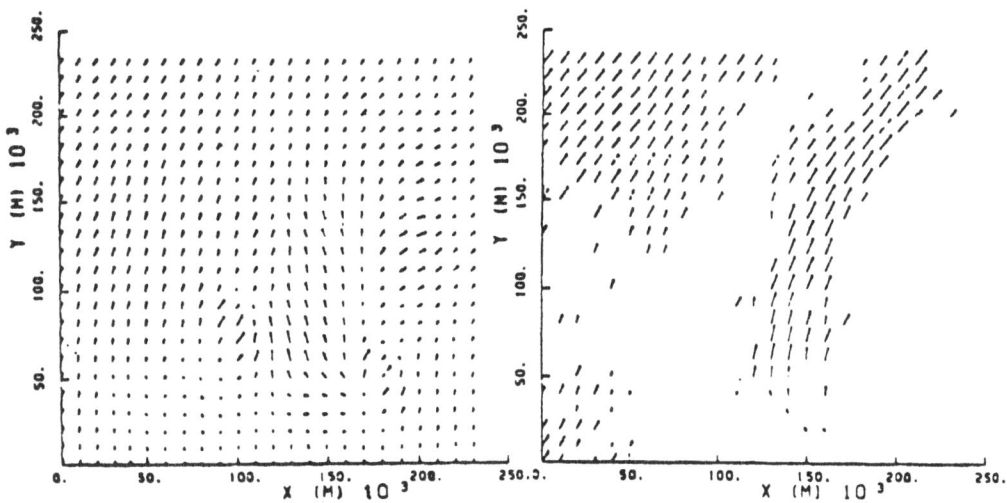

3-a. Surface winds

3-b. Horizontal cross-section
at z = 300 m

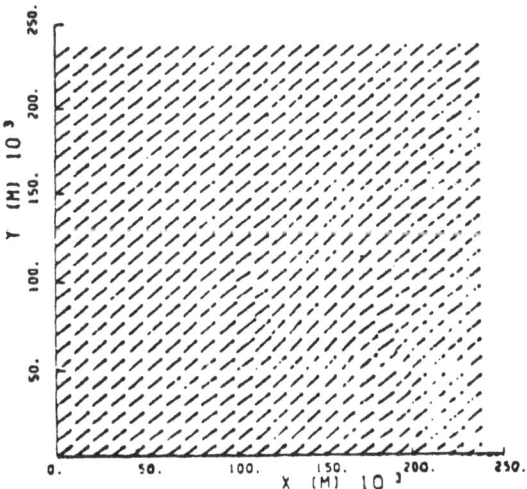

3-c. Horizontal cross-section
at z = 1800 m

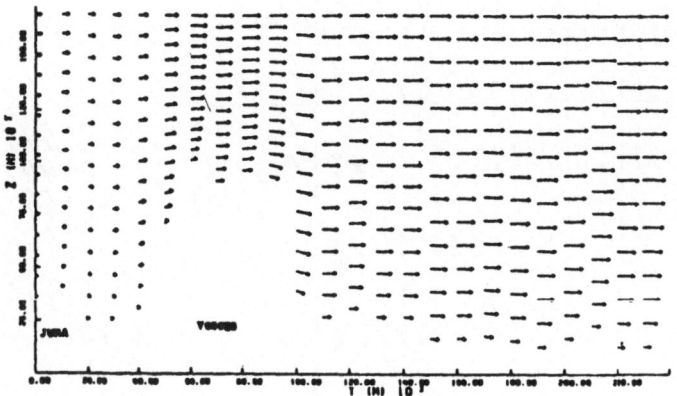

3-d. Vertical cross-section at x = 100 km

Fig. 3. Simulation of the situation for october 6th 1976
        (12 GMT). Southwesterly synoptic flow.

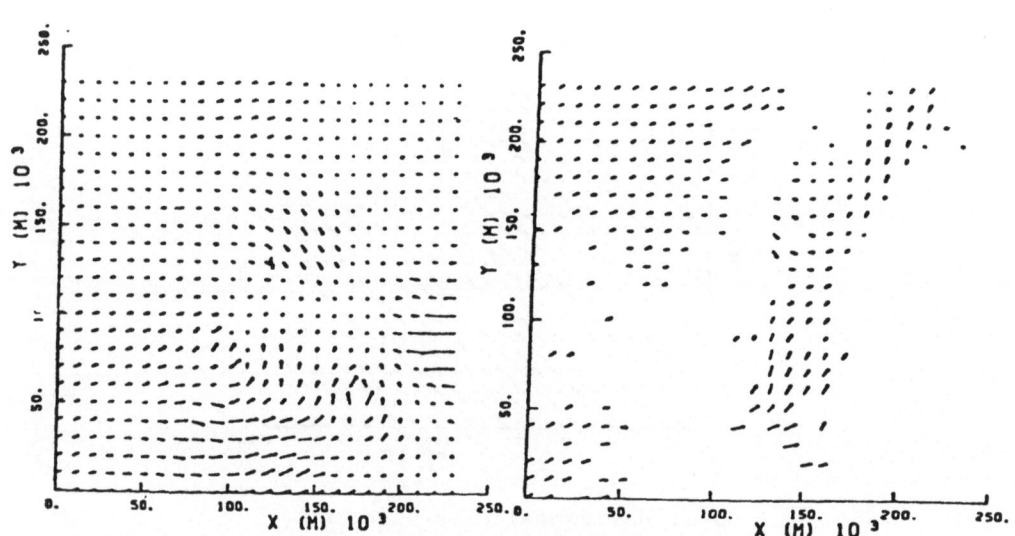

4-a. Surface winds   4-b. Horizontal cross-section
                                        at z = 300 m

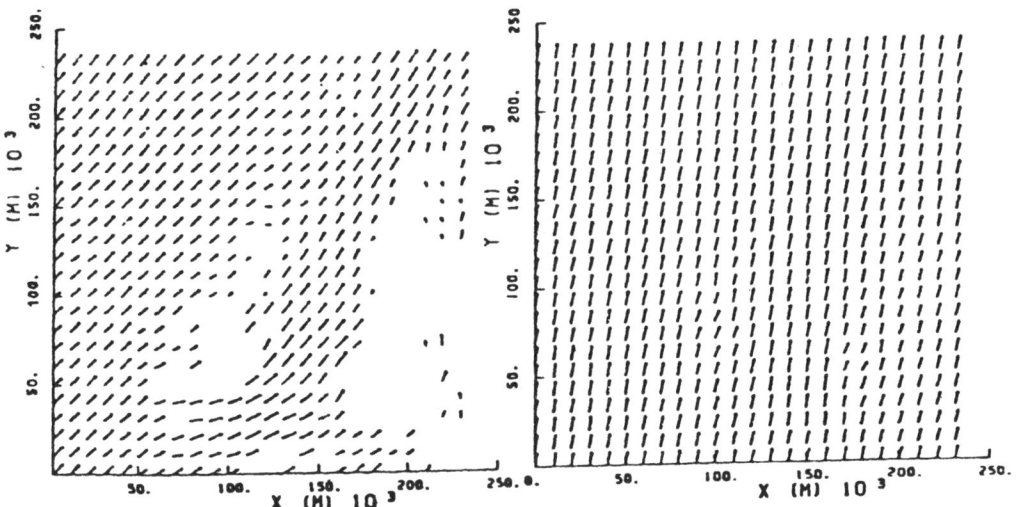

4-c. Horizontal cross-   4-d. Horizontal cross-section
     section at z = 600 m   at z = 1800 m

Fig. 4. Simulation of the situation for october 12th, 1976
        (12 GMT). Strong southerly synoptic flux.

CONCLUSIONS

We have applied a three-dimensional variational objective
analysis procedure to the computation of wind fields over the
mid-Rhine valley and the surrounding mountains. Our primary
goal was to accomodate most of the meteorological information
available in the mesoscale domain of study : in this respect,
the model has proven to be operationally effective with a low
cost. Present improvements bear upon the incorporation of the
surface roughness and thermal effects together with the recasting
of the numerical code into a finite element formulation.

In this present state, our model is part of a more extensive study devoted to the transport and diffusion of airborne
pollutants. We now plan to incorporate within it the lagrangian scheme of turbulent diffusion,[7] that we have developed
previously.

REFERENCES

1. C. E. Sherman, A mass-consistent model for wind fields over
   complex terrain, Journal Appl. Met., 17 : 312 (1978).

2. J. Sasaki, Some basic formalisms in numerical variational

analysis, <u>Month. Weath. Rev.</u>, 98 : 875 (1970).

3. G. J. Haltiner, "Numerical weather prediction", John Wiley
   & sons Inc. (1971).

4. G. P. Cressman, An operational objective analysis system,
   <u>Month. Weath. Rev.</u>, 87 : 367 (1959).

5. D. H. Mac Lain, Two dimensional interpolation from random
   data, <u>Comput. Journal</u>, 19 : 178 (1976).

6. A. Wün-Nielsen, Vorticity, divergence and vertical velocity
   in a baroclinic boundary layer with a linear variation of the
   geostrophic wind, <u>Bound. Layer Meteo.</u>, 6 : 459 (1974).

7. F. X. Le Dimet and R. Rosset, Un schéma lagrangien de diffusion turbulente, <u>Journal Rech. Atmos.</u>, 12 : 35 (1978).

DISCUSSION

P. HECQ   Did you compare your results
                with the model presented by C. Blondin ?
                And did you notice significant differences ?

A. ROSSET   This comparison has not yet
                been fully performed, but we intend to do
                it soon - there appears to be no major dif-
                ferences in the case of strong SW flows,
                but discrepances arise in the presence of
                low-wind regimes with noticeable thermal ef-
                fects. In fact such a comparison is one of
                our objectives with the concept of comple-
                mentary models in mind.
                The model we have presented here is much
                cheaper to run and more easy to accomodate
                with smaller grid mesks than that referred
                to in your question, though the present model
                is sophisticated.

V. FERRARA   Are you familiar with the wind
                field model of the Larence Livermore Labo-
                ratory ? and if you are, how is it in com-
                parison with your model ?

R. ROSSET   The answer to your first
                question is yes.
                With respect to your 2 question two points
                must be made clear - Firstly, the model we
                have used is not by itself very much diffe-
                rent from the one you refer to. We use a
                $\sim$ coordinate system instead of a J-S coor-
                dinate system and different boundary con-

ditions. One difference comes too from the
treatment of the input data, with emphasis
being put upon the objective analysis of the
surface wind measurements.
Then on the specifiation of the vertical
wind profiles :

This is crucial for the field obtained and
is not only a matter of detail. As I told
in my introduction, our final aim is to
have several complementary models able to
be used on different scales in our field.

A. VENKATRAM   Is your model sensitive
to the vertical variation of the wind speed
and the height of the boundary layer ?

R. ROSSET   Yes it is, and it is one
of its weakest points since the usual net-
work of radiosounding stations is much
looser than the surface stations network.
Furthermore it is difficult to compare different profiles in a given situation since
real profiles sometimes depart from power
law profiles especially of low wind regimes.

A. BERGER   As the windfield is probably the most important meteorological
factor in air pollution modeling, I would
like to suggest to Panel 2 of the New Pilot
Study to start a comparison between the
results obtained from different wind field
models when applied to the same data base.

A MESOSCALE NUMERICAL MODEL OF

ATMOSPHERIC FLOW OVER THE ALSACE PLAIN

A.E. Saab - C. Rolin - V. Villouvier

Electricité de France
Direction des Etudes et Recherches
Chatou, France 78400

INTRODUCTION

The problem of modeling the atmosphere at the mesoscale may be stated as drawing up two types of approaches to simulate the atmospheric boundary layer pattern.

Firstly, there are multi-level primitive equation models which can simulate complex mesoscale flow patterns satisfactorily, but require large amounts of computer storage and time.

Secondly, there are one-layer models which represent the boundary layer by assuming efficient mixing up to the capping inversion (heat, momentum and moisture are distributed uniformly in the vertical) and parameterizing the interactions of the mixed layer with the underlying surface layer and the overlying stable layer. These assumptions eliminate the dependence of the pronostic variables on height, and reduce considerably computing cost.

An early model of this latter kind was developped by Lavoie (1972) and has been used for simulating the Föhn effect and the heat island effect on the mesoscale airflow in the Alsace Plain. Using a case study and simulated heat island characteristics, we present the results of the mixed-layer model and discuss its sensitivity to some basic variables : topography, roughness, heat and moisture fluxes, strength and size of heat sources.

THE MIXED-LAYER MODEL

Structure of the model atmosphere

The atmosphere is assumed to consist of three layers defined by distinctive lapse rates of potential temperature [Figure 1].

The lowest layer (z_0 , z_s) consists of a parameterized surface layer of fixed depth ($\simeq 50$ m) which contains most of the wind shear and superadiabatic lapse rate and follows the variable terrain.

The middle layer (z_s , h) is well mixed so that the horizontal wind V, the potential temperature Θ, and the specific humidity q are assumed homogeneous in the vertical. Its depth varies locally due to topography, heat, momentum and moisture inputs. Its upper limit is defined by a first-order discontinuity in potential temperature.

The upper stable layer (h, H) contains a constant lapse rate of Θ. The wind is assumed to be geostrophic and may include a constant shear in the vertical.

Model equation

The basic equations are the horizontal momentum equation, the first law of thermodynamics, the mixed-layer height equation, the conservation of water substance, the continuity equation, plus the equation of state and the definition of hydrostatic balance.

By parametrizing the interactions of the mixed-layer with both the surface layer and the stable layer, the time-dependent calculations are limited to the mixed-layer. Assuming the vertical homogeneity, we can integrate the pronostic equations between z_s and h and write them as :

$$
\begin{cases}
\dfrac{\partial \vec{V}}{\partial t} = -\vec{V}\cdot(\vec{\nabla}\cdot\vec{V}) - \vec{k}\times(f\vec{V}) - \vec{F}_i - (h_i-h)f\vec{V} + \left.\dfrac{\partial \vec{V}}{\partial t}\right|_{ent} + K_{MH}\nabla^2\cdot\vec{V} \\[3mm]
\qquad + \dfrac{g}{\theta_h}\left(\theta - \theta_h - \dfrac{\Gamma}{4}(h_m-h_i)\right)\vec{V}(h) + \dfrac{g}{2\theta}(h-z_s)\vec{V}(\theta) - \dfrac{C_D}{(h-z_s)}|\vec{V}|\vec{V} \\[3mm]
\dfrac{\partial \theta}{\partial t} = -\vec{V}\cdot\vec{\nabla}(\theta) + \dfrac{C_H}{(h-z_s)}|\vec{V}|(\theta_0-\theta) + \dfrac{L\theta_{\overline{\alpha}}}{C_pT}\dfrac{M}{(h-z_s)} + \left.\dfrac{\partial \theta}{\partial t}\right|_{ent} + \left.\dfrac{\partial \theta}{\partial t}\right|_{ca} + K_{\theta H}\nabla^2\cdot\vec{V} \\[3mm]
\dfrac{\partial h}{\partial t} = -\vec{V}\cdot\vec{\nabla}(h) + w_h + \left.\dfrac{\partial h}{\partial t}\right|_{ent} + \left.\dfrac{\partial h}{\partial t}\right|_{ca} + K_{hH}\nabla^2\cdot h \\[3mm]
\dfrac{\partial q}{\partial t} = -\vec{V}\cdot\vec{\nabla}(q) + \dfrac{C_E|\vec{V}|}{(h-z_s)}(q_0-q) - \dfrac{\alpha M}{(h-z_s)} + \left.\dfrac{\partial q}{\partial t}\right|_{ent} + K_{qH}\nabla^2\cdot q
\end{cases}
$$

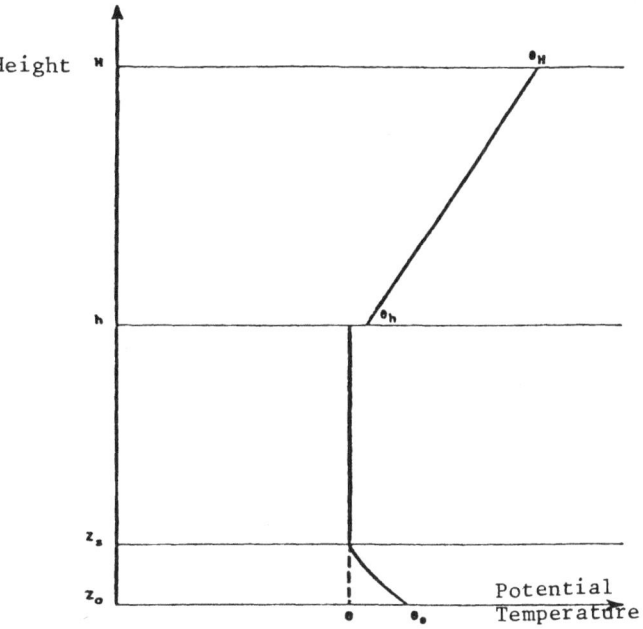

Figure 1:   Thermal structure of the model

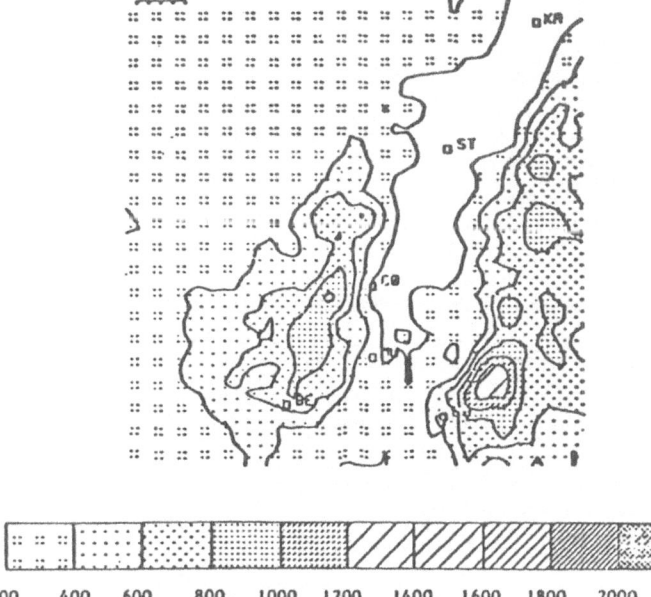

Figure 2 : Terrain elevation in the Alsace Plain (meters)

The present equations presented here contain some modifications to those proposed by Lavoie (1972) :

1/ We adopt Keyser's formulation (1977) for horizontal mixing and entrainment of monumentum, heat and moisture across the mixed layer. K_{MH}, $K_{\Theta H}$ and K_{qH} are the eddy coefficients for the horizontal transport of monumentum, heat and moisture. $\frac{\partial \vec{V}}{\partial t}\bigg|_{ent}$, $\frac{\partial \Theta}{\partial t}\bigg|_{ent}$ and $\frac{\partial h}{\partial t}\bigg|_{ent}$ represent the effects of entrainment across h, using a parameterization proposed by Tennekes (1973) and Keyser-Anthes (1977).

2/ We adopt also the convective adjustment parameterization which removes super adiabatic lapse rates at the top of the mixed layer. $\frac{\partial \Theta}{\partial t}\bigg|_{ca}$ and $\frac{\partial h}{\partial t}\bigg|_{ca}$ represent this effect.

3/ The rainfall rate M is given by $M = (\varepsilon W_{CB} + 1) N$ mm/hr with $N = a(D-k)$. W_{CB} is the mesoscale vertical velocity (cm s^{-1}) and the parameter ε is set equal to 0.5 cm s^{-1} so that the mesoscale subsidence of 2 cm $^{-1}$ is sufficient to prevent precipitation. D is the depth of the cloud and k is a critical value of cloud depth above which precipitation occurs. The parameter a is a measure of the efficiency of production of precipitation within the clouds and is set equal to 1 [Raddatz – Khandekar (1977)]. A sensitivity test shows us that the cloud depth and the rainfall rate are highly dependent on k value so that k must be adjusted in terms of meteorological conditions varying from 250 m to 1000 m.

Numerical procedure

The set of equations is solved by the forward upstream finite-difference technique as used earlier by Lavoie (1972). The advective terms are approximated by upstream differencing, and the non-advective terms by centered differencing. The system is integrated forward from a specified set of initial conditions over a grid (56x58 grid points) which is stretched at the boundaries. In order to reduce the influence of boundary conditions, the grid spacing of 5 km in the interior of the domain is increased regularly by 50 per cent between two grids on the 5 peripheral grids.

The terrain elevation at each grid point is obtained from topographic maps of scale 1 : 10^6. A spatial smoothing using linear filtering is introduced in order to damp short wavelength noise.

The model needs about 200 seconds of C.P.U. time to reach permanent conditions on an I.B.M. 370/168. This is obtained after 6 hours of integration time using a 100 s – time step.

AIRFLOW OVER THE ALSACE PLAIN

    Before discussing the results of the sensitivity tests of the
impact of large heat releases on the mesoscale airflow, we present
a case study on the Föhn effect in the Alsace Plain. This case demonstrates some of the physical capabilities of the model and serves as
a control experiment for the following sensitivity tests.

A case study on the Föhn effect

    The Föhn effect in the Alsace Plain, created by the Vosges Moun-
tains, is mainly defined by its impact on the rainfall rates. This
effect is so that the city of Colmar has the lowest rainfall rate in
France (less than 500 mm per year) while the Ballon d'Alsace located
at 40 km east Colmar has the highest rate (over 2300 mm per year).

    Generally, this effect is more marked when the wind in the west
direction (200°-350°) is strong. It happens during winter and the related rainfall rates are much lower than those observed in the Vosges
area and in the north of the Alsace Plain.[Figure 2]

    The model run contains the folowing specifications and inputs
for the case study of March 27, 1967 :

Surface layer height	z_s = 50 m (at hour 15.00)
Mixed layer height	h = 2500 m
Wind speed and direction	V = 15 m/s, 220°
Wind gradient in the stable layer	$\frac{\partial Vg}{\partial z}$ = 0
Humidity at ground level	q = 5 g/kg
Humidity at the stable layer base	q_h = 3 g/kg
Potential temperature lapse rate in the stable layer	Γ = 10° C/km
Initial potential temperature in the mixed layer	Θ_i = 2 83 °k
Initial potential temperature at the stable layer base	Θ_{hi} = 285°k

    The results of this simulation are given in Figure 3 and can be
summarized in the following terms :

- The maximum height of inversion base is 2750 m and the minimum
  height 2200 m
- The behaviour of the horizontal wind component shows a 6 % acceleration upwind of Colmar ($\simeq$ 20 km), and 10 % deceleration 100 km
  west of Belfort.
- The Föhn effect in Colmar area can be specified by an increase of
  horizontal wind component, a minimum of inversion base and negative
  vertical wind components which prevent precipitation.

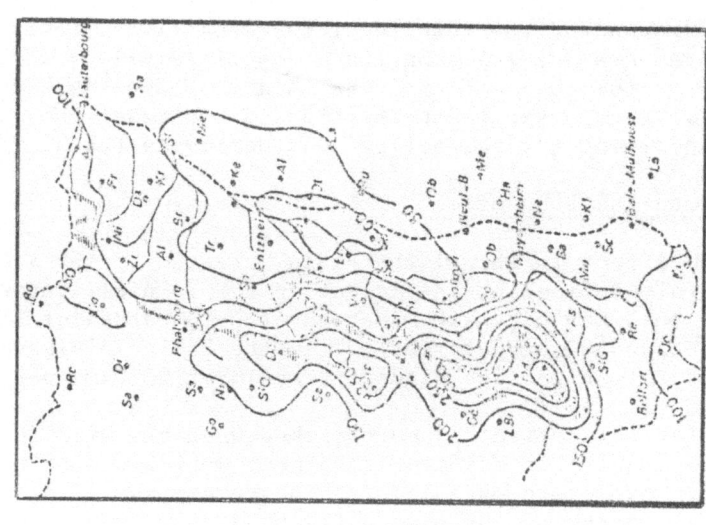

Figure 4: Observed precipitation pattern –
March 27, 1967 (Isocontours 1/10
mm/hour)

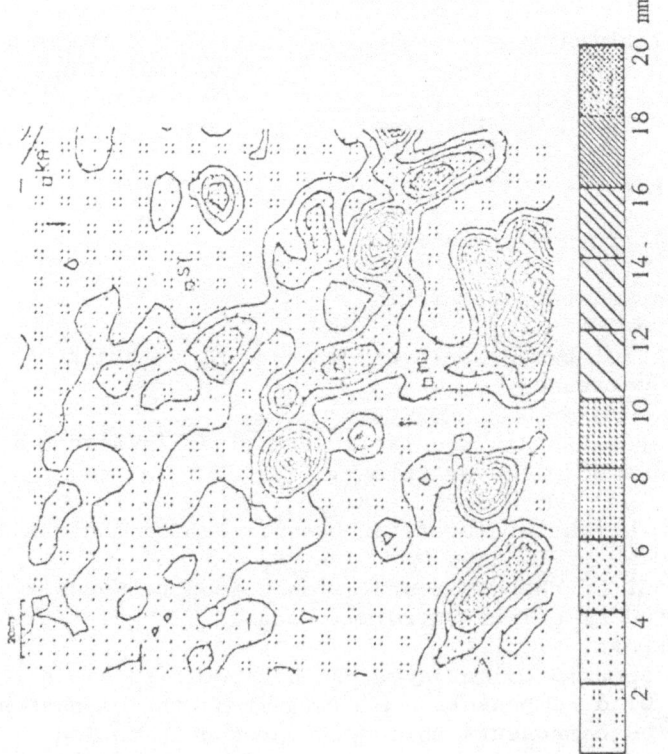

Figure 3: Calculated precipitation pattern –
March 27, 1967

The results can be compared with the observed precipitation
pattern [Chappaz (1978)]. This pattern points out that the average
rainfall rates in the Alsace Plain are 60 % less the surrounding
ones (Lorraine area) during strong wind episodes. The order of
magnitude of these rainfall rates 24 hour averaged (35 mm per day)
are quite close to those predicted (1,4 x 24 = 33,6 mm per day).
The model successfully predicts the two areas of maximum precipitation (near Belfort, 60 km north-west of Colmar) and the three
areas of minimum precipitation (10 km west of Colmar, mid-distance
between Strasbourg and Colmar, and Strasbourg area) [Figure 4].

Sensitivity tests on heat island characteristics

We investigate the model's response to various combinations of
heat island characteristics simulating large heat releases from an
industrial area or large power parks.(Villouvier et Rolin 1980).

The first variation to be discussed is that the relative importance of large heat releases on the mesoscale airflow in the absence
of terrain elevation effects (hypothetical flat terrain). It comprises
different runs varying the source strength from $P = 2 \times 10^3$ MW to
$P = 2 \times 10^5$ MW for a specified source area (S = 100 km2). The results
show a significant perturbation of the initial flow pattern (U = 12 m/s)
when the source strength is increased : the maximum horizontal wind
deformation varies from 1,5 cm/s ($P = 2 \times 10^3$ MW) to 90 cm/s
$P = 2 \times 10^5$ MW) ; the furthest distance from the source where temperature deformation $\Delta\Theta > 0.1°C$ varies from 20 km to 150 km; and the
maximum rainfall rate M increases from 0.6×10^{-2} mm/h to 15×10^{-2} mm/h.

The second variation is related to the combined effects of a
large heat release and terrain elevation (mainly Vosges Mountains)
on the airflow over the Alsace Plain. It discusses the influence of
source strength, source area and its location with regard to terrain
elevation and atmospheric conditions. Table 1 summarizes the specifications for the various runs.

The source location corresponds to a hypothetical power park in
the Strasbourg area within the Alsace Plain. Run 1 considers the
reference case without heat perturbation, and allows us to analyze the deformations in the mesoscale flow for the following runs
in terms of contribution of the assumed heat releases.

Runs 2, 3 and 4 explore the effect of source strength for the
north-west wind where the influence of the Vosges Mountains is weak.
The results show that the induced effects are rather important yielding to a temperature deformation over the source area varying respectively from 0.0 1°C over 300 km2 (Run 2) to 0.03°C over 400 km2
(Run 3) and 0.15°C over 600 km2 (Run 4) ; the increase in the precipitation pattern as compared with Run 1 is more sensitive varying
respectively from 0.2 mm/h over 100 km2 (Run 2) to 0.7 mm/h over
200 km2 (Run 3), and 1 mm/h over 400 km2 (Run 4).

Table 1 : Specifications for the various runs

Run	Source characteristics		Atmospheric Characteristics		
	Strength (MW)	Area (km2)	Wind-speed (m/s)	Wind (deg) Direction	Inversion height (m)
1	–	–	15	315	2000
2	2000	100	15	315	2000
3	10000	100	15	315	2000
4	50000	100	15	315	2000
5	50000	25	15	315	2000
6	10000	100	15	225	2000
7	50000	100	15	225	2000

Run 5 explores the effect of a smaller source area (S = 25 km2) as compared with Run 4. We can see an important increase in the temperature pattern – 0.3°C over 600 km2 –, and in the precipitation pattern – 2,2 mm/h over 400 km2.

Runs 6 and 7 explore the effect of source strength for the southwest wind where the influence of the Vosges Mountains is predominant. We can see a similar effect on the temperature pattern as compared with Runs 3 and 4, while the deformation in the precipitation pattern is definitely weaker – 0.2 mm/h over 100 km2 – independently of source strength.

CONCLUSION

A simulation based on an observed case study on the mesoscale airflow in the Alsace Plain showed that a mixed layer model is capable of reproducing a reasonable precipitation pattern induced by the Fôhn effect.

The model has been used to simulate the effects of heat releases similar to those generated by assumed power parks in the mid-Rhine Valley. While it is too early to draw firm conclusions on what amounts of heat releases would be acceptable, the model provided a reasonable orientation concerning the influence of source strength and area and its location in terms of atmospheric conditions (Koenig 1979).

REFERENCES

Chappaz R., 1978, L'effet de foëhn dans la plaine d'Alsace. La Météorologie, n° 12, pp 129-147

Keyser D, Anthes R., 1977, The applicability of a mixed layer model of the planetary boundary layer to real data forecasting, Month Weather Rev, Vol 15, n° 14, pp 1351-1371

Koenig L.R., 1979, Anomalous cloudiness and precipitation caused by industrial heat rejection, Rard Corp. R-2465

Lavoie R, 1972, A mesoscale numerical model of lake effect storms, J. Atm. Sc. Vol 29, pp 1025-1040.

Raddatz R, Kandekar M; 1977, Numerical Simulation of cold easterly circulations over the canadian western plains using a mesoscale model, Bound. Layer Met. 11, pp. 307-327.

Rolin C., 1980, Modelisation numérique bidimensionnelle de l'écoulement atmosphérique à moyenne échelle. Application à la plaine d'Alsace. Thèse de 3é cycle, Paris VI.

Tennekes H., 1973, A model for the dynamics of inversion above a convective boundary layer, J. Atm. Sc, N° 30, pp. 558-567.

Villouvier V.,1980, Validation et application aux ilôts de chaleur d'un model numérique bidimensionnel à moyenne échelle, Rapport DEA, Electricité de France, Chatou.

DISCUSSION

D.P. EPPEL
In your finite difference scheme you apply upstream differencing. Do you have any numerical problems such as numerical diffusion.

A. SAAB
With the simplified microphysics scheme we don't have really problems as artificial diffusion is weak.
Therefore if we consider a more elaborated .microphysics parametrisation using Kessler's approach - with two additional equations for suspended droplets and precipitating drops -

we have to revise the numerical procedure to accomodate time dependent solutions. Thus we may use centered time and centered space differences.

THE APPLICATION OF A STOCHASTIC WIND MODEL TO THE METEOROLOGY

OF NORTH WEST ENGLAND

J.W. Bacon
B. Henderson-Sellers
Dept. Civil Engineering
University of Salford
Salford, U.K.

A. Henderson-Sellers

Dept. Geography
University of Liverpool
Liverpool, U.K.

INTRODUCTION

Many air pollution models utilise a wind speed given by a mean value over the time period under investigation (see e.g. discussions in references 1 and 2). This provides a valuable tool for many investigations since for successful utilisation of these models it is necessary to understand the individual effects (e.g. source strength, atmospheric stratification and chimney height) before attempting to analyse the inter-relationships and feedbacks. Wind speed variation at surface type boundaries (e.g. land-sea) have been analysed by Jensen[3] in terms of the surface roughness and these ideas have been utilised in an urban mixing layer model[4] to predict the influence of surface roughness on plume trajectories[5].

Probability distributions of wind speed have been discussed by e.g. Fortak[6], Stewart and Essenwanger[7] but the dependence of speed at time t +1 on speed at time t was not considered. Bacon and Henderson-Sellers[8] have described in detail a bivariate, first order Markov chain model, which can synthesise a realistic time series of wind speed over a lengthy period of time. Here this model is utilised to describe the wind fields using data from two stations in North West England.

STOCHASTIC MODEL

In the U.K. data on the wind field are collected as hourly mean values of speed and direction, measured to the nearest knot ($=0.5148$ ms^{-1}) and 10° respectively. This permits a stochastic simulation by modelling the wind vector as a first order Markov chain:

$$\Pr \left[\underline{X}_{n+1} = \underline{x}_{n+1} \mid \underline{X}_n = \underline{x}_n, \ \underline{X}_{n-1} = \underline{x}_{n-1} \cdots \underline{X}_0 = \underline{x}_0 \right]$$

$$= \Pr \left[\underline{X}_{n+1} = \underline{x}_{n+1} \mid \underline{X}_n = \underline{x}_n \right] \qquad (1)$$

(see e.g. Takács[9]), where the vector $\underline{X}_n$ has a value $\underline{x}_n$ at time n. To simplify the computation, it was assumed that the process was time homogeneous within each calendar month and that there was no inter-annual trend. A transition array (stochastic matrix) is then calculated. Random sampling from the four dimensional area permits the construction of a time series simulation over any selected period (this method is fully discussed in reference 8).

Data Storage

To minimise computer storage requirements the largest simulated wind speed was taken to be equal to the maximum observed at Manchester Weather Centre (40 knots) during the period of the data base (1971 - 1980). The resultant stochastic matrices have thus 36 x 40 x 36 x 40 = 2073600 elements - uneconomically large. Two methods for reducing storage requirements were investigated:

(i) a coarser grouping of states with a resolution of 2 knots and 20°;
(ii) collapsing the matrix.

The former method restricts the precision of the generated vectors and was thus rejected after initial trials. However, the second method was found to be applicable since many of the elements in the full 4-D matrix had a value of zero.

Matrix Reduction

Using conditional probability techniques[10], a bivariate frequency function $f(x,y)$ can be expressed as the product of the conditional frequency function $f_c(y \mid x)$ and the marginal frequency function $f_m(x)$ i.e.

$$f(x,y) = f_c(y \mid x) \ f_m(x) \qquad (2)$$

This can be applied equally to the stochastic matrices as required here.

$$f \left(\underline{X}_{n+1} = \underline{x}_{n+1} \mid \underline{X}_n = \underline{x}_n \right)$$

$$= f(D_{n+1} = d_{n+1}, \ S_{n+1} = s_{n+1} \mid D_n = d_n, \ S_n = s_n)$$

$$= f_c \ (S_{n+1} = s_{n+1} \mid D_{n+1} = d_{n+1}, \ D_n = d_n, \ S_n = s_n)$$

$$\times f_m \ (D_{n+1} = d_{n+1} \mid D_n = d_n, \ S_n = s_n) \qquad (3)$$

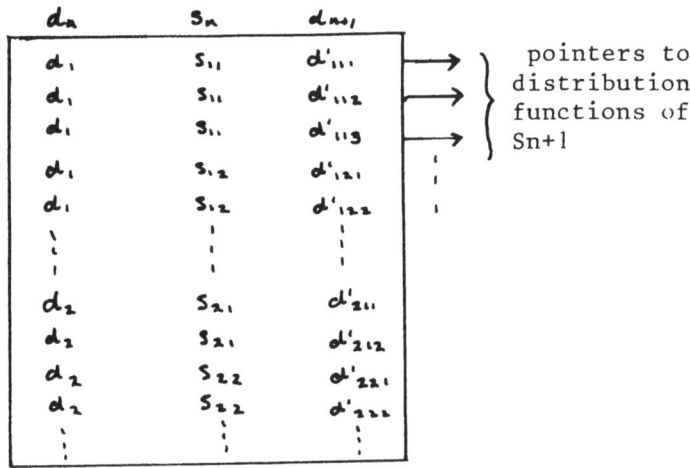

Figure 1. Matrix of all occurring d_n, s_n and d_{n+1}.

in which D_n, S_n are the wind direction and speed at time n (The lower case representations refer to realizations of these variables).

For each calendar month a matrix, A, for all occurring values of d_n and s_n was constructed. Associated with each pair of values is a frequency function for D_{n+1} which corresponds to f_m in equation 3. In addition a matrix, B, for all occurring values of d_n, s_n and d_{n+1} (see Figure 1) was made with an associated value for the frequency function S_{n+1} (thus corresponding to f_c in equation 3).

Sampling was undertaken by searching the (d_n, s_n) matrix (A) for the wind vector at time n and then a random sample taken from the associated distribution of D_{n+1}. Using this generated value, d_{n+1}, the matrix (B) was searched for this calculated wind triple and similarly a sample taken from the associated distribution of S_{n+1}. Thus the wind vector at time n+1 viz (d_{n+1}, s_{n+1}) is derived.

However, the matrix searching is time consuming and a faster method is currently under development. This is outlined in Appendix I.

Sampling

The "top hat" sampling procedure, described by Tocher[10] was utilised based on a pseudo-random number generator. This uses the congruential formulae

$$r_i = \rho_i k^{-z} \qquad (4)$$

where $\rho_i = a^t \rho_{i-1} \bmod k^z$ (5)

$$\rho_o = 1$$

Suitable constants are given[11] as

$$a = 5, \quad t = 17, \quad k = 2, \quad z = 40 \tag{6}$$

Other values for the constants are possible (see e.g. Atkinson[12]) and alternative pseudo-random number generators are currently under investigation by the authors.

This method is found to give a 20-fold reduction in storage requirement, at the expense of an increase in execution time. This was acceptable although methods of reducing execution time are currently under review.

SIMULATIONS

Wind Speed Model

Using data from two sites in North West England (Manchester Weather Centre and Riversdale College, Liverpool), synthetic time series of the hourly mean wind vectors were generated. To compare these sites, samples of the vectors were obtained (as above) from the stochastic matrix for the wind vector at each site, taking an identical initial value for the wind vector and using the same set of random numbers (see Table 1, which shows the results for January). Since the pseudo-random numbers were generated by a congruential method, this set of random numbers is in effect deterministic and thus readily repeatable[11] for comparison simulations. Thus the differences between Liverpool and Manchester are depicted in terms of the divergence between these two data sets. Using the January simulations given in Table 1, it can be seen that there are many differences in detail, (e.g. Manchester has more winds in the first quadrant (N-E)), although both data sets contain a high percentage of winds in the second and third quadrants (90º - 270º). This comparison is further demonstrated graphically in Figure 2. Here the initial vector is a calm. The two series depict the winds over a six hour period. Although the winds are predominantly from the third sector (180º - 270º), the difference evident in these two simulations suggest that the wind fields for Manchester and Liverpool cannot be regarded as identical for many investigations in the field of air pollution meteorology.

The use of such a wind vector series is indicated briefly in the following two sections: for box models and plume rise models.

Table 1. Simulated wind vectors (tabulated in the format (direction$^\circ$
          /speed ms^{-1}) for Manchester and Liverpool using January data.

Hrs from start	Man	Liv	Hrs from start	Man	Liv
0	0/0	0/0	42	40/2.6	120/1.5
1	210/2.1	0/0	43	40/2.6	120/1.0
2	210/2.1	270/2.6	44	50/2.1	120/1.0
3	200/2.1	260/1.0	45	60/1.5	120/1.0
4	200/3.6	260/1.5	46	270/1.0	130/2.1
5	170/3.1	300/2.6	47	210/1.0	150/2.1
6	170/3.6	300/2.6	48	150/1.5	170/2.1
7	170/3.1	300/2.6	49	100/2.1	170/2.6
8	160/3.1	0/0	50	90/2.6	150/3.1
9	170/4.1	0/0	51	90/1.5	140/2.6
10	140/3.6	0/0	52	60/1.0	130/2.6
11	140/4.1	160/1.0	53	70/1.0	130/2.1
12	130/3.6	120/1.0	* 54	230/1.0	130/1.0
13	120/3.1	120/2.6	55	270/1.0	130/3.1
14	140/4.1	120/2.6	56	240/1.5	150/2.6
15	150/5.1	120/2.6	57	240/1.5	160/2.6
16	150/5.7	160/4.1	58	240/2.1	150/2.6
17	150/6.7	170/5.1	59	270/2.6	160/1.5
18	140/6.2	160/5.7	60	250/2.1	170/2.6
19	130/6.2	160/5.7	61	250/2.6	290/5.1
20	140/6.7	150/4.6	62	240/1.5	240/2.1
21	140/4.6	130/5.1	63	240/1.5	200/3.1
22	140/5.1	130/6.2	64	220/1.5	180/3.6
23	120/5.1	130/5.7	65	210/1.0	180/3.6
24	120/5.1	130/6.2	66	280/1.0	170/4.6
25	110/4.6	120/6.7	67	20/1.0	160/4.1
26	110/4.6	120/6.7	68	10/2.1	180/4.1
27	110/5.1	110/6.2	69	360/2.1	130/3.6
28	110/6.7	120/6.2	70	350/1.0	110/4.1
29	100/6.2	120/6./	71	360/2.1	150/6.7
30	110/6.2	120/7.2	72	360/1.5	150/6.2
31	100/5.7	100/5.7	73	340/1.0	180/9.3
32	100/6.7	110/5.1	74	340/1.0	210/10.8
33	100/6.7	110/4.1	75	270/1.0	220/11.8
34	90/7.7	110/4.6	76	260/1.0	220/12.9
35	80/6.2	120/4.1	77	290/1.0	210/12.4
36	70/4.6	120/4.6	78	240/1.0	220/10.3
37	60/4.1	120/4.6	79	330/1.0	240/11.8
38	60/3.6	120/4.1	80	300/1.0	240/9.3
39	60/4.6	130/3.6	81	300/1.0	240/9.8
40	30/3.6	150/2.1	82	320/1.5	240/11.3
41	30/3.1	150/2.1	83	360/2.1	240/11.3

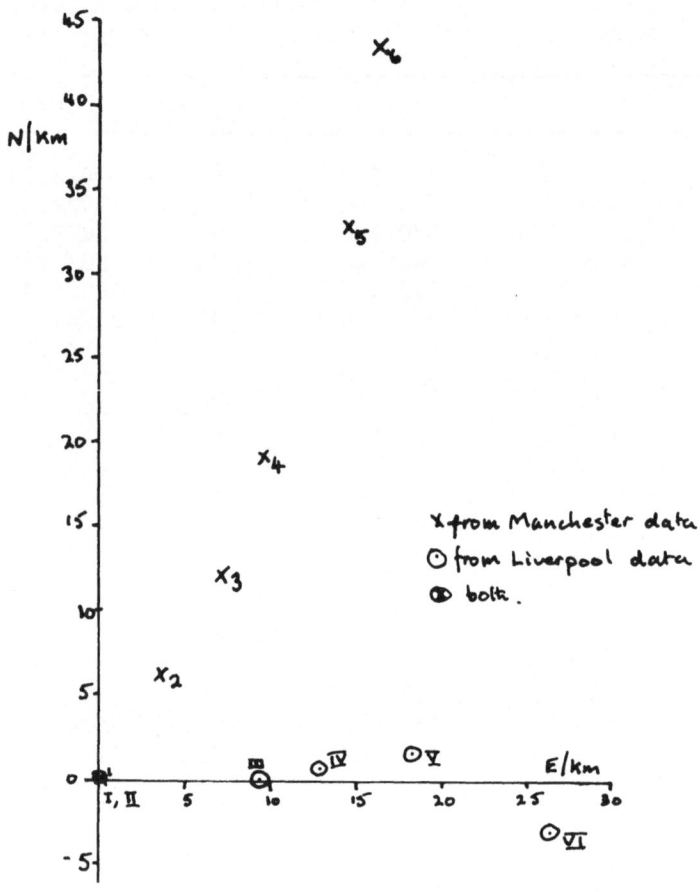

Figure 2.   Series of wind vectors over a six hour period beginning
            with a calm, for Manchester and Liverpool data.

Box Model

     As an example, consider the slug-type box-model of pollutant
dispersion[13] in which the mean wind direction and speed are assumed to
be constant or to vary according to a simple relationship (e.g.
direction varying linearly between SSW and W and speed decreasing
linearly with time). If the plot of the wind vectors is considered
as the centre of a box of pollutant, a different spatial and temporal
distribution of pollutants is obtained with this stochastic model in

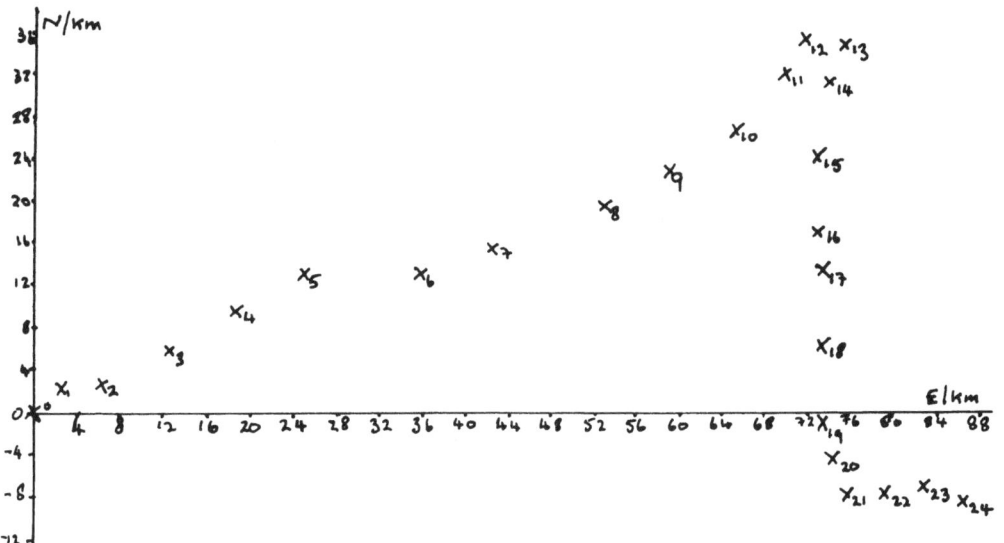

Figure 3.   Trajectory of "slug" of pollution for 24 hours simulation
            of wind vectors from Manchester data.

comparison with, say, a linear variation of wind speed and
direction.

     Figure 3 illustrates a 24 hour simulation for Manchester
(January) - the initial wind vector is indicated * in Table 1. For
the first 12 hours, the wind blows from the same quadrant, the
direction varying between 210° and 270° - roughly SSW to W. During
this period, a comparison can be made with a box model (or similar)
using a constant wind direction. It is readily seen that the
difference between the trajectory given by the stochastic model and
for that using a constant wind direction may be of the order of 2-3 km.
For a "slug" of pollution following this trajectory, the resultant
differences in ground level concentrations will be of importance.
Furthermore, the stochastic model simulates quite rapid direction
changes (e.g. from SSW to N in Figure 3). The simple box model is
totally unable to predict such a change. However, any box or slug
model requiring hourly averaged wind input (both direction and speed)
could benefit from the type of stochastic simulation described here.

     The method of generation allows many possible series of wind
vectors, the frequency of occurrence of each series being proportional
to its likelihood. By calculating the pollutant dispersion predicted
by a box model using different series of wind vectors as inputs, it is

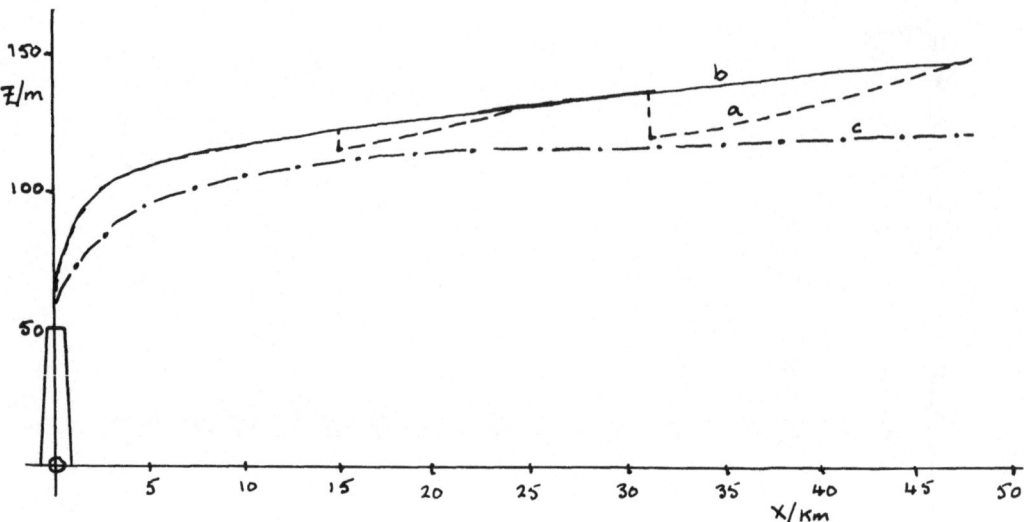

Figure 4.   Plume shape changes as a result of changing wind speed.

thus possible to build up a probabalistic picture of the spatial and
temporal variation of pollutant concentration.

Plume Rise Model

     A second use of this stochastic wind model is in plume rise
calculations. Using a model described elsewhere in this volume[14],
the effect of a stochastically changing (rather than constant) wind
speed (the directional variation is neglected for ease of graphical
representation) is shown in Figure 4. Using a sequence of discrete
hour-long "puffs" the effect of changing wind speed is observed in the
discontinuous trajectory (curve a - dotted line). However, since
most effluent discharges are continuous the smooth curve b (solid line)
will more accurately describe the plume centreline. In the case
depicted here, the wind time series is monotonic decreasing and this
is seen in restricted rise during periods of stronger winds. Curve c
shows the result of the plume rise simulation for a constant wind
speed (of 6.2 ms[-1]). The difference between the constant wind speed
model and the stochastic model is large enough to give an appreciable
difference (of the order of several tens of percent) in ground level
concentrations.
CONCLUSIONS

     The use of a stochastic model for wind vector generation has been
demonstrated for analysing local meteorologies in North West England.
Furthermore, it can provide a useful and realistic input to many types
of air pollution model - its applicability in two such cases has been

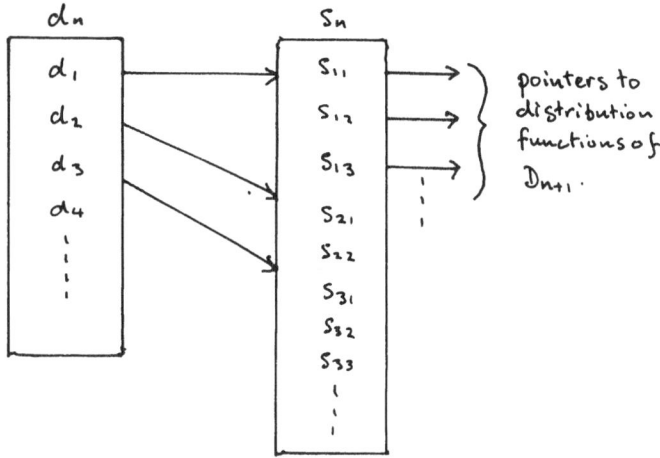

Figure 5. Lists of occurring values of d_n and s_n.

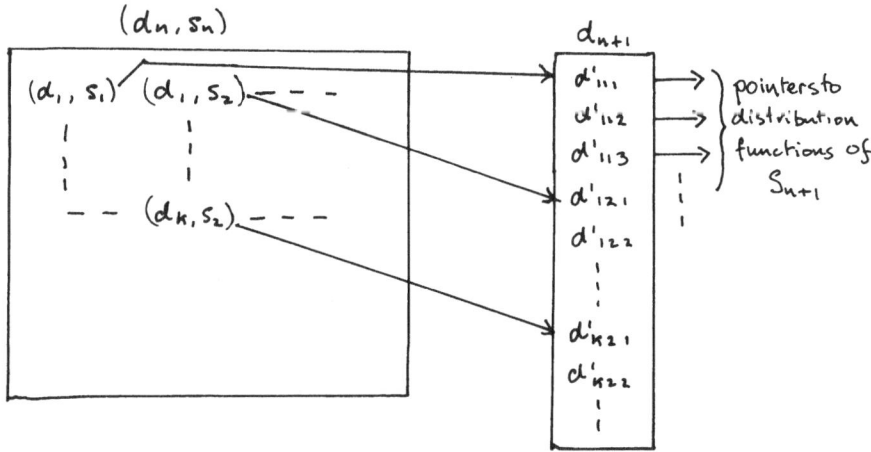

Figure 6. Matrix of values of d_n, s_n and list of values of d_{n+1}.

exemplified. Using a box model, changes in both wind speed and direction can now be successfully simulated; as can the effect in plume rise
models (although only speed changes have been considered here).

Acknowledgements

We wish to thank Captain H. Lawson of the Department of Navigation
Riversdale College of Technology, Liverpool for supplying data for
Liverpool, and the Manchester Weather Centre for Manchester data.

APPENDIX I

To decrease execution time, the following storage and search
routine is being developed. The information corresponding to f_m in
equation 3 is stored as a list of directions, d_n, and a list of
speeds, s_n. For each value of d_n a pointer gives the address in the
s_n list for the start of the speeds associated with that d_n value
(see Figure 5).

From each value of s_n a second pointer gives the address of the
associated distribution function of D_{n+1}. Similarly for f_c, data are
stored as (d_n, s_n) values in a 36 x 40 matrix, together with a list
of directions d_{n+1}. Pointers are then again used to find the relevant
address in the list of directions d_{n+1} and subsequently the distribution function for S_{n+1} (see Figure 6). Although some elements of
the (d_n, s_n) matrix will be zero and hence the store required greater
than that of the currently used method, the reduction in execution
time should make this second method more attractive.

REFERENCES

1. M.I. Hoffert, Atmospheric transport, dispersion, and chemical
    reactions in air pollution: a review, AIAA Journal, 10:
    377 (1972).
2. F.A. Gifford, Atmospheric dispersion models for environmental
    pollution applications, in "Lectures in Air Pollution
    and Environmental Impact Analysis", 35-58, AMS, Boston,
    Ma. (1975).
3. N.O. Jensen, Change of surface roughness and the planetary
    boundary layer, Quart. J. Roy. Met. Soc., 104, 351(1978).
4. A. Henderson-Sellers, A simple numerical simulation of urban
    mixing depths, J. Appl. Meteor., 19:215 (1980).
5. B. Henderson-Sellers, The inclusion of wind and wind shear
    effects in plume rise models, Symposium on Plumes and
    Visibility: Measurements and Model Components, Grand
    Canyon, 10-14 November, 1980.
6. H.G. Fortak, Numerical simulation of the temporal and spatial
    distributions of urban air pollution concentration in
    Proc. Symposium on Multiple-Source Urban Diffusion
    Models, USEPA, APCO, Publ. No. AP86 (1970).

7. D.A. Stewart and O.M. Essenwanger, Frequency distribution of
    wind speed near the surface, J. Appl. Meteor., 17: 1633
    (1978).

8. J.W. Bacon and B. Henderson-Sellers, The generation of hourly
    average wind vectors using a Markov process, in
    "Atmospheric Pollution, 1980, ed. M. Benarie, Studies in
    Environmental Science", 8: 195 (1980).

9. L. Takács, "Stochastic Processes", Chapman & Hall Ltd.
    London (1978).

10. K.D. Tocher, "The Art of Simulation", English Universities
    Press Ltd., London (1963).

11. D. Teichrow, A history of distribution sampling prior to the
    era of the computer, and its relevance to simulation,
    American Statistical Association Journal, March:27 (1965).

12. A.C. Atkinson, Tests of pseudo-random numbers, Appl. Statist.
    29: 164 (1980).

13. A. Venkatram, An examination of box models for air quality
    simulation, Atmos. Environ. 12: 2243 (1978).

14. A. Henderson-Sellers and B. Henderson-Sellers, Numerical
    modelling of stack plumes within a city environment at
    distances of several kilometres downstream, 11th Inter-
    national Technical Meeting in Air Pollution Modeling
    and its Application, Amsterdam, 24-27 November, 1980.

ATMOSPHERIC CIRCULATION ON THE REGIONAL SCALE AND ISENTROPIC

TRAJECTORIES AS SUPPORT TO THE LONG RANGE TRANSPORT (LRT) OF

AIR POLLUTION

Sergio Borghi

Osservatorio Meteorologico di Brera
Via Brera, 28
20121 Milano, Italy

ABSTRACT

Atmospheric circulation over northern Italy has been examined for three winter and three summer periods up to 500 millibar-level, to characterize behavior and frequencies of currents, mainly regarding to the effects connected to orographic obstacles. From this point of view the importance of southern winds has been emphasized. For what is concerning the transport of air pollutants through long distances (several hundreds of kilometres), a method has been suggested to follow the individual motion of air particles, necessary to take into account also vertical displacements. By means of the isentropic analysis, under suitable conditions – besides typical of the persistence of air pollution – the method allows to consider tridimensional trajectories, which generally cannot be obtained by using isobaric meteorological maps, normally available.

INTRODUCTION

Local diffusion of air pollutants is evaluated generally through statistical or thermodynamical interpretation of turbulent airflow.

On the contrary, the medium or long range transport of air pollutants is mainly a synoptical problem. A complete investigation on it can carried on only if two fundamental viewpoints, strictly connected with the problem, are taken into account, namely: (1) the knowledge of atmospheric circulation over the region considered, and (2) the related possibility to find individual trajectories of the air particles, which have to be followed as vehicles of air pollutants.

For these purposes a particular region has been chosen, Po Valley, where many urban areas, arterial roads and industrial plants
can be found. Also from a geographical point of view this   area
shows some important characters, because it is particularly sheltered by orographic obstacles, which can influence the incoming air
masses and modify also their movements.

BEHAVIOR OF THE WIND OVER PO VALLEY

To point out some characteristics of atmospheric circulation
over the chosen area, three winter periods (December, January and
February 1971 - 72, 1972 - 73 and 1973-74) and three summer periods
(June, July and August 1972, 1973 and 1974) have been considered,
through the examination of two meteorological situations a day, at 0
and 12 GMT. Altogether 542 winter situations and 552 summer situations have been taken into account.

To approach the problem, frequencies of wind speed values have
been analysed, by distinguishing southern from northern currents,
and cold from warm season. Four classes of wind speed, besides the
calm, have been considered for both levels of 850 and 500 millibars
(tab. 1 and tab.2).

The choice of these two levels was suggested because over Po
Valley, at the higher level, corresponding to about 5500 m above
s. l., airflow is sufficiently not disturbed by the orography, while
at the lower level, corresponding to about 1500 m above s. l., although currents are influenced by the orography, they are not strongly affected by the friction stresses due to the ground.

As we can observe, at 500 mbar (tab.1) the calm frequencies are
very small (0.7% in winter and 1.8% in summer), and northern circulation is prevailing, especially during summer. The range of speeds
between 10 and 30 knots includes most cases and, disregarding the
wind directions, the sum of frequencies is 70.4% during winter, and
76.2% during summer.

At 850 mbar (tab.2) frequencies have a behavior quite different
from above. The calm frequencies, greater than 12%, indicate that
over Po Valley the lower layer considered does not often take part
in the motion of the atmosphere on the large scale. During both
seasons southern winds are widely prevailing, but this result could
be expected for the strong blocking action, made by the Alps against
northern currents.

Southern winds have been analysed with more details, by the
examination of wind speed frequencies, separately for six sectors of
30 degrees of amplitude. For both levels considered, in fig. 1, a, b
wind speed classifications are shown, corresponding to winter and
summer periods, and to cyclonic, rectilinear and anticyclonic flows.

Table I. 500 millibars level: Frequencies of calm and four classes of wind speed, separately considered for season and meridional component. The classes of wind speed are the following: (A) $5 \leqslant V < I0$ kt, (B) $I0 \leqslant V < 20$ kt, (C) $20 \leqslant V < 30$ kt, (D) $V \geqslant 30$ kt, calm $V < 5$ kt. Total values do not comprehend the cases of calm.

	Wind speed	(A)	(B)	(C)	(D)	Total	Calm
Winter	Northern wind	2,8	I6,4	I5,I	I7,2	5I,5	0,7
	Southern wind	2,6	22,5	I6,4	6,3	47,8	
Summer	Northern wind	2,0	28,3	I8,I	8,9	57,3	I,8
	Southern wind	3,4	I5,9	I3,9	7,7	40,9	

Table 2. 850 millibars level. For explanation see table I.

	Wind speed	(A)	(B)	(C)	(D)	Total	Calm
Winter	Northern wind	II,I	I6,2	4,2	4,I	35,6	I2,4
	Southern wind	I4,0	24,6	9,6	3,8	52,0	
Summer	Northern wind	I7,I	I6,3	4,3	0,7	38,4	I2,3
	Southern wind	25,9	I9,I	3,4	0,9	49,3	

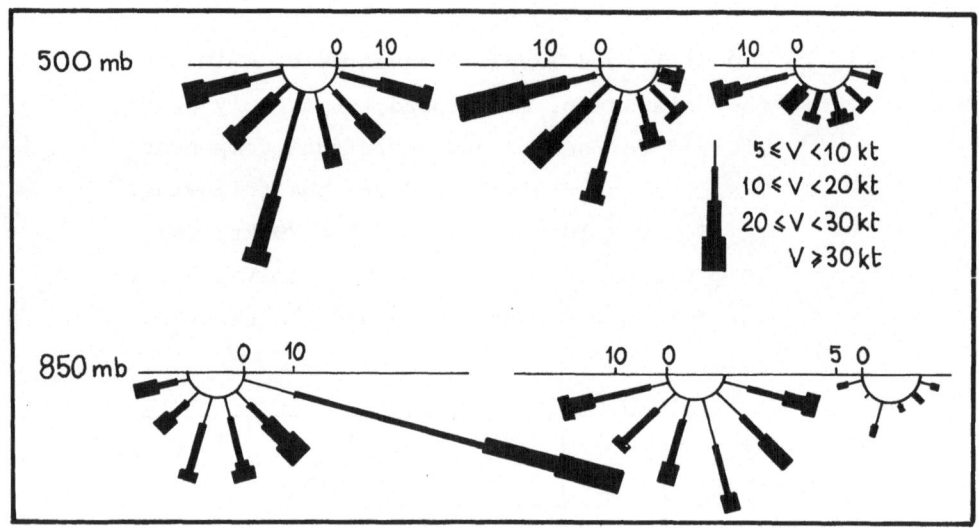

Fig.1,a. Winter. Southern currents at pressure levels
of 500 and 850 mbar. Number of cases of cyclonic (C), rectilinear (R) and anticyclonic
(A) circulation referred to four classes of
wind intensities and to 6 southern sectors.

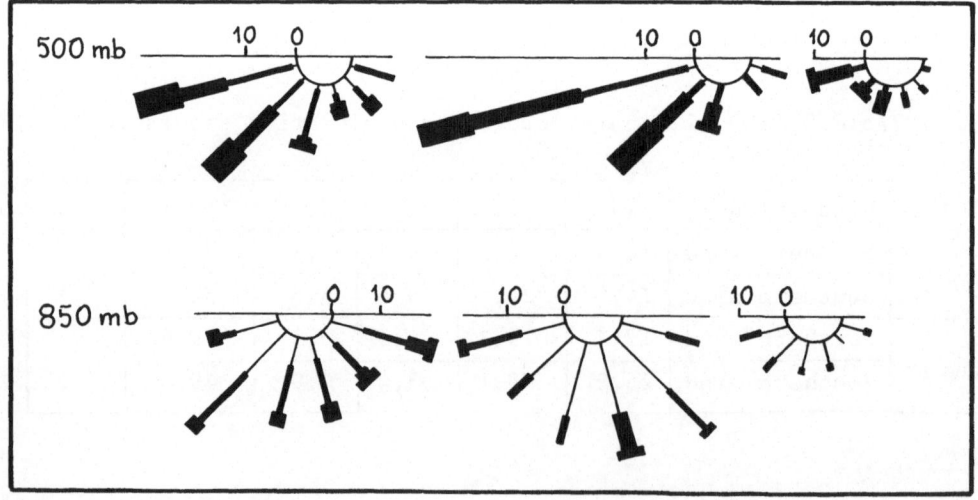

Fig.1,b. Summer. For the explanations see fig.1,a.

At 500 mbar level, cyclonic and rectilinear winds are prevailing in both seasons; frequencies and intensities of western currents have higher values with respect to eastern currents.

But it is at the lower layer (fig. 1,b) that the behavior of currents becomes more significant; here we can observe some singularities, among which the most important is perhaps the high frequency of the winds coming from ESE during winter (fig.1,a), corresponding to the irruption of cold continental air masses from Balkan area.

These results, even though referred to only three years and to only two seasonal periods, are in good agreement with some results pointed out in two papers, concerning the wind over Milan (RIMA, 1963) and Cagliari (MATTANA et al., 1967).

Under the hypothesis of an eventual connection between atmospheric circulation and LRT, for two couples of levels (850 - 700 mbar, and 700-500 mbar) the persistence of the wind direction has been tested along the vertical of Milan, and some results are shown in fig.2,a (winter) and fig. 2,b (summer).

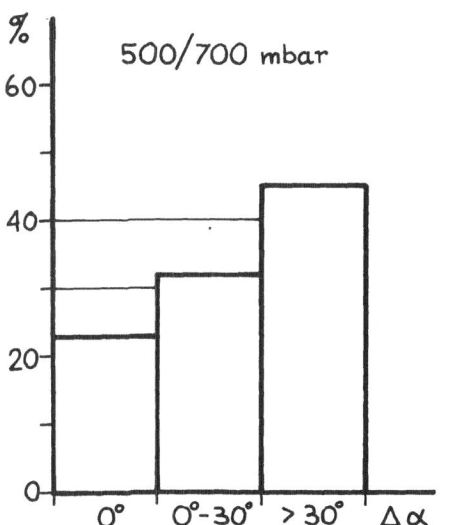
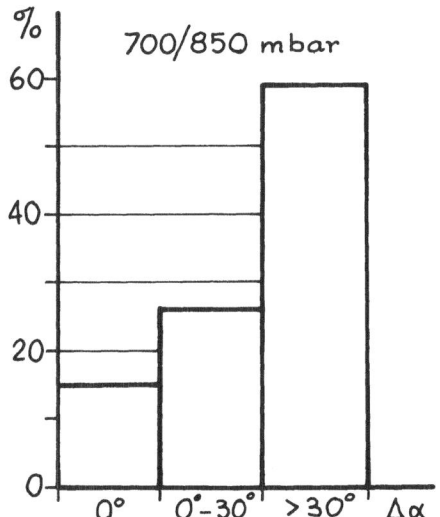

Fig.2,a. Winter. Percentage of cases with indicated differences between wind directions (Δα) for two couples of levels (500/700 and 700/850 mbar).

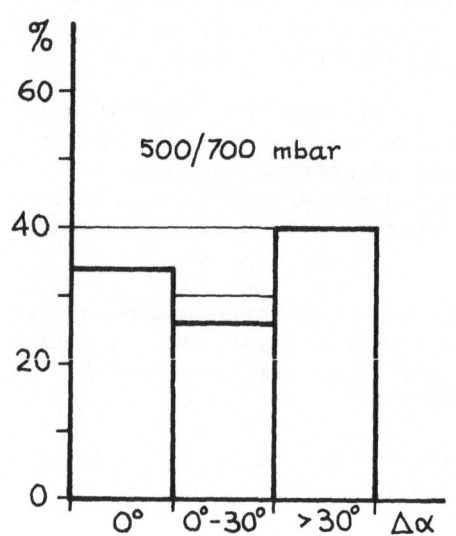
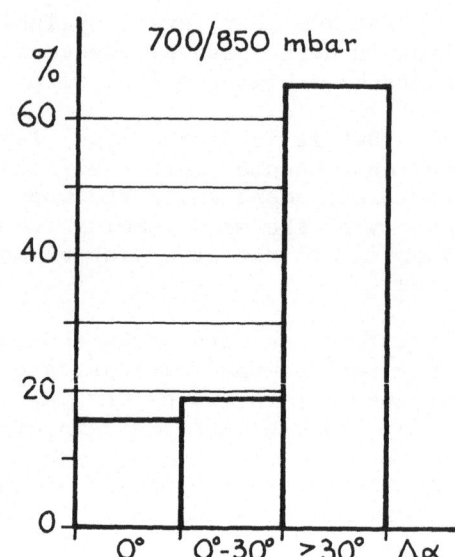

Fig.2,b. For explanations see fig.2,a.

It is rather interesting to observe that during summer, and especially for the lower layer considered, a lot of cases (65%) corresponds to a difference between the directions of the two winds, which is not less than 30 degrees.

That is also important, as it will be seen later, because our major aim is the determination of individual atmospheric trajectories, which can be obtained only by following an invariant property for the chosen sample of air.

But we cannot find, and not even follow any invariant property of the air particle on the isobaric maps, normally available. These maps reproduce the actual atmospheric motion on quasi-horizontal surfaces, and vertical displacements of air particles cannot be deduced.

THE METHOD OF ISENTROPIC TRAJECTORIES

If the atmospheric conditions are favorable to persistence of atmospheric pollution, to obtain a significant determination of trajectories, the property which can be chosen is the entropy of the air mass, if its movements are adiabatic. This is reasonably true under these two following assumptions:

- there are neither release of latent heat by water vapor condensation, nor wash-out;

- the period considered has to be short enough to have a negligible transfer of sensible heat from the environment to the air sample.

The first condition is generally verified because, with an abundant condensation and wash-out, air pollution is drastically reduced, and its transport does not occur.

For the second condition we have to choose a period during which the potential temperature of the air mass can be considered as a constant; this condition is normally guaranteed by a period not exceeding 24 hours.

According to the Thermodynamics First Law, the virtual potential temperature

$$\theta_v = T_v \left(\frac{1000}{p}\right)^k$$

can be considered as a conservative property during an adiabatic, i.e. isentropic transformation of the air particle.

The isentropic trajectories can be obtained if motion of air masses is referred to a system (x, y, z, t) where the axes are respectively directed eastward, northward and upward, and t is the time.

On a surface where θ_v = Const., represented horizontally on a map, bidimensional equation which describes the variability of the horizontal component $\vec{V}_H$ of the velocity has the following form

(1) $$\frac{d\vec{V}_H}{dt} = -\nabla_\theta\left(c_p T + gz\right) - f\,\vec{k}\times\vec{V}_H$$

where ∇_θ = horizontal del-operator for θ_v = const.
          g = 9.81 m sec^{-2} is the gravity
          f = $2\Omega \cdot \sin\varphi$ is the Coriolis parameter
          $\bar{k}$ = unit vector along the z - axis
          Ω = 7.29 10^{-5} rad sec^{-1} is the rotation speed of the earth
          φ = latitude.

Using the stream function

$$\Psi = c_p T + gz$$

equation (1) can take the form

$$(2) \qquad \frac{d\vec{V}_H}{dt} + f\,\vec{k} \times \vec{V}_H = \nabla_\Theta \Psi$$

On an isentropic map, with respect to the wind, the Ψ - field has the same significance as the geopotential field has on the iso- baric maps. Equation (2) gives us the geostrophic wind (on an isen- tropic map) when we put $d\vec{V}_H/dt = 0$ (geostrophic equilibrium).

Taking the total derivative

$$(3) \qquad \frac{d\Psi}{dt} = \frac{\partial \Psi}{\partial t} + \vec{V}_H \cdot \nabla_\Theta \Psi + \frac{\partial \Psi}{\partial \theta_v} \cdot \frac{d\theta_v}{dt}$$

and the dot-product between $\vec{V}_H$ and equation (2)

$$(4) \qquad \frac{d}{dt}\left(\frac{V_H^2}{2}\right) = -\,\vec{V}_H \cdot \nabla_\Theta \Psi$$

the advective term $-\vec{V}_H \cdot \nabla_\Theta \Psi$ can be eliminated between equations (3) and (4). The result is the following equation

$$(5) \qquad \frac{d}{dt}\left(\Psi + \frac{V_H^2}{2}\right) = \frac{\partial \Psi}{\partial t} + \frac{\partial \Psi}{\partial \theta_v} \cdot \frac{d\theta_v}{dt}$$

To construct trajectories, equation (5) has to be integrated following the individual motion of the air particle (DANIELSEN, 1966). In general we have

$$(6) \qquad \Psi_f - \Psi_i + \frac{1}{2}\left(V_{Hf}^2 - V_{Hi}^2\right) = \int_{t_1}^{t_2} \frac{\partial \Psi}{\partial t}\,dt + \int_{\theta_1}^{\theta_2} \frac{\partial \Psi}{\partial \theta_v}\,d\theta_v$$

where the subscripts i, f denote the initial and final values of Ψ

and V_H^2 respectively assumed at time t_1 , when $\theta_v = \theta_1$, and at time t_2 , when $\theta_v = \theta_2$.

With the adiabatic approximation the last term on the right side of equation (6) vanishes, because of the constance of θ_v (i.e. $\theta_1 = \theta_2$).

Replacing the integration of the local change $\partial\Psi/dt$ with a finite difference for an interval Δt , equation (6) becomes

$$(7) \qquad \Psi_f = \Psi_i - \frac{1}{2}\left(V_{Hf}^2 - V_{Hi}^2\right) + \frac{\Delta\Psi_i + 2\Delta\Psi_m + \Delta\Psi_f}{4}$$

where the subscript m denotes the mid-time position, which in general differs from the mid-point of the trajectory.

On the synoptic scale, under the approximation of linear speed variations, the air particle displacement is given by

$$(8) \qquad D = \frac{V_{Hf} + V_{Hi}}{2} \cdot \Delta t \qquad \text{or by}$$

$$(8') \qquad D = \frac{V_{Hf} + 2 V_{Hm} + V_{Hi}}{4} \cdot \Delta t$$

where $\Delta t \leqslant 24$ h and the subscript m has the meaning just specified.

The trajectories can be determined by choosing a point on the initial isentropic map, and then exploring the down-stream area of the final map to select a point which simultaneously satisfies equations (7) and (8) or (7) and (8'). The uniqueness of the solutions has been demonstrated (DANIELSEN, 1966), so that trajectories can be obtained graphically.

INVESTIGATED AREA AND BASIC DATA EMPLOYED

As regards medium or long range transport of air pollutants, Italian regions and in particular Po Valley are interested by the meteorological conditions of the Mediterranean area. As pointed out previously, the frequency of southern air-flow over Po Valley is prevalent; the northern winds, in fact, rarely interest the LRT phenomena because of the vigorous obstacle represented by the Alps.

A practical determination of isentropic trajectories for three levels, namely

$$\theta_v = 298, \ 303 \ \text{and} \ 308 \ \text{K}$$

has been made by choosing the site of Milan as the final point for the adiabatic displacement of air masses moving in the Mediterranean area. A 36 h – period for the investigation has been chosen, from 12 GMT June 17th to 24 GMT June, 1977.

The basic meteorological data employed have been obtained from the observations of the radiosounding stations, available for the chosen investigated area (fig. 3), and where daily two soundings are made at 0 and 12 GMT.

Fig.3. Network of radiosounding stations employed.

USE OF ISENTROPIC MAPS. DETERMINATION OF SOME INDIVIDUAL TRAJECTORIES

The single isentropic maps (each one is characterized by a time and by a θ_v level) are the results of the analyses of simultaneous observations made at radiosounding stations. Each sounding has

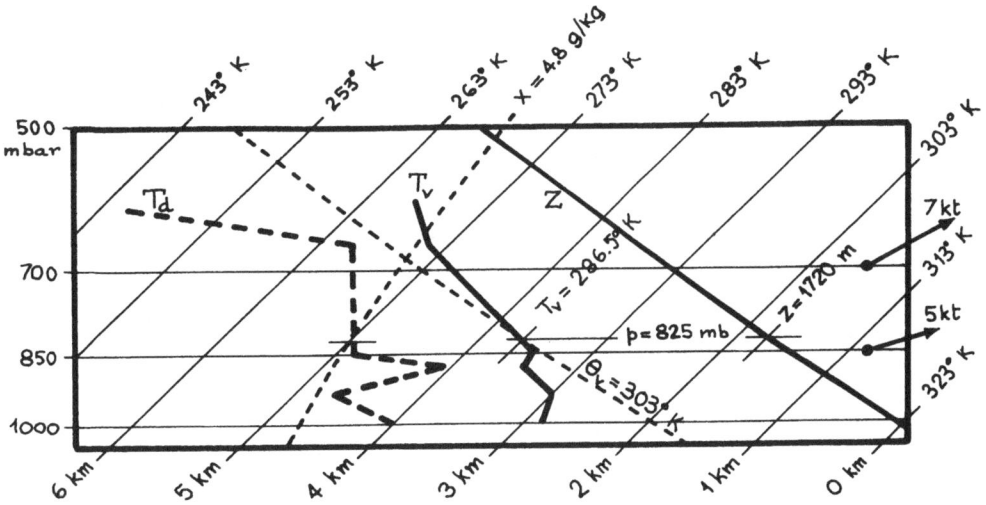

Fig.4. Example of determination of the virtual temperature θ_v, the pressure p, the level z and the wind $\vec{V}_H$ for the level θ_v = 303 K, by using a Herlofson Thermodynamic Diagram.

been considered on a Herlofson Thermodynamic Diagram to have the values of Ψ, p and V_H (see fig. 4).

Three series of four isentropic maps was used, one for each level of constant entropy, and the Ψ- field was analysed for successive 12 h - intervals, to have the possibility to solve equations (7) and (8).

The method of the isentropic trajectories was applied to the entire period considered, by starting backwards from the site of Milan, and inverting the role of Ψ_f and Ψ_i in equation (7).

For a 12 h - step, initial and final Ψ- fields and pressure - fields are shown in fig. 5, a, b respectively; both maps correspond to the level θ_v = 303 K.

In fig. 6 is shown the $\Delta\Psi$ - field related to the situations above.

The three trajectories determined are sketched in fig. 7, for the three levels considered. It is very interesting to observe that, for each trajectory, the vertical displacements are the more important the higher the θ_v - levels considered are, also for the wind intensities, which generally rise with the altitude and produce longer horizontal displacements.

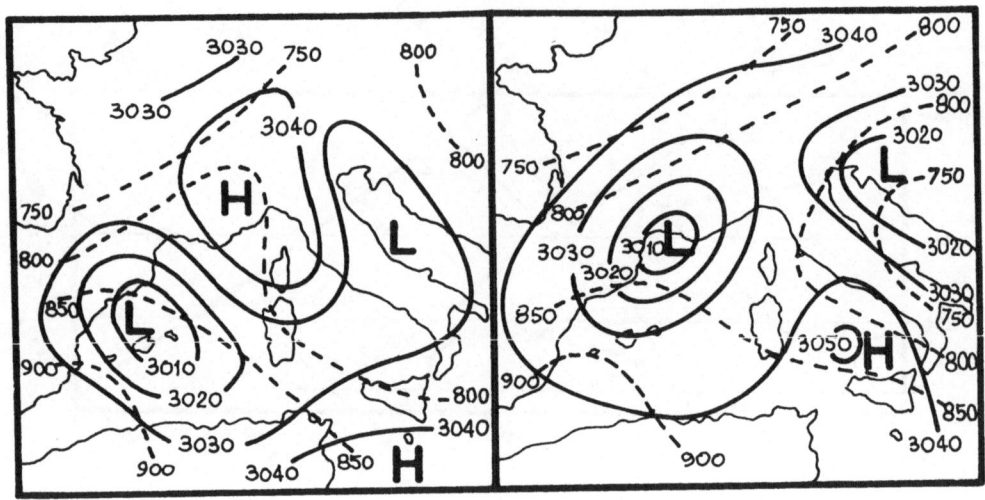

Fig.5. a) Ψ - field (solid lines, and values in 10^2
m² sec^{-2}) and pressure field (dashed lines
and values in mbar) at 12 GMT, June 17th,
1977. b) The same at 0 GMT, June 18th, 1977.

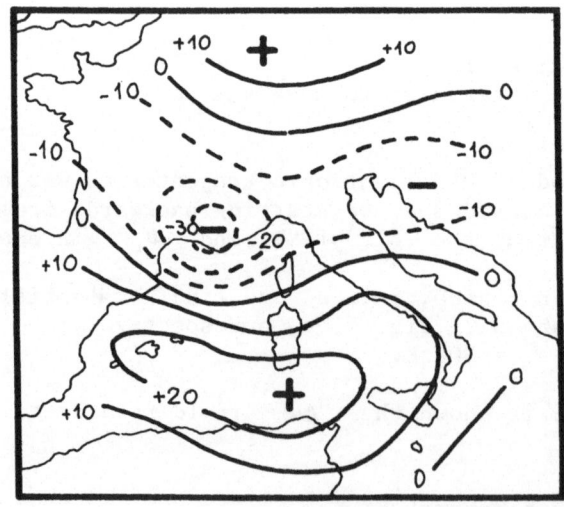

Fig.6. ΔΨ - field (values in 10^2 m² sec^{-2}) at 0 GMT,
June 18th, 1977, for Δt = 12 h and θ$_v$ = 303 K.

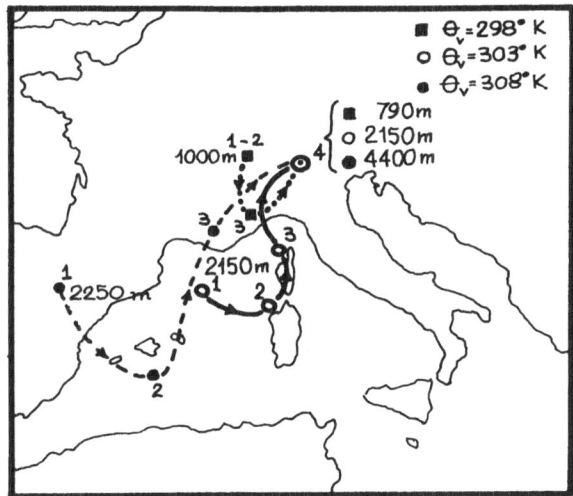

Fig.7. Three evaluated trajectories, corresponding to
three values of θ_v. Each one (1 to 4) starts
on June 17th, 1977 at 12 GMT (point no.1) and
ends on June 19th, 1977 at 0 GMT (point no.4).

CONCLUSIONS

After a convenient examination of prevailing atmospheric currents over the chosen area of Po Valley up to about 5000 m above sea
level, a method has been pointed out to approach some meteorological
problems connected to the medium and long range transport of air
pollutants, especially by considering the single air particle as a
vehicle of the same pollutants.

The method suggested, applied to some situations, operates
through the determination of isentropic maps, so that under suitable
conditions (besides typical of pollution transport), objective trajectories can be evaluated to take into account also the vertical
displacements of air masses, whose magnitude generally is not negligible.

BIBLIOGRAPHY

Danielsen, E. F., 1966, Research in Four-dimensional Diagnosis of
        Cyclonic Storm Cloud Systems, Air Force Cambrige Research
        Laboratories, Bedford, Massachussets.
Mattana, N., Sanna S., and Serra, A., 1967, Le Correnti Atmosferiche
        in Quota sulla Sardegna, nel Periodo 1948-1963 (Confronto
        con una Serie di Dati in Valpadana), Riv. Met. Aer., 27,4.

Rima, A., 1963, Variazioni Anemologiche a Milano-Linate ai Livelli
        Standard dal 1952 al 1961, nel Semestre Caldo e nel Seme-
        stre Freddo, Riv. Met. Aer., 23, 4.

ACKNOWLEDGEMENTS

    This work was partly supported by grant of "Progetto Finalizzato
per la Promozione della Qualità dell'Ambiente", C.N.R..

DISCUSSION

H. VAN DOP   Have you tested your
                      assumption that the potential temperature of an
                      air parcel is approximately constant over a
                      period of 24 hours ?

S. BORGHI   The control was made
                      for a certain number of situations of weak cir-
                      culation with absence of clouds (anticyclonic
                      conditions), when locally we can have a relati-
                      vely strong transfer of sensible heat from the
                      surface to the atmosphere. This transfer be-
                      comes negligible in the free atmosphere over
                      Po Valley from about 1000 m above ground and for
                      upper levels, especially for the period used of
                      12 h. In these conditions $\delta\theta v/\delta t \simeq d\theta v/dt \simeq 0$
                      which is our assumption to construct isentropic
                      trajectories.

P.J. SAMSON   How is the interpreta-
                      tion of your model in the boundary layer when
                       $d\theta/dz$ becomes constant.

S. BORGHI   The complete model that
                      we are trying to define is a fluidodynamical
                      model for Po Valley region. The trajectories
                      obtained with the isentropic analysis are used
                      to describe the upper boundary of the atmospheric
                      fluid, when we connect the properties of the
                      boundary layer with the properties of the free
                      atmosphere. The method has its significance
                      from a synoptical point of view, at the levels
                      where the behaviour of the atmosphere is the
                      typical one of the free atmosphere. Isentropic
                      trajectories take into account the vertical dis-
                      placements and the consequent direction of the
                      wind in the free atmosphere, which could not be
                      deduced by isobaric maps, normally available,
                      where atmospheric notion is represented on

quasi-horizontal surfaces. The fluidodynamical
model operates in the lower layers with different assumptions : vertical profiles of the
temperature and of the velocity are considered
together with the vertical fluxes of sensible
heat.

DIFFUSIVITY PROFILES DEDUCED FROM SYNOPTIC DATA

G. Schayes and M. Cravatte
Institute of Astronomy and Geophysics
Catholic University of Louvain
Louvain-la-Neuve
Belgium

Abstract. In order to be able to use
K-formulation dispersion models with the
most commonly available meteorological
data (i.e. synoptic data), a simple physical model has been built.

The radiative balance is computed at
the ground level with the date of year,
time of day and cloudiness data. Assuming
a given Bowen ratio and Monin-Obukov
similarity, surface layer parameters are
deduced and applied to find the K_z profile
with the Wippermann stationnary planetary
boundary layer formulation.

The results are compared with a PBL
model and with available data.

Introduction

Mathematical models of pollution dispersion based on the
K-theory need more elaborate input data than the classical Gaussian plume model to be run. These additional data consist mainly
of diffusivity profile K(Z) and eventually wind speed profile
U(Z), both taken to be constant with height in the Gaussian model.

Although an approximation of the wind profile can be derived
easily from the wind speed at anemometer level and stability classes a good K(Z) profile determination must be obtained by rather
sophisticated modelling of the PBL, which need many input data
generally not available and a long computing time.

The present more simple formulation allows the eddy diffusivity profile to be calculated from the data widely available in the synoptic report, namely air temperature and dew point, wind speed, cloudiness, time of day and day of year.

The model is not at all intended to yield an accurate representation of the K(Z) profile in all meteorological situations, but it will be used as a substitution scheme of the σ_z, for use in advanced pollution modelling. Moreover, the computation time is very shorter than PBL models, and this allows the present formulation to be applied to longer periods of time.

However, the application to real cases requires an estimation of the roughness length Z_0 the albedo A and the Bowen ratio B of the place in question. A sensitivity study of the model on these parameters is presented in section 4.

The validity of the present formulation has been tested against the results of an one-dimensional Atmospheric Boundary Layer (ABL) model based on the turbulent energy, which is considered here as giving the "correct" K(Z) answer. This last ABL model has been itself tested against quality data (e.g. Wangara) and compared favorably with the results of other investigators[1,2].

The model

In this model, the K(Z) profile is deduced when u_* the friction velocity and H the turbulent heat flux are known, by using a formulation devised by Wippermann[3].

Energy fluxes

The turbulent heat flux H can be found from the energy balance at ground level :

$$H + E = R_S(1 - A) - I - G \tag{1}$$

where E = latent heat flux
  R_S = shortwave radiative flux reaching the ground
  I = longwave outgoing radiative flux
  G = heat flux going into the ground
  A = albedo of the surface.

Every component of this balance must be calculated separately from the available synoptic data.

First we compute the zenital distance θ of the sun by well known astronomical relationships from the day of the year and time of the day. If S_0 is the solar constant (here taken as 1353 Wm^{-2}), then the visible radiative flux R_S reaching the ground surface may be expressed as :

$$R_S = S_0 \cos \theta \underbrace{(\tau - A_w)}_{(a)} \times \underbrace{N}_{(b)} \qquad (2)$$

where the factor (a) is the global transmissivity of the atmosphere taking into account the transmissivity τ of dry air including scattering and the absorption A_w due to water vapor, and the factor (b) takes into account the effect of clouds when they are present.

The absorption by water vapor A_w may be calculated with the relation given by McDonald[4] :

$$A_w = 0.077 \left(\frac{w}{\cos \theta}\right)^{0.3} \qquad (3)$$

where w is the water vapor path length in cm. According to Atwater and Brown[5], the transmissivity coefficient τ in clear air including scattering may be expressed as :

$$\tau = 1.021 - 0.0824 \left[\frac{949 \times 10^{-6} \times P + 0.05}{\cos \theta}\right]^{0.5} \qquad (4)$$

where P is the surface pressure in mb. We must mention that aerosol absorption is not taken into account by lack of regular data about it.

When present, the absorption of radiation by clouds is written after Haurwitz[6] :

$$N = \prod_{i=1}^{3} [1 - C_i (1 - T_i)] \qquad (5)$$

Here i = 1,2,3 represents respectively low, medium and high cloud types, the C_i are the cloud cover of each category (C_i = 1 means totally overcast) and the T_i are the mean transmissivity of each cloud category. As shown in (1), a part of the shortwave radiative flux is reflected by the ground surface depending on its albedo A.

The net outgoing longwave radiation leaving the surface is approximated by a Brunt type formula : (Arnfield[7]) :

$$I = \sigma T^4 [1-(0;65+0.045\sqrt{e}) \times (1+a_i)] \qquad (6)$$

Here σ is the Stefan-Boltzmann constant, T is the surface temperature and a_i takes into account the effect of clouds on the infrared flux in a manner similar to equ.(5).

Finally, the ground heat flux G can be estimated from the surface air temperature T if we make the following assumptions.

Let us define the mean temperature $\tilde{T}_G$ in a ground layer of depth d by :

$$\tilde{T}_G(t) = \frac{1}{d} \int_0^d T_S(\zeta,t)d\zeta \tag{7}$$

where T_S is the soil temperature at depth ζ and time t. Then the heat balance of that ground slice may be expressed by :

$$\frac{G(0) - G(d)}{d} = - \rho_G \, C_G \, \frac{\partial \tilde{T}_G}{\partial t} \tag{8}$$

where G(d) = heat flux through depth d
   ρ_G = density of soil
   C_G = specific heat of soil

If we choose a depth d such as G(d) becomes small compared to surface flux G(0), then we can neglect it and

$$G(0) = -d \, \rho_G \, C_G \, \frac{\partial \tilde{T}_G}{\partial t} \tag{9}$$

The only data with which we can guess $\partial \tilde{T}_G/\partial t$ in our model is the air temperature T. We assume the relation between the two quantities is :

$$\frac{\partial \tilde{T}_G(t)}{\partial t} = \eta \, \frac{\partial T(t+\Delta t)}{\partial t} \tag{10}$$

allowing the two variations to be out of phase by Δt. This relation has been tested over many months of hourly data available and we found a best correlation for :

$$\eta \approx 0.35, \; \Delta t = 0 \; ; \; d \approx 0.30 \text{ m}$$

Returning now to equ.(1) all right hand side terms have been calculated, and the remaining energy fluxes must be divided between H and E, respectively sensible and latent turbulent heat fluxes. If we now assume that the Bowen Ration B (defined by B = H/E) may be considered as a constant for a given place and has a known value, the flux H can be deduced by :

$$H = \frac{R_S(1-A) - I - G}{1 + \frac{1}{B}} \tag{11}$$

Friction velocity

The friction velocity u_* can be calculated from the anemometer wind speed U_a and the heat flux H by using the surface layer similarity formulation (Businger[8]) if an approximate roughness length Z_0 can be attributed for the place under investigation.

If Z_a is the anemometer level (usually 10m), then by the similarity theory, we have :

$$u_* = \frac{k \, U_a}{(\log \frac{Z_a}{Z_0} - \psi_1)} \tag{12}$$

where $k = 0.35$ is Von Karman constant and ψ_1 is the integrated form of the surface layer universal function ϕ_m (cfr Businger[8]). This function ϕ depends on the Monin-Obukov length L defined by :

$$L = - \frac{T \, u_*^3 \, \rho \, c_p}{k \, g \, H} \tag{13}$$

(ρ is the air density, c_p the specific heat of air at constant pressure and g is the acceleration of gravity).

The neutral atmospheric boundary layer height Z_E is then :

$$Z_E = \frac{ku_*}{f} \qquad ; \qquad f = 2 \, \Omega \, \sin \phi, \text{ the Coriolis parameter}$$

and the global atmospheric boundary layer stability parameter μ can now be calculated as :

$$\mu = \frac{Z_E}{L} = - \frac{k^2 \, g \, H}{f \, T \, u_*^2 \, \rho \, c_p} \tag{14}$$

The K(Z) profile

Wippermann[3] devised a stationary barotropic planetary boundary layer model based on the Rossby similarity relationship in the Ekman layer. He finds that with his closure scheme, the non dimensional $\hat{K}(Z)$ profile is well approximated by a simple relation like :

$$\hat{K}(\hat{Z}) = \hat{Z} \exp (-C \, \hat{Z}^{0.764}) \tag{15}$$

where all $\hat{}$ symbols are dimensionless quantities defined by :

$$\hat{K} = \frac{K \, f}{k^2 \, u_*^2} \qquad \text{and} \qquad \hat{Z} = \frac{Z \, f}{ku_*}$$

The parameter C is dependent on the global stability state of the boundary layer. Wippermann gives the correspondance between and C and this table is reproduced here below :

μ	C	stability
+ 20	27.8	very stable
+ 10	14.8	stable
0	7.8	neutral
− 20	3.0	unstable
− 40	1.6	very unstable

We found a connexion between μ and C by fitting a second degree polynomial in μ with ln C, or :

$$C = \exp\ (0.264 + 0.0162\ \mu' + 0.000396\ \mu'^2) \tag{16}$$

where $\mu' = \mu + 50$.

Now μ is given by (14), C is found by (16) and K(Z) is given by the dimensional form of (15).

Atmospheric Boundary Layer Model (ABL)

In this section we will describe breefly the ABL model that is used to simulate the evolution of the boundary layer in speci-fied meteorological situations and also to give a "reference" diffusivity profile. The different kinds of such models that can be formed depend on the turbulence closure scheme adopted. In our case, the turbulent energy equation (also called 1.5 order closure) is thought to give a better physical simulation than simple first order closure, without having the complexity of higher order models[9,10,11].

Since the horizontal homogeneity is assumed, a one dimensional model (only in z) is used. Table I gives the main equations of the model namely those of the horizontal wind u and v, the poten-tial temperature θ, the specific humidity q and the specific tur-bulent energy b.

This model gives the vertical profiles of those quantities if initial profiles of wind, temperature and humidity are given as well as time evolution of temperature and humidity at ground level and at the upper boundary.

In order to test the efficiency of this model we have to com-pare the computed values of meteorological quantities with observed ones in an actual situation. We have chosen the day 33 of the Wangara Boundary Layer Experiment, often considered in the litte-rature as a test day for models ([1,2] and Clarke[12]).

The conditions observed during day 33 and the night from day 33 to day 34 are of particular interest because they are caracté-ristic of a clear sunny day with moderate wind (3 to 4 m/sec),

Table 1. Equations of the unidimensional and unsteady model of the atmospheric boundary layer.

EQUATIONS

$$\frac{\partial u}{\partial t} = f(v-v_g) + \frac{\partial}{\partial z}(K_m \frac{\partial u}{\partial z}) - w\frac{\partial u}{\partial z}$$

$$\frac{\partial v}{\partial t} = f(u_g-u) + \frac{\partial}{\partial z}(K_m \frac{\partial v}{\partial z}) - w\frac{\partial v}{\partial z}$$

$$\frac{\partial \theta}{\partial t} = \frac{\partial}{\partial z}(K_\theta \frac{\partial \theta}{\partial z}) - w\frac{\partial \theta}{\partial z} + R$$

$$\frac{\partial q}{\partial t} = \frac{\partial}{\partial z}(K_q \frac{\partial q}{\partial z}) - w\frac{\partial q}{\partial z}$$

$$\frac{\partial b}{\partial t} = -w\frac{\partial b}{\partial z} + K_m [(\frac{\partial u}{\partial z})^2 + (\frac{\partial v}{\partial z})^2 - 1,35 \frac{g}{\theta}(\frac{\partial \theta}{\partial z} - \gamma)] +$$

$$1,2 \frac{\partial}{\partial z}(K_m \frac{\partial b}{\partial z}) - \frac{(0,2b)^{3/2}}{1}$$

$$K_m = 1(0,2b)^{1/2} \qquad K_\theta = K_q = 1,35\ K_m$$

INDEPENDENT VARIABLES

t : time, in seconds
z : vertical coordinate, in meters

DEPENDENT VARIABLES

u,v : wind horizontal components, in meters per second
θ : air potential temperature, in degrees Kelvin
q : specific humidity, in kg per kg
b : turbulent energy, in Joule per kilogramme

PARAMETERS

u_g, v_g : geostrophic wind components, in meters per second
K_m, K_θ, K_q : turbulent diffusivities, in square meter per second
f : Coriolis parameter, in second minus one
w : wind vertical component, in meters per second
l : mixing length, in meter
g : gravitational constant, in meters per square second
γ : counter gradient heat flux term, in degrees Kelvin per meter
R : radiative heating or cooling rate, in degrees Kelvin per second

thus close to the free convection case. Our simulation begins at
6 a.m. of day 33 and lasts for 24 hours. The results are close
to reality for temperature profiles and wind profiles[13].

The figure 1 shows the evolution of potential temperature
profiles during the day every 3 hours (i.e. at 9, 12, 15 and 18),
computed by the model (Fig. 1a) and observed at the same time
(Fig. 1b). The mixing layer temperature and height are well reproduced during day hours. Computed and observed night time temperature profiles are similarly presented in the figure 2. During the
first part of the night, the model reproduces fairly well the
observed inversion formation. But early in the morning the measured inversion rising is not simulated by the computation. Probably,
radiative phenomena that are not included in the model play a important role and should be taken in consideration.

The figure 3 shows the computed diffusivity profiles generated
by the model during the day (Fig. 3a) and during the night (Fig.
3b). We notice the very high values of K(Z) produced during the
most convective hours of the day because of a high thermal production of turbulence. During the night, a first maximum near the
ground is due to mechanical friction of the air on the surface,
and is separated of the second maximum aloft by a region of very
small turbulence in the inversion.

Test Case

In December 1979, a campaign took place in Belgium and the
meteorological situation has been examined in detail. Profiles
of temperature and wind velocity have been obtained by tethered
balloon up to 500m above ground, and were completed by the aerological sounding of Uccle up to 2 000 m. These data were fed in
the ABL model described in the preceeding section which yielded
the K(Z) profile. The figure 4 shows the initial temperature
profile with a marked inversion at about 300m though the high wind
present (about 5-6 m s^{-1}). The resulting K(Z) profile is presented in figure 5 (dotted line) and is thus considered here as a
"reference" profile.

The synoptic observations of a nearby station were available
for the same period and were introduced in the simple model developed in section 2. The parameters used in the model are :
Z_0 = 0.05m, Bowen ratio B = 0.6 and albedo A = 0.2 (typical values
over land, mid-latitudes). This gives the K(Z) profile as shown
in figure 5 (solid line). The comparison of the two curves shows
a good agreement for the lower part of the profile and for the
magnitude and height of the maximum. Of course, the Wippermann
formulation cannot "feel" the presence of the strong inversion at
350m and also because of the exponential function, K(Z) decreases
more slowly than the one computed with the ABL model.

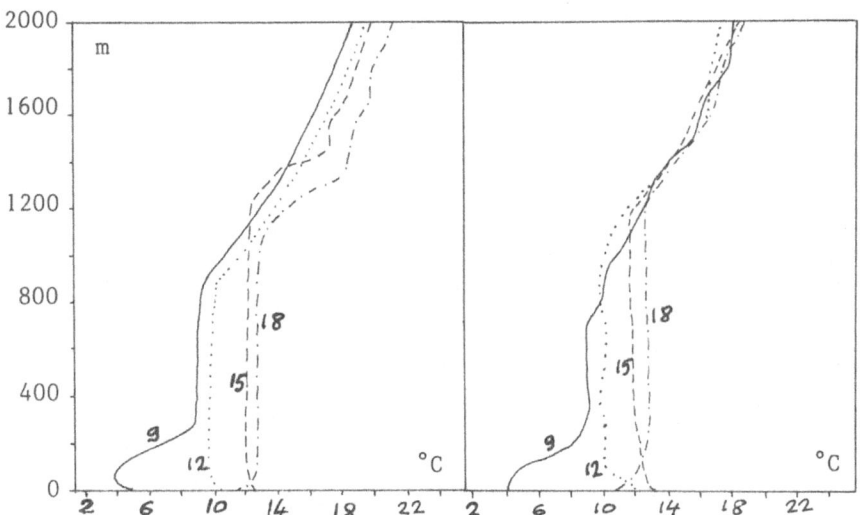

Figure 1.   Computed (left) and observed (right) pro-
            files of potential temperature during
            Wangara day 33.   (figures show time of day)

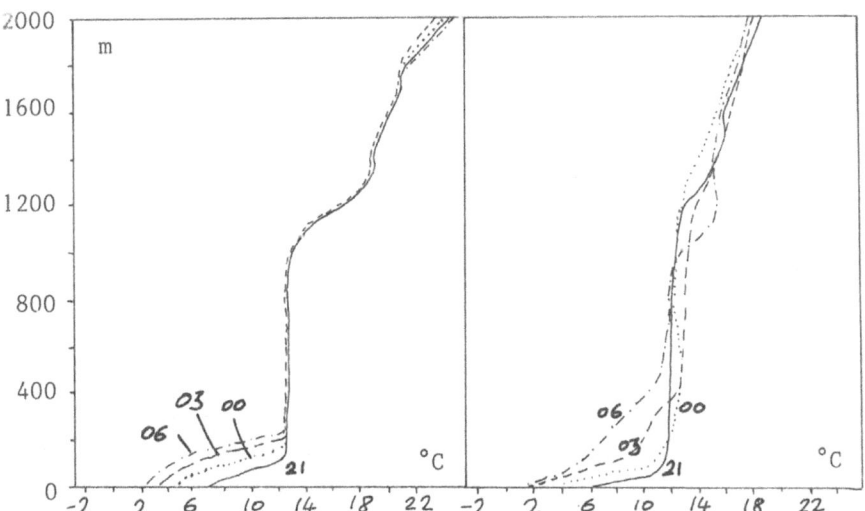

Figure 2.   Computed (left) and observed (right) profiles
            of potential temperature during Wangara
            night 33-34.

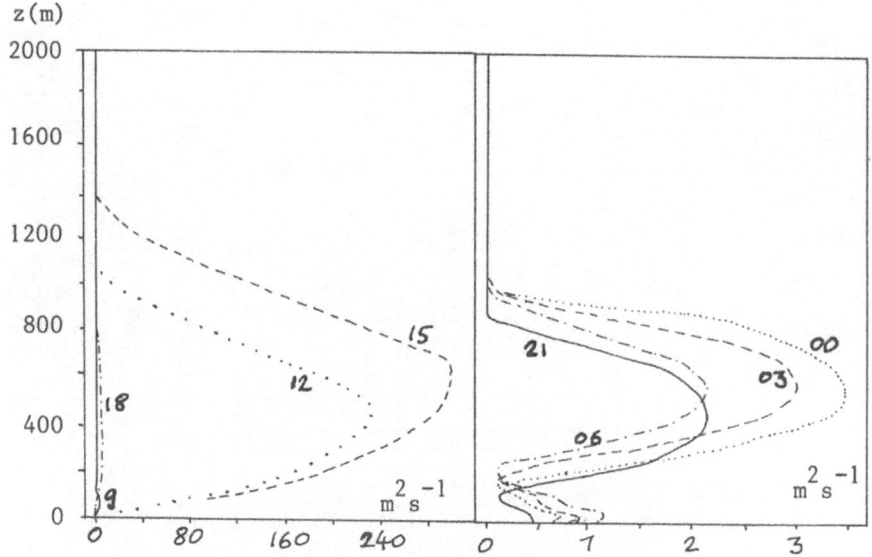

Figure 3.   Computed profiles of diffusivity K(Z)
            during day 33 (left) and night 33-34
            (right) with Wangara data.

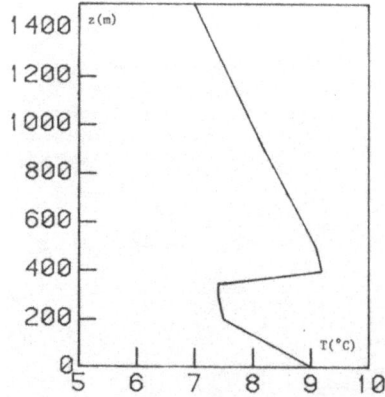

Figure 4.   Observed temperature profile in December
            1979 campaign (Belgium).

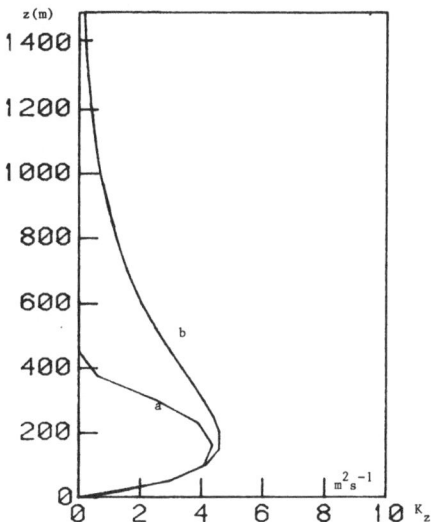

Figure 5. Computed K(Z) diffusivity profiles from data of Figure 4. (a) with ABL model, (b) with simple model.

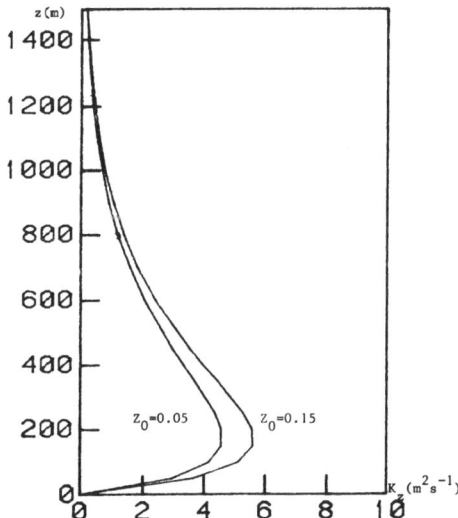

Figure 6. Sensitivity of the simple model of the Z_0 estimated.

Sensitivity

It was pointed out at the beginning of section 2 that the
most important parameters of the simple model were u_* and H, respectively the friction velocity and the turbulent heat flux.
The determination of u_* requires the prior knowledge of z_0. As
this parameter is not easy to determine accurately, (it can only
be guessed after the typical values listed in well known textbooks
(e.g. [14])) we must ensure that the model is not too sensitive to
say a factor 2 or 3 of error in the estimated z_0. The figure 6
shows the results of the test case when z_0 is 0.05 m (curve (a))
and z_0 = 0.15 m (curve (b)). We   note in the second case an
increase of about 20% in the K(Z) when z_0 is tripled and thus we
may conclude that the sensitivity to an error in z_0 is rather low,
which is an interesting feature.

The same procedure is used to test the response to an error
in the turbulent heat flux H. We must point out that in the model,
the calculated H is a result of a series of steps and assumptions
and therefore, a factor two of error in H is not unprobable. Here
we modify H by changing the Bowen ratio B. The figure 7 shows
three curves :

    7a is the same as fig. 5 with the Bowen ratio B = 0.6
    7b is the K(Z) when B = 0.3
    7c is the K(Z) when B = 1.2

It is obvious here that the sensitivity of the K(Z) to B (or H) is
much greater than in the case of Z_0, since a doubling of B gives
an variation of about 35% on the maximum value of K(Z).

Fortunately, all possible K(Z) values vary in a considerable
range if we take into account all possible meteorological situations, for example, a factor of 100 in K(Z) is not unusual between
stable and unstable situations. On the other hand, some similar
experiment on K-diffusion model showed a small-sensitivity of these
latter models to a reasonable variation in K(Z), for example, a 20%
forced variation in K(Z) produced a few percent variation in the
resulting concentrations.

Conclusions

Although simple to use, the present formulation of K(Z) seems
promising. An example of the use of it in a K-dispersion model
will be found in Demuth et al.[15]. More comparisons with available
data must be performed to ensure that on the average the model
works properly in all situations. However, we must keep in mind
that this simple formulation is unlikely to give good results in
situations where the hypotheses are in default, that is mainly :

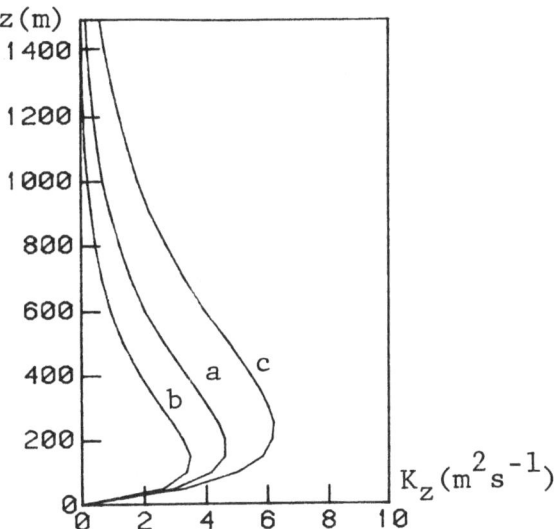

Figure 7. Sensitivity of the simple model to the
Bowen ratio estimated.
(a : B=0.6 ; b: B=0.3 ; c : B=1.2).

(a) in calm wind situation (day or night)
(b) in rapidly varying situations such as at sunset and
    sunrise, and in frontal situations.

Acknowledgements

This research is a part of the National R-D Programme on
Environment-Air, financed by the Services for Science Policy Programming, Belgium.

References

1. Pielke, R., Mahrer, Y., 1975 : Representation of a Heated PBL
   in Mesoscale models with coarse vertical resolution.
   J. Atm. Sci. 32, pp. 2288-2308.
2. Yamada, Mellor, 1975 : A simulation of the Wangara atmospheric
   BL data. J. Atm. Sci. 32, pp. 2309-2329.
3. Wippermann, F., 1972 : Universal profiles in the barotropic
   Boundary Layer. Contr. Atm. Phys. 45, pp. 148-163.
4. McDonald, J., 1960 : Direct absorption of solar radiation by
   atmospheric water vapor. J. of Met. 17, pp. 319.
5. Atwater, Brown, 1974 : Numerical computation of the latitudi-
   nal variation of solar radiation for an atmosphere of
   varying opacity. J. Appl. Met. 13, pp. 289-297.
6. Haurwitz, G., 1948 : Insolation in relation to cloud type.

J. Meteorl. 5, pp. 110-113.
7. Arnfield, A., 1979 : Evaluation of empirical expressions for
    the estimation of hourly and daily totals of atmospheric
    longwave emission under all sky conditions. Quat. J. Roy.
    Met. Soc. 105, pp. 1041-1053.
8. Businger, J., 1973 : Turbulent transfer in the atmospheric
    surface layer. Workshop on Micrometeorology. Am. Met.
    Soc., Haugen ed., pp. 77.
9. Delage, Y., 1974 : A numerical study of nocturnal BL. Quat.
    J. Roy. Met. Soc. 100, pp. 351-364.
10. Bodin, S., 1979 : A predictive numerical model of the Atm.
    BL based on the turbulent Energy Equation. SMMI Rapporter
    no RMK 13.
11. Yu, T., 1977 : A comparative study on parametrization of ver-
    tical turbulent exchange process. Monthly Weat. Rev. 105,
    pp. 57-66.
12. Clarke et al., 1971 : The Wangara experiment. Boundary Layer
    Data. Techn. Paper no 19 CSIRO, Australia.
13. Cravatte, M., Schayes, G., 1980 : Simulation des journées
    Wangara 33 et 34 par le modèle de CLA en énergie turbulente.
    Scientific Report 1980/5, Inst. Astr. Geophys., UCL.
14. Haltiner and Martin, 1957 : Dynamical and Physical meteorology.
    McGraw-Hill, pp. 229.
15. Demuth et al., 1980 : Numerical computation of high air pollu-
    tion levels. 11th ITM on air pollution modeling and its
    applications. Amsterdam, Nov. 1980.

DISCUSSION

J. KRETZSCHMAR   The specific site where
the december 1979 experiment took place is
characterized by the presence of grouped pine
trees ($\sim$20 m), some factories with high chimneys and some important excavation (up to 10 m
deep over 100 m x 200 m areas). Personally
I think that this is a (medium) rough terrain
with a Z_0 between 10 cm and 50 cm (see paper
by Wieringa, figure 3.) On what criteria is
your choice of 5 to 15 cm based ?

G. SCHAYES   These Z_0 were taken on
a subjective basis just to show that the K_z
calculated does not depend strongly of the Z_0
parameter. I agree that probably a larger Z_0
would be more realistic. This would not change
the conclusions.

J. KRETZSCHMAR   The simple method is
intended to be used with synoptic data. This
means that no exact information (e.g. by means

of balloon soundings) is available for the
height of the mixing layer. As the dec. 79
experiments took place in neutral conditions
one normally would say H $\tilde{}$ 1000 m.
This involves that in practice you would accept
the calculated K_z profile as being realistic.
What are the implications of this assumption ?

G. SCHAYES   If you are interested
in a short horizontal range of pollutant dis-
persion (about max. 5 km) only the lower part
of the K_z profile is relevant (up to the maximum). For the dispersion to larger downwind
distance, the presence of a capping inversion
becomes more important. In some cases, a formulation taking the inversion height Z_i into
account would be preferable (e.g. O'Brien
profile). But then an additional equation on
$z_i(t)$ is needed.

R. BERKOWICZ   The Bowen ratio varies
from a negative value to a large positive value.
It will influence the K-profile substantially.

G. SCHAYES   Yes this is the weakest
point of the model. An appropriate parameteri-
zation of B should be introduced to take the
night/day and soilwater availability conditions;
but which one ?

WIND VELOCITY VARIANCES IN THE ATMOSPHERIC BOUNDARY LAYER

R. Berkowicz[+] and L. P. Prahm

National Agency of Environmental Portection
Air Pollution Laboratory
DK-4000 Roskilde, Denmark

[+]Institute of Mathematical Statistics and Operations
Research, Technical University of Denmark,
2800 Lyngby, Denmark

INTRODUCTION

Velocity fluctuation in a turbulent flow is the major parameter determining the strength of the turbulent diffusion. It
is desirable to have evaluations of wind velocity variances at
particular site and time, when modelling the dispersion of pollutants in the atmosphere is attempted. However, measurements
are usually very difficult and costly. Therefore, it is of great
value if the wind velocity variances can be modelled relating
them to easily measured or predicted parameters. Numerical models based on solution of the boundary layer equations, e.g.
Deardorff (1972)[1], seem still to be too complicated. The behaviour of wind velocity variances in the surface layer is well
described in terms of the similarity theory. A review of the
semi-empirical expressions valid in the surface layer is given
by Dutton et al. (1979)[2]. In the case of the unstable stratification, the so-called free-convection surface-layer formulation
can be used (Wyngaard et al., 1971)[3].

We present here a simple model of wind velocity variances
in the entire atmospheric boundary layer in the case of unstable
stratification. The constant parameters of the model are estimated from the surface-layer data of Kansas (Izumi, 1971)[4] and
Minnesota (Izumi and Caughey, 1976)[5] experiments.

We will discuss here the vertical structure of the wind
velocity variances and horizontally homogeneous and stationary
turbulence is assumed.

The model results are compared with the water-tank laboratory experiments of Willis and Deardorff (1974)[6] and aircraft measurements of Lenschow[6].

ENERGY SPECTRUM AND VELOCITY VARIANCES FOR A HOMOGENEOUS TURBULENCE

Before proceeding with discussion of the vertical structure of the velocity variances, we recapitulate the well-known relations for a homogeneous turbulent flow.

We may express any instantaneous velocity distribution in terms of a Fourier integral.

$$V(\underline{r}) = \int a(\underline{k}) \, e^{i\underline{k}\underline{r}}d\underline{k} \tag{1}$$

where $\underline{r}$ is a space vector and the Fourier coefficients $a(\underline{k})$ are random functions of time. Assuming that the average value of V is zero, the variance becomes

$$\sigma_V^2 = \overline{V^2} = \iint \overline{a(\underline{k})a(\underline{k}')} \, e^{i(\underline{k}+\underline{k}')\underline{r}} \, d\underline{k}d\underline{k}' \tag{2}$$

where the bar means time averaging. Because of the homogeneity, the variance can not be a function of the space coordinates and therefore we must have

$$\overline{a(\underline{k})a(\underline{k}')} = E(\underline{k}) \; \delta(\underline{k}+\underline{k}') \tag{3}$$

Substituting (3) into (2), we obtain

$$\sigma^2_V = \int E(\underline{k}) \, d\underline{k} \tag{4}$$

$E(k)$ is here the energy spectrum of the turbulent velocity V.

The right hand side of (1) can be considered as a superposition of eddies of different size represented by the wave vector $\underline{k}$. The expression (3) means thus that only eddies represented by equally large wave vectors are correlated with each other and thus contribute to the variance of V.

VERTICAL STRUCTURE OF THE VARIANCE OF THE VERTICAL WIND VELOCITY COMPONENT

We now consider the vertical distribution of the vertical component w(z) of the turbulent wind field in the atmospheric boundary layer. An instantaneous realization of the w(z) distribution can also be expressed in terms of a Fourier integral. However, the existence of a solid boundary, the earth's surface, implies that w(z) must be zero at z = O. This again implies that w(z) must be expressed in terms of a sinus Fourier integral

$$w(z) = \int_0^\infty a(k) \sin kz\,dk \tag{5}$$

The variance of w is

$$\sigma_w^2(z) = \overline{w^2(z)} = \int_0^\infty \int_0^\infty \overline{a(k)\,a(k')} \sin(kz)\sin(k'z)dkdk' \tag{6}$$

In the previous section, we have shown, that in the case of a homogeneous turbulence, only eddies of the same size are correlated. We suppose that this is also the case for the vertical velocity fluctuation. We thus write

$$\sigma_w^2(z) = \int_0^\infty E_w(k) \sin^2(kz)dk \tag{7}$$

where

$$\overline{a(k)a(k')} = E_w(k)\,\delta\,(k-k') \tag{8}$$

Equation (8) expresses the principle of weak interaction of eddies of non-equal size.

The vertical structure of $\sigma_w^2(z)$ is determined through Eq.(7) when the spectral function $E_w(k)$ is given. In the following, we will discuss how $E_w(k)$ can be determined.

Eric Convective Boundary Layer

The Convective Boundary Layer

In the atmospheric boundary layer, there exist two mechanisms responsible for generation of the turbulence. The one is wind shear and the other is bouyancy. In the following discussion, we will use the terms shear- and heat-generated turbulence for the two types of production mechanisms. In spite of the fact that the two processes are strongly influenced by each other, we may consider their contribution to the turbulent energy separately.

In the case of an unstable stratified boundary layer, the heat-generated turbulence is usually much stronger than the shear-generated turbulence, creating the so-called convective boundary layer. The turbulence manifests itself by existence of thermal updraughts and downdraughts penetrating throughout the whole boundary layer (Kaimal et al., 1976)[7].

We denote the thickness of the convective boundary layer by h. The dominating eddy motion is also of the size h. The turbulence is generated by eddies of this size and is transferred to smaller eddies. From the theory of the homogeneous turbulence (Lumley and Panofsky, 1964)[8], it is known that when there is a sufficient separation between the energy production and energy dissipation range, the energy spectrum can be described by the inertial sub-range law, i.e. by the $k^{-5/3}$ dependence. The use of the $k^{-5/3}$ dependence for the whole energy spectrum requires furthermore that the band of the energy producing eddies is small compared to the width of the inertial sub-range band. This requirement is quite well-fulfilled for the heat-generated turbulence in a convective boundary layer. However, for the vertical component of the wind velocity fluctuations, a further modification must be made. When the unstable layer is capped by a stable layer, the turbulence is strongly attenuated above the inversion height. It is, therefore, to be expected that those Fourier components of the vertical velocity spectrum which have a maximum at the top of the boundary layer will have a smaller strength than the Fourier modes which exhibit a minimum at the top of the boundary layer. One can say that the spectrum of $w(z)$ exhibits a resonance phenomenon. We have decided to use the $\cos^2 kh$ function in order to account for this behaviour.

The heat-generated turbulent energy scales with the so-called <u>convective velocity scale</u> w_* (Deardorff, 1970)[9],

$$w_*^3 = \frac{g}{\bar{\theta}} \, \overline{(\theta'w')}_0 h \tag{9}$$

where $\overline{(\theta'w')}_0$ is the surface turbulent heat flux, g is the earth's acceleration, $\bar{\theta}$ is the average potential temperature of the boundary layer and θ' and w' are temperature and velocity fluctuations, respectively.

Further discussion of the appropriate velocity scales for the planetary boundary layer is given elsewhere (Berkowicz and Prahm, 1980)[10].

The height-dependent energy spectrum of the vertical wind velocity component, due to the heat-generated turbulence in a convective boundary layer applied in our model, is thus

$$\varepsilon_w(k,z) = E_w(k)\sin^2 kz = \begin{cases} \alpha w_*^2 (k/k_m)^{-5/3} k_m^{-1} \cos^2(kh)\sin^2(kz) \\ \qquad\qquad\qquad \text{for } k \geqslant k_m \\ \\ 0 \qquad\qquad\qquad \text{for } k < k_m \end{cases} \tag{10}$$

where $k_m = \pi/h$ is the wave number associated with the largest energy containing eddies and α is a proportionality constant to be determined.

The contribution to the energy spectrum due to the shear-generated turbulence will depend on the actual wind profile.

During the convective conditions, almost all the wind shear is confined to a very shallow region close to the ground[7]. There is therefore reason to believe that the shear-generated turbulence will in this case produce vertical velocity fluctuations which are almost constant with height.

The appropriate velocity scale for the shear-generated turbulence is discussed in ref. 10. However, we will here use the conventional scaling with the friction velocity u_* in order to make the comparison with other existing models easier. Considering the contribution from both the heat-generated and shear-generated turbulence, the model expression for the variance of the vertical wind velocity component is

$$\underline{\sigma_w^2(z)} = \alpha w_*^2 \int_1^\infty x^{-5/3} \cos^2(\pi x) \sin^2(\pi x \tfrac{z}{h}) dx + a_w u_*^2 \quad (11)$$

where $x = k/k_m$ and a_w is a constant.

The Surface Layer Relations

We will discuss here the behaviour of σ_w^2 for $z \ll h$, where the predictions of our model can be compared with the similarity theory predictions.

We make, in the integral of (11), the substitution

$$s = x\frac{z}{h} .$$

Now,

$$\sigma_w^2(z) = \alpha w_*^2 \left(\frac{z}{h}\right)^{2/3} \int_{z/h}^\infty s^{-5/3} \cos^2(\pi \frac{sh}{z}) \sin^2(\pi s) ds + a_w u_*^2 . \quad (12)$$

For $z \ll h$, the integral in (12) tends to a constant and we have

$$\sigma_w^2(z) = b_w w_*^2 \left(\frac{z}{h}\right)^{2/3} + a_w u_*^2 \qquad \text{for } z \ll h . \quad (13)$$

Dividing both sides of (13) by u_*^2, we obtain

$$\sigma_w^2 / u_*^2 = (b_w/\kappa^{2/3})\left(\frac{z}{-L}\right)^{2/3} + a_w \quad (14)$$

where L is the Monin-Obukhov length and κ is the von Karman constant (k = 0.35). We use here the relationship

$$\frac{w_*^2}{u_*^2} = \left(\frac{h}{-\kappa L}\right)^{2/3} \quad (15)$$

Equation (14) confirms that the present model gives results in

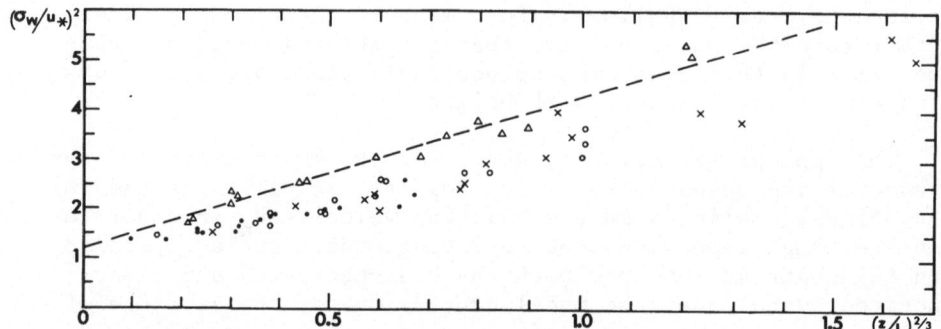

Figure 1: Variance of the vertical wind velocity fluctuations in the
surface layer.
Data from Kansas: • (z= 5.66m); ○ (z=11.31m); x (z=22.63m).
Data from Minnesota: Δ (z=4m); ∇ (z=32m). z is the height
above the ground. The dashed line is the fit to the Minnesota data.

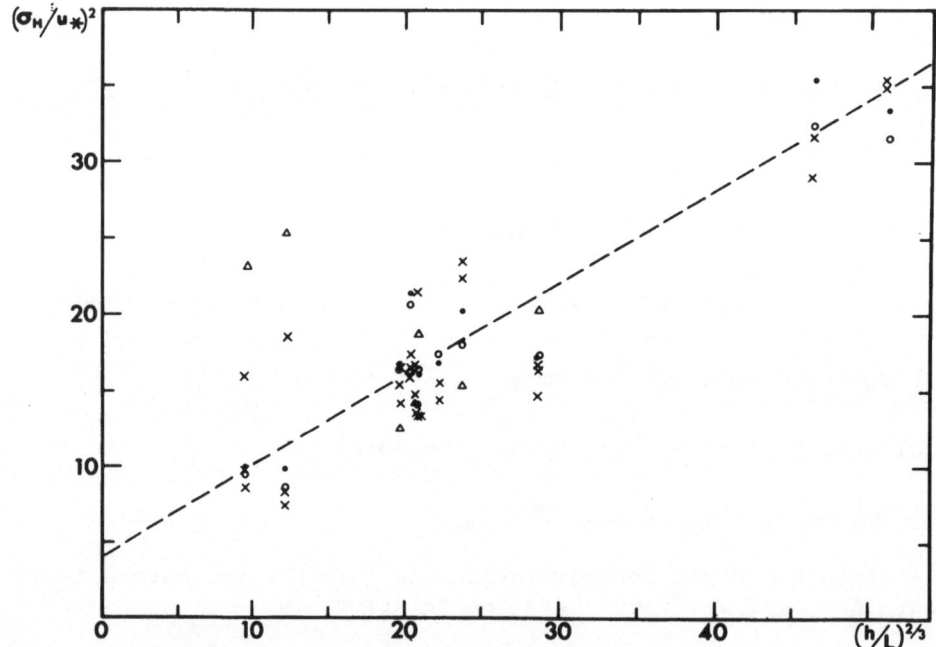

Figure 2: Variance of the horizontal wind velocity fluctuations in
the surface layer.
Data from Minnesota: v-component: • (z=4m); ○ (z=32m);
Δ (z=61m); u-component: x (z=4m, 32m, 61m).

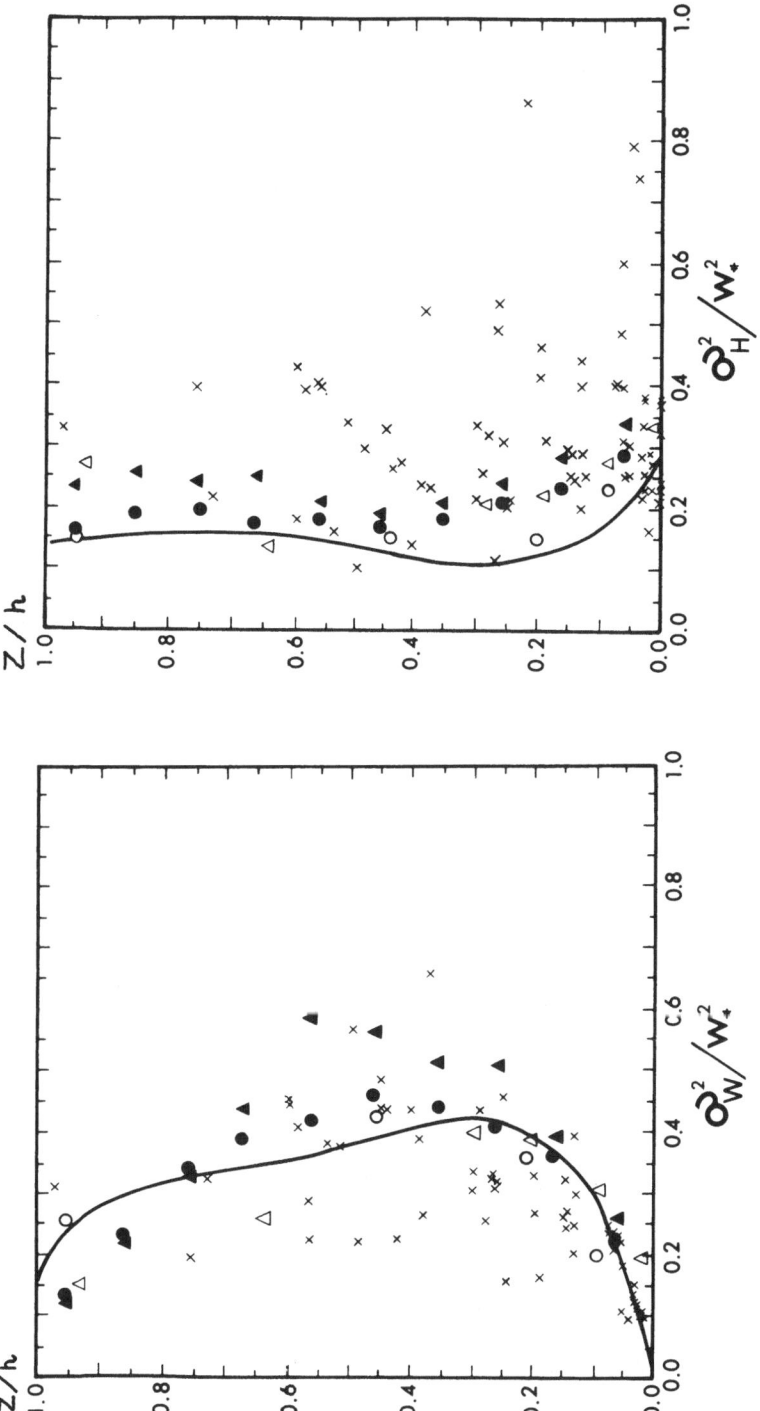

Fig. 3: The normalisation variance of the vertical velocity fluctuations in the atmospheric boundary layer. x: Minesota experiments; ● and ▲: water-tank experiments; ∘ and △: aircraft measurements. The solid line represents the theoretical expression from the present study.

Fig. 4: The normalized variance of the horizontal wind velocity fluctuations in the atmospheric boundary layer. All the symbols are as in Fig. 3.

accordance with the similarity theory in the surface layer. Panofsky et al. (1977)[11] proposed an expression identical to (14)
as an empirical expression. Here, however, we have derived it
from eq. (12).

Priestley (1954)[12] has predicted the $(z/L)^{2/3}$ -behaviour of
$(\sigma_w/u_*)^2$ in the case of free convection conditions (the wind-
shear-generated turbulence can be neglected) by dimensional
reasoning. Wyngaard et al. (1971)[3] verified this relationship on
the basis of experimental data. From eqs. (11) and (12), it
appears that the $z^{2/3}$-relationship is connected with the $k^{-5/3}$ -
behaviour of the energy spectrum.

VERTICAL STRUCTURE OF THE VARIANCE OF THE HORIZONTAL WIND VELOCITY COMPONENT

The vertical structure of the variance of the horizontal
component of the wind velocity fluctuations can be modelled in a
way similar to that described in the case of the vertical component. Here, however, instead of using a sine Fourier representation, we have to use a cosine representation because the
horizontal component of the velocity fluctuation has a maximum
at the ground. We have thus

$$\sigma_H^2 (z) = \int_0^\infty E_H(k)\cos^2(kz)dk \qquad (16)$$

Another significant difference is that there is no reason
to expect a resonance phenomenon with the thickness of the
boundary layer and therefore no $\cos^2 kh$ term is present in the
expression for $E_H(k)$. The energy spectrum used here is

$$\varepsilon_H(k,z) =$$

$$E_H(k)\cos^2 kz = \begin{cases} \beta w_*^2 (k/k_m)^{-5/3} k_m^{-1} \cos^2 kz & \text{for } k \geqslant k_m \\ \\ 0 & \text{for } k < k_m \end{cases} \qquad (17)$$

We again assume a constant contribution from the shear-generated
turbulence and the expression for the variance of the horizontal
component of the wind velocity fluctuations is

$$\sigma_H^2 (z) = \beta w_*^2 \int_0^\infty x^{-5/3} \cos^2(x\frac{z}{h})dx + a_H u_*^2 \qquad (18)$$

where β and a_H are constants and $x = k/k_m$.

The Surface Layer Relations

For z/h approaching zero, the integral in (18) tends to a
constant value and we have

$$\sigma_H^2 (z) = b_H w_*^2 + a_H u_*^2 \qquad\qquad \text{for } z << h \qquad (19)$$

where $b_H = 3/2\ \beta$ is a constant.

Dividing both sides of (19) by u_*^2, we obtain

$$\sigma_H^2 /u_*^2 = (b_H/\kappa^{2/3})(\frac{h}{-L})^{2/3} + a_H . \qquad (20)$$

It was already recognized by Lumley and Panofsky (1964)[8] the surface layer, σ_u^2 /u_*^2 scales with (h/-L) and not with (z/-L). Panofsky et al. (1977)[11] proposed an equation similar to (20) but on a basis of semi-empirical arguments. We have derived eq. (20) from eq. (18).

DETERMINATION OF THE MODEL PARAMETERS

Comparing eqs. (14) and (20) with the experimental data in the surface layer, the constants b_w, a_w, b_H, and a_H can be determined. We use here Kansas[4] and Minnesota[5] experiments because they contain both the necessary turbulent parameters. However, only Minnesota data are used for the horizontal component because the boundary layer height was not measured during the Kansas experiment.

Fig. 1 shows σ_w^2 /u_*^2 as function of (z/-L)$^{2/3}$. The dashed line is the best fit-by-eye to the Minnesota data. The Kansas data indicate somewhat smaller slope, but about the same intercept. We obtain the following values for a_w and b_w:

$$\begin{aligned} a_w &= 1.20 \\ b_w &= 1.54 \end{aligned} \qquad\qquad (21)$$

In Fig. 2, $\sigma_{v,u}^2 /u_*^2$ is plotted against (h/-L)$^{2/3}$. Here, σ_v^2 and σ_u^2 are variances of the two horizontal components measured at Minnesota. No significant difference can be seen between these two components. The large scatter of the data prevents, however, an exact analysis. Following estimates of a_H and b_H are made :

$$\begin{aligned} a_H &= 4.0 \\ b_H &= 0.3. \end{aligned} \qquad\qquad (22)$$

The values of   and   are

$$\begin{aligned} \alpha &= 0.9 \\ \beta &= 0.2 \end{aligned}$$

COMPARISON WITH ATMOSPHERIC BOUNDARY LAYER DATA

The model results are compared with the experimental data obtained at Minnesota, Willis and Deardorff's water-tank experiments and Lenschow's aircraft experiments. According to our model, the only height variation of the wind variances is due to the convective term. The shear term makes a constant contribution. We subtract, therefore, from the Minnesota and aircraft data the value of $1.2(u_*/w_*)^2$ in the case of the vertical component and $4.0 (u_*/w_*)^2$ in the case of the horizontal component. The relevant boundary layer parameters for the aircraft measurements are from Lenschow (1974)[13]. The water-tank experiment is made without wind shear and therefore no correction is necessary. In the case of the vertical variance, the contribution by the shear term is usually very small compared to the total variance, apart from the points close to the earth boundary. In the case of the horizontal component, both the convective and the shear terms are of the same order of magnitude.

The model results and the experimental data for the vertical variance are shown in Fig. 3., while in Fig. 4, the horizontal component is presented.

DISCUSSION

It appears from Figs. 3 and 4 that our model is able to predict the essential features of the vertical structure of the variances of the wind velocity fluctuations in convective conditions. The model predictions show that the vertical variance exhibits a rather broad maximum at $z/h \sim 0.3$. The water-tank experiments indicate a maximum at $z/h \sim 0.5$ but the numerical results of Deardorff[9] also suggest a maximum at $z/h \sim 0.3$. The relatively large scatter of the Minnesota data makes a quantitative comparison rather difficult. Only few measurements are made above $z/h = 0.6$.

In the case of the horizontal component, the model predicts a minimum at $z/h \sim 0.3$. and a very shallow maximum at $z/h \sim 0.8$. The water-tank results exhibit a similar behaviour, but the minimum is here again situated higher, i.e. at $z/h \sim 0.5$. It is interesting to note that the aircraft data, after being corrected for the shear contribution, coincide now quite well with both the model and water-tank data. Willis and Deardorff[6] explain the difference between the aircraft measurements and the water-tank results as due to large scale motions present in the atmosphere but absent in the laboratory model. In view of our results, it seems however that the shear-generated turbulence alone can account for the difference.

CONCLUSIONS

A simple model of the vertical structure of wind velocity variances in the boundary layer is developed. The model is based on spectral (Fourier) representation of the velocity fluctuations. The model gives results in accordance with the similarity theory in the surface layer. Comparison with the existing boundary layer data is also satisfactory. At the present stage, only the unstable conditions can be accounted for by the model. The contribution to the turbulence by heat and wind shear is treated separately. The shear-generated turbulence is described by a constant term, but in the case of neutral and stable conditions, a more realistic treatment is necessary for this term.

REFERENCES

1. J. W. Deardorff, 1972 : Numerical investigation of neutral and unstable planetary boundary layers. J. Atmos. Sci., 29, 91-115.

2. J. A. Dutton, H. A. Panofsky, D. Larko, H. N. Shirer, G. Stone and M. Vilardo, 1979 : Statistics of wind fluctuations over complex terrain. Final report. Department of Meteorology, Pennsylvania State University, University Park, PA 16802.

3. J. C. Wyngaard, O. R. Coté and Y. Izumi, 1971 : Local free convection similarity, and the budgets of shear stress and heat flux. J. Atmos. Sci., 28, 1171-1182.

4. Y. Izumi, 1971 : Kansas 1968 Field Program Data Report. AFCRL-72-0041, ERP No. 379, Air Force Cambridge Research Laboratories, Hanscom AFB, Mass., USA.

5. Y. Izumi and J. S. Caughey, 1976 : Minnesota 1973 Atmospheric Boundary Layer Experiment Data Report. AFCRL-TR-76-0038, ERP No. 547, Air Force Cambridge Research Laboratories, Hanscom AFB, Mass., USA.

6. G. E. Willis and J. W. Deardorff, 1974 : A laboratory model of the unstable planetary boundary layer. J. Atmos. Sci., 31, 1297-1307.

7. J. C. Kaimal, J. C. Wyngaard, D. A. Haugen, O. R. Coté, Y. Izumi, S. J. Caughey and C. J. Readings, 1976 : Turbulence structure in the convective boundary layer. J. Atmos. Sci., 33, 2152-2169.

8. J. Lumley and H. A. Panofsky, 1964 : The structure of Atmospheric Turbulence. Interscience, New York, 239 pp.

9. J. W. Deardorff, 1970 : Convective velocity and temperature scales for the unstable planetary boundary layer and for Rayleigh Convection. J. Atmos. Sci., 27, 1211-1213.

10. R. Berkowicz and L.P. Prahm, 1980 : Note on turbulent scaling parameters for the planetary boundary layer. (Submitted for publication to Boundary-Layer Meteorology).

11. H. A. Panofsky, H. Tennekes, D. H. Lenschow and J. C. Wyngaard, 1977 :The characteristics of turbulent velocity components in the surface layer under convection conditions. Boundary-Layer Met., 11, 355-361.

12. C. H. B. Priestley, 1954 : Convection from a large horizontal surface. Australian J. Phys., 7, 176-201.

13. D. H. Lenschow, 1974 : Model of the height variation of the turbulence kinetic energy budget in the unstable planetary boundary layer. J. Atmos. Sci., 31, 465-474.

DISCUSSION

J. D. REID   Where the scales which are responsible for production included in those that contribute to the variance ? Why is it reasonable to assume the − 5/3-1aw for all scales ? Is it true, that the contribution of energy at scales ≥ 100 m to the local variance is very small ?

R. BERKOWICZ, L. P. PRAHM   It makes very little difference to σ_w profile what is the exact shape of the spectrum at large wave-length.

ESTIMATION OF MESOSCALE AND LOCAL-SCALE ROUGHNESS

FOR ATMOSPHERIC TRANSPORT MODELING

Jon Wieringa

Royal Netherlands Meteorological Institute
De Bilt
The Netherlands

INTRODUCTION

The rough surface of the earth acts as a sink for horizontal mean flow and as a generator of turbulence. Both for vertical turbulent diffusion calculation and for estimation of horizontal transport some specification of the degree of roughness of the underlying surface is necessary.

This paper deals with the choice of roughness specification parameter: drag coefficient, roughness length or power law exponent. Additionally a discussion is given of the effectiveness and usefulness of various methods to determine roughness parameters: turbulence measurements, wind profile measurements, and visual terrain evaluation. The discussion is restricted to generally level terrain at scales varying from local surface-layer description below 100 m height over a few kilometers horizontal distance to mesoscale description of the planetary boundary layer (PBL) below ∿1 km height over maybe 50 km horizontal distance. Corresponding time scales range from 10 minutes to a few hours.

A major conclusion is, that for roughness evaluation in non-homogeneous terrain wind profile measurements are less suitable than gustiness determination, particularly at mesoscale. Roughness classifications are outlined for the assessment and description of terrain roughness.

ROUGHNESS SPECIFICATION ALTERNATIVES

The major effect of local-scale roughness is the transformation of horizontal motion into turbulence. The effectiveness of

this process is a function of thermal stratification: in a stable
atmosphere the vertical exchange of motion is suppressed to the
extent that roughness specification has little practical use for
the flow above a few meters height. In adiabatic and unstable
stratification roughness effects will be important in the lower
part of the PBL. As a reference value of roughness parameters
their magnitude in near-neutral stratification is used. In spite
of many published remarks to the contrary, near-neutral stratification does occur frequently in the lowest 50 to 100 m, where for
surface wind velocities u exceeding ∿6 m/s the mechanical turbulence dominates the turbulence of thermal origin quite markedly.
On the other hand, above the level z ∿100 m the buoyancy and earth
rotation effects tend to dominate the PBL behavior.

Geostrophic drag coefficient

A possible descriptor of the surface roughness is the magnitude of the turbulence generated in the surface layer by a given
horizontal pressure gradient, the latter being specified as
geostrophic wind velocity G. The turbulence level is generally
specified by the friction velocity $u_* \equiv \sqrt{\tau/\rho}$, where τ is the turbulent stress and ρ the atmospheric density. However, the actual value
of the geostrophic drag coefficient u_*/G is virtually always strongly influenced by the presence of horizontal temperature gradients
and non-adiabatic vertical temperature gradients somewhere in the
PBL. The adiabatic barotropic situation, which would be required to
make u_*/G solely determined by the roughness, hardly ever occurs
(McBean, 1979). In fact, for the similarity-theory relation

$$C_G \equiv u_*/G = \kappa \left(\left(\ln \frac{h}{z_0} \right) - A \right)^2 + B^2 \right)^{-\frac{1}{2}}$$

(where h is the PBL height, κ the Kármán constant and z_0 the surface
roughness length) it has so far been impossible to determine the
values of the empirical constants A and B with even moderate accuracy for a stationary, homogeneous, barotropic, non-equatorial and
fully adiabatic PBL, because the situation has not occurred in experimental practice. Similarly, the angle between geostrophic and
surface wind directions is highly variable and seldom takes the
value which would be appropriate in an adiabatic Ekman layer over
terrain of given roughness (Marshall, 1954; Manier and Weingärtner,
1978) - even in strong-wind conditions the observed standard
deviation of that angle is ∿10°.

Therefore the geostrophic drag coefficient and the surface
cross-isobar angle are unsuitable for initial specification of
surface roughness. When geostrophic drag has to be determined, the
proper approach is to determine a roughness parameter for the
surface layer first.

Roughness length and surface drag coefficient

In the surface layer a steady-valued roughness-dependent wind structure description can be obtained for the inertial sublayer (Tennekes, 1973). This is the layer where the velocity-defect law of wind variation valid in the upper part of the PBL (above ~ 0.1 h) matches with the wall law which is appropriate to the lowest few meters ($z < 20\ z_o$) of the atmosphere. Over the full height of this intermediate layer (typically from 6 to 60 m height) a logarithmic wind profile

$$u_{z_2}/u_{z_1} = \ln(z_2/z_o)/\ln(z_1/z_o) \tag{1}$$

applies in near-neutral conditions (i.e. in strong winds), even though the stress τ decreases with height in the inertial sublayer.

The roughness length z_o determined over this layer is an exclusively terrain-dependent parameter. The single-level formulation of the logarithmic profile law is

$$u_z/u_* = (^1/_\kappa)\ \ln(z/z_o)$$

This shows, that the surface drag coefficient $C_{d(z)} \equiv (u_*/u_z)^2_{\text{neutral}}$ also must be an exclusively terrain-dependent parameter when defined for near-neutral stability at a predetermined reference level, in view of the change of u_* with height. The standard WMO wind measuring height of 10 m is, luckily, an excellent choice of reference level, just in the lower part of the inertial sublayer in most terrain situations. Through the relation

$$C_{d(10)} = (\kappa/\ln(10/z_o))^2 \tag{2}$$

the surface drag coefficient and the roughness length can be considered as equivalent alternative terrain roughness descriptions.

The reliability of (2) has recently been strengthened through research on the value of the Kármán constant κ, which was disputed to range between 0.35 and 0.41. Primarily Garratt (1977), comparing 1200 wind profiles and 369 eddy correlation stresses measured over the sea by a dozen different reliable investigators, found $\overline{\kappa} = 0.41 \pm 0.025$. Additionally, Wieringa (1980a) found that the reputable anomalous value $\kappa = 0.35$ obtained in the Kansas project (Businger et al., 1971) can probably be traced back to a mast influence error, and that after its correction the Kansas data give $\kappa = 0.41$ as well.

Therefore the exchangeability of $C_{d(10)}$ and z_o as roughness parameters is assured. In certain cases, where determination of z_o from profiles is less viable and $C_d(10)$ can still be obtained, the roughness will still be specified in terms of z_o by way of (2) for reasons of consistency.

Power "law" exponent

The logarithmic wind profile (1) is a theoretically and experimentally well-founded description of wind behavior in the near-neutral surface layer. However, traditionally often the wind profile is fitted to an exponential curve:

$$u_{z_2}/u_{z_1} = (z_2/z_1)^p \qquad (3)$$

This approach may sometimes have mathematical advantages, but it has no physical relevance (which is the reason why quotation marks seem appropriate to use in discussing power "laws"). Various atmospheric structure parameters such as turbulence level and stress, which are directly determinable from e.g. z_o, must in the exponential methodology be obtained from lengthy specifications in tables or nomograms (e.g. Irwin, 1979). It is therefore often an illusion that using (3) leads to simpler calculations than using (1), even in the diabatic case not discussed here. Another ancient argument for preferring (3) over (1) was the notion, that the logarithmic profile was restricted to a very shallow "constant-stress layer". However, this notion has been proven invalid (Tennekes, 1973). In experiments the application of a logarithmic profile with a properly chosen z_o-value will give more realistic estimates of high-level winds than exponential extrapolations.

Finally, specification of roughness (as apparent from the wind profile slope) in terms of power "law" exponents leads to large uncertainties, because the fitting of an exponential curve to a profile curve depends strongly on the height interval considered. For example, an exponent value $p = 0.2$ for an adiabatic profile between 2 m and 20 m corresponds to $z_o = 0.04$ m, but if the height interval is 10 m - 300 m the corresponding z_o-value is 0.30 m. Specification of the geometrical mean height $\sqrt{z_1 z_2}$ is a tolerable, though not exact, way to remove this uncertainty. But in practice publications on the power law never specify height ranges of application (e.g. Touma, 1977). An accurate evaluation of exponentially specified relations is therefore often impossible.

For these reasons the use of double-logarithmic graph paper in wind structure evaluations ought to be rejected in favour of single-logarithmic paper. Only by a logarithmic approach an internally consistent representation of the flow structure is achieved. Those who are addicted to the power "law" can always derive consistent expo-

nent values, for any desired height range, from given roughness
lengths or surface drag coefficients.

ROUGHNESS PARAMETER DETERMINATION PROBLEMS

    The effect of surface roughness is the slowing down of the
near-surface flow and the transformation of its kinetic energy into
turbulent energy. This gives the following possibilities for
quantization of roughness:
(1) Roughness $\sim \partial \overline{u}/\partial z$, the local-scale decrease of flow speed with
    decreasing height (wind profile evaluation).
(2) Roughness $\sim \partial \overline{u}/\partial x$, the mesoscale decrease of flow speed as the
    flow advances over rough terrain - e.g. the fact that in a
    homogeneous pressure gradient along a coast the winds over
    water will exceed those over land by $\sim 20\%$.
(3) Roughness $\sim$ relative turbulence level. The choices are:
    (a) Eddy correlation measurement of the local stress
        $\tau \equiv \rho \, u_*^2$.
    (b) Measurement of the standard deviation of the horizontal
        wind velocity, σ_u, or its estimation from gust data.
    (c) Measurement of the turbulent dissipation ε (Khalsa and
        Businger, 1977).
Eddy correlation and dissipation measurements require fast-
responding sensors and extensive data handling equipment. For
stress determination one also requires a vertical velocity sensor,
and alignment of the sensing direction within half a degree to a
well-defined horizontal flow direction. The last requirement poses
problems in nonhomogeneous terrain, and the instrumentation effort
restricts this type of approach to research projects.

    Measurements of the gustiness level $\sigma_u/\overline{u}$ are easier to achieve,
and should become more and more attainable as signal processing
techniques become cheaper and more reliable. A poor-man's version
of this approach is the evaluation of maximum recorded gusts u_{mx}
since the gust factor $G \equiv u_{mx}/\overline{u}$ is statistically related to $\sigma_u/\overline{u}$
(Wieringa, 1973 and 1977). For ordinary synoptic wind stations the
surrounding roughness can thus be evaluated objectively, provided
the response of the instruments is known.

    Wind profile measurements, though simple in principle, have
some severe experimental problems, like the necessity to measure
well away from the flow-distorting profile mast, and the requirement for high calibration accuracy. If these conditions are met,
the profile evaluation in terms of z_o is straightforward and gives
reliable results, provided the conditions are stationary (Stearns,
1971). However, in nonhomogeneous terrain the application of (1)
may no longer be justified, because the adjustment of profiles to
roughness changes is much faster than the turbulence adjustment.
The equilibrium between profiles and turbulence downwind of a

roughness change is only regained at a distance x larger than 100
times the height of the largest roughness elements (Brooks, 1961;
Jackson, 1976). Closer to the roughness jump the profile still will
be non-logarithmic, sometimes S-shaped and sometimes bent, as seen
e.g. in the model results of Rao et al. (1974) and in the experiments of Peterson et al. (1976). Therefore z_o-values derived from
such profiles will vary with the fetch and the evaluated height
interval.

Cabauw wind profiles

The Cabauw 200 m mast (Driedonks et al., 1978) is during
easterly winds only 300 m downwind of a 10 m high orchard, and for
that reason the near-adiabatic wind profile shows profile slope
variations with height (Figure 1). In the lowest few meters the

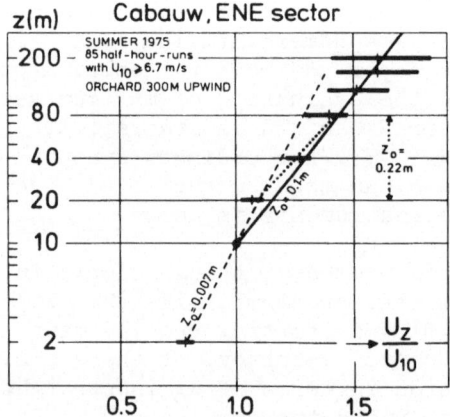

Figure 1.   Wind profile measurements for the inhomogeneous sector
           East of the Cabauw mast. Measurements are normalized
           by division through the simultaneous wind speed at 10 m.
           Variation bars indicate $\pm\,\sigma$.

profile is already adjusted to the grassy open space surrounding
the mast, and therefore evaluation of the profile below 20 m leads
to z_0-values $\lesssim 0.01$ m. These values are much too low to account for
the orchard presence. On the other hand, from the orchard edge a
plume of increased turbulence extends downwind and upward (Pasquill,
1972), causing the wind profile between 20 m and 80 m to show the
rapid increase with height which is typical of large roughness
$(z_{0(20-80)} = 0.22$ m).

Noteworthy is that the evaluation of gustiness of the wind
measured at 10 m height gives $z_{0(G10)} = 0.23$ m for easterly winds
(Wieringa, 1980b). The G-derived roughness length thus is a suitable
parameter to relate u_{10} to the wind speed at the top of the surface
layer, 60-80 m. Incidentally, at 80 m G hardly varies with azimuth,
showing that at that level individual roughness plumes are no longer
noticeable and only the mesoscale roughness counts (see Munn and
Reimer (1968) for a similar observation).

Above 80 m the profile can well be represented by $z_0 = 0.10$ m -
0.15 m until the 160 m level, indicating that in these strong-wind
conditions the properties of an inertial sublayer are retained up
to high levels, at least as far as the logarithmic behavior of the
profile is concerned. At 80 m, measured values of σ_u have already
decreased to 75% of simultaneous σ_u-values at 10 m, and at 200 m
to 65%. This agrees with the conclusion of Tennekes (1973), that
the accuracy of the logarithmic law is not dependent on the accuracy
of the assumption that stress should be height-independent in the
inertial sublayer.

Independently of the strong-wind profile measurements at
Cabauw discussed above, Nieuwstadt (1978) compared eddy correlation
measurements for easterly winds in diabatic conditions with fluxes
evaluated from simultaneous wind and temperature profiles. He found
good agreement between the two, using $z_0 = 0.3$ m as evaluated from
gustiness measurements for that sector at that time (with more
trees). Next to this, Van Ulden en Holtslag (1980) modelled
satisfactorily a full year of Cabauw wind profiles from all directions, using only the geostrophic wind and $z_0 = 0.15$ m, which is the
all-azimuth averaged value of the roughness lengths derived from
the gustiness at 10 m height. Generally at Cabauw more consistent
boundary layer structure descriptions are obtained through using
gustiness-derived roughness $z_0 \geqslant 0.1$ m than through using the much
smaller z_0-values resulting from low-level wind profile measurements.

Similar conclusions are presented by Teunissen (1979) from a
similar experimental situation – a mast on an airport downwind of
a suburb – where also the roughness lengths derived from a 33 m-mast
profile were $\leqslant 0.01$ m, obviously unsuitable for application to the

available mesoscale data. Simultaneously Teunissen observed
$\sigma_u/\overline{u} = 0.23$ along the mast, giving $z_o \sim 0.25$ m which agrees better
with the data.

Effective roughness length

For mesoscale modeling of boundary layer fluxes over a
heterogenous region it was proposed by Fiedler and Panofsky (1972)
to define an effective roughness length, being the z_o-value which
homogeneous terrain would need to have in order to produce the
same space-average downward flux of momentum near the ground as is
observed over the actual heterogeneous terrain.

For the moderately heterogeneous Cabauw region the gustiness-
derived roughness parameters have proved themselves consistently
superior to profile-derived roughness lengths for PBL modeling
purposes over a layer of 200 m height. The properties of this layer
are determined by the upwind terrain over $\sim$10 km distance. The
reason for this may be the fact, that turbulence intensity adjusts
much slower to roughness changes than profiles. Therefore observed
values of $\sigma_u/\overline{u}$ are integrated results of turbulent exchange over a
rather large upwind area, and therefore rather likely to approach
mesoscale average values. In this respect it is interesting, that
longitudinal velocity spectra do not follow strictly the Monin-
Obukhov laws of local similarity, but that their spectral shape is
dependent on location and only moderately height-dependent
(Panofsky, 1974).

For similar heterogeneous rural terrain Wamser and Müller
(1977) found an effective roughness length of 0.27 to 0.36 m, while
Guyot and Seguin (1978) found regional roughness lengths of similar
or slightly larger magnitude for a shelterbelt-infested area. From
the comparability of the gustiness-derived z_o-values at Cabauw and
from Teunissen's $\sigma_u/\overline{u}$-values it may be concluded, that low-level
gustiness measurements in the surface layer furnish an suitable
estimate of the effective mesoscale roughness length.

On the other hand, low-level wind profiles only furnish
consistent roughness length values if over an upwind distance of
$\sim$100 times the mast height the terrain is sufficiently homogeneous
that equilibrium between profile and turbulence is established. If
this conditions is not approximately met, the application of (1)
to profile data is a purely algebraic exercise, the result of which
can only be accidentally relevant to local turbulent exchange.
Since therefore measuring sites in a heterogeneous region are
smoother than their surroundings, it is not surprising that the
roughness value required for regional mesoscale modeling is larger
than typical site profile roughness.

Unfortunately, tables of "typical" roughnesses used by PBL
modellers like e.g. Goodin and McRae (1980) are usually compiled
mainly from published wind profile experiments. Such tables usually
feature hardly any z_0-values between 0.1 m and 1 m, and cannot be
expected to give a suitable description of surface roughness for
mesoscale model application.

AERODYNAMIC DRAG IN VARIOUS TERRAIN TYPES

In order to structurize the discussion on terain, its general
features are systematized here in a manner which originated in
pipe flow investigations and has since been used by Perrier et al.
(1972) to categorize turbulent flow above agricultural crops. The
following flow categories are given:
(A) <u>Smooth</u> turbulent flow occurs over a flat rough surface, without
obstacles that are sufficiently prominent to produce wakes.
(B) <u>Semi-smooth</u> turbulent flow occurs over a surface with isolated
obstacles, which are sufficiently far apart that individual
wakes are almost dissipated before the next obstacle is reached.
In that case the total drag exerted by the surface on the flow
results from addition of wake drag and surface drag in the
wake-free areas - sometimes called form drag and friction,
respectively (Marshall, 1971).
(C) <u>Wake-interference flow</u> occurs when average obstacle intervals
are of the same order of magnitude as the average length of
their direct wakes - i.e. 10 to 15 obstacle heights, depending
on obstacle shape and porosity. In that case the obstacle
effects are not simply additive, and close to the ground the
turbulent flow will generally not be in equilibrium.
(D) <u>Skimming flow</u> occurs when the obstacles are so close together,
that the flow in the cavities between the obstacles no longer
directly relates to the bulk flow above. Asymptotically this
last case approaches to vertically displaced regular turbulent
flow, as the interspaces between the obstacles disappear.
The aspect of the corresponding terrain types is pictured in
Figure 2, except for smooth terrain of which the aspect is evident.

In the extreme flow categories, A and D, we have a homogeneous
roughness distribution, therefore consistent inertial sublayer
values of roughness parameters can be defined. For skimming flow
the inertial sublayer has a lower limit level lying at least one
obstacle interval distance above the obstacle tops (Dubov and
Marunich, 1973; Raupach et al., 1980), at least 5 m above a typical
forest and 10 to 15 m above the roof tops of a homogeneously
built-up suburb. In this case the profile origin is also displaced
upward by ~0.7 h (Brutsaert, 1975) but this is unimportant if we
are chiefly interested in evaluating turbulent exchange. For these
categories profile data are a suitable source of roughness information.

The intermediate flow categories, B and C, suffer to some
degree from inhomogenuity problems, so profile shapes become
location-dependent and are less suitable for the evaluation of
roughness parameters. In these cases it is better to evaluate an
mesoscale average roughness from the gustiness level, specified by
σ_u/u or $u_{mx}/\overline{u}$. For these categories the ratio between roughness
length and obstacle height, z_0/h, is highly dependent on obstacle
density and distribution (Lettau, 1969; Wooding et al., 1973).
z_0/h has a maximum of ~0.2 at obstacle densities of $\sim10\%$ (depending
on the density definition) and falls rapidly to ~0.02 for both
lower and higher densities. Therefore the exclusive use of
vegetation height as a roughness parameter, e.g. by Baumgartner
et al. (1977) necessarily leads to unreliable roughness estimates.

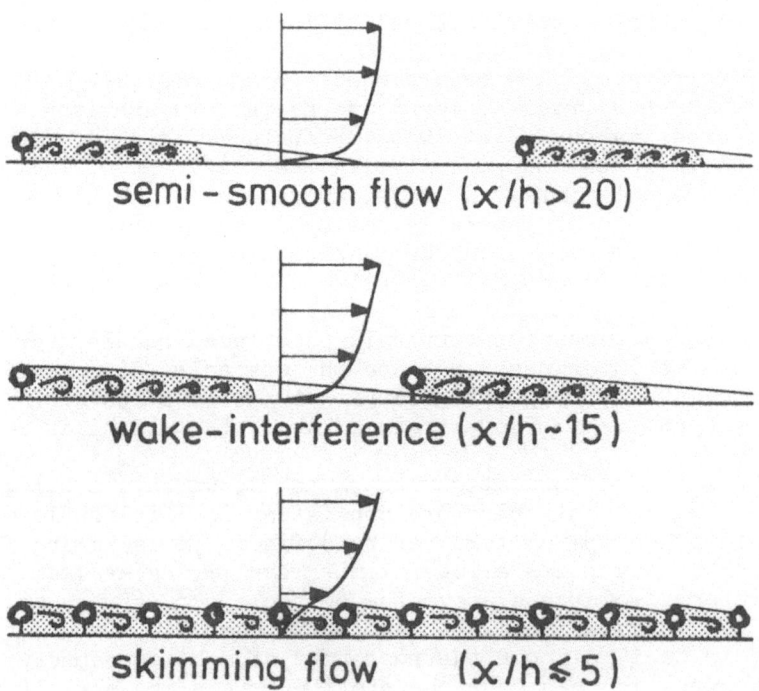

Figure 2.   Appropriate distribution of terrain obstacles and
            wakes for various flow categories, with typical wind
            profile shapes indicated.

Table 1. Davenport roughness classification
regrouped by flow category.

Identification Flow category Davenport classes	Terrain type Obstacle situation	Z_0 -range (m) $C_{d(10)}$ $(\times 10^{-3})$
sea A = smooth 1, 2	open water, mud flats , snow no obstacles	0.0002 - 0.005 1.5 - 3
open B = semi-smooth 3, 4	grassland and low crops few isolated obstacles (x/h > 20)	0.03 - 0.10 5 - 8
rough C = wake interference 5, 6	high crops , bushland , parks scattered obstacles (20 > x/h ⩾ 10)	0.25 - 0.5 12 - 20
closed D = skimming 7	homogeneous forest , suburb full obstacle coverage (10 ≫ x/h)	~ 1.0 ~ 30
chaotic - 8	city centre with high - rise buildings , tropical forest with large clearings	~ 2.0 ~ 60

Roughness classification

The fact that Baumgartner and others try to achieve such areal
z_0-estimates, however, indicates the desirability of translating
visible and geographically-mapped terrain information into a specification of surface roughness. To this purpose the pioneering
classification effort of Davenport (1960), originally loosely
defined in terms of power "law" exponents, has been tentatively
translated into z_0 terms (Wieringa, 1980b). His eight classes can
also be combined towards a practically manageable description of
the four flow categories defined above, and this has been done in
Table 1.

Further investigation and literature search is being made to
designate the roughness of various terrain types with more accuracy,
and to describe them in such a manner that they can be applied to
reality available geographical data without much misunderstanding.
In this respect it is for instance of interest, that satellite
photographing (LANDSAT) can at least describe the following
categories: water, open country, forests and built-up areas
(Loats et al., 1978). Maps give the same information, or more.
When it is necessary to evaluate an area with variable terrain
types, the mesoscale roughness will be dominated by the largest
roughness available (Smith and Carson, 1977).

Horizontal transformation of surface wind speed

For increasing roughness (and increasing gustiness level)
the surface layer has an increasing sink capacity for horizontal
flow velocity because of increased downward turbulent transport.
In other words horizontal differences in surface wind velocity
between neighbouring terrain areas of different roughness aspect
can be estimated from a good estimate of the mesoscale roughness.
A model for the horizontal transformation of wind speed to terrain
of different roughness has been developed by Wieringa (1976, 1980b)
and was successfully checked against field data. From this can be
deduced that the four flow categories ABCD defined above have a
relative difference of ∿15% between surface wind velocities of
subsequent categories. Such a differentiation corresponds to
practically discernible differences, -witness the fact that the
higher Beaufort classes differ 15% to 20% in class wind speed.

In a nomogram description of the just-mentioned model for
horizontal transformation of surface wind speed (Figure 3) the
z_0-adaptation of Davenport's roughness classification has been
used, since this degree of classification refinement is necessary
for local-scale terrain description. Less and larger classes bring
too many different terrain types in one class - a typical example
is the class "moderately rough" of Counihan (1975), ranging from

z_o = 0.001 m (mud flats, snow) to z_o = 0.2 m (farmland). Such coarse classification invites the use of double designation ("class 2/3") or subdivision for descriptive purposes.

On the other hand, for the purpose of accounting for frictional drag in large-scale modeling it may be sufficient to describe the terrain summarily in the four (or five) flow categories described and tabulated above. Then it will be necessary to integrate this with the growing knowledge on form drag, caused by elevation variations and mountains (e.g. Bowen and Lindley, 1977; Sacré, 1979; Jackson and Hunt, 1975; Taylor and Gent, 1980). The final goal is the satisfactory estimation of drag effects in atmospheric flows on all scales, from local to global.

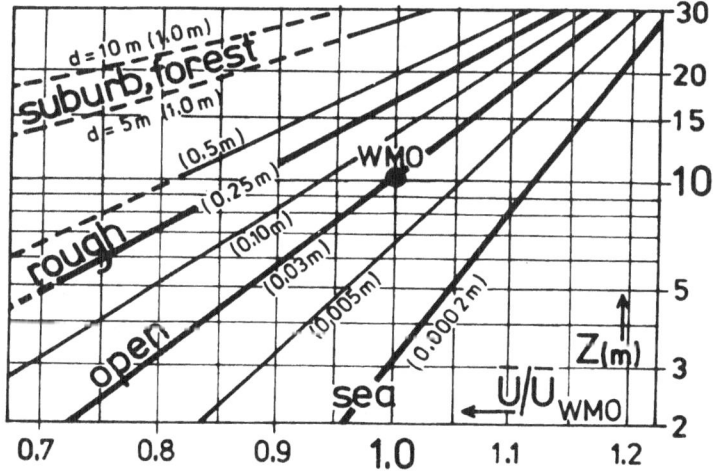

Figure 3. Nomogram for transforming WMO-standardized wind speed (10 m above open terrain) to wind speed at height z above nearby terrain with specified roughness length z_o (z_o-value indicated in parentheses). Broken lines at the left-hand side indicate that below z ∿ 20 z_o the transformation validity is doubtful because of non-logarithmicity and inhomogenuity of the wind field close to the roughness-inducing obstacles.

REFERENCES

Baumgartner, A., Mayer, H., Metz, W., 1977, Weltweite Verteilung
    des Rauhigkeitsparameters z_0 mit Anwendung auf die Energie-
    dissipation an der Erdoberfläche, Met. Rdschau, 30, 43-48.
Bowen, A.J., Lindley, D., 1977, A windtunnel investigation of the
    wind speed and turbulence characteristics close to the ground
    over various escarpment shapes, Bound. Layer Meteor., 12,
    259-271.
Brooks, F.A., 1961, Need for measuring horizontal gradients in
    determining vertical eddy transfers of heat and moisture,
    J. Meteor., 18, 589-596.
Brutsaert, W., 1975, Comments on surface roughness parameters and
    the height of dense vegetation, J. Met. Soc. Japan, 53, 96-97.
Businger, J.A., Wyngaard, J.C., Izumi, Y., Bradley, E.F., 1971,
    Flux-profile relationships in the atmospheric surface layer,
    J. Atm. Sci., 28, 190-201.
Counihan, J., 1975, Adiabatic atmospheric boundary layers: a review
    and analysis of data from the period 1880-1972, Atm. Env.,
    9, 871-905.
Davenport, A.G., 1960, Rationale for determining design wind velo-
    cities, J. Struct. Div. Am. Soc. Civ. Eng., 86, 39-68.
Driedonks, A.G.M., van Dop, H., Kohsiek, W.H., 1978, Meteorological
    observations on the 213 mast at Cabauw, in the Netherlands,
    Prepr. 4th AMS Symp. on Meteorol. Obs. and Instr., Denver,
    (publ. Am. Meteor. Soc., Boston, 41-46).
Dubov, A.S. Marunich, S.V., 1973, Structure of the air flow above
    forested areas, SSSR Atm. Oc. Phys. (Engl. transl), 9, 362-365.
Fiedler, F., Panofsky, H.A., 1972, The geostrophic drag coefficient
    and the "effective" roughness length, Qu. J. Roy. Meteor. Soc.,
    98, 213-220.
Garratt, J.R., 1977, Review of drag coefficients over oceans and
    continents, Month. Wea. Rev., 105, 915-929.
Goodin, W.R., McRae, G.J., 1980, A procedure for wind field con-
    struction from measured data which utilizes surface rough-
    ness, Prepr. 2nd AMS Conf. on Coastal Meteorology, Am. Meteor.
    Soc., Boston, 233-239.
Guyot, G., Seguin, B, 1978, Influence du bocage sur le climat d'une
    petite région: resultats des mesures effectuées en Bretagne,
    Agric. Meteor., 19, 411-430.
Irwin, J.S., 1979, A theoretical variation of the wind profile
    power-law exponent as a function of surface roughness and
    stability, Atm. Env., 13, 191-194.
Jackson, P.S., Hunt, J.C.R., 1975, Turbulent wind flow over a low
    hill, Qu. J. Roy. Meteor. Soc., 101, 929-955.
Jackson, N.A., The propagation of modified flow downstream of a
    change in roughness, Qu. J. Roy. Meteor. Soc., 102, 924-933.
Khalsa, S.J.S., Businger, J.A., 1977, The drag coefficient as
    determined by the dissipation method and its relation to the
    intermittent convection in the surface layer, Bound. Layer

Meteor., 12, 273-297.

Lettau, H.H., 1969, Note on aerodynamic roughness-parameter estimation on the basis of roughness-clement description, J. Appl.
Meteor., 8, 828-832.

Loats, H., Fowler, F., Castruccio, P., 1978, Applications of remote
sensing to hydrologic planning, NASA-CR-3041.

Manier, G., Weingärtner, H., 1978, Untersuchungen zum Einfluss von
Orographie, Stabilität und Baroklinität auf den Zusammenhang
zwischen Bodenwind und geostrophischen Wind, Meteor. Rundschau
31, 145-154.

Marshall, W.A.L., 1954, Comparison of the wind recorded by anemo-
graph with the geostrophic wind. Meteor. Off. Prof. Note, 108.
(H.M. Stationary Off., Londen), 26 pp.

Marshall, J.K., 1971, Drag measurements in roughness arrays of
varying density and distribution, Agr. Met., 8, 269-292.

McBean, G.A. (ed.), 1979, The planetary boundary layer, WMO Techn.
Note, 165 (WMO-No. 530).

Munn, R.E., Reimer, A., 1968, Turbulence statistics at 30 and 200
ft at Pinawa, Manitoba, Atm. Env., 2, 409-417.

Nieuwstadt, F.T.M., 1978, The computation of the friction velocity
u_* and the temperature scale T_* from temperature and wind
velocity profiles by least-square methods, Bound. Layer
Meteor., 14, 235-246.

Panofsky, H.A., 1974, The atmospheric boundary layer below 150
meters, Ann. Rev. Fluid Mech., 6, 147-177.

Pasquill, F., 1972, Some aspects of boundary layer description,
Qu. J. Roy. Meteor. Soc., 98, 469-494.

Perrier, E.R., Robertson, J.M., Millington, R.J., Peters, D.B.,
1972, Spatial and temporal variations of wind above and
within a soybean canopy. Agric. Meteor., 10, 421-442.

Peterson, E.W., Kristensen, L., Su, C.C., 1976, Some observations
and analysis of wind over non-uniform terrain. Qu. J. Roy.
Meteor. Soc., 102, 857-869.

Rao, K.S. Wyngaard, J.C., Coté, O.R., 1974, The structure of the
two-dimensional internal boundary layer over a sudden charge
in surface roughness, J. Atm. Sci., 31, 738-746.

Raupach, M.R., Thom, A.S., Edwards, I., 1980, Wind-tunnel study of
turbulent flow close to regularly arrayed rough surfaces,
Bound. Layer Meteor., 18, 373-397.

Sacré, C., 1979, An experimental study of the air flow over a hill
in the atmospheric boundary layer, Bound. Layer Meteor., 17,
381-401.

Smith, F.B., Carson, D.J., 1974, A scheme for deriving boundary-
layer wind profiles, Met. Mag., 103, 241-255.

Smith, F.B., Carson, D.J., 1977, Some thoughts on the specification
of the boundary layer relevant to numerical modelling. Bound.
Layer Meteor., 12, 307-330.

Stearns, C.R., 1971, The effect of time-variable fluxes on mean
wind and temperature profile structure. Bound. Layer. Meteor.,
1, 389-398.

Taylor, P.A., Gent, R.R., 1980, Modification of the boundary layer by orography, "Orographic effects in planetary flows", R. Hide and P.W. White, eds., WMO-GARP Publ. Series No. 23, 143-165.

Tennekes, H., 1973, The logarithmic windprofile, J. Atm. Sci., 30, 234-238.

Teunissen, H.W., 1979, Measurements of planetary boundary layer winds and turbulence characteristics over a small suburban airport, J. Industr. Aerodyn., 4, 1-34.

Touma, J.S., 1977, Dependence of the wind profile power law on stability for various locations, J. Air Poll. Contr. Ass., 27, 363-366.

Van Ulden, A.P., Holtslag, A.A.M., 1980, The wind at heights between 10 m and 200 m in comparison with the geostrophic wind. Risø Seminar on Radiative releases and their dispersion in the atmosphere following a hypothetical reactor accident (publ. Comm. European Communities, Luxembourg), vol. 1, 83-92.

Wamser, C., Müller, H., 1977, On the spectral scale of wind fluctuations within and above the surface layer. Qu. J. Roy. Met. Soc., 103, 721-730.

Wieringa, J., 1973, Gust factors over open water and built-up country, Bound. Layer Meteor., 3, 424-441.

Wieringa, J., 1976, An objective exposure correction method for average wind speeds measured at a sheltered location, Qu. J. Roy. Met. Soc., 102, 241-253.

Wieringa, J., 1977, Wind representativity increase due to an exposure convection, obtainable from past analog station wind records, WMO-No. 480, 39-44.

Wieringa, J., 1980a, A revaluation of the Kansas mast influence on measurements of stress and cup anemometer overspeeding, Bound. Layer Meteor., 18, 411-430.

Wieringa, J., 1980b, Representativeness of wind observations at airports, Bull. Am. Meteor. Soc., 61, 962-971.

Wooding, R.A., Bradley, E.F., Marshall, J.K., 1973, Drag due to regular arrays of roughness elements of varying geometry, Bound. Layer Meteor., 5, 285-308.

DISCUSSION

A. VENKATRAM   Why do you expect the turbulence to adjust more slowly to the underlying surface than the wind profile ?

J. WIERINGA   Generally profile adjustment is speeded up by additional non-turbulent exchange mechanisms, such as the pressure deficit occurring immediately downwind of

a rough area. On the other hand, excess turbulence has to dissipate all by itself. The issue
has been analysed in a more serious fashion by
diverse authors (e.g. P. TAYLOR, J. HUNT),and
they all agree that the turbulence has a longer
memory than the profile. This is particularly
the case for rough-to-smooth changes, where at
a downwind distance of ~ 30 x the height of the
large roughness we find, that the greater part
of the excess turbulence is still present, while the profile velocity deficits are already
diminished to less than 10 % of their original
value. This must be one of the basic reasons
why for a heterogeneous site the upwind roughness averages over mesoscale distances can
better be evaluated from turbulence level indicators (e.g. gustiness) than from profiles.
A. VAN ULDEN suggests another viewpoint: the
horizontal velocity variance is directly linked
to the behaviour of the largest boundary layer
eddies (TENNEKES and LUMLEY 1972). Therefore
σ_u may be an optimal starting point for boun-
dary-layer-averaged parameter determination.

R. YAMARTINO   Does the solid line of
Figure 1 represent a fit to these data ?

J. WIERINGA   No ! The solid line
(z_o = 0.1 m) corresponds to a typical regional
roughness as estimated visually (DAVENPORT-
class 4; see Bull. AMS. 61,967).
It is entered in the way it would be applied
in practice, with reference to the wind measured
at 10 m, and is seen to give a reasonable overall representation of the profile up to 160 m,
with exception of the near-surface behaviour.

A CONVECTIVE PLUME MODEL FOR PBL DISPERSION

John D. Reid

Boundary Layer Research Division
Atmospheric Environment Service
Downsview, Ontario, Canada

INTRODUCTION

It is now clear that dispersion of air pollutants within the
unstable planetary boundary layer needs to be handled somewhat
differently from dispersion in less unstable conditions. This
realization has come about largely on the basis of physical
simulation studies by Willis and Deardorff (1974, 1976, 1978,
1979), numerical studies by Deardorff (1972, 1974a, 1974b), and
continued analysis of Deardorff's numerical experiments by Lamb
(1978, 1979).

In their 1979 paper Willis and Deardorff review the physical
manifestations of convective structure causing the peculiar
unstable dispersion regime. Deardorff (1976) discussed the
evolution of convective structure with z_i/L, where z_i is the
height of the mixed layer and L is the Monin-Obukhov length. For
the most unstable conditions, $-z_i/L > 45$, random patterns of
convective plumes are found. As the atmosphere becomes less
unstable groupings of plumes into rings are found, and when
$-z_i/L \sim 25$ ring patterns of plumes with their most prominent axis
oriented with the mean wind are observed. For slightly unstable
conditions, $-z_i/L \lesssim 10$, a pattern of plumes aligned in longitudinal rolls along the wind is found.

An important common aspect of each of the above structures
is the presence of convective plumes. It would be reasonable to
expect that these are important contributors to the convective
dispersion regime. Indeed, this is just what Willis and
Deardorff have observed in their water tank. In this study a
highly simplified model of a single convective plume and its

297

associated subsidence field is examined to investigate convective boundary layer dispersion.

THE MODEL

Consider an elementary convective plume consisting of an updraft region, radius r_u surrounded by a compensating subsidence region concentric with the updraft of radius r_s. With cylindrical symmetry the equation of continuity for incompressible flow in cylindrical polar coordinates is

$$\frac{u}{r} + \frac{\partial u}{\partial r} = \frac{\partial w}{\partial z} \qquad (1)$$

Where the u is the radial velocity component and w the vertical velocity component. Telford (1970) achieved some success with a cylindrically symmetric model of a convective plume. One of his results was that, at least in some conditions, the radius of the plume did not vary greatly with height. This gives motivation for attempting a solution of (1) by the method of separation of variables

$$u = u(z) \cdot u(r) \qquad (2)$$

$$w = w(z) \cdot w(r)$$

which yields

$$\frac{u(r)}{r} + \frac{du(r)}{dr} = M\,w(r) \qquad (3)$$

and

$$\frac{dw(z)}{dz} = -M\,u(z) \qquad (4)$$

where M is a separation constant. Notice that the plume has been assumed to have a vertical axis in agreement with the most probable orientation observed by Hall et al (1975).

To proceed profiles of u(z), u(r), w(z), w(r) need to be specified, preferably simple analytic expressions meeting physically determined boundary conditions. For radial components some appropriate boundary conditions are:

1. $w(r)$ a maximum at $r = 0$
2. $w'(r) = 0$ at the outer edge of the subsidence region, $r = r_s$
3. $w(r) = 0$ at the updraft edge, $r = r_u$
4. $u(r) = 0$ at the plume centre, $r = 0$
5. $u(r) = 0$ at $r = r_s$

A natural solution of (3) which satisfies these boundary conditions is

$$w(r) = J_0(\alpha r)$$

$$0 > \alpha r > 3.83171 \qquad (5)$$

$$u(r) = J_1(\alpha r)$$

where the J's are Bessel functions and constant multipliers are to be considered with the z terms. Substitution gives

$$M = \alpha$$

and

$$r_s = 3.83171/\alpha.$$

In addition

$$r_u = 2.40482/\alpha.$$

The ratio $r_u/r_s = 0.63$, and the ratio of area of downdrafts to total area of the domain is 0.61. Lamb (1978) gives a figure of the vertical profile of fraction of area occupied by downdrafts, based on Deardorff's numerical model, with values ranging from less than 0.5 at the surface and the inversion to 0.63 at $z/z_i \sim 0.8$. Telford (1970) has a result which gives the fractional area of downdrafts ~ 0.70. Thus the value which arises from the assumed solution (5) is within the range of other investigators values. Because of the requirement for continuity (5) also specifies the ratio of mean updraft to downdraft speed as 1.56.

The value α can be determined to give a best fit to observed plume dimensions. A number of studies are available. Fitzjarrald (1978) suggests that for the most unstable conditions (for which the single plume model is most appropriate) with $-z_i/L > 100$, $\ell/z_i \sim 1.5$, where ℓ is a horizontal length scale. Willis and Deardorff (1979) find in their water tank an average horizontal dimension of 1.2 z_i. Kaimal et al (1976) observed a peak in their vertical velocity spectra through most of the unstable boundary layer at a wave length $\sim 1.5 z_i$ which they state corresponds to the length scale of large thermals dominating the boundary layer circulation. The question then arises regarding the relation between the length scales above and the plume dimension, say r_s. Since the length scales are generally area averages we might expect $r_s < \ell$. For convenience take $r_s = z_i$, which gives $\alpha = 3.83171/z_i$ and $r_u = 0.63 z_i$. Equation (5) becomes

$$w(\hat{r}) = J_o (3.83171\hat{r})$$

$$0 \le \hat{r} \le 1 \tag{6}$$

$$u(\hat{r}) = J_1 (3.83171\hat{r})$$

and

$$\hat{r} = r/z_i .$$

Some interesting properties of the model plume may now be derived. Combining (2) and (6) gives a total vertical velocity

$$w = w(z) J_o (3.83171\hat{r}) \tag{7}$$

Thus: (a) The maximum upward vertical velocity is $w(z)$,
         (b) The mean updraft velocity is $0.43 \, w(z)$,
         (c) The maximum downdraft velocity is $0.40 \, w(z)$,
         (d) The mean downdraft velocity is $0.28 \, w(z)$,
         (e) The contribution of the plume circulation to the vertical velocity variance is

$$\sigma_{w,p}^2 = 0.16 \, w^2(z) \tag{8}$$

         (f) For a horizontal component of velocity, the velocity variance due to the plume circulation is

$$\sigma_{v,p}^2 = 0.51 \, u^2(z) \tag{9}$$

As regards the vertical profile a recent analysis of data from the AMTEX experiment by Lenschow and Stephens (1980) led them to propose

$$\frac{\bar{w}_u}{w_*} = 1.0 \, \hat{z}^{1/3} (1 - 1.1 \, \hat{z}) \tag{10}$$

where $\bar{w}_u$ is the mean updraft velocity, a function of $\hat{z}(=z/z_i)$. This gives the updraft velocity increasing from zero at the surface to a maximum of $0.46 \, w_*$ at $\hat{z} = 0.23$, decreasing to zero at $\hat{z} = 0.91$, and having only relatively small values above. However, combining equations (4) and (10) indicates $u(\hat{z}) \to \infty$ as $\hat{z} \to 0$ which is physically unreasonable.

Some additional support for the general profile form of (10) arises from studies of the vertical variations of σ_w^2 and of the and of the vertical eddy diffusivity. Deardorff (1974) shows σ_w^2 from his numerical model reaching a maximum at 0.3 to 0.4 z_i with a value near $0.44 \, w_*^2$. Deardorff and Willis (1974) show a similar behaviour for their water tank simulation although the maximum is somewhat higher. They show two aircraft observations confirming the existence of a mid-level maximum. Crane et al (1977) defined the vertical eddy diffusivity for material by dividing a measured pollutant flux by the pollutant gradient and found the diffusivity decreasing with height in the upper part of the convective boundary layer.

Since sub-plume as well as plume-scale motions contribute to the variance and diffusivity, it is not possible to make direct inferences about the vertical structure of $w(\hat{z})$ from the information in the paragraph above. It does demonstrate the uncertainty in the vertical profile. In lieu of any better form that of equation (10) is adopted, with the expectation that the infinite surface velocity will not be important to the early particle height evolution. This yields

$$w(\hat{z}) = 2.33 \, w_* \, \hat{z}^{1/3} \, (1 - 1.1 \, \hat{z}) \tag{11}$$

$$u(\hat{z}) = -0.61 \, w_* \, (\hat{z}^{-2/3}/3 - 1.47 \, \hat{z}^{1/3})$$

and combining with (6)

$$w(\hat{r},\hat{z}) = 2.33 \, w_* \, \hat{z}^{1/3} \, (1 - 1.1 \, \hat{z}) \, J_0(3.83r) \tag{12}$$

$$u(\hat{r},\hat{z}) = -0.61 \, w_* \, (\hat{z}^{-2/3}/3 - 1.47 \, \hat{z}^{1/3}) \, J_1(3.83 \, \hat{r}).$$

COMPUTATIONAL TECHNIQUE

Trajectories of pollutant particles through the plume domain are computed. Forward differencing with a very small time step, $t_* = 0.0001$, is adopted. For a given release height trajectories of ten particles were followed. The initial radii for the particles were chosen so that the ensemble was representative of the entire plume. To do this the plume domain was divided into ten equal area concentric annuli. An initial particle radius was located within each annulus so that there was equal annulus area inside and outside that radius. Thus, the radius of the n'th particle is given by

$$r_n = 0.31623 \, (n - \tfrac{1}{2})^{1/2} \tag{13}$$

Computations for initial source non-dimensional heights, 0.1, 0.25, 0.5, 0.75 were made.

Computation results are presented in terms of mean particle height

$$\bar{z} = \sum_{n=1}^{10} z_n/10. \tag{14}$$

RESULTS

Figures 1 - 4 show the variation of mean particle height as a function of downwind displacement $X = t_* = xw_*/z_iu$ for four different source heights. In each figure the solid line is the variation for the present model. Predictions by other investigations for approximately the same release height are also shown.

In figure 1, for release at 0.1, the present simulation shows that initially the centre of mass remains at that height. This is a result of the continuity requirement for no net mass flux through any level. Radial motion and vertical variability force entrainment into updrafts and the centre of mass rises to reach a maximum near 0.6 at X = 2.0. The variation is in excellent agreement with the numerical predictions of Deardorff and Willis (1974), shown as crosses, who used a numerically computed flow field with $-z_i/L = 45$. Also shown as a dashed line is the variation from the Willis - Deardorff water tank experiments for 0.12 release height. The indicated rise is more rapid but the level of the maximum very similar.

For particles released near 0.25, figure 2, the present model shows an initial minimum at X = 0.7 at height 0.23 and a subsequent maximum at X = 2.3 at height 0.49. Shown as a dashed line are results obtained by Lamb (personal communication) using Deardorffs' numerically simulated boundary layer for Wangara Day 33. For this case $-z_i/L = 1100$ so that the convection should be dominated by isolated plumes. Lambs' calculations show a much earlier rise to a higher level but with a later maximum, at X = 2.9, compared to the present simulation. Shown as dots are results from the Willis - Deardorff tank simulation which suggest an early minimum, a rise in good agreement with Lamb, and a maximum near X = 2.0 in better agreement with the present simulation than with Lamb.

Figure 3 is for releases near 0.5. The present simulation shows $\bar{z}$ dipping to 0.26 at X = 1.5 and then attaining source height again near 3.0. Lambs simulation shows an earlier and higher minimum and return to $\bar{z} = 0.5$ at X = 2.3. The dots are from a recent Willis - Deardorff tank simulation (personal communication) and show a minimum even earlier and higher than Lambs'. The trend for the present simulation to show more extreme but slower oscillations than Lambs' is evident again for the 0.75 release height case shown in figure 4.

DISCUSSION

It is clear from the results above that, although the qualitative agreement between these predictions is good, the quantitative

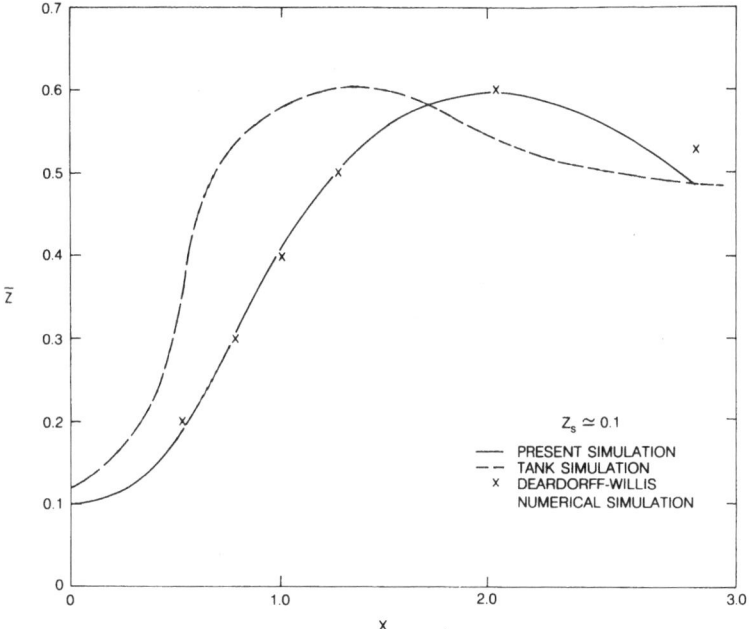

Figure 1: Variation of $\bar{z}$ with x for source height 0.1

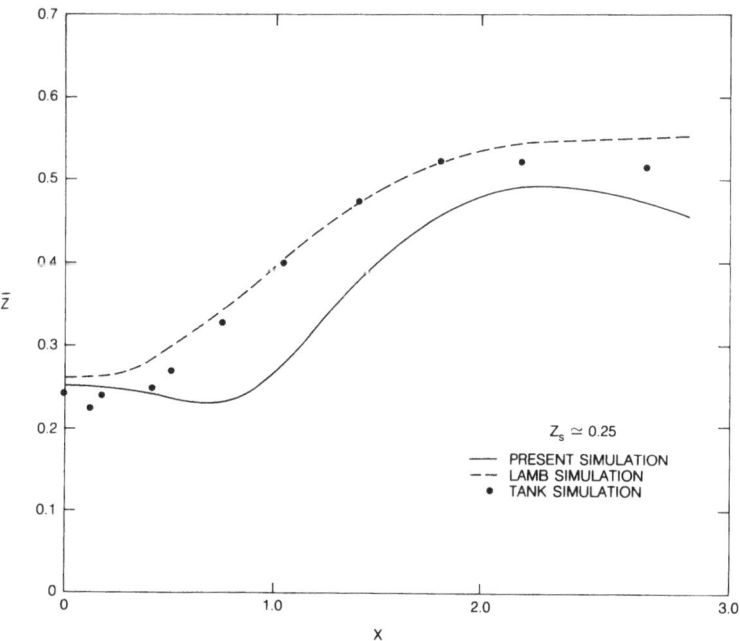

Figure 2: As figure 1, but $z_s = 0.25$

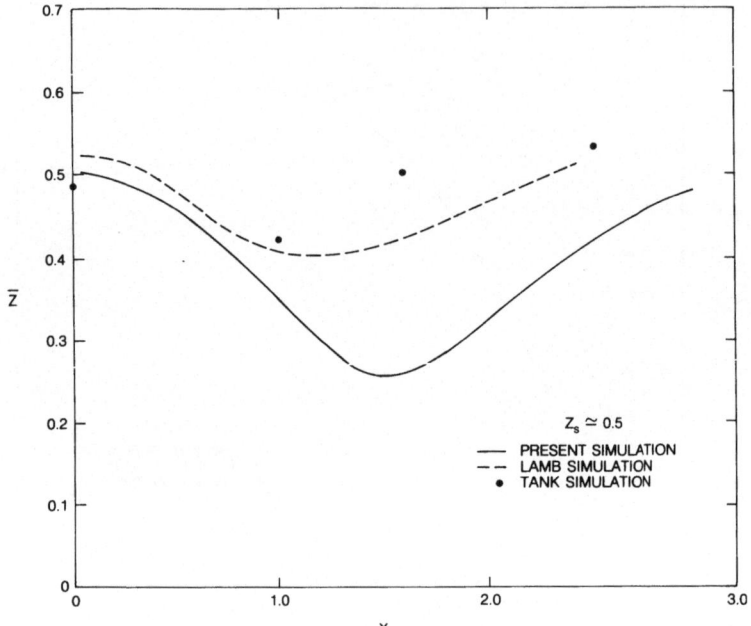

Figure 3: As figure 1, but $z_s = 0.5$

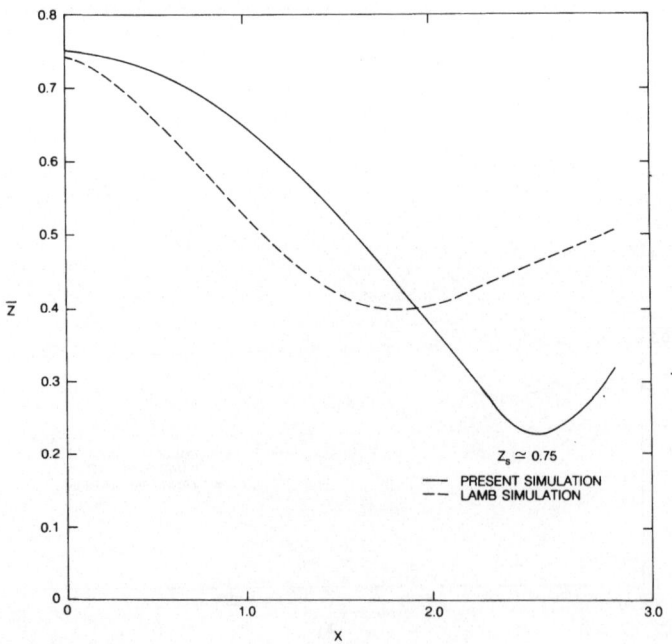

Figure 4: As figure 1, but $z_s = 0.75$

agreement is far from satisfactory. It is also not clear which prediction is the most representative of atmospheric conditions.
Within the context of the present model at least five sources of
error can be identified.

1. The small source height differences. The effect here should be
   insignificant.

2. Different equilibrium levels. The $\hat{z}$ depedence specified by (10)
   does not permit the plume circulation to mix particles above
   0.91 so that, with through mixing the equilibrium level is
   0.455. Of course, through mixing below the inversion gives an
   equilibrium level 0.5. There is a suggestion that this difference is appearing in the variations near X = 3.0.

3. Plume radius scaling. It is likely that by fine tuning the radius
   scaling of the present model, that is taking r_s/z_i different from
   one, the agreement could be improved.

4. Vigour of the circulation. The leading constant in (10) can be
   varied to scale the downwind displacement so that the same ex-
   tremum would be attained earlier or later. In the cases of the
   more elevated releases the agrement with the Lamb simulations
   could be dramatically improved by increasing the constant.

5. Effect of sub-plume scale motions. These will mix particles
   across their plume flow trajectories and this will tend to reduce the amplitude of $\bar{z}$ oscillations, the more so for longer
   downwind distances. The sub-plume scale contribution to the
   total velocity variance is typically of the same order as the
   plume-scale contribution. This is likely a major source of
   error.

ACKNOWLEDGEMENTS

        The author would like to acknowledge Drs. J.W. Deardorff and
R.G. Lamb for helpful discusions and permission to reproduce data
from their models.

REFERENCES

Crane G., H.A. Panofsky and O. Zeman, 1977 : A Model for Dispersion
        from Area Sources in Convective Turbulence. Atmos. Environ-
        ment 11, 893 - 900

Deardorff, J.W., 1972 : Numerical Investigation of Neutral and Un-
        stable Planetary Boundary Layers. J. Atmos. Sci. 29, 91-115

Deardorff, J.W., 1974 : Three-Dimensional Numerical Study of the
        Height and Mean Structure of a Heated Planetary Boudary
        Layer. Boundary-Layer Meteor. 7, 81-106

Deardorff, J.W., 1974 : Three-Dimensional Numerical Study of Turbu-
        lence in an Entraining Mixed Layer. Boundary-Layer Meteor.,
        7, 199-226

Deardorff, J.W., 1976 : Discussion of "Thermals over the Sea and
        Gulf Flight Behavior" by A.H. Woodcock, Boundary-Layer
        Meteor., 10, 241-246

Deardorff, J.W. and G.E. Willis, 1974 : Computer and Laboratory Mo-
        deling of the Vertical Diffusion of Non-Buoyant Particles
        in the Mixed Layer. Adv. Geophys, 18B, 187-200

Fitzjarrald, D.E., 1978 : Horizontal Scales of Motion in Atmospheric
        Free Convection Observed During the Gate Experiment.
        J. Appl. Meteor., 17, 213-221

Hall, F.F., J.G. Edinger and W.D. Neff, 1975 : Convective Plumes in
        the Planetary Boundary Layer, Investigated with an Acoustic
        Echo Sounder. J. Appl . Meteor., 14, 513-523

Kaimal, J.C., J.C. Wyngaard, D.A. Haugen, O.R. Coté, Y. Izumi,
        S.J. Caughey and C.J. Readings, 1976 : Turbulence Structure
        in the Convective Boundary Layer. J. Atmos. Sci., 33,
        2152-2169

Lamb,R.G., 1978 : A Numerical Simulation of Dispersion from an Ele-
        vated Point Source in the Convective Planetary Boundary
        Layer. Atmospheric Environment,12, 1297-1304

Lamb, R.G., 1979 : The Effects of Release Height on Material Dis-
        persion in the Convective Planetary Boundary Layer. Pre-
        prints, Fourth Symposium on Turbulence, Diffusion, and
        Air Pollution, (Reno, Nevada), AMS, Boston, pp 27-33

Lenschow, D.H. and P.L. Stephens, 1980 : The Role of Thermals in the
        Convective Boundary Layers. To appear in Boundary Layer
        Meteor.

Telford, J.W., 1970 : Convective Plumes in a Convective Field.
        J. Atmos. Sci., 27, 347-358

Willis, G.E. and J.W. Deardorff, 1974 : A Laboratory Model of
        the Unstable Planetary Boundary Layer. J. Atmos. Sci,

31, 1297-1307

Willis, G.E. and J.W. Deardorff, 1976 : A Laboratory Model of Diffusion into the Convective Planetary Boundary Layer. Quart. J. Roy. Meteor. SocK, 102, 427-445

Willis, G.E. and J.W. Deardorff, 1979 : Laboratory Observations of Turbulent Penetrative - Convection Planforms. J. Geophys. Res., 84, 295-302

Willis, G.E. and J.W. Deardorff, 1978 : A Laboratory Study of Dispersion from an Elevated Source Within a Modeled Convective Planetary Boundary Layer. Atmospheric Environment, 12, 1305-1311.

DISCUSSION

R. BERKOWICZ   In our group we have modelled convective boundary layer effects using the spectral turbulent diffusivity approach.For this we use same height-dependent energy spectrum as those we use to determine the K-profile.We have obtained qualitative agreement with results reported by Willis and Deardorff.
1° I would ask you to show how the source of particles was located with respect to the convective plume in your numerical experiment?
2° In our opinion the experiments reported by Willis and Deardorff show too much symmetry because of presence of the walls in the water-tank. Is the convective turbulence in the atmosphere as much organized as it is in a water-tank?

J.D. REID   Question 1°
The total plume cross- sectional area at the source height was divided into 10 equal area annuli. One particle was located in each annulus so that it represented that area.

Question 2°
There are always doubts about the validity of these types of physical model despite the best efforts of the experimenters.Good field experiments to resolve these doubts must certainly be regarded as a high priority.

NUMERICAL MODELLING OF STACK PLUMES WITHIN A CITY ENVIRONMENT

AT DISTANCES OF SEVERAL KILOMETRES DOWNSTREAM

Ann Henderson-Sellers   Brian Henderson-Sellers

Dept. of Geography   Dept. of Civil Engineering
University of Liverpool   University of Salford
U.K.   U.K.

INTRODUCTION

The buoyant rise of effluent emitted from a tall stack can be severely restricted by the presence of an inversion at any level. Whether the plume will penetrate into or through the stable layer depends upon the magnitude of its remnant buoyancy when it encounters the inversion. This is a function of the characteristics of both the emitted effluent (e.g. momentum, buoyancy) and the atmosphere (e.g. environmental lapse rate, height of inversion above stack top, turbulent mixing induced by topography). Some of these effects have been considered[1,2] by including the mixing effects of a city in the third phase of an extended three phase model based on work by Slawson and Csanady[3,4] - for details see references 2 and 5. However these simulations considered a marginally buoyant plume where the rise was insufficient to lead to concern about the effects of the vertical temperature profile of the atmosphere.

This paper aims to discuss the temperature effects of an urban mixing dome by "coupling" the plume rise model with a numerical description of an urban mixing layer, originally derived for flat terrain[6] and later extended to cases of irregular terrain[7]. A short discussion is included on the necessary height for a chimney so that the effluent may be able to "punch through" the inversion - an important concept in the regulation of ground level concentrations.

THEORY

Plume Model

The forced plume simulation uses a gross parameter numerical model which takes parametric values integrated across the plume, although the assumed profile permits concentrations etc. to be determined at any point in the plume. Differential equations for flux variables (of mass, momentum, buoyancy) are formulated[1] together with a relationship for the angle of the plume trajectory to the horizontal.

The entrainment velocity[8] concept permits the modelling of self-induced turbulent mixing and thus defines this stage of the plume rise as the "first phase". The second and third phases, defined by Slawson and Csanady[3,4] relate the turbulence levels to the eddies present in the plume, and the largescale energy containing atmospheric eddies respectively. When the plume enters the third phase, it can be shown[9] that the entrainment velocity is identical to that associated with a non-buoyant Gaussian plume model, even whilst the buoyant effluent may still be rising. (The mathematical formulation for these entrainment velocities in the three phases is detailed in references 1 and 9).

City Mixing Height

In many areas overnight cooling of the rural ground around a city produces a well defined ground level inversion. This may be enhanced by the formation of fog, which can accumulate following radiative cooling plus drainage from the surrounding hills. This latter phenomenon is especially noticeable in many European conurbations where the history of industrialization has led to many cities being built effectively in a "bowl" (e.g. Leeds) or a valley (e.g. Liege/Seraing). If a wind now blows the air mass containing the rural ground level inversion over the centre of the city, the urban heat island tends to distort the streamlines. Relatively warm air rising from the city forms a well mixed layer beneath the incoming air mass and forces it to rise over the urban heat dome. The model described here simulates the origin and development of this typical temperature profile as the air mass moves all the way across the city forming an urban plume downwind. The model is directly applicable to any times when there is a ground level inversion upwind of the city (i.e. mainly in the early mornings). As radiative heating destroys the inversion then the heat dome is effectively strengthened. It has been shown that the model can be equally well applied in this case too for as long as an inversion exists at some height, although if the depth below the inversion is great assumptions about its well-mixed nature become dubious.

By considering the energy balance of a column of air advected across the city, Leahey and Friend[6] calculated a value for the mixing height, $h(x,y)$. This is the height of the dome above a zero level defined arbitrarily (e.g. $z = 0$ at the ground level upwind of the city at $x = 0$). The model, which assumes that the wind speed is uniform and steady and that all properties (e.g. pollutant concentrations, heat, momentum) are well-mixed within the urban dome, has been applied successfully e.g. in reference 10, for cities in Canada.

However, for non-planar topography, the change in potential energy must also be taken into account. This gives an expression for the (adjusted) height of the mixing layer, h', above $z = 0$, as[7]

$$\Gamma h' = T_o - (T_o - \Gamma h)((T_o + \Gamma z')/T_o)^{-\gamma/\alpha} \tag{1}$$

where Γ is the dry adiabatic lapse rate, z' the height of the ground below $z = 0$, γ the lapse rate in stable air and $\alpha = \Gamma - \gamma$. The total depth of the mixing layer is thus $h' + z'$.

Urban Plume Model

It is now possible to integrate the urban mixing layer model with the plume model (both outlined above) in order to undertake a theoretical study of the effects of the urban heat island on stack plumes emitted at different heights and in different city locations. This model then provides a useful predictive tool for the location and heights of proposed new chimneys.

Numerically the linkage is straightforward. The plume model requires a vertical temperature profile as a function of downwind distance, x. This is provided by the advection model. Mixing heights are calculated as mean values over a grid square. There is thus either a small discontinuity in parametric values as the plume crosses the grid square boundary; or the grid value is taken as the centre value and linear interpolation is employed - results given here use the former method.

SIMULATIONS

The behaviour typical of plumes of efflux temperature 290K and 350K emitted from a chimney at three specific locations in Leeds with a prevailing Westerly wind of 6 m s^{-1} is summarised in Table 1. The points of emission are as follows: location A is 5 kilometres downwind from the Western boundary of the city; locations B and C are 5 and 10 kilometres due South of A. In all cases the chimney is assumed to be 50 metres in height. The 290K plumes (emitted at a low velocity of 0.01 m s^{-1}) are only marginally buoyant. They enter

Table 1.   Inversion Height and Plume Rise and Plumes of 290K
$(w_O = 0.01 \text{ m s}^{-1})$ and 350K $(w_O = 1.0 \text{ m s}^{-1})$ at
Three Locations in Leeds.

Location	Height of base of inversion above stack top		Plume rise at 5km downstream	
	At chimney	5km downstream	290K,0.01m s^{-1}	350K, 1.0m s^{-1}
A	4	4	0.58	68
B	104	175	0.50	234
C	87	210	0.41	211

(all values are in metres)

the atmosphere below the inversion base and level off rapidly;
remaining trapped beneath the stable layer, thus adding grossly
(possibly) to the pollutant level of the urban mixing layer. For the
plumes of higher temperature (350K) and emitted at 1 m s^{-1}, the rise
is substantial: the plume from chimney B is seen to penetrate into
the stable layer; the plume from C hits the inversion base and its
rise is terminated at this level by the inversion. At location A the
dampening effect of the inversion layer is still seen, restricting
plume rise to less than 70m. These figures thus demonstrate the
effect of heat content on the plume behaviour; although this aspect
is·discussed in more detail elsewhere[11] where the effect of the
varying mixing due to changing city topography is also considered.

     To illustrate the effects of chimney height, we consider an
effluent emitted at a temperature of 300K at 0.5 m s^{-1} at location A
(where the choice of chimney height for such an effluent is found to
be important). Table 2 gives the height of rise (at a distance of
5 km downstream for chimneys of height 10, 25, 50, 100 metres (at
location A). For the smaller chimney, there is little remnant
buoyancy in the plume when it reaches the inversion level. It remains
at this level and can thus be considered to exhibit "trapping" with
the possible consequences of "fumigation". With a higher chimney,
effluent emitted just below (25m chimney) or above the base of the
stable layer (50m and 100m chimney) does reach an equilibrium height
above the inversion base ("lofting") thus resulting in comparatively
lower ground level concentration. However it is important to note
that by increasing the chimney height, the total rise (above ground)
is not increased by an identical amount. Rise through the stable
layer is restricted so that, for instance, doubling the chimney
height from 50m to 100m only increases the total plume height by

Table 2.   Heights of Rise for Plumes Emitted at Location A with
            a Temperature of 300K and Velocity of 0.5 m s^{-1}

Chimney Height	Inversion Height (above stack top) (approximately constant over 5 km)	Height of Rise (at 5 km downstream)	Total Height above Ground	Average Spread Rate dR/dx
10	44	44.8	54.8	0.013
25	29	43.7	68.7	0.006
50	4	35.9	85.9	0.002
100	-46	25.0	125.0	0.001

40.9 metres. This dampening is readily noticeable in the values for
dR/dx in Table 2. For the plume trapped within the mixed layer the
spread rate of 0.013 is good agreement with observations and other
model results[12] for a plume dominated by third phase mixing. A much
smaller value for dR/dx is to be expected for plumes within the
stable layer (see Table 2).

DISCUSSION

       Comparison of these results with predictions of plume rise in a
neutral environment, derived from semi-empirical equations for total
plume rise, suggest that empirical calculations may underestimate
final rise. However such formulae often assume that the rise is
terminated at a distance of ten stack heights downstream. This is
neither borne out by observations nor models (as discussed above).
An alternative formulation given by Briggs[13]:

$$\Delta H = \frac{1.6}{U} \left[g \, w_o \, R^2 \left(\frac{T_s - T_\infty}{T_\infty} \right) \right]^{1/3} x^{2/3} \qquad (2)$$

which gives plume rise as a function of distance - an equation
devised for use within ten stack heights downstream. Evaluating
equation (2) at ten stack heights it can be seen from Table 3 that
agreement between equation (2) and the numerical model described
here at a distance of 500m downstream is reasonable - to within about
50%: a value well within the error bars of the models. In fact
Briggs[14] comments that plume rise model predictions can often differ
from each other by a factor of 10! Although Briggs does not advocate
use of this formula (equation (2)) for long-range transport,

Table 3.   Plume Rise (in metres) at 500 m downstream from Source.
            (Efflux Temperature 350K; Velocity 1 m s^{-1}; Wind Speed
            6 m s^{-1}; Chimney Height 50m)

| T_s | T_∞ | Locations | | | Equation (2) |
		A	B	C	
350	281	51.2	57.3	56.8	22.5

it is interesting to note that at 5000m, it gives ΔH (for a 350K
effluent emitted at 1 m s^{-1}) as 105 m - in tolerable agreement with
the numerical calculations displayed in Table 1.

CONCLUSIONS

     The amalgamation of numerical models of plume rise and the urban
atmospheric environment has been shown to be able to predict the
plume trajectory under the influence of the urban mixing layer
(although no advective effects of the displacement of streamlines
has been included in the discussion here). The importance of topography, mixing depth and chimney height have been highlighted.
Increased stack heights do not necessarily lead to an equal increase
in total rise (for emissions above or just below the inversion level).
The main criterion seems to be the relevant magnitudes of chimney
height and inversion height for an emission of given efflux characteristics. The inter-relationships of the many variables involved
(stack emission and meteorology) cannot be simply analysed - it is
suggested that the numerical model described here will be able to
offer guidance on matching emissions and chimney heights to local
meteorological conditions.

REFERENCES

1.   B. Henderson-Sellers, A gross parameter plume model: theory and
     application for non-planar topography and for negatively buoyant
     effluents, 10th International Technical Meeting on Air Pollution
     Modeling and its Application, Rome, 23-26 October, 1979.
2.   B. Henderson-Sellers, The behaviour of marginally buoyant plumes
     in an urban environment, Ecological Modelling, 9: 43 (1980).
3.   P.R. Slawson and G.T. Csanady, On the mean path of buoyant bent
     over chimney plumes, J. Fluid Mech., 28: 311 (1967).
4.   P.R. Slawson and G.T. Csanady, The effect of atmospheric
     conditions on plume rise, J. Fluid Mech., 37: 33 (1971).

5.   B. Henderson-Sellers, Forced plumes in a stratified reservoir,
     Procs. ASCE., J. Hyd. Div., 104, HY4: 487 (1978).
6.   D.M. Leahey and J.P. Friend, A model for predicting the depth
     of the mixing layer over an urban heat island with application to
     New York City, J. Appl. Meteor., 10: 1162 (1971).
7.   A. Henderson-Sellers, A simple numerical simulation of urban
     mixing depths, J. Appl. Meteor., 19: 215 (1980).
8.   B.R. Morton, G.I. Taylor and J.S. Turner, Turbulent gravitational
     convection from maintained and instantaneous sources, Procs.
     Roy. Soc., A234:1(1956).
9.   B. Henderson-Sellers, Plume rise and Fickian diffusion models,
     Boundary Layer Meteorology, in press (1980).
10. D.M. Leahey, An application of a simple advective pollution
     model to the City of Edmonton, Atmos. Environ., 9: 817 (1975).
11. A. Henderson-Sellers and B. Henderson-Sellers, The effects of
     the urban environment on the trajectories of stack plumes,
     submitted to Applied Mathematical Modelling.
12. B. Henderson-Sellers, Rates of spread for a bent-over chimney
     plume, Ecological Modelling, in press (1980).
13. G. Briggs, Some recent analyses of plume rise observations, in
     "Procs.2nd Intern. Clean Air Conf.", Washington, D.C., 1979,
     Academic Press, N.Y. (1971).
14. G. Briggs, Plume rise predictions, in "Lectures in Air Pollution
     and Environmental Impact Analysis", 59-111, AMS, Boston, Ma.(1975).

DISCUSSION

B. E. A. FISHER   Does the model show that the
                     presence of an elevated inversion causes a
                     doubling of the maximum ground level concentra-
                     tion ?

B. HENDERSON-SELLERS   Ground level concentrations
                     were not calculated, but could easily be eva-
                     luated.

RESULTS OF LIDAR MEASUREMENTS OF

ATMOSPHERIC BARRIER LAYERS

Josef Giebel

Landesanstalt für Immissionsschutz
Essen, Federal Republic of Germany

ABSTRACT

The diurnal and annual cycles of atmospheric barrier layers
were determined by soundings of the lower atmosphere with a ruby
laser. In summer the degree of cloud cover had a great influence
on the height of the barrier layers. The influence increased with
the time of day and the degree of cloud cover. In winter there was
only a significant influence, when the degree of cloud cover was
more than 7o %. But then windspeed had a greater influence. The
measured heights of the barrier layer, which was nearly identical
to the top of the mixing layer, were ordered to the corresponding
stability classes. In each class, appearing in daytime, heights
between 2oo and 2ooo m were measured. The mean heights were 97o m
above sea level for class 2 (unstable atmospheric conditions)
and 76o m for class 3 (slightly unstable conditions). The diurnal
mean between sunrise and sunset was calculated to about 8oo to
9oo m above sea level.

INTRODUCTION

Soundings of the lower atmosphere were carried out with a
ruby laser, to get information about the behaviour of atmospheric
barrier layers.

The method is based on the reflection of Lidar signals from
aerosols present in the atmosphere.

The Lidar signals show relative maxima and sudden changes in
the detectable particle concentration, as they are caused by
barrier layers in the atmosphere. These layers usually cause an

accumulation of the aerosols between surface and barrier layer and
(or) in the barrier layer itself. At the bottom of the barrier
layer, in the barrier layer or at its top the aerosol concentration
then decreases abruptly.

The measurements were carried out in the south-west of the
Ruhr-area, about 15o m above sea level. Measurements wer performed
on 73 days throughout the year between 8.3o a.m. and 3.3o p.m..
The interval between the measurements usually was half an hour.

RESULTS OF LIDAR MEASUREMENTS

Lidar Measurements and the Temperature Profile

Figure 1 shows the relation between abrupt decreases of
aerosol concentrations - as it was measured by Lidar in November
between 9 a.m. and 3 p.m. - and the temperature profile at noon.
Up to a height of 8oo m every change of the temperature gradient
was connected with an abrupt decrease of the aerosol concentration.

But not every decrease of the aerosol concentration was
connected with a change of the temperature gradient. Sometimes,
however, there was a change of the temperature gradient in the
night before. Above 8oo m not every alteration of the temperature
gradient was connected with a decrease of the aerosol concentration
detected by Lidar.

The Barrier Layer with the lowest Transmission value in
comparison with the Temperature Gradient.The barrier layer which
is most important for the dispersion of air pollution is the
barrier layer with the strongest decrease of the aerosol concentration. This layer can be defined as the barrier layer with the
lowest transmission value. The transmission value T is the quotient
of the Laser energy I, which is backscattered from the points
P_1 and P_2 with point P_1 near to the Lidar system.It is a measure
for the decrease of the aerosol concentration with height.

$$T = \frac{I(P_2)}{I(P_1)}$$

T :   Transmission value

$I(P_2)$ Laser energy backscattered

$I(P_1)$ from the points P_1 resp.

P_2

As a comparison with the temperature profile at noon shows.
(figure 2), at this time the barrier layer with the least transmission value is usually identical with the top of the mixing
layer. The Lidar Measurements show that in the average it is also
nearly identical with the highest barrier layer influenced by the
turbulence from the surface. Therefore in the following the barrier

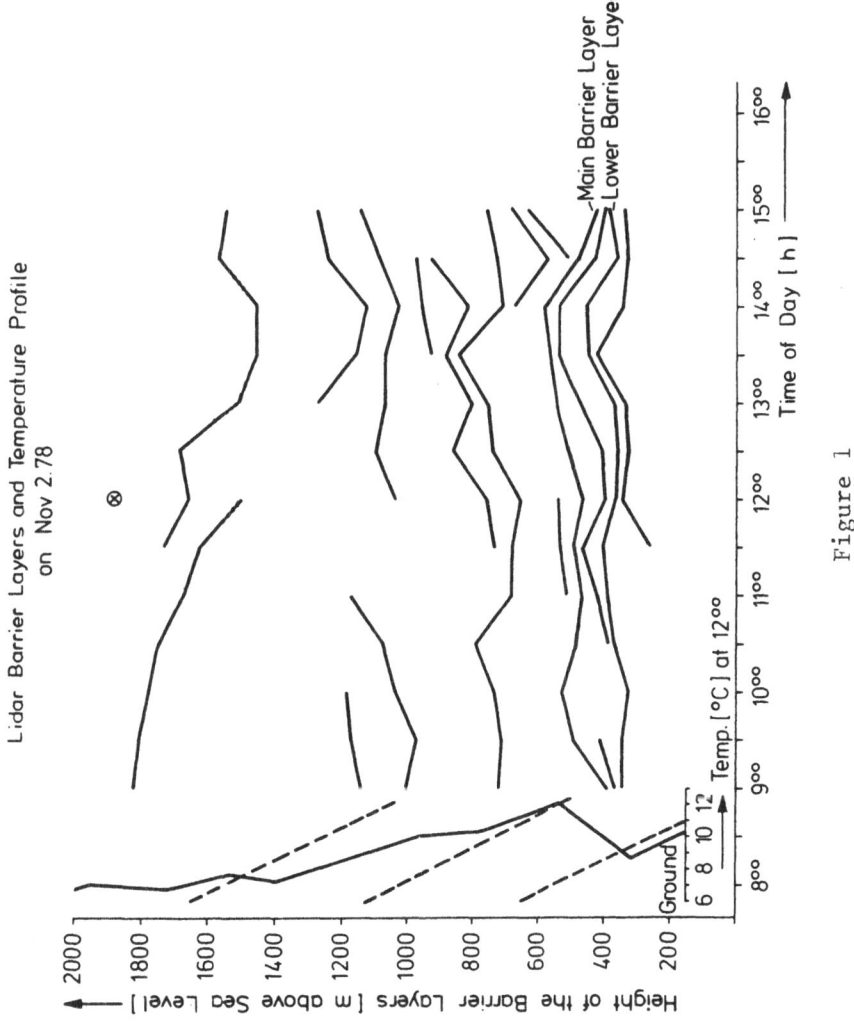

Figure 1

layer with the least transmission value is called 'the main barrier layer'.

Figure 2 shows the height of the main barrier layer at noon in dependence on the height of the nearest layer with a change of the temperature gradient. The figure depends on 69 couples of values. The correlation coefficient is o.96. Near the height of the main barrier layer the following changes of temperature gradient occured:

Main Barrier Layer and Temperature Gradient

Type of temperature gradient	Number of Cases
Base of an inversion	34
Base of a stable layer but no inversion (temperature gradient $>-o.6^{\circ}C/1oo$ m)	18
Stable layer in the night before	6
Top of an inversion	6
Top of a stable layer but no inversion	2
Base of an unstable layer (temperature gradient $<-o.6^{\circ}C/1oo$ m)	3
	69

In 76 % of the cases the main barrier layer was near the base of an inversion(50 %) or near the base of a stable layer that was no inversion(26 %).

In 9 % of the cases it was near the base of an inversion or a stable layer measured in the night before.

In the remaining 16 % however, the main barrier layer was near the top of an inversion or stable layer or base of an unstable layer (4 %).

The mean height difference between the main barrier layer and the nearest layer with a change of the temperature gradient was 88m. One of the reasons for this difference is that the Lidar measurements don't agree exactly in time and space with the measurements of the temperature gradient. As the Lidar measurements show, the barrier layers can vary in height up to 100 m within minutes. Another reason for the difference is probably the strong wind which is found within the inversion near its top. In this wind layer the aerosol concentration should decrease abruptly.

Characterization of the different kinds of Barrier Layers

As figure 1 shows, it is possible to measure by Lidar barrier layers between 200 and 2000 m above sea level. According to their

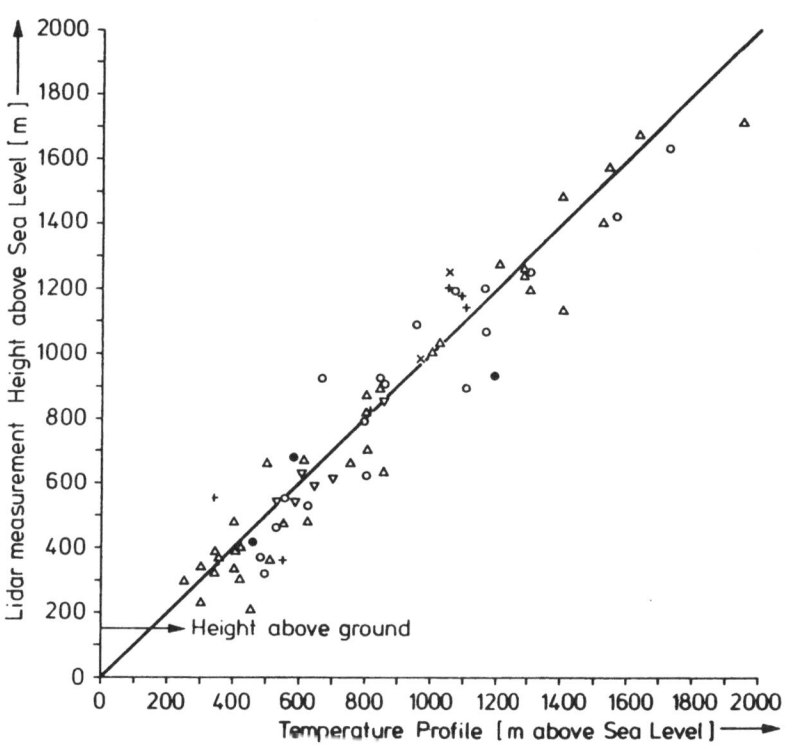

Comparison of Barrier Layers
Temperature Profile – Lidar Measurement
(12°° Noon)

Δ Base of an inversion   ∇ Top of an inversion

o Base of a stable layer, but   x Top of a stable layer,
  no inversion   but no inversion

+ Base of stable layers in the   ● Base of an unstable layer
  night before

Figure 2

behaviour during the day, the different barrier layers can be
attached to one of the following three groups:

In greater heights are layers which are not influenced by
fluxes of heat and momentum from the surface. Near the ground are
often found weaker barrier layers which disappear during the
morning. They form new in the afternoon and evening and probably
also during the night. The Lidar measurements show this group of
barrier layers mainly in summer where they may appear up to a height
of about 1ooo m. The third group of barrier layers is between the
weaker ones near the ground and the layers in greater heights, ·
which are not influenced by the turbulence from the surface. The
highest of them is usually the top of the mixing layer. These layers
usually don't disappear during the day, but are gradually pushed
upwards by the growing turbulence and show a diurnal and annual
cycle. They are mostly the base of a stable layer, which the turbulence from the surface partly dissolves. When they sometimes break
down,an upper layer takes over their function. The further investigations are concerned with this group of barrier layers. The most
important of them is the main barrier layer. As the Lidar measurements show, in the average it is nearly equivalent to the highest
barrier layer influenced by the surface. Therefore it is also nearly
equivalent to the top of the mixing layer, which in general is
defined in that way. Another important layer of this group is the
lowest barrier layer. Between it and the surface the mixing is
greatest. In the following it is called 'the lower barrier layer'.

Diurnal and Annual Cycles of the main Barrier Layer

Figure 3 shows the mean height of the main barrier layer for
the different months in dependence on the time of day. Each curve
represents the diurnal cycle of the two months with the same
distance from the solstice at the 21 of December. The curves form
two different groups. The winter-months belong to the one the
summer-months to the other. In summer the height of the main barrier
layer increased rapidly from 9.oo a.m. to 1.oo p.m.. Then the
increase diminished. The maximum was reached at 3.oo p.m.. The mean
height at 9.oo a.m. was about 55o m above sea level, the maximum
at 3.oo p.m. was between 13oo and 165o m.

The main barrier layer had the lowest height in the two months
with the least distance from the solstice. The main height was about
4oo m above sea level at 9.oo a.m. and about 530 m at 3.oo p.m..
Two to three months away from the solstice the mean height of the
main barrier layer was about 45o m at 9.oo a.m. and about 7oo m at
3.oo p.m.. All values are heights above sea level.

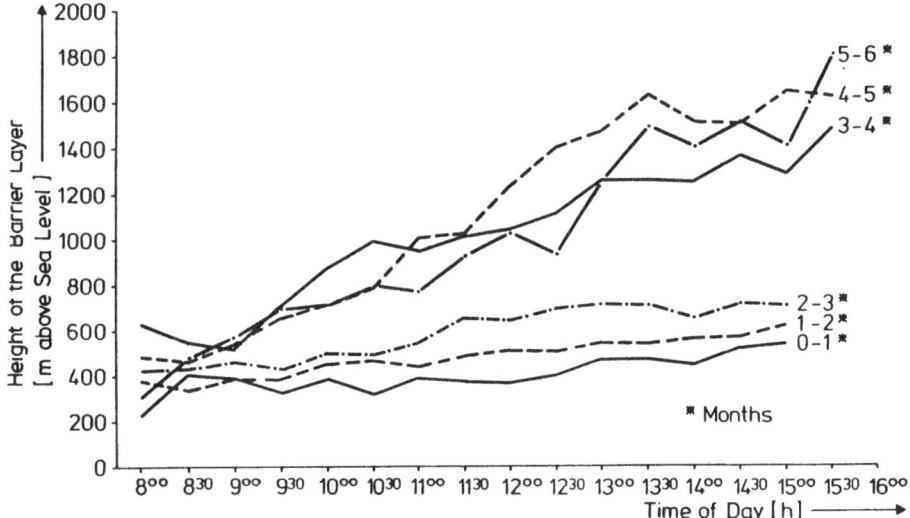

Figure 3. Diurnal variation of the main Lidar Barrier Layer for months with the same distance from the solstice of the 21st of December.

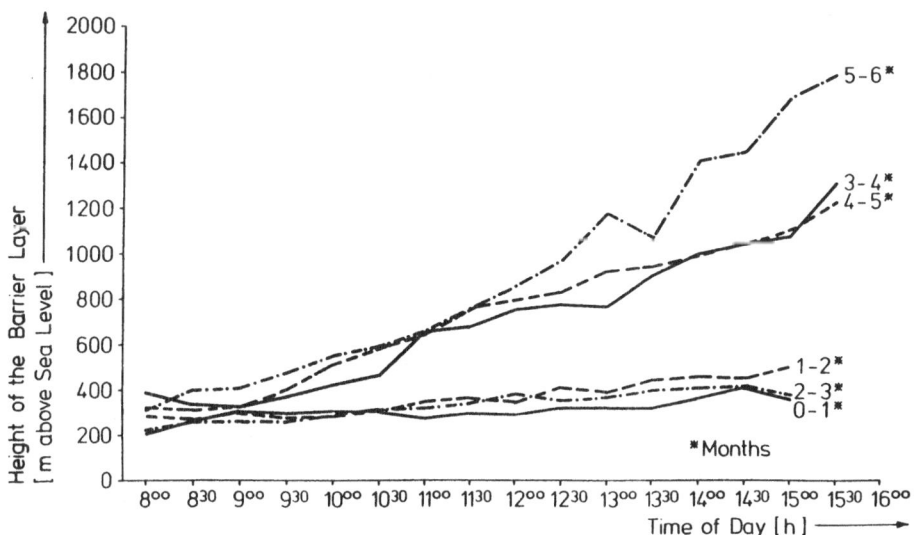

Figure 4. Diurnal variation of the lower Lidar Barrier Layer for months with some distance from the solstice of the 21st of December.

Diurnal and Annual Cycles of the lower Barrier Layer

Figure 4 shows the diurnal and annual variation of the height of
the lower barrier layer measured by Lidar.

In winter the mean height of this layer increased from about
3oo m above sea level at 9.oo a.m. to about 4oo m at 3.oo p.m.. It
was about 1oo m lower than the main barrier layer.

Three to five months away from the solstice the lower barrier
layer increased from about 35o m at 9.oo a.m. to about 125o m at
3.oo p.m.. In these months the distance to the main barrier layer
was 2oo to 3oo m. Five to six months away from the 2? of December
the lower barrier layer increased from about 35o m at 9.oo a.m. to
about 18oo m at 3 p.m.. Between 8.oo a.m. and 1.3o p.m. it was
1oo to 2oo m lower than the main barrier layer, but from 2.oo p.m.
until 3.3o p.m. it had nearly the same height as the main barrier
layer. This is an indication that in the afternoon of these two
months with the highest solar radiation the boundary layer is best
mixed by strong convective turbulence.

The Influence of meteorological Parameters on the Height of
the Barrier Layers

The Influence of the Temperature. In summer there was a positive
correlation between the height of the barrier layers and the surface
temperature at the moment of the Lidar measurement. There was also
a positive correlation when instead of the surface temperature
at the moment of the Lidar measurement the difference of the surface
temperature at 9.oo a.m. and the surface temperature at the moment
of the Lidar measurement was taken. The greater the temperature or
the temperature difference the higher the barrier layers. The correlation was stronger, when instead of the temperature the degree of
cloud cover was taken. Regression analysis was performed. As a
measure for the convective turbulence the average degree of cloud
cover between one hour after sunrise and the moment of the Lidar
measurement was taken.

Beside the cloud cover the windspeed at the surface (mean over
the three hours before the moment of the Lidar measurement) was
taken as a measure for the mechanical turbulence.

The Influence of Cloud Cover and Surface Windspeed on the main
Barrier Layer. The Influence in Summer *) Table 1 contains regression
equations for every hour in summer between 9.oo a.m. and 3.oo p.m.

*) In this paper summer is the time between the 21 of March and the
22 of September and winter is the time between the 23 of September
and 2o of March.

Table 1:   Regression Equations

Height of Barrier Layers in Dependence on Windspeed
and Cloud Cover

y :   Barrier Layer   [m]

u: Windspeed   [m/s]

           (mean windspeed at the surface over the 3 hours
           before the moment at the Lidar measurement)

B :   Degree of Cloud cover [%]
           (Mean cloud cover between 1 hour after sunrise
           and the moment of the Lidar measurement)

The main Barrier Layer in Summer

Time of Day	Regression Equation
9.oo a.m.	y = 69.3 u − 1.o B + 365
1o.oo a.m.	y = 15.2 u − o.4o B + 688
11.oo a.m.	y = 4.4 u − 2.4 B + 1o36
12.oo	y = − 2o.9 u − 3.9 B + 14oo
1.oo p.m.	y = − 41.8 u − 3.8 B + 1673
2.oo p.m.	y = − 52.7 u − 8.9 B + 2o89
3.oo p.m.	y = − 36.1 u − 8.7 B + 211o

The main Barrier Layer in Winter

9.oo a.m.	y = 8.9 u + o.59 B + 354
1o.oo a.m.	y = 3o.1 u + o,6o B + 297
11.oo a.m.	y = 26.3 u + o.1o B + 349
12.oo	y = 68.1 u − o.79 B + 28o
1.oo p.m.	y = 81.2 u − 1.4 B + 314
2.oo p.m.	y = 39.3 u + o.15 B + 4o1
3.oo p.m.	y = 16.6 u + o.63 B + 517

Independent variables are degree of cloud cover and windspeed, dependent variable is the height of the main barrier layer.

As the gradient shows, in the course of day the cloud cover gains increasing influence on the height of the main barrier layer. At 3.oo p.m. it was about loo m higher, when the cloud cover was lo % lower. At 9.oo a.m. the difference was only lo m. The main barrier layer reached a maximum, when there were no clouds and no wind.

With the regression equations one can calculate the mean height of the main barrier layer in summer for the average of cloud cover and windspeed. In the Ruhr-area it was 136o m at 3 p.m..

Figure 5 shows in semi logarithmic paper the gradient due to cloud cover multiplied by loo in dependence on the time of day. The result is a nearly straight line. The gradient due to cloud cover multiplied by loo is the mean height of the main barrier layer on cloudless days minus the height of the mean height of that layer on days with loo % cloud cover. At 9.oo a.m. the height difference was about loo m, at noon about 3oo m and at 3 p.m. about looo m.

The influence of cloud cover on the height of the main barrier layer was not only dependent on the time of day, but also on the degree of cloud cover.

When the cloud cover diminished from 9o to 8o %, the increase of height of the main barrier layer was much stronger as when the cloud cover diminished from 5o to 4o %.

Figure 6 shows the gradient due to cloud cover in dependence on the degree of cloud cover. Winter and summer are separated, but not the different hours of the day. Regression analyses were performed for degrees of cloud cover of more than lo, 2o, 4o, 6o, 7o, 8o and 9o %. The gradient multiplied by lo is the mean height increase of the main barrier layer corresponding to lo % decrease in cloud cover. When the cloud cover was more than 5o instead of more than 6o %, the height difference was about loom, but when the cloud cover was more than 8o instead of more than 9o % the height difference was about 25o m. The different influence of the cloud cover depends probably on the formation of inversions and the release of latent heat through condensation in dependence on the degree of cloud cover.

The influence of windspeed was less than the influence of cloud cover. At 9.oo a.m. and lo.oo a.m. the main barrier layer still increased with increasing windspeed, but between 11.oo a.m. and 3.oo p.m. it descended with increasing windspeed (Table 1). Probably the windspeed reduced the convective turbulence, which at this

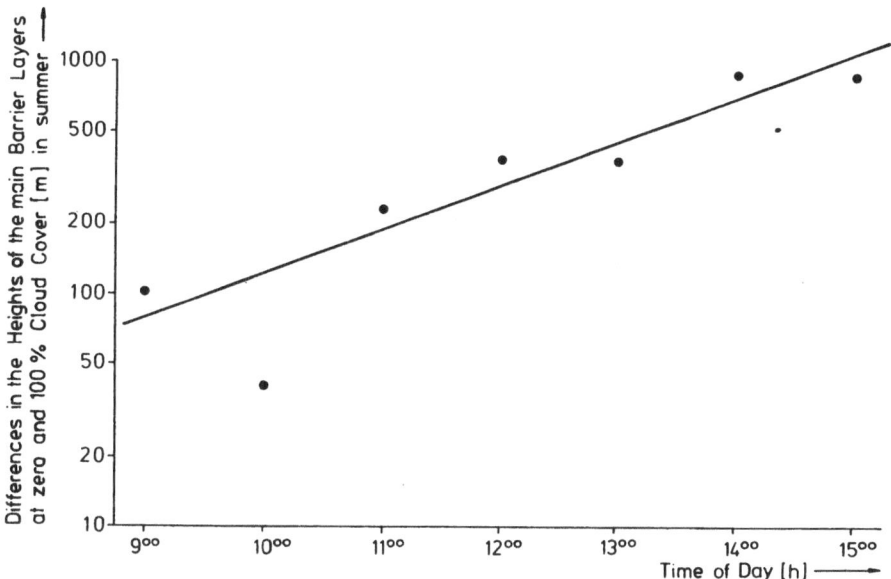

Fig. 5. Dependence of the height of the main barrier layer on the degree of the cloud cover at different times of the day.

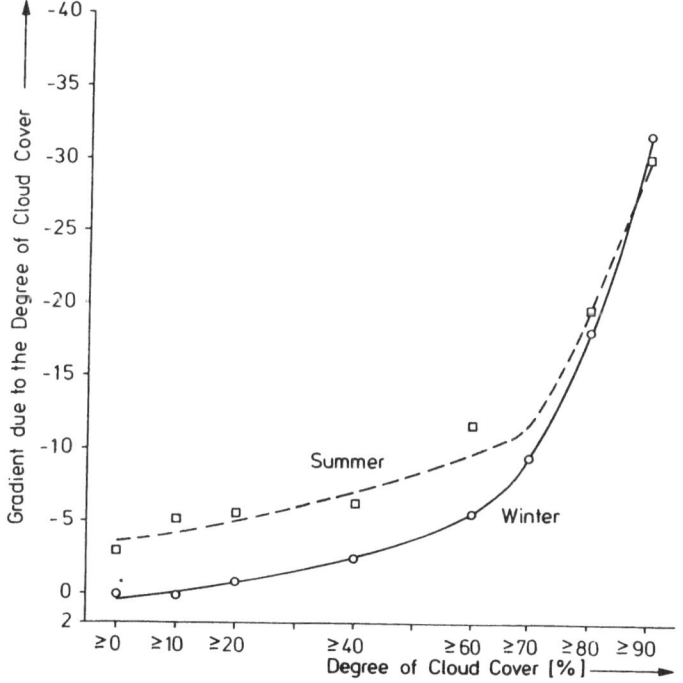

Fig. 6. Influence of the cloud cover on the height of the main barrier layer (mean over the timeperiod between 9:00 am and 3 3:00 pm) at different degrees of cloud cover.

time of the day is mainly responsible for the height of the main barrier layer.

Figure 7 shows the influence of windspeed on the height of the
main barrier layer in dependence on the degree of cloud cover. When
the degree of cloud cover was less than 7o %, the influence of windspeed was low, but it increased considerably with increasing cloud
cover, when the degree of cloud cover was more than 7o %.

The influence of Cloud Cover and Surface Windspeed on the main
Barrier Layer in Winter. In winter the influence of cloud cover on
the height of the main barrier layer was smaller than in summer and
regarding all degrees of cloud cover it had the opposite effect.
The height of the main barrier layer was decreased, when the cloud
cover was decreased. Figure 6 shows the gradient due to cloud cover
in dependence on the degree of cloud cover. Up to a degree of about
7o % the influence of cloud cover was only low, but with degrees of
cloud cover of more than 7o % it was in the right direction and grew
strongly with increasing cloud cover. When the degree of cloud cover
was more than 8o % instead of more than 9o % the main barrier layer
was about 2oo m higher.

The influence of the windspeed on the height of the main barrier layer was in winter stronger than in summer (Table 1; Figure 7).
It increased from morning to noon and than diminished. When at 1.oo
p.m. the windspeed was 1 m/s higher (average over the 3 hours before
the moment of the Lidar measurement), the main barrier layer was a-
bout 8o higher. At 9.oo a.m. and 3.oo p.m. the influence of the windspeed was least. As Figure 7 shows, the influence of windspeed grew
with increasing cloud cover.

Weather Conditions during the Performance of the Lidar Measurements

Usually the Lidar measurements were performed when the cloud
cover was low. On the days of Lidar measurements the average clouc
cover was 46 % (Table 2) while in the average the cloud cover in
Essen is 7o %. On the days of Lidar measurements the windspeed wa's
also lower than the long term mean. The difference was about o.5 m/s.
The regression equations can be used to adjust the mean heights of
barrier layers measured in summer to mean windspeed and cloud cover.

To adjust the heights of the layers measured in winter is doubtful because of the different influence of cloud cover in dependence
on the degree of cloud cover. Figure 8 shows the measured and the
adjusted annual cycle of the height of the main barrier layer for
3.oo p.m.. The adjusted cycle is in summer more symmetrical to the
solstice at the 21 of December than the uncorrected one.

Table 2: Average Cloud Cover and Windspeed on the Days with Lidar Measurements and in the long Mean

Month	Number of Days with Lidar Measurements	Cloud Cover		Windspeed in Anemometerlevel	
		monthly Average	on the Days with Lidar Measurements	monthly Average	on the Days with Lidar Measurements
		[%]	[%]	[m/s]	[m/s]
January	2	78	49	4.3	5.0
February	5	75	34	4.0	4.6
March	8	69	61	4.1	4.4
April	8	66	18	3.9	4.o
May	4	66	5	2.5	3.6
June	1	66	59	3.o	3.4
July	7	71	62	2.8	3.4
August	4	67	52	2.8	3.6
September	4	62	81	3.6	3.8
October	5	66	58	3.1	4.o
November	9	77	48	3.6	4.6
December	6	79	19	3.9	4.8
Annal Average		70	46	3.5	4.1

Influence of the Wind Speed on the Height of the main
Barrier Layer

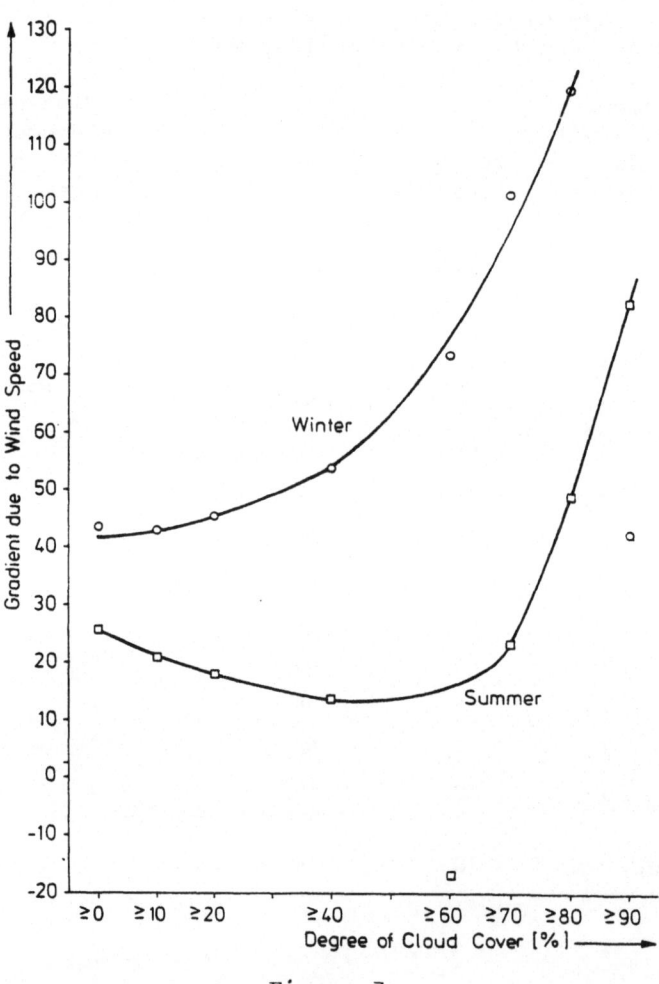

Figure 7

Annual Variation of the Height of the main Barrier Layer at 3 p.m.
Original and with Correction over Cloud Cover and Wind Speed

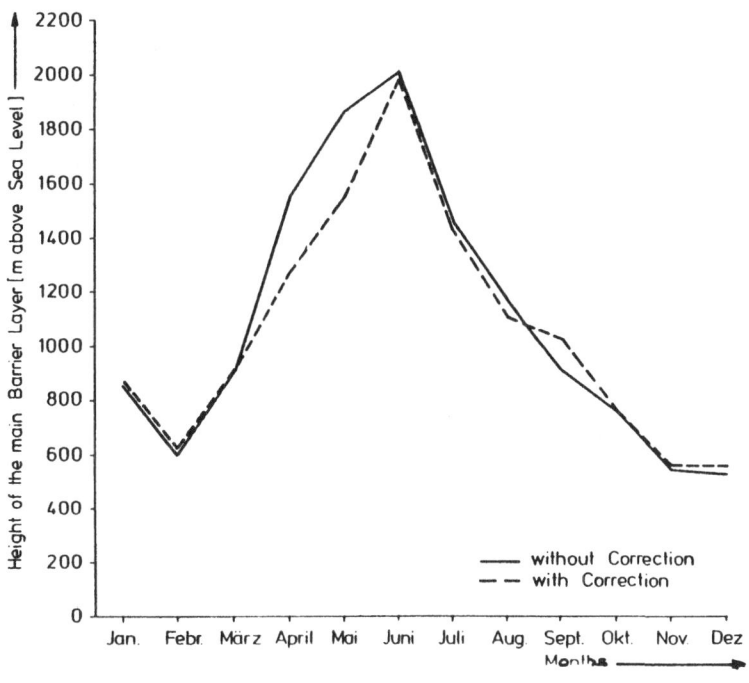

Figure 8

The Height of the main Barrier Layer in Dependence on the Stability Classes

For modeling air pollution the mean height of the mixing layer for the different stability classes is needed. Therefore the measured heights of the main barrier layers were ordered to their corresponding stability classes.

The stability classes are those defined by Turner.[1] It is the system described in the Raffinerie Richtlinie.

Unstable atmospheric conditions are represented by stability class 2, stable conditions by stability class 5. The classes 1 and 6 do not exist.

Lidar measurements were performed in the classes 2, 3 and 4 but not in class 5 which represents (stable conditions).

The percentage of the Lidar measurements falling to the different stability classes is shown by table 3. The mean height of of the main barrier layer was 965 m above sea level in stability class 2, 7o3 m in stability class 3 and 733 m in stability class 4.

In winter the mean height of the main barrier layer was 393 m in stability class 2 (unstable conditions) and 519 m in stability class 4 (slightly stable conditions). This is the opposite of what is assumed, namely that the mixing layer is higher in class 2 than in class 4.

The reason for that is probably that due to the formation of inversions in the case of low cloud cover, in winter the atmosphere is more stable in stability class 2 than in stability class 4.

Figure 9 shows the frequency distribution of the different heights of the main barrier layer in dependence on the stability classes. In each stability class heights between 2oo an 2ooo m were measured. The Lidar measurements were carried out between 8.oo a.m. and 3.oo p.m.. On account of that, the measurements don't show the mean height of the main barrier layer in the different stability classes. To get those heights for the stability classes 2 and 3, which only appear in the day, the heights of the main barrier layer at 9.oo a.m. were extrapolated until sunrise and the heights at 3.oo p.m. were extrapolated until sunset.

The mean heights received by that method were 97o m above sea level for stability class 2 and 76o m for stability class 3.

In class 4 the result was 8o3 m. But this is not the mean value for this class, because class 4 appears also in the neight.

Table 3:   Mean Heights of the Main Barrier Layer in Dependence on the Stability Classes

Stability Class	Frequency of the Stability Class in the long Mean [%]	Frequency of the Lidar Measurements in the different Stability Classes [%]	Height above Sealevel in the annual Mean [m]	Standard Deviation measurements [%]	Number of measurements	Height above Sealevel in winter resp.in summer	Standard Deviation [%]	No. of measurements
5	22.3	0	–	–	–	–	–	–
4	63.6	58.6	733	57	276	519/1o16	45/43	157/119
3	9.6	22.9	7o3	59	1o8	492/1o34	28/47	66/42
2	4.5	18.5	965	53	87	393/1o84	25/44	15/72
								238/233

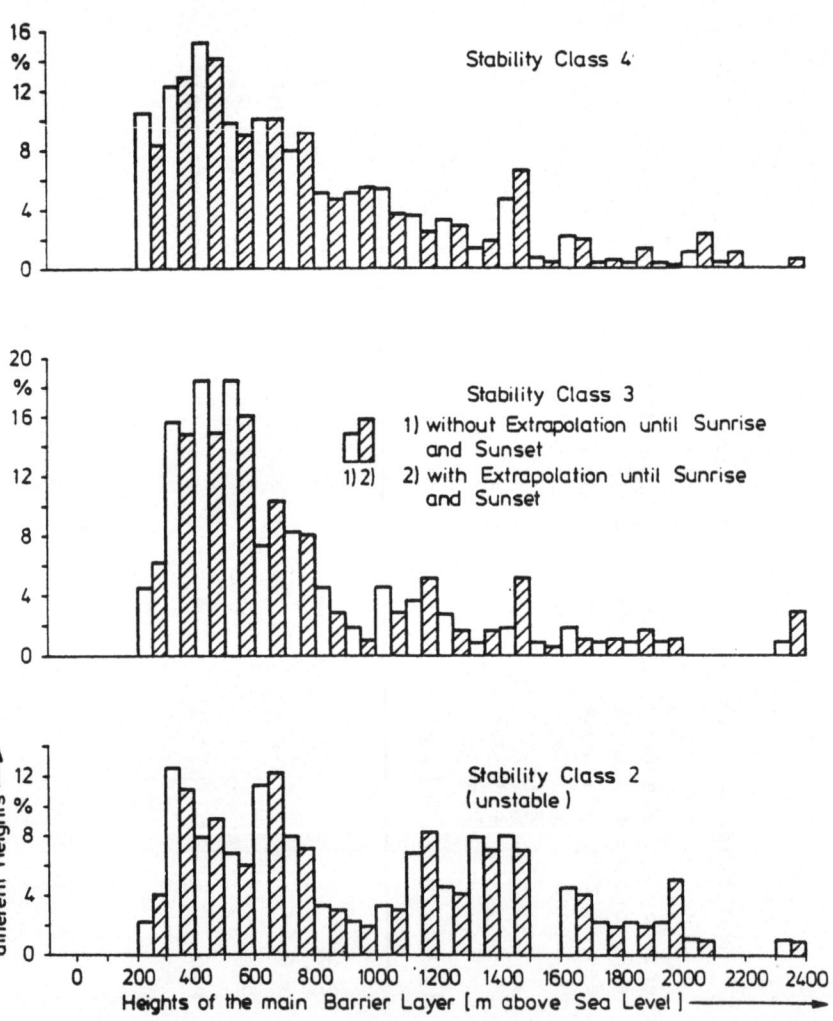

Figure 9

The average height of the main barrier layer over the day from sunrise to sunset was 806 m.

The adjusting to mean cloud cover and windspeed with the use of the regression equations changes the result only slightly.

The calculated mean heights of the main barrier layers may be somewhat lower than the real means because the Lidar measurements were performed mainly on days with low cloud cover, and in winter when - regarding all degrees of cloud cover - the main barrier decreased with decreasing cloud cover, the adjusting to mean cloud cover and windspeed was not well possible. But the height difference is probably less than 5o m.

REFERENCES

1   Verwaltungsvorschriften zum Genehmigungsverfahren nach
    § 6, 15 Bundes-Immissionsschutzgesetz (BImSchG) für
    Mineralölraffinerien und petrochemische Anlagen zur
    Kohlenwasserstoffherstellung.
    RdErl. d. Ministers für Arbeit, Gesundheit und Soziales
    des Landes NW - III B 4/III B 6 - 8856.4 (III Nr. 13/
    1975) v.14.4.1975 Ministerialblatt NW, 28 (1975), 65
    S.966-982

DISCUSSION

R. V. PORTELLI   Did you preselect the 73 days
                        for which you presented data or did you monitor
                        every day and utilize data of those days for
                        which aerosol barriers were detected ?

J. GIEBEL   We preferred to make measure-
                        ments on days with small amounts of clouds,
                        since it is more difficult to interput the data
                        with clouds present. However, we did perform
                        measurements in between clouds on partly clou-
                        ded days. Aerosols were always present in the
                        air and did not cause problems.

R. V. PORTELLI   I mention this since it is
                        important in the interpretation of the seasonal
                        variations in Barrier Height (or Mixing Height)
                        you presented. Account must be taken of those
                        conditions excluded from your data sample.

M. L. WILLIAMS   I have a comment concerning
                        our work at WSL on SODAR-derived barrier layers.

We have derived barrier layer frequency distributions for the full range of Pasquill-Smith
stability conditions from a year's continuous
measurements. To fill in a gap in your presentation, we find that in stable conditions (F,G)
that the maximum in the distributions lies at
a height of ∿ 100 m above ground level. One question I have is that, you measure barrier layers
in unstable conditions between ∿ 100 m up to
∿ 2000 m. Do you think that this is because the
stability you measure is a surface value only
and not representative for the full boundary
layer ?

J. GIEBEL   The minimum barrier height
that may be recorded by the lidar is approxi-
mately 200 m above sealevel or 50 m above the
ground on our observations side. Inversions
closer to the ground may occur specially in
rural areas. We considered mainly the barrier-
layers that were influenced by the ground.
Above these   layers other barrier layers were
recorded in accordance with the temperature
profiles.

EFFECTS OF RELEASE HEIGHT ON σ_y AND σ_z IN DAYTIME CONDITIONS

Steven R. Hanna

Air Resources
Atmospheric Turbulence and Diffusion Laboratory
National Oceanic and Atmospheric Administration
Oak Ridge, Tennessee 37830

ABSTRACT

A statistical diffusion model is used to estimate the variation of σ_y and σ_z with height in daytime conditions. The existing literature is reviewed, showing inconsistencies among the few field data. The statistical Monte Carlo-type model is described in detail and some asymptotic predictions of the theory are outlined. For example, the usefulness of the formulas $\sigma_y = \sigma_\theta x$ and $\sigma_z = \sigma_e x$ are shown, where σ_θ and σ_e are the standard deviations of the horizontal and vertical wind directions.

Results of the model show that σ_y varies little with release height in daytime conditions, but that it can depart significantly from the standard Pasquill-Gifford-Turner (PGT) curves for roughnesses outside their range of derivation. The vertical component σ_z typically increases by a factor of two as heights increase from the surface layer to mid-planetary boundary layer height.

1. INTRODUCTION AND BACKGROUND

The standard Pasquill-Gifford-Turner curves used for σ_y and σ_z, the crosswind standard deviations of pollutant concentration distributions, are not functions of height. Gifford (1976) provides graphs and recent reference materials on these graphs. They have been shown to work satisfactorily when compared with observed pollutant concentration data. Yet we know from theory that σ_y and σ_z should depend on height. For example, the initial crosswind spread of a plume is given by the formula,

$$\sigma_y = \sigma_\theta x, \tag{1}$$

where x is downwind distance and σ_θ is the standard deviation of wind direction fluctuations. The parameter σ_θ is a strong function of height in the surface layer :

$$\sigma_\theta \overset{\sim}{=} \sigma_v/u = 2u_x/2.5u_x \ln(z/z_o) = 0.8/\ln(z/z_o) \tag{2}$$

The neutral formula for wind speed is used in deriving equation (2). Note that roughness length, z_o, is an important parameter. Combining equations (1) and (2) we find the solution

$$\sigma_y = 0.8x/\ln(z/z_o), \tag{3}$$

which implies, contrary to common usage, that σ_y is indeed a function of height. For example, as height doubles from five to ten meters with a roughness of 0.1 m, this theory predicts that σ_y should decrease by about 18 %. Proponents of equation (1) generally recommend that the appropriate height for determination of σ_θ should be the effective release height of the plume.

There have been a few observations and theories reported in the literature on the effect of release height on σ_y and σ_z. Vogt et al (1978) released neutrally-buoyant tracers on several occasions from heights, z_r, equal to 50 m and 100 m on a tower in Julich, West Germany. In 11 experiments releases were made simultaneously from both heights, using different tracers. Ground level observations of concentrations at 150 locations were used to derive the variation of σ_y and σ_z with downwind distance out to distances on the order of 10 km. Underlying terrain was mixed farmland and forests with a roughness of 1.8 m at the central tower. The data were separated according to Pasquill stability classes (A through F), and power laws of the form $\sigma_y = ax^b$ were fitted to the data. Consequently the ratios σ_y (100 m)/σ_y (50 m) and σ_z (100 m)/σ_z (50) are also simple power laws, and they have been plotted in Figures 1 and 2 for different stability classes. In general the ratios are within a factor of two of unity.

The results in the figures can be compared with some theoretical predictions. We assume that the proportionalities in equations (1) and (2) are valid for the vertical component of diffusion, also. In neutral and unstable conditions, σ_y is known to be nearly constant with height in the surface layer (Panofsky et al, 1977). The wind shear will be slightly greater in neutral conditions than in unstable conditions. Thus the ratio σ_y (100 m)/σ_y (50 m) in A, B, C, and D stabilities should theoretically be less than unity and should decrease going from stability class A to stability class D. The first prediction is not verified by the data in Figure 1 but the second is verified quite nicely at distances

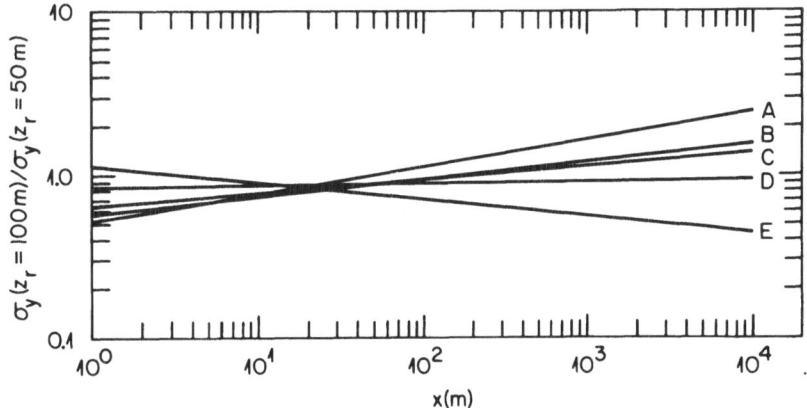

Fig. 1 : Vogt et al's (1978) observed ratio $\sigma_y(z_r = 100$ m$)/$
$\sigma_y(z_r = 50$ m$)$ for several stability classes.

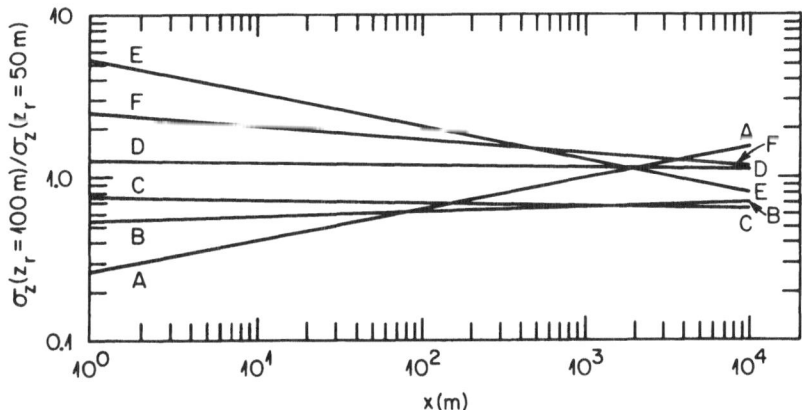

Fig. 2 : Vogt et al's (1978) observed ratio $\sigma_z(z_r = 100$ m$)$
$/\sigma_z(z_r = 50$ m$)$ for several stability classes.

greater than 20 m from the source. Predictions for stability class
E are uncertain because of the erratic behavior of σ_v with height
in very stable conditions.

In unstable conditions, σ_w is known to increase slightly
with height in the surface layer (Panofsky et al, 1977). Thus the
ratio σ_w/u theoretically should not vary much between heights of
50 m and 100 m in unstable conditions. On the other hand, σ_w is
known to decrease and u is known to increase with height in stable
conditions, thus implying that $\sigma_z(100 \text{ m})/\sigma_z(50 \text{ m})$ should be less
than unity. This theoretical prediction is not verified by the
curves for E and F stabilities in Figure 2, which are greater than
unity over most of their range. At downwind distances from 10 m
to 1000 m, the variation of $\sigma_z(100 \text{ m})/\sigma_z(50 \text{ m})$ with stability is
the opposite of what one would expect from physical intuition.
But it should be recognized that there is a certain amount of error
in the power law coefficients given by Vogt et al (1978), implying
that the ratios $\sigma(100 \text{ m})/\sigma(50 \text{ m})$ contain even more error.

Another series of diffusion experiments in which the effects
of source on height on σ_y and σ_z were studied was reported by Doran
et al (1978). This program was called the Hanford-67 series and
consisted of over 50 tests made over flat desert terrain with a
roughness of 3 cm. Releases of non-buoyant tracers were made at
heights of 2, 26, 56, and 111 m under neutral or stable conditions
(no unstable run included). A network of surface stations and
towers was used to determine σ_y and σ_z. Vertical distributions of
material were measured up to a height of about 40 m on the towers,
thus permitting a direct observation of σ_z in many cases. In contrast, σ_z at the Julich site reported by Vogt et al (1978) had to
be inferred from ground level measurements. It was found at Hanford that σ_z generally increases as release height increases for
these stability classes, in agreement with the results of Vogt et
al (1978), but in disagreement with what we would expect from the
physics in equations (1) and (2).

A theoretical analysis was done by Reid (1979), who used a
Monte Carlo procedure to calculate the trajectories of thousands of
particles in the surface layer and then calculated σ_z from the
resulting distribution. He considered release heights of 0, 2.5,
5, and 10 m and found that there was very little difference in σ_z
for these release heights. There was evidence that σ_z is slightly
higher for higher release heights near the source ($x < 30$ m), but
slightly lower for higher release heights at greater distances
($x > 70$ m).

Lamb's numerical (1979) study of the effect of release height
on σ_y and σ_z used still another technique. He followed the trajectories of thousands of particles but used a three dimensional turbulent velocity field produced by Deardorff's (1974) numerical mo-

del of the convective planetary boundary layer. Release height s of
.025, .25, .5, and .75 times the mixing depth were used. His results are consistent with what one would expect from planetary
boundary layer theory for daytime conditions, assuming the validity
of equations (1) and (2). Namely, σ_y is about 50 % greater for
near-surface releases ($z_r \sim 25$ m) than for mid-PBL releases and
σ_z is about 50 % less for near-surface releases than for mid-PBL
releases.

It is difficult to get a consistent picture from the above
references on the effect of release height on σ_y and σ_z. The
experiments yield scattered data points and the theories do not
always agree on the sense and magnitude of the σ_y and σ_z variation.
If we argue from the premise that $\sigma_y \propto \sigma_v/u$ and $\sigma_z \propto \sigma_w/u$ we can
use the observed behavior of σ_v, σ_w, and u over the depth of the
mixed layer to make some qualitative predictions concerning σ_y and
σ_z :

Very stable :	σ_y: uncertain, depends on meanders
	σ_z: strong decrease with height
Slightly stable:	σ_y: decrease with height
	σ_z: decrease with height
Neutral :	σ_y: slight decrease with height
	σ_z: slight decrease with height
Unstable :	σ_y: slight decrease with height in surface layer, constant above.
	σ_z: increase with height

These simple predictions will be tested by the model described in
the following sections.

2. MODEL DESCRIPTION

The transport and diffusion of a cloud of contaminants is
assumed to be determined by the time and space distribution of
mean velocity and the three components of turbulent energy and
Lagrangian time scale. In this section these quantities are specified for a homogeneous and stationary daytime planetary boundary layer using results of an experiment in Minnesota (Kaimal et
al, 1977). Trajectories of 1000 particles all released at the

same point are followed by computer using an equation for the turbulent speed fluctuation u'. This equation was first suggested by
Smith (1968) and further studied by Hanna (1978) and Reid (1979) :

$$u' (t + dt) = u' (t)R(dt) + u'' (t + dt), \qquad (4)$$

where t is time, dt is time step, R is the autocorrelation coefficient, and u'' is a random speed component chosen from a Gaussian
distribution with mean zero and variance σ^2_u given by

$$\sigma^2_{u''} = \sigma^2_{u'} (1 - R^2) \qquad (5)$$

Equations (4) and (5) were verified in an earlier paper by Hanna
(1978) by looking at observed velocity fluctuations of neutral
balloons.

Smith (1968) used equation (4) to investigate the conditioned
motion of particles. Initially, all particles were released with
the same speed, and so the results simulated relative diffusion
(a puff or an instantaneous plume). In Hanna's (1978) study, the
particles were initially released with a speed randomly picked
from a Gaussian distribution with mean zero and variance $\sigma^2_{u'}$.
The resulting diffusion corresponds to a plume sampled over a time
period of about one hour. The same procedure was used by Reid
(1979). Gifford (1980) converted equation (4) into a differential
equation equivalent to Langevin's equation, and obtained a general
analytical solution from which both the instantaneous and time-
averaged plume could be calculated. It is important to note that
there is a significant difference between the cross-wind standard
deviation σ_y of the particle distribution for the instantaneous
and time-averaged plumes. Initially, σ_y (instantaneous) is much
less than σ_y (time-averaged plume), but since σ_y (inst.) grows
as $t^{3/2}$ and σ_y (time avg) grows as t at intermediate times, the
gap decreases with time, and disappears altogether as both curves
approach a $\sigma_y \propto t^{1/2}$ law at large times. Clearly the proper
specification of the distribution of initial particle speed is
very important, depending on whether one is interested in simulating puffs or plumes.

To calculate diffusion in three dimensions, equation (5) is
written for the three components of turbulence. The functional
form of the autocorrelation coefficient R(dt) is not important in
equations (5) or (6), but for convenience the exponential form is
assumed :

$$R(dt) = e^{-dt/T_L}, \qquad (6)$$

where T_L is the Lagrangian time scale. This function gives a reasonable fit to most observed R(dt) curves (Pasquill, 1974).

In most realistic situations a mean wind component $\bar{u}$ must be accounted for, so that the total wind speed is given by the formula :

$$u = \bar{u} + u' \qquad (7)$$

If the mean wind represents an average over, for example, one hour then the turbulent component u' represents fluctuations with periods less than about one hour. The mean wind field can be measured or can be estimated using standard boundary layer formulas for flow over flat terrain. For complex terrain the mean wind field can be given by the output of a dynamic mesoscale model. McNider et al, (1980) and Ohmstede (1980) have used the model in this mode to estimate diffusion in complex terrain.

The diffusing particles in the model are immersed in a flow field in which turbulent energy, mean wind velocity, and Lagrangian time scales determine the particle trajectories. These variables are not well known in the stable and neutral PBL. Recent experiments in the full depth of the PBL by Caughey et al. (1979) in Minnesota have shown that the neutral PBL is transient, giving way soon after sunrise to the daytime unstable PBL. For these reasons, the daytime unstable PBL was studied first in this project both because of its frequency of occurrence and the accuracy of its empirical parameterizations.

The problem is simplified if we work with nondimensional variables. Suitable scaling parameters for the daytime PBL have been shown to be the mixing depth, z_i, and the convective velocity, $w_\ast$, defined by the relation :

$$w_\ast = u_\ast (- z_i/.4L)^{1/3}, \qquad (8)$$

where $u_\ast$ is the friction velocity and L is the Monin-Obukhov length. The parameter $w_\ast$ is essentially proportional to the cube root of the product of the surface heat flux, $\overline{- w'T'}$, and the mixing depth, z_i :

$$w_\ast = (z_i (g/T) \overline{w'T'})^{1/3} \qquad (9)$$

Using these scaling parameters, the following dimensionless equations can be written for the daytime PBL over flat terrain :

Mean Winds

Paulson's (1970) integration of the Businger et al (1971) flux profile equation is used for heights z/z_i less than 0.1 :

$$u/w_* = 2.85 \ (u_*/w_*) \ (\ln \ (z/z_o) - 2 \ \ln \ (0.5(1 + 1/\phi_m))$$

$$- \ \ln \ (0.5(1 + 1/\phi_m^2)) + 2 \ \tan^{-1}(1/\phi_m) - \pi/2) \qquad (10)$$

The dimensionless wind shear ϕ_m in an unstable boundary layer is defined by :

$$\phi_m(z/L) = (1 - 15 \ z/L)^{-1/4} \qquad (11)$$

In this lower layer we also assume that

$$v/w_* = w/w_* = 0 \qquad (12)$$

In the upper part of the daytime PBL, the Minnesota observations show that the wind speed remains nearly constant, but the wind direction veers by about 10°. (Kaimal et al, 1977). This observation is parameterized in the model by the assumptions (for $0.1 < z/z_i < 1$) :

$$v/w_* = -(\sin \alpha) \ (z/z_i - .1)(u/w_*(\text{at } z/z_i = .1))/.9 \qquad (13)$$

$$u/w_* = (u^2/w_*^2(\text{at } z/z_i = .1) - v^2/w_*^2)^{1/2} \qquad (14)$$

$$w/w_* = 0 \qquad (15)$$

Note that the mean vertical motions are assumed to be zero at all heights in this application.

Turbulence

The horizontal components of turbulent energy in the daytime PBL were shown by Panofsky et al. (1977) to be functions only of z_i/L :

$$\sigma_u/w_* = \sigma_v/w_* = (u_*/w_*)(12. - 0.5 \ z_i/L)^{1/3} \qquad (16)$$

Irwin (1979) suggests power law formulas for the vertical component σ_w/w_*. Since his formulas let σ_w/w_* approach zero as z/z_i approaches zero, the surface layer formula suggested by Panofsky et al. (1977) is used near the ground :

$$\sigma_w/w_* = 0.96(3z/z_i - L/z_i)^{1/3} \qquad z/z_i < .03 \qquad (17)$$

$$\sigma_w/w_* = \min \ (0.96(3z/z_i - L/z_i)^{1/3}, 0.763(z/z_i)^{.175}) \qquad .03 < z/z_i .4$$
$$(18)$$

$$\sigma_w/w_* = .722(1-z/z_i)^{.207} \qquad .4 < z/z < .96 \qquad (19)$$

$$\sigma_w/w_* = .37 \qquad .96 < z/z_i < 1 \qquad (20)$$

The "min" function in equation (18) is used to provide a continuous variation in σ_w/w_* between the Panofsky et al. (1977) and Irwin (1979) functions.

Lagrangian Time Scales

The Minnesota PBL experiment gives values of T_{Ep}, or the time period at which peak energy occurs in the Eulerian turbulent energy spectrum. Hanna (1980) showed that the Eulerian time scale T_E is approximately equal to $T_{Ep}/6$. To calculate the Lagrangian time scale, T_L, the following relation is used :

$$T_L/T_E = \beta = 0.6/i = 0.6 \ u/\sigma_v, \tag{21}$$

Hay and Pasquill (1961) suggested that the ratio β was inversely proportional to the intensity of turbulence, $i = \sigma_v/u$, and it is found that the constant 0.6 provides a best fit to observations. Equation (21) and the relation $T_E/T_{Ep} = 1/6$ are applied to the formulas for T_{Ep} given by Kaimal et al. (1977) to yield the following formulas :

$$T_{Lu} \ w_*/z_i = T_{LV} \ w_*/z_i = 0.15/(\sigma_u/w_*) \tag{22}$$

$$T_{Lw} \ w_*/z_i = 0.1 \ ((z/z_i)/(\sigma_w/w_*))/ \qquad z/z_i < 0.1$$
$$(0.55 + 0.38((z-z_o)/L)) \quad -(z-z_o)/L > 1 \tag{23}$$

$$T_{Lw} w_*/z_i = 0.59(z/z_i)/(\sigma_w/w_*) \qquad \begin{matrix} z/z_i < 0.1 \\ -(z-z_o)/L > 1 \end{matrix} \tag{24}$$

$$T_{Lw} w_*/z_i = 0.15(1-\exp(-5z/z_i))/(\sigma_w/w_*) \quad z/z_i > 0.1 \tag{25}$$

As an added restriction, $T_{Lw} \ w_*/z_i$ is not allowed to drop below its value at $z/z_i = 10^{-4}$. Otherwise, problems arise with the computational procedure, which necessarily employs a finite time step.

3. EXPECTED RESULTS BASED ON FIELD EQUATIONS

For times t less than the Lagrangian time scale T_L, it is expected that diffusion will follow equations (4), which can be rewritten with time as the independent variable :

$$\sigma_y/z_i = \sigma_v t/z_i \tag{26}$$

$$\sigma_z/z_i = \sigma_w t/z_i \tag{27}$$

Note from equation (16) that σ_v/w_* is a function only of z_i/L.

Thus equation (26) becomes :

$$\sigma_y/z_i = w_*(u_*/w_*)(12-.5z_i/L)^{1/3}t/z_i \qquad (28)$$

If equation (8) is substituted into equation (28), the following relation is obtained :

$$\sigma_y/z_i = (w_*t/z_i)(-z_i/.4L)^{-1/3}(12-.5z_i/L)^{1/3} \qquad (29)$$

For large z_i/L, this equation reduces to the simple relation :

$$\sigma_y/z_i \simeq 0.6\, w_*t/z_i \quad (z_i/L> -10,\ t/T_L> 1) \qquad (30)$$

Converting this simple formula from t to x yields :

$$\sigma_y/z_i \simeq 0.6\, (w_*/u)(x/z_i), \qquad (31)$$

thus introducing a new variable w_*/u, which is a function of z_i/z_o, z_i/L, and z_r/z_i (see equations (8) and (10)). We have the paradox of having a very simple formula in the time system, which isn't of much practical use, and a complex formula in the distance system, which is really the system we want.

For releases at heights such that $z_r/z_i > 0.1$, the wind speed can be assumed constant, and the nomogram in Figure 3 can be used to estimate w_*/u. Diffusion (σ_y/z_i) increases as z_i/z_o decreases or $|z_i/L|$ increases. Another way of saying this is that diffusion increases as roughness increases or instability increases. As an example of the application of this simple method for estimating σ_y/z_i, consider Pasquill-Gifford-Turner stability classes A,B, and C. Assume a mixing depth z_i of 1000 m and a roughness z_o of 0.01 m.
According to Golder (1972), classes A,B, and C correspond to Monin Obukhov lengths of $L > -7$ m, -20 m $< L < -7$ m, and $L < -20$ m, respectively, for an assumed roughness of 0.01 m. Let's arbitrarily assume that these criteria give values of z_i/L of about -300, -100, and -20, respectively. Figure 3 then provides values of about 0.50, 0.33, and 0.16 for w_*/u for classes A,B, and C. Substituting into equation (31), we obtain the predictions :

$$\sigma_y/z_i = \begin{cases} 0.3\ x/z_i & \text{class A} \\ 0.2\ x/z_i & \text{class B} \\ 0.1\ x/z_i & \text{class C} \end{cases} \qquad (32)$$

These predictions agree with the Pasquill-Gifford-Turner σ_y curves within a factor of two for downwind distances x in the range from the source to 10 km. Thus for release heights z_r such that $z_r/z_i > 0.1$, this analysis predicts that σ_y is not a function of release height and is in fact fairly well-simulated during unstable conditions by the Pasquill-Gifford-Turner curves. At $z_r/z_i < 0.1$,

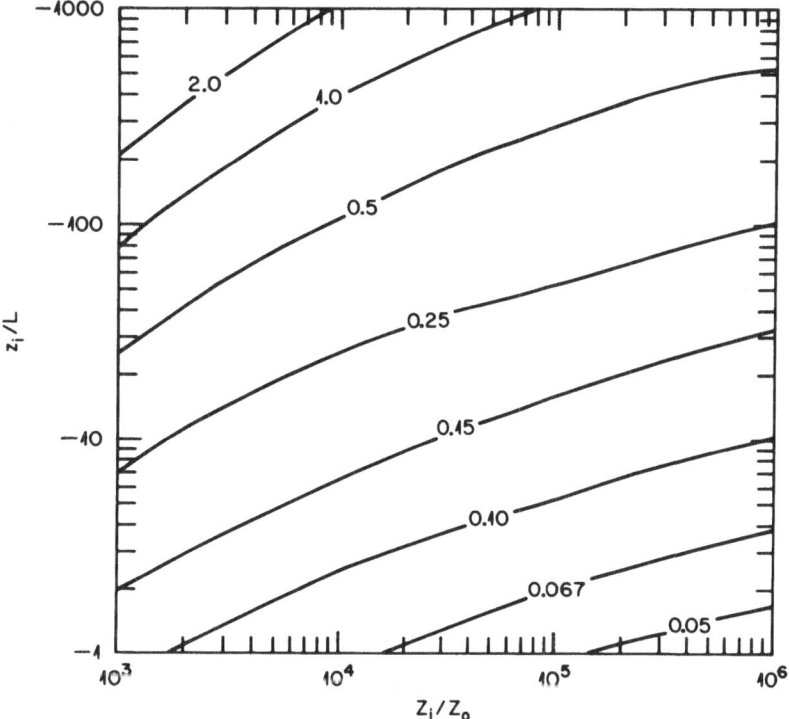

Fig. 3 : Nomogram of w_*/u as a function of z_i/z and z_i/L based on equations (8) and (10) and assuming $z/z_i \geqslant 0.1$.

release height enters the picture because of the dependence of wind speed (in equation (31)) on height.

Vertical diffusion σ_z, on the other hand, is predicted to be a function of release height at all heights, since equations (17) through (20) show that σ_w/w_* continuously varies with height. But midway in the mixed layer ($z/z_i = 0.5$), σ_w/w_* holds a value close to 0.6 and it results that equation (31) is valid for σ_z/z_i also. In otherwords, for releases in the middle of the mixed layer during unstable conditions, $\sigma_w \approx \sigma_v$, and consequently $\sigma_z \approx \sigma_v$. As the surface is approached and σ_w drops off relative to σ_v, σ_z will drop off relative to σ_y also.

4. RESULTS

The model was first run with input parameters that would permit comparison with other models. If this comparison showed that the model was valid, it could then be run for a large variety of input parameters. In all cases, 1000 particles are released from a point with initial velocities randomly picked from a Gaussian distribution of velocities with mean zero and standard deviation σ_u, σ_v, σ_w.

4.1 Comparison With Other Models

The solution to Taylor's (1921) analytical diffusion equation for homogeneous, stationary turbulence is :

$$\sigma_y^2 = 2\sigma_v^2 T_L (t/T_L - 1 + \exp(-t/T_L)), \qquad (33)$$

where an exponential autocorrelogram ($R = \exp(-t/T_L)$) is assumed. Our computer model was run for these conditions, yielding results that differed by less than 1 % from the analytical equation.

Comparisons were also made with Lamb's (1979) numerical predictions and Willis and Deardorff's (1978) laboratory simulations, using input parameters typical of the "Wangara" PBL experiments. The following parameters were used :

$$z_r/z_i = 0.26 \qquad z_i/z_0 = 10^6 \qquad z_i/L = -1100 \qquad \alpha = 0,$$

where z_r is the release height. To conform with the methods used by Lamb (1979) and Willis and Deardorff (1978), the standard deviations σ_y and σ_z are normalized by z_i, and downwind distance x is normalized by $z_i u/w_*$. A time step $dt = .01\ T_{Lu}$ was used, and time was converted to distance through the simple assumption $t = x/u$. There is fairly good agreement (± 50 %) among the predictions of the two numerical models and the laboratory model. These differences can be explained by the slightly different turbulent energy profiles used by the models.

4.2 Results of Computer Program

The computer model was applied to all combinations of z_i/L equal to -10 and -100; z_i/z_o equal to 10^3, 10^4, and 10^6; and z_r/z_i equal to .01, .025, .05, .18, .25, .50, .75 and .95. Wind direction turning angle, α, is set equal to zero in all these calculations. Gates, or y-z planes are set up at ten specified downwind distances and y and z positions of each particle are noted at the instant they pass through each gate. For a time step equal to $0.1T_{Lw}$, distance gates of x/z_i equal to .025, .05, .075, .10, .15, .20, .30, .50, .70, and 1.0, and 1000 particles, a typical run takes about 30 seconds on an IBM 3033 computer.

The last distance gate in these calculations is close enough to the source that the diffusion is usually within the initial $\sigma \propto t$ regime of Taylor diffusion. That is, the travel times are never much greater than the Lagrangian time scale. Diffusion at longer times or distances will be studied in the future. As a result, the output of the computer model should be compared with the expected results outlined in section 3.

As mentioned at the beginning of this section, three values of z_i/z_o (10^6, 10^4, and 10^3) and two values of z_i/L (-10,-100) were used. If the mixing height is about 1000 m, these values correspond to roughness lengths z_o of about .001, .1, and 1 m, respectively ; and Monin-Obukhov lengths L of about -100 m and - 10m, respectively. Golder's (1972) method of estimating Pasquill-Gifford-Turner turbulence types form z_o and L give the results in Table 1.

Table 1
Estimates of Pasquill-Gifford-Turner Turbulence Types
(Golder, 1972), Assuming z_i = 1000 m.

z_i/L \ z_i/z_o	10^6	10^4	10^3
- 10	D	C	B
- 100	B	A	A

The classes for z_i/z_o equal to 10^6 and 10^3 are only estimates, since they fall off the edges of Golder's nomogram. Furthermore, it must be emphasized that it is assumed that mixing depth, z_i, equals 1000 m. Despite these uncertainties, it is interesting to compare the σ_y and σ_z curves predicted by this model with the P-G-T σ_y and σ_z curves.

4.2.1 σ_y/z_i Variation With x/z_i For Set Values of z_i/z_o
and z_i/L, and Various Release Heights.

For each set of meteorological conditions, non-dimensional
release heights z_r/z_i equal to .01, .025, .05, .1, .25, .5, .75,
and .95 were used in the computer calculations. Figure 4 contains an example of curves of predicted σ_y/z_i versus x/z_i for
the case z_i/L = -10 and z_i/z_o = 10^4. The PGT curve from Table 1
is also drawn on the figure, using Briggs (1973) analytical formulas. It is seen that the PGT curve for C conditions is a factor
of two below the predicted curve. The predicted curves all
follow a + 1 power law, in agreement with theory for travel times
less than the Lagrangian time scale. In addition, the predicted
curves all follow the proper coefficients as given by equation (32).
Thus, the disagreement with the PGT curves is indeed a matter of
concern, since the predicted curves are based on sound theory and
excellent PBL field observations. It appears that σ_y should be
made a function of roughness in any empirical scheme. Through use
of the formula $\sigma_y = \sigma_\theta x$, and field observations of σ_θ, difficulties with the PGT curves are circumvented.

Figure 4 also shows the variation of σ_y/z_i with release height
Because σ_v is nearly constant with height, the variation of σ_y
depends solely on the variation of wind speed, u, with height.
Consequently σ_y/z_i varies by about 30 % between release heights
z_r/z_o of .01 and .25. The ratio $\sigma_y(z_r/z_i$ = .5)/ $\sigma_y(z_r/z_i$ = .01)
is given in Table 2 for the five other classes of roughness and
stability that were studied. We can tentatively conclude that
the assumption that σ_y is independent of release height is fairly
good (+ 30 %) for the daytime boundary layer with typical surface
roughness. Since the variation of σ_y with height in unstable conditions is mainly due to the variation of wind speed with height,
the ratios in Table 2 are mainly functions of roughness length,
with the greatest variation of σ_y predicted for the largest roughness.

Table 2

The Ratio $\sigma_y(z_r/z_i$ = .5)/ $\sigma_y(z_r/z_i$ = .01) at x/z_i = .1
for the Cases in Table 1. PGT Letter Class is Also
Given.

z_i/L \ z_i/z_o	10^6		10^4		10^3	
-10	.91	D	.76	C	.47	B
-100	.90	B	.76	A	.69	A

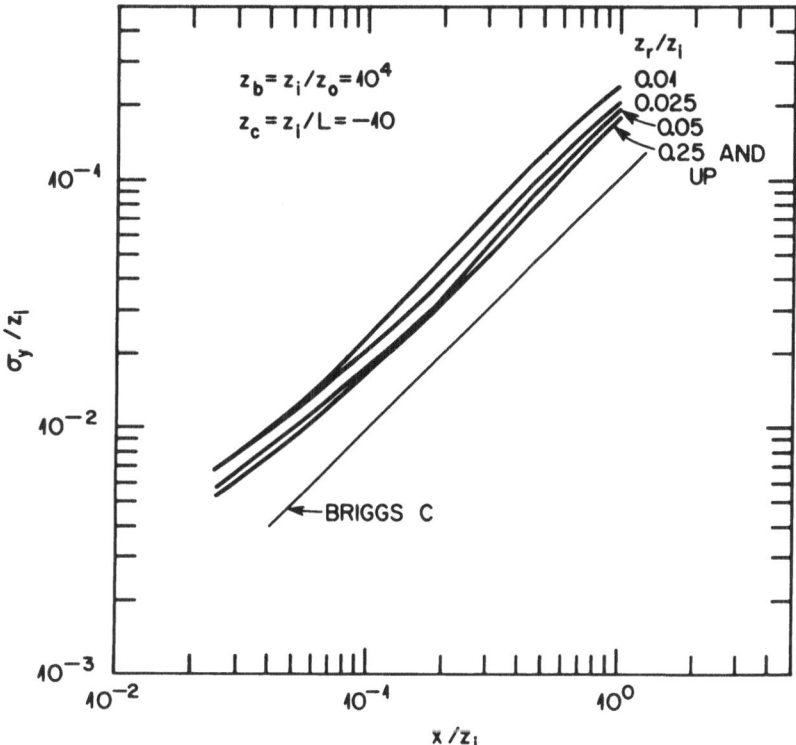

Fig. 4 : Model predictions of σ_y/z_i as a function of x/z_i, for $z_i/z_0 = 10^4$, $z_i/L = -10$, and several release heights. The corresponding PGT curve is also sketched in, using Briggs (1973) empirical formulas.

4.3.2 σ_z/z_i Variation With x/z_i for Set Values of z_i/z_o and z_i/L, and Various Release Heights.

The sets of conditions in Table 1 were also used to study σ_z. As an example of the results, Figure 5 contains curves of predicted σ_z/z_i versus x/z_i for the case $z_i/L = -10$ and $z_i/z_o = 10^4$. There is seen to be much more variation of σ_z than σ_y with release height, due to the variability of σ_w with height. The PGT curves are in the middle of the computer generated curves on this figure.

Since both σ_w and u increase with height between the surface and z/z_i equal to .5, it is not so intuitively obvious how σ_z will vary with height. Table 3 gives the ratio of σ_z at z_r/z_i = .5 to σ_z at z_r/z_i = .01 from Figure 5 and the data from the five other classes.
Largest variation occurs at moderate roughness and greatest instability. Variation of a factor of two is common. We can tentatively conclude that it may be necessary to account for the variation of σ_z with release height in future empirical models. This can be done through use of theoretical estimates or observations of the ratio σ_w/u.

Table 3

The Ratio $\sigma_z(z_r/z_i = .5)/\sigma_z(z_r/z_i = .01)$ at x/z_i = .1 for Cases in Table 1.

z_i/L \ z_i/z_o	10^6	10^4	10^3
-10	1.70	1.83	1.67
-100	2.33	2.33	1.41

Curves of σ_y and σ_z are plotted in this paper for only one combination of z_i/L and z_i/z_o. The other ten figures are contained in a larger annual report to the Nuclear Regulatory Commission. Copies of this report are available from the author.

ACKNOWLEDGEMENT : This research was performed under an agreement among the Nuclear Regulatory Commission, the Department of Energy, and the National Oceanic and Atmospheric Administration.

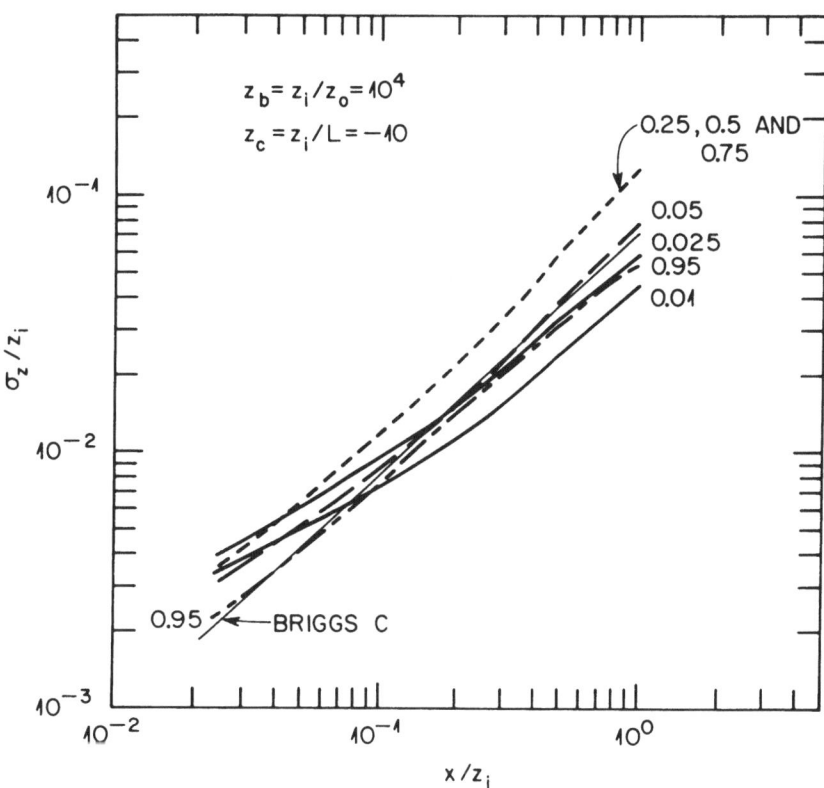

Fig. 5 : Model predictions of σ_z/z_i as a function of x/z_i,
for $z_i/z_o = 10^6$, $z_i/L = -10$, and several release
heights.
The corresponding PGT curve is also sketched in,
using Briggs (1973) empirical formulas.

REFERENCES :

Briggs, G. A., 1973, Diffusion Estimation for Small Emissions, in
    Environmental Research Laboratories, Air Resources Atmosphere
    Turbulence and Diffusion Laboratory 1973 Annual Report, USAEC
    Report ATDL-106, National Oceanic and Atmospheric Administra-
    tion, December 1974.

Businger, J. A., J. C. Wyngaard, Y. Izumi and E. F. Bradley, 1971,
    Flux profile relationships in the atmosphere surface layer,
    J. Atmos Sci., 28 : 181 - 189.

Caughey, S. J., J.C. Wyngaard and J. C. Kaimal, 1979, Turbulence in
    the evolving stable boundary layer, J. Atm. Sc. 36 : 1041 -
    1052.

Deardorff, J.W., 1974, Three dimensional numerical study of the
    height and mean structure of a heated planetary boundary
    layer, Bound. Layer Meteorol., 7 : 81 - 106.

Doran, J. C., T. W. Horst and P. Nickola, 1978, Experimental ob-
    servations of the dependence of lateral and vertical disper-
    sion characteristics on source height, Atmos. Environ., 12 :
    2259 - 2263.

Gifford, F. A., 1976, Turbulent diffusion typing schemes-a re-
    view. Nuclear Safety, 17 : 68 - 86.

Gifford, F. A., 1980, Study of the horizontal diffusion length.,
    Annual Report to NRC, ATDL, P.O. Box E, Oak Ridge, TN.37830.

Golder, D, 1972, Relations among stability parameters in the sur-
    face layer, Boundary-Layer Meteor., 3 : 47 - 58.

Hanna, S. R., 1978, Some statistics of Lagrangian and Eulerian
    wind fluctuations, J. Appl. Meteorol., 18 : 518 - 525.

Hanna, S. R., 1980, Lagrangian and Eulerian time scale relations
    in the daytime boundary layer, To be published in J. Applied
    Meteo..

Hay, J. S. and F. Pasquill, 1959, Diffusion from a continuous
    source in relation to the spectrum and scale of turbulence,
    Advances in Geophysics, 6, Academic Press : 345 - 365.

Irwin, J. S., 1979, Estimating plume dispersion - a recommended
    generalized scheme, Preprints, Fourth Symposium on Turbulence
    Diffusion, and Air Pollution. Am. Meteorol. Soc., 45 Beacon
    St., Boston, MA 02108 : 62 - 69.

Kaimal, J. S., J. C. Wyngaard, D. A. Haugen, O. R. Cote, Y. Izumi, J. J. Caughey and C. J. Readings, 1977, Turbulence structure in the convective boundary layer, J. Atmos. Sci., 33 : 2152 - 2169.

Lamb, R. G., 1979, The effects of release height on material dispersion in the convective planetary boundary layer. Proceedings, Fourth Symp. on Turb., Diff., and Air Poll., AMS, 45 Beacon St., Boston, MA : 27 - 33.

McNider, R. T., and S. R. Hanna, and R. A. Pielke, 1980, Sub-grid scale plume dispersion in coarse resolution mesoscale models. Proceedings, Second Joint Conference on Applications of Air Pollution Meteorology, Am. Meteorol. Soc., 45 Beacon St., Boston, MA, 02108 , 424 - 429.

Ohmstede, W.D. and E. B. Stenmark, 1980 : A model for characterizing transport and diffusion of air pollution in the battlefield environment. Proceedings of Second Joint Conf. on Applic. of Air Poll. Meteorol., New Orleans, AMS, 45 Beacon St., Boston, Mass 02108, 416 - 423.

Panofsky,H. A., H. Tennekes, D. H. Lenschow, and J. C. Wyngaard, 1977, The characteristics of turbulent velocity components in the surface layer under convective conditions, Bound. Lay. Meteorol., 11 : 355 -361.

Pasquill, F., 1974, Atmospheric Diffusion, 2nd ed., J. Wiley and Sons, New York, 429 pp.

Paulson, C. A., 1970, The mathematical representation of wind speed and temperature profiles in the unstable atmospheric surface layer, J. Appl. Meteorol. 9 : 857 - 861.

Reid, J., 1979, Markov chain simulations of vertical dispersion in the neutral surface layer for surface and elevated releases, Bound. Lay. Meteorol., 16 : 3 - 22.

Smith, F. B., 1968 : Conditioned particle motion in a homogeneous turbulent field. Atmos. Environ., 2, 491 - 508.

Taylor, G. I., 1921, Diffusion by continuous movements. Proc. London Math. Soc. Ser 2., 20 : 196.

Vogt, K. J., H. Geiss and G. Polster, 1978, New sets of diffusion parameters resulting from tracer experiments with 50 and 100 m release height, Proc. of Ninth Int. Tech. Meeting on Air Poll. Modeling and its Applic., No. 103, NATO Comm. on Challenges to Modern Society : 221 - 239.

Willis,G. E. and J. W. Deardorff, 1978, A laboratory study of dispersion from an elevated source in a convective mixed layer. Atmos. Environ., 12 : 1305 - 1313.

DISCUSSION

F. NIEUWSTADT   In your calculations you use a symmetric probability distribution for the vertical velocity fluctuations. In convective conditions, however, this probability distribution is very skew . How does this influence your calculations ?

S.R. HANNA   I do not believe that the skewed distribution function would influence σ_y and σ_z calculations very much. However, the height of the maximum concentration or the mean height of the particles would be changed if a skewed distribution were used.

L. JANICKE   A numerical simulation model of turbulent diffusion has to meet certain conditions to be consistent with the physical processes it is supposed to simulate. For example, it should reproduce a constant concentration field as an equilibrium state. Models of the type you used, however, often have the tendency to accumulate particles in regions where σ_w is small. Did you check your model in this respect ?

S. R. HANNA   With a Gaussian probability distribution function assumed for vertical velocity fluctuations and a value of $\delta\sigma_w / \delta z$ not equal to zero, we do have problems in the model with accumulation of particles at heights with low σ_w . In unstable condition, particle accumulation tends to occur near the ground surface. This problem can be partially solved by allowing the particles to reflect off $z z_o$ rather than z_o, and by not permitting the Lagrangian time scale to drop below some minimum value. A more physical solution would be to assume a skewness in the vertical velocity probability distribution function, where the skewness is related to $\delta\sigma_w / \delta z$.

DISPERSION NEAR TO A TALL STACK

R. Steenkist

N.V. KEMA
Environmental Research Department
Arnhem, The Netherlands

F.T.M. Nieuwstadt

Royal Netherlands Meteorological Institute
De Bilt, The Netherlands

ABSTRACT

In many places data for σ_y have been measured. By combining these data, we found that σ_y for non-buoyant plumes from tall stacks can be given by

$$\sigma_y = 0.079 \ x^{0.86} \sigma_A^{0.57}$$

independent of roughness, stability and source height. For a rising buoyant plume

$$\sigma_z = 0.6 \ \Delta h$$

has been found. Using this, the maximum ground concentration for a rising plume is found when the plume rise is 2 x stack height at a distance $x = (0.6 \ h_s \ u \ Q_H^{0.33})^{1.5}$, so that this distance depends on the wind velocity. A formula for the maximum concentration is given.

INTRODUCTION

The σ_y and σ_z values are essential in the calculation of the concentration in the Gaussian plume model. Of these two parameters σ_y can be measured easiest, whereas σ_z is often a derived value. It will be tried to find values for these parameters for a buoyant plume. The distance from source to acceptor point has been kept restricted.

σ_y FOR NON-BUOYANT PLUMES

A number of σ_y measuring results from sources without buoyancy are known. These σ_y values have been used by considering them statistically. The σ_y figures used here originate from Hanford[1], Jülich[2], Copenhagen[3], Möl[4], and our measurements in Cabouw[5]. Nearly 300 data were available. A list with the information on the conditions during the various measurements is given in Table 1. The ranges for the various parameters for all the measurements are shown in Table 2. When processing the data, the following relation between σ_y and the various parameters was assumed:

$$\sigma_y = t^{\alpha} \sigma_A^{\beta} u^{\gamma} \cdot (\text{source height})$$

in which t = x/u. In succession, the values of α, β and γ and the relation with the source height were determined using the multi-regression analysis on the logarithms of the parameters, which give the following result:

$$\sigma_y = t^{0.86 \pm 0.04} \cdot \sigma_A^{0.57 \pm 0.08} \cdot u^{0.80 \pm 0.15} \cdot \exp(-2.35 \pm 0.45)$$

The tolerances given are the 95% limits. σ_y appears not to depend significantly on the source height. The coefficients of u and t are not significantly deviating so that for the model under consideration the equation x = t.u can be used. A new calculation gives the following result:

$$\sigma_y = \exp(-2.55 \pm 0.08)\ x^{0.86 \pm 0.04} \sigma_A^{0.57 \pm 0.08}$$

or rather

$$\sigma_y = 0.079\ x^{0.86} \sigma_A^{0.57} \qquad\qquad (1)$$

(x in m, σ_A at source height in degrees). σ_y appears to be independent of wind velocity and source height directly. Measurements at Karlsruhe[6] showed a relation between σ_y and x according to $\sigma_y = C\ x^{0.871}$, which agrees with our results. These measurements did not show the relation with σ_A as we found it. Our investigation was based on the assumption that σ_y depends on σ_A by means of a power function. This was checked in Fig. 1. It appears that the power function provides a reasonable description of the relation.

The standard deviation of ln σ_y in the original data can be explained for 85 % by the relation with x and for 7 % by the relation with σ_A. The standard error of estimate is 0.39 so that 95 % of the measured values holds ln (σ_y(measured)/ σ_y(predicted) 2.0.39; σ_y(m)/ σ_y (p) is between .5 and 2.

In the foregoing it was assumed that the observations used, were drawn from the same population. As a check, the average value of ln(σ_y(m)/ σ_y(p)) for each set is presented in Fig. 2. These average values appear to differ so little that the deviation could easily be explained from differences in measuring equipment; there

Table 1. Conditions when measuring σ_y

location	stability	local situation	roughness	wind velocity	source height	σ_A	distance
Hanford	stable and neutral	sagabush 1+2 m		2.5-10 m/s	56 and 111 m	2.5-11°	400-12800 m
Jülich	neutral and instable		1 - 2 m	3.5-12 m/s	50 and 100 m	-16°	200-10000 m
Copenhagen	instable and neutral	suburban		3.4-1.6 m/s	115 m	-16°	1900- 6000 m
Mol	stable and neutral	meadows and forest		1.8- 9 m/s	69 m	-20°	200- 8000 m
Cabouw	neutral and slightly instable	meadows	5 cm	4.2-10 m/s	80 and 200 m	4.5- 7°	3100- 4300 m

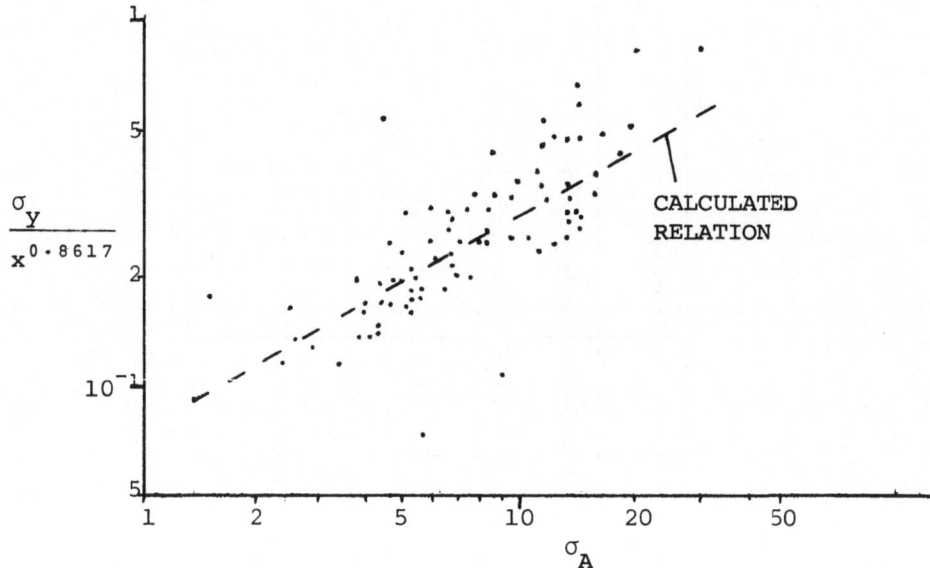

Fig. 1.   Check of the exponential relation between σ_y and the standard deviation of the horizontal fluctuations of wind directions at source height σ_A. (The number of dots has been reduced by combining measurement results from one point of time but taken at various distances).

Table 2. List of measuring conditions

stability	in-stable – stable
σ_A	4–16°
distance	200–13000 m
wind velocity	4–11 m/s
source height	50–200 m
roughness	0.05–2 m

σ_A and u at source height

is no visible difference between the measurements at the various locations nor between the various source heights.

Summarizing within the limits of Table 2 the following applies for sources without buoyancy:

$$\sigma_y = 0.079 \, x^{0.86} \sigma_A^{0.57}$$

in which σ_y appears not to depend on source height, nature of the local ground surface and wind velocity. As for tall stacks especially the independence of σ_y of the source height is essential.

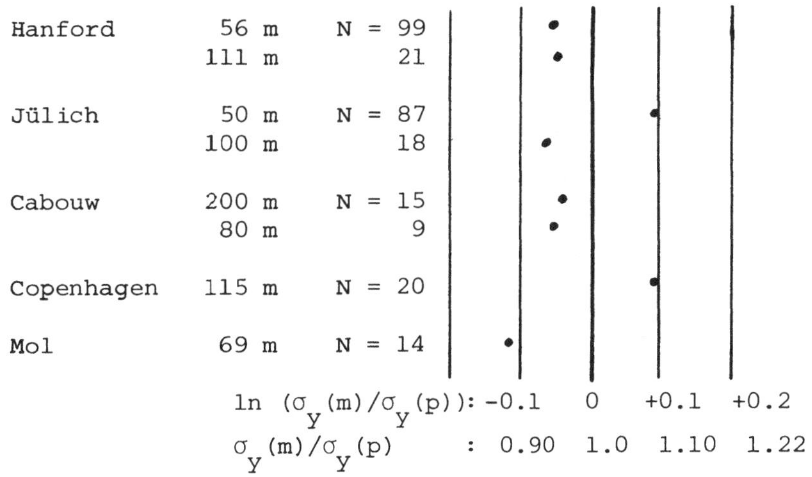

Hanford 56 m N = 99
111 m 21
Jülich 50 m N = 87
100 m 18
Cabouw 200 m N = 15
80 m 9
Copenhagen 115 m N = 20
Mol 69 m N = 14
ln (σy(m)/σy(p)): -0.1 0 +0.1 +0.2
σy(m)/σy(p) : 0.90 1.0 1.10 1.22

Fig. 2. Average values of ln $(\sigma_y(m)/\sigma_y(p))$ for the various measuring series.

This aspect can also be seen with regard to turbulence. The lateral dispersion is caused by lateral wind fluctuations. Kaimal[8] measured the power spectrum of these fluctuations up to a height of 22 m. In the stable situation the normalized spectrum was found to depend on $f/4f_m$, in which $f = nz/u$ and for f_m (at L positive) was found $f_m = 0.65\ z/L + 0.2$. For $z/L > 0.3$ (stable situation) f/f_m can be approximated by:

$$f/4f_m = 0.38\ n\ L/u \text{ (see Fig. 3.)}$$

1.0
.5
.2
.1
.05
.02
n sv(n)/v²
.1
1.0
10
100
f/fo = f/4fm = .38 n L/u

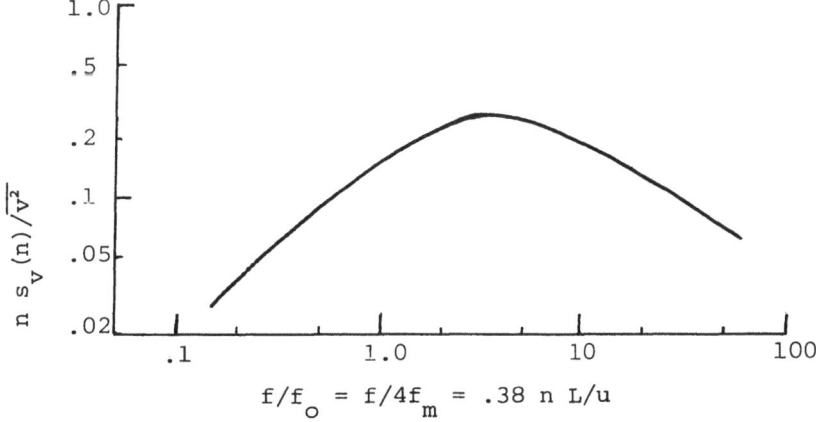

Fig. 3. Logarithmic spectra of v, normalized by the variance, plotted against modified f scale[8].

From this normalized power spectrum σ_y can be found by [9]:

$$\sigma_y^2 = \overline{v^2}\ T^2 \int_0^\infty (s_v/\overline{v^2})(1 - \sin^2 \pi n\tau/(\pi n\tau)^2)\sin^2 (\pi nT/\beta)/(\pi nT/\beta)^2\ dn$$

$$= \sigma_A^2\ x^2 \int_0^\infty (s_v/\overline{v^2})(1-\sin^2 \pi n\tau/(\pi n\tau)^2)\sin^2 (\pi nT/\beta)/(\pi nT/\beta)^2\ dn$$

$$\cong \sigma_A^2\ x^2 \int_{n\cong 0.1/\tau}^{n\cong\beta/T} s_v(n)/\overline{v^2}\ dn$$

$$= \sigma_A^2\ x^2\ u/(0.38L) \int_{n\cong 0.1/\tau}^{n\cong\beta/T} s_v(n)/\overline{v^2}\ d(0.38nL/u)$$

$$\sigma_y^2 \cong \sigma_A^2\ x^2\ u/(0.38L) \int_{N=0.38.0.1L/u\tau}^{N=0.38\beta L/(u T)} s_v(N)/\overline{v^2}\ dN$$

An estimation of the lower limit gives, with $z/L > 0.3$ as assumed before, $z < 200$ m for the mixing layer in stable situation and $\tau \cong 3600$ s, $N \cong 0.02$. As in Fig. 3 can be seen this limit can be replaced by zero. In the upper limit $u.T = x$

$$\sigma_y^2 \cong \sigma_A^2\ x^2\ u.\ 2.6/L \int_0^{0.38\beta L/x} s_v(N)/\overline{v^2}\ dN$$

The integral is a function of L or σ_A and x and independent of z. σ_y expressed as a function of σ_A and x is not directly dependent on z, but since u is a function of z, σ_y is, via u, a weak function of z (about $z^{.15}$). This dependancy is not found in the measurements used before.

In the convective situation Kaimal[10] found measuring at 60 to 1200 m, that the normalized spectrum depends on nz_i/u, in which z_i stands for mixing height. From this, it follows, as in the stable situation, that σ_y is only via u a very weak function of z ($\cong z^{0.05}$).

No satisfactory explanation was found for the neutral situation.

From the foregoing it appears that equation (1) is likely to apply to the entire mixing layer.

σ_z FOR BUOYANT PLUMES

The power spectrum of the vertical wind velocity in the convective situation appears to be dependent on height[10]. The value of σ_z for a non-buoyant plume will depend on height.

Generally tall stacks have buoyant plumes. Briggs' formula[11] is used rather generally for the calculation of the plume rise for these sources. For the derivation of this formula the relation $\sigma_z = c.\Delta h$ is used. For "radius" of the instantaneous plume holds $r = (0.5$ à $0.6)\Delta h$[11]. (Briggs uses a value of 0.6 for the derivation of the plume rise formula).

This radius has been determined by the measurements of photographs of plumes; therefore it can be expected that the edges of the plume have been measured. The radius will be about $2 \times \sigma$ (instantaneous): $\sigma_z(i) \cong 0.3 \cdot \Delta h$. Pasquill[12] suggests $\sigma(i) = 0.33 \Delta h$. Pasquill[13] gives a factor of 2 for the ratio $\sigma_z(\tau)/\sigma_z(i)$ near the stack (τ of the order of tens of minutes); $\sigma_z(\tau) = 0.6 \Delta h$. Lidar measurements resulted in[14] $\sigma_z(\tau) = -9.4 + 0.56 \Delta h$ respectively[15] $\sigma_z(\tau) = 0.62 \Delta h$, which corresponds very well with our estimation.

Generally it is assumed that the plume keeps rising until the dispersion due to atmospheric turbulences becomes faster than that due to buoyancy of the plume; as long as the plume rises it will hardly be affected by atmospheric turbulences. The instantaneous dispersion of the rising plume hardly depends on stability. $\sigma_z(\tau) = 0.62 \Delta h$ has been measured[15] in neutral and convective circumstances. In this situation also $\sigma_z(\tau)$ is independent of stability.

MAXIMUM GROUND CONCENTRATION

If the maximum ground concentration occurs while the plume is rising, this concentration, as well as its location, can be found from:

$$\chi = \frac{\sigma_z(\tau)}{\sigma_y(\tau)} \quad \frac{Q}{\pi u \sigma_z^2(\tau)} \exp \frac{-(h_s + \Delta h)^2}{2 \sigma_z^2(\tau)}$$

$$\sigma_z(\tau) = a \Delta h$$

With $d\chi/d(\Delta h) = 0$, this gives (if $\sigma_z(\tau)/\sigma_y(\tau)$ is assumed to be independent of Δh):

$$\Delta h/h_s = 2/((1 + 8a^2)^{0.5} - 1)$$

which at $a = 0.6$ becomes $\Delta h = 2.06 \, h_s$.

The maximum concentration occurs if the height of the plume is 3 times the height of the stack.

A combination of this value with Briggs' formula ($\Delta h = 3.34 \, Q_H^{0.33} u^{-1} x^{0.66}$) gives for the location of the maximum concentration:

$$x(\text{max}) = \left(\frac{0.6}{(\sqrt{1+8a^2} - 1)} \quad \frac{h_s u}{Q_H^{0.33}} \right)^{1.5}$$

which for $a = 0.6$ becomes $x(\text{max}) = (0.6 \, h_s u/Q_H^{0.33})^{1.5}$. The distance to the maximum concentration is proportional to $u^{1.5}$; at low wind velocities the maximum concentration is found near the stack, at higher wind velocities the distance is larger. When $a = 0.6$, the following equation applies for the maximum concentration:

$$\frac{\chi(\text{max})u}{Q} \cdot \frac{h_s^2 \pi}{2} \cdot \frac{\sigma_y}{\sigma_z} = 1.5 \; 10^{-2}$$

Since $\sigma_z = a \Delta h$ has been measured in neutral and convective conditions[15], the formula for $\Delta h/h_s$ and χ (max) are also applicable for this conditions. The formula for x(max) depends on the Briggs' formula and is for the neutral situation only.

The χ(max) equation contains $\sigma_y(\tau)/\sigma_z(\tau)$, which refers to a buoyant plume, whereas the formula for σ_y derived before is for a plume without buoyancy. Assuming that this σ_y is only caused by meandering of the wind, σ_y for a buoyant plume can be estimated by adding the instanteneous dispersion for the buoyant plume to σ_y derived before. Approximately $\sigma_y(i) = \sigma_z(i) = 0.5\sigma_z(\tau)$ applies to the buoyant plume.

$$\frac{\sigma_y(\tau)}{\sigma_z(\tau)} = \left(\frac{\sigma_y^2(\text{meander}) + \sigma_y^2(i)}{4\sigma_y^2(i)}\right)^{.5} = 0.5(1 + \sigma_y^2(m)/\sigma_y^2(i))^{0.5}$$

This results in:

$$\sigma_y(\tau)/\sigma_z(\tau) = 0.5\ (1 + 0.0022\ x^{0.40}\ \sigma_A^{1.14} Q_H^{-0.66} u^2 a^{-2})^{0.5}$$

When a = 0.6, the combination of this result with the formula for χ(max) results in:

$$\frac{\chi(\text{max})u}{Q} \cdot h_s^2(1 + 4.58\ 10^{-3}\ h_s^{0.6} Q_H^{-0.86} \sigma_A^{1.14} u^{2.6})^{0.5} = 1.9\ 10^{-2}$$

on the location x(max) = $(0.6\ h_s u/Q_H^{0.33})^{1.5}$.

NUMERICAL EXAMPLE

As a numerical example χ(max)/Q and x(max) were determined with the method as stated above, and in accordance with the Dutch national model[16] in which the σ_z values of Singer/Smith are used, supplied with the σ_y values of Singer/Smith, for a source with h_s = 150 m, Q_H = 100 MW and σ_A = 7° (neutral). For u has been taken u = 5 respectively 9, and 15 m/s, 9 m/s is considered to be the normal value at the plume height at x(max) ($\approx$ 450 m).

The following results were found:

	u = 5 m/s		u = 9 m/s		u = 15 m/s	
	χ/Q	x(max)	χ/Q	x(max)	χ/Q	x(max)
this model :	$1.3\ 10^{-7}$	950 m	$4.2\ 10^{-8}$	2400 m	$1.3\ 10^{-8}$	5200 m
Dutch model:	$2.0\ 10^{-7}$	3100 m	$1.1\ 10^{-7}$	3100 m	$6.5\ 10^{-8}$	3100 m

The range of x(max) found with this model agrees fairly well with x(max) of the Dutch model. The concentrations calculated with this model are less than calculated with the Dutch model.

For the foregoing calculations it was assumed that a = 0.6. Both locations and maximum concentrations depend on the value of a. The relation for the 3 situations considered above are shown in Fig. 4; when a decreases, the distance at which the maximum is found will

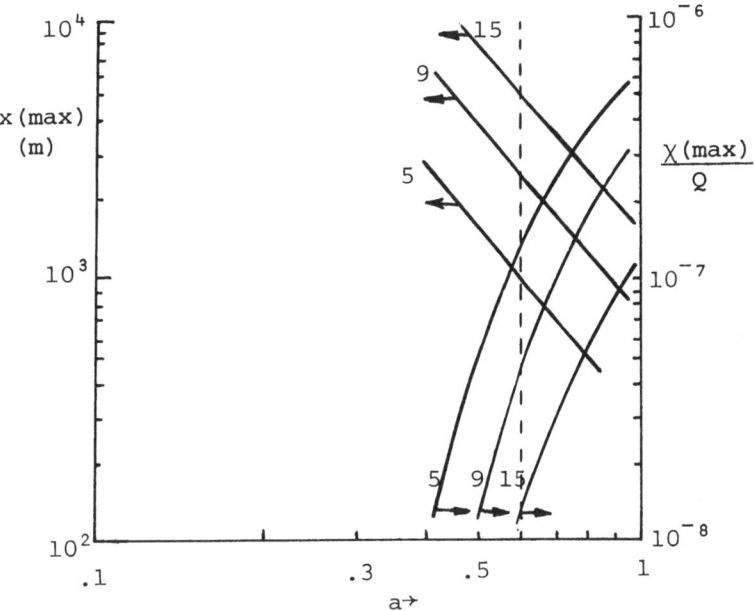

Fig. 4. Maximum concentrations with corresponding distances as a
function of a (σ_z = aΔh) at h_s = 150 m, Q_H = 100 MW,
u = 5, respectively 9 and 15 m/s.

increase and the concentration will decrease. As to x(max), a = 0.6
seems to be a good value.

CONCLUSION

From the model used here it is concluded that if a plume can
rise up to 3 times the stack height, the maximum ground concentration
is reached while the plume is rising. This maximum concentration is
found at a distance from the stack at x(max) = $(0.6 \ h_s .u/Q_H^{0.33})^{1.5}$.
The maximum concentration can be calculated with:

$$\chi(max)uQ^{-1}h_s^2(1 + 4.58 \ 10^{-3}h_s^{0.6}Q_H^{-0.86}\sigma_A^{1.14}u^{2.6})^{0.5} = 1.9 \ 10^{-2}.$$

The above equations are based on $\sigma_z(\tau)$ = aΔh using a = 0.6. This
value of a seems to give good results for x(max).

Advantages of the method, described here, are
- in the calculations, the final plume rise, which is difficult to detect, is not used.
- the distance, the maximum concentration is found, depends
  on the wind velocity. In the theoretical limit of u = o
  (although the method cannot be used in this situation) the
  maximum is found at the base of the stack as it should be.

REFERENCES

1. P.W. Nickola, The Hanford 67 series, Batelle PNL 2433 UC 11
   (1977).
2. K.J. Vogt, H. Geifs, H. Nordsieck, G. Polster and F. Rohloff,
   Untersuchungen zur Ausbreitung von Abluftfahnen in der
   Atmosphäre, Jül. 998-ST (1973).
3. S.E. Gryming, E.L. Petersen, E. Lyck, Elevated source SF_6 tracer
   dispersion experiments in the Copenhagen area, NATO-CCMS
   108 : 119 (1979).
4. H. Bultynck, L. Malet, Diffusion turbulente des effluents émis
   dans l'atmosphère par une source élevée a émission continue
   en relation avec la stabilité de l'air, Studiecentrum voor
   Kernenergie BLG 434 (1969).
5. H. van Duuren and F.T.M. Nieuwstadt, Dispersion experiments from
   the 213 m high meteorological mast at Cabouw in The Netherlands, Atmospheric Pollution 1980, Proc. 14th Intern. Coll.,
   Paris : 77 (1980).
6. W.G. Hübschman, K. Nester, P. Thomas, Diffusion of atmospheric
   pollutants being emitted from tall stacks, NATO-CCMS 108 :
   339 (1979).
7. M.J. Meroney, Facts from figures, Penguin : 296 (1964).
8. J.C. Kaimal, J.C. Wijngaard, IJ. Izumi, O.R. Coté, Spectral
   characteristics of surface layer turbulence, Quart. J.R.
   Met. Soc. 98 : 563 (1972).
9. F. Pasquill, Atmospheric diffusion, John Wiley and sons ltd : 13
   (1972).
10.J.C. Kaimal, J.C. Wijngaard, D.A. Hangen, O.R. Coté, IJ. Izumi,
   Turbulence structure in the convective boundary layer, J.
   Atm. Sc. 33 : 2152 (1976).
11.G.A. Briggs, in : "Lectures on air pollution and environmental
   impact analysis", AMS : 73 (1975).
12.S.R. Hanna, G.A. Briggs, J. Deardorff, B.A. Egan, F.A. Gifford
   and F. Pasquill, Meeting review, BAMS 58, 12 : 1307 (1977).
13.F. Pasquill, Atmospheric diffusion, John Wiley : 282 (1974).
14.R.M. Hoff,F.A. Froude, Lidar observations of plume dispersion
   in Northern Alberte, Atm. Env, 13 : 35 (1979).
15.J.C. Weil, J.L. Altman, Stack plume characterization and model
   assessment with lidar data, NATO-CCMS 103 : 197 (1978).
16.Modellen voor de berekening van de verspreiding van luchtverontreiniging inclusief aanbevelingen voor de waarden van de
   parameters in het lange termijnmodel (Models for the calculation of the dispersion of air pollution including re-
   commandations for the values of parameters in the long
   term model), Staatsuitgeverij, The Hague (1976).

DISCUSSION

E. RUNCA   Is your model related
                to Gifford's fluctuating plume model ?

R. STEENKIST   No, it is not.

A STATISTICAL APPROACH FOR ESTIMATING ATMOSPHERIC

STABILITY CLASSES FROM NEAR-GROUND OBSERVATIONS

S. Cieslik, H. Bultynck and J.G. Kretzschmar

Nuclear Energy Centre
Boeretang 200
B-2400   Mol, Belgium

INTRODUCTION

The controlling role of the stability of the atmospheric boundary layer (ABL) in the process of pollutant dispersal is commonly recognized and is the rationale of the present study.

The ABL stability is a rather ill-defined concept and its expression varies from author to author. Physically, it is a combination of mechanical and thermal stability; mathematically, it generally involves a function of the vertical temperature gradient and of the wind speed. Many proposals have been made to quantify this concept. Among them, the Richardson number, proportional to the ratio of the temperature and wind speed vertical gradients, is often used, but it does not provide a satisfactory representation of the stability of the global ABL, because it refers to a particular height. Several authors expressed the ABL stability in classes (Pasquill, 1961, Singer and Smith, 1966; Klug, 1969; U.S.NRC, 1972; Bultynck and Malet, 1972; Gifford, 1976). The correspondence between these various typing schemes was examined by Kretzschmar and Mertens (1980). In the present work, we adopted the classification scheme of Bultynck and Malet (1972), which is used in all dispersion calculations which are made in Belgium.

The classes used here are defined by the value of the parameter

$$S = \frac{G}{u_{69}^2} \qquad (1)$$

with

$$G = \frac{\theta_{114} - \theta_8}{114 - 8} \qquad (2)$$

where u_{69} is the half-hourly average of the wind speed at a height of 69 m; θ_8 and θ_{114} are half-hourly averages of the potential temperatures at heights of 8 and 114 m, respectively. The definition of the classes (E_1, E_2 ... E_i, or E_i) is given in table 1 and illustrated in fig. 1.

In Belgium, dispersion calculations for impact assessment studies are mostly based on the so-called IFDM (Immission Frequency Distribution Model), which has been developped by Kretzschmar et.al. (1977, 1978). Although whatever diffusion-typing scheme can be used as an input to this model, routine calculations made in Belgium make use of the E_i classification scheme, since the E_i classes are determined on a routine basis from observations made on the Belgian meteorological towers. The location of these is shown in fig. 2.

Table 1. Definition of the E_i stability classes

Class	Qualitative description	Criteria defining the class		
		Lower limit for S $(Km^{-3}s^2)$	Upper limit for S $(Km^{-3}s^2)$	Limit for u_{69} (ms^{-1})
E_1	very stable	5.6×10^{-4}	–	< 11.5
E_2	stable	5.6×10^{-5}	5.6×10^{-4}	< 11.5
E_3	near-neutral	$- 10^{-4}$	5.6×10^{-5}	< 11.5
E_4	slightly unstable	$- 5.6 \times 10^{-4}$	$- 10^{-4}$	< 11.5
E_5	unstable	$- 2 \times 10^{-3}$	$- 5.6 \times 10^{-4}$	< 11.5
E_6	very unstable	–	2×10^{-3}	< 11.5
E_7	neutral by strong wind	–	–	$\geqslant$ 11.5

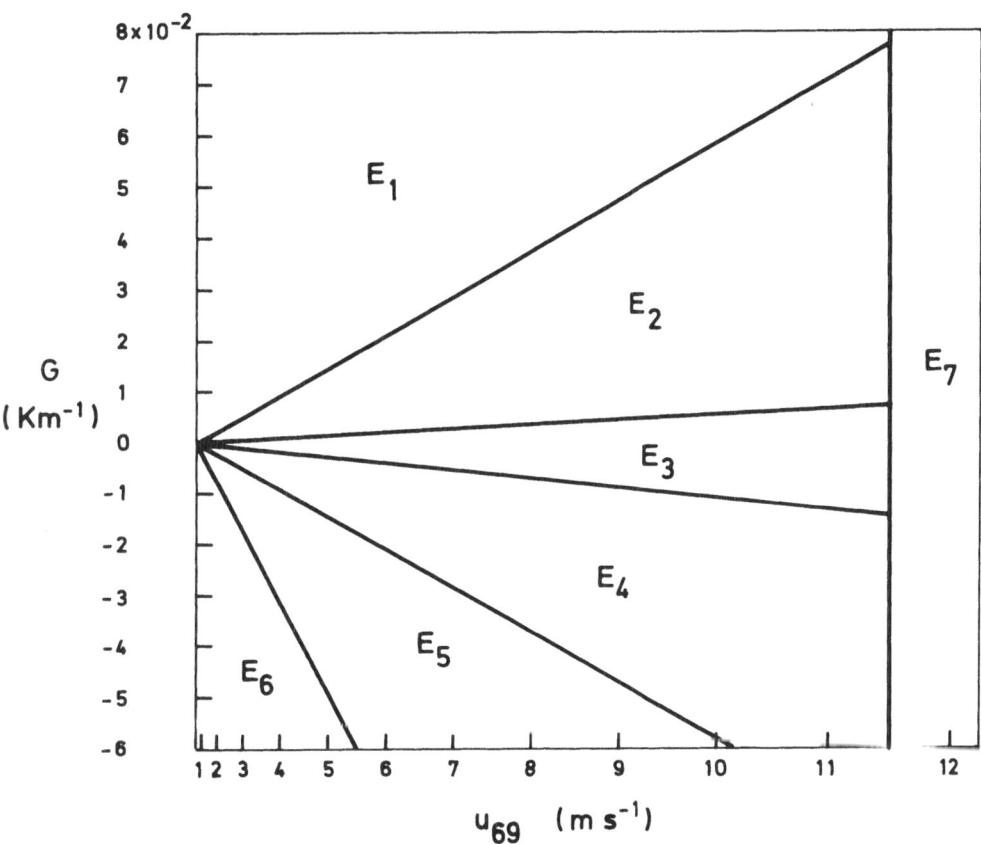

Fig.1. Location of the E_i stability classes in a $G - u_{69}$ space
(see text)

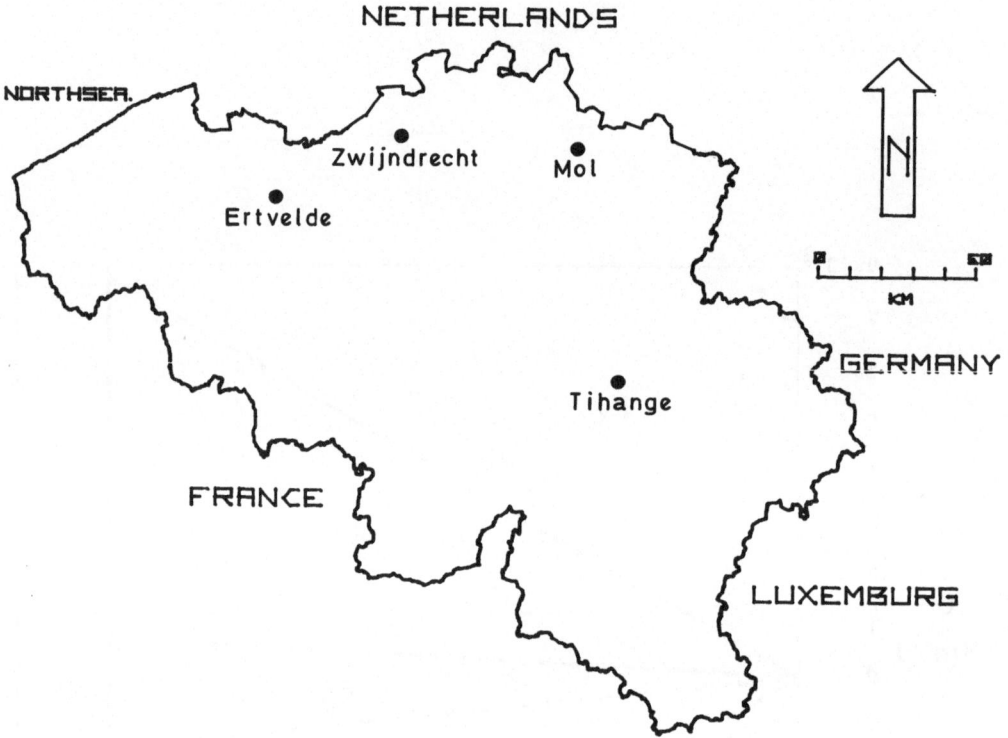

Fig. 2.   Location of the four Belgian meteorological towers

In absence of any tower, however, the E_i classes cannot be determined observationally. A computational scheme has therefore been
worked out to determine these classes approximately by means of
near-ground temperature (NGT) and near-ground wind speed (NGWS)
observations.

PHILOSOPHY OF THE METHOD

As stated before, the E_i class is defined by the value of the
parameter S, which consists of tho factors : the mean vertical
temperature gradient between the heights of 8 m and 114 m, and the
reciprocal squared wind speed at 69 m. These factors both involve
meteorological quantities existing at a certain distance from the

ground level. These quantities are in some way related with the similar quantities near the ground level. The method developed here is based on the determination of a statistical relationship between the NGT on one side and the quantities G (as defined by eq. 2) and u_{69}^2 on the other side.

DESCRIPTION OF THE METHOD

Obviously, the NGT is not in itself directly correlated with G and u_{69}^2. But an examination of the shape of the time variations of all these quantities leads to the conclusion that, in most cases, a quasi-periodical variation occurs almost simultaneously, if we consider the three following variables : T_8, G, and the exponent m appearing in the wind speed vertical profile power law

$$\frac{u_2}{u_1} = \left(\frac{z_2}{z_1}\right)^m \qquad (3)$$

where z is the height and the subscripts 1 and 2 refer to the lower and upper observational levels, respectively. In the present situation (Mol tower), z_1 and z_2 are equal to 24 m and 69 m, respectively. The main characteristics of the simultaneous variations mentioned above are summarized in table 2

Table 2. General shape of the simultaneous variations of T_8, G and m

Time of the day	T_8	G	m
Short after sunrise	Minimum	Largest positive value	Largest value ($\approx$ 0.6 to 1)
Early afternoon	Maximum	Largest negative value (not exceeding $- 0.02$ Km^{-1})	Near zero

The NGT consists in fact of two terms : the first one is the non-oscillatory synoptical background, and the second one is purely oscillatory and is generated by the radiative processes which occur near the ground. This may be expressed by

$$T_8(t) = T_s(t) + D(t) \qquad (4)$$

where $T_s(t)$ is the synoptical term and $D(t)$ is the radiative contribution (Figure 3). In order to become independent of the synoptical
situation we shall examine the relationship between D and G, on one
hand, and between D and m, on the other hand.

It is practically impossible to separate the two terms appearing
in (4) exactly from each other and some approximating assumptions
had therefore to be made. For this purpose three characteristic
moments of the time oscillation of T_8 are defined. The time when the
T_8 maximum occurs at a given day is called t_1; t_2 corresponds to the
T_8 maximum of the following day, and t_3 corresponds to the T_8 minimum occuring between t_1 and t_2. For clarity, later in this work,
the period extending from t_1 to t_2 will be called "a day".

The set of assumptions which lead to an approximative separation
of T_s and D are as follows :

$$T_s(t) = at + b \qquad\qquad (5)$$

$$\int_{t_1}^{t_2} D(t) \, dt = 0 \qquad\qquad (6)$$

$$T_s(t_2) - T_s(t_1) = T_8(t_2) - T_8(t_1) \qquad\qquad (7)$$

The synoptical component of the NGT is assumed to vary linearly
with the time between two successive daily maxima; the assumptions
(5) - (7) are illustrated on fig. 3. These relations must be used
in a discretized form, owing to the fact that the data are available
on a half-hourly, thus discrete basis. The coefficients a and b
can be deduced from eqs.(6) and (7); these two coefficients yield
the linear function $T_s(t)$, and the function $D(t)$ is calculated by
means of eq (4). Plots of the dependence between G and D, and
between m and D (figs. 4 and 5) show a hysteresis - like behaviour
during the near - 24h - cycle. The crosses correspond to successive
half-hourly values. The upper part of the "hysteresis" cycle corresponds to its cooling period (from t_1 to t_3); the lower part corresponds to the warming period (from t_3 to t_2). The two parts are very
satisfactorily approximated by quadratic trinomials of the following
form :

$$G = A_0 + A_1 D + \Delta_2 D^2 \qquad\qquad (8)$$

$$m = U_0 + U_1 D + U_2 D^2 \qquad\qquad (9)$$

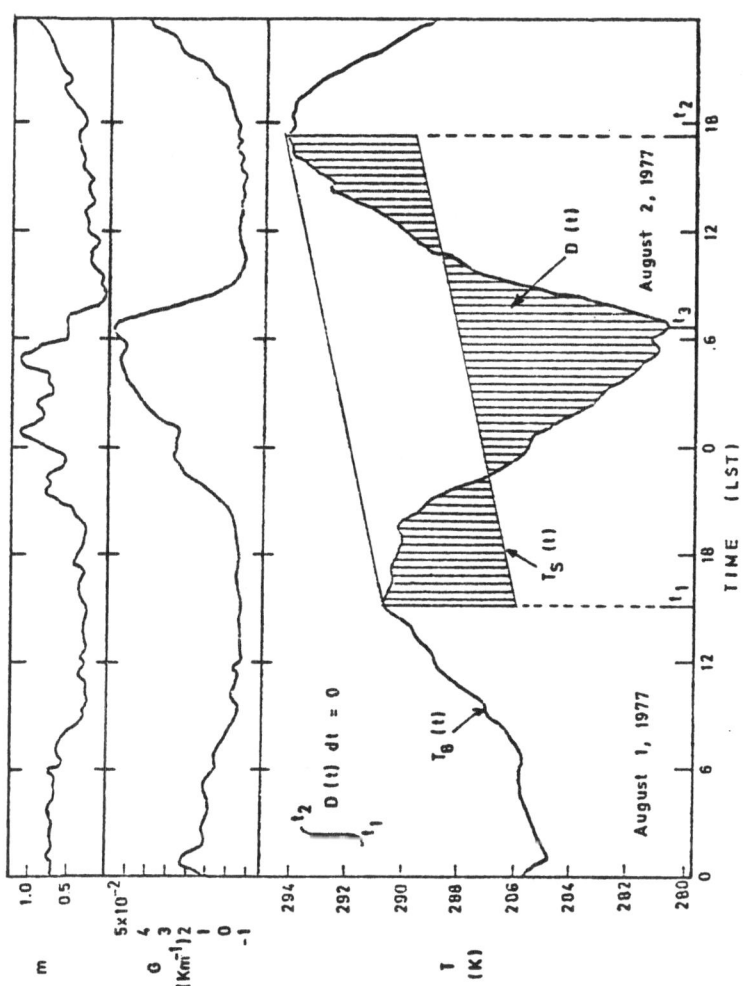

Fig. 3. Diurnal oscillation of (a) the near-ground temperature, (b) the mean vertical potential temperature gradient, and (c) the exponent of the vertical wind profile power law, as deduced from the Mol tower data, for August 1-2, 1977.

for the cooling period, and

$$G = B_0 + B_1 D + B_2 D^2 \qquad\qquad (10)$$

$$m = V_0 + V_1 D + V_2 D^2 \qquad\qquad (11)$$

for the warming period. For each case, the correlation coefficient
between the observed values and those calculated by means of eqs.
(8) - (11) has been calculated; these coefficients are also represented on figs. 4 and 5; their values which are often close to
unity, show that the quadratic approximation is a good one.

The fitting procedure outlined above has been applied to all
the days of the period August 1975 - December 1978 which satisfy
to the following set of conditions :
 - the data set must be complete from t_1 to t_2;
 - t_1 and t_2 must lie between 4 a.m. and 12 a.m.;
 - t_3 must lie between 1 p.m. and 10 p.m.
If the latter two conditions are not fulfilled, the typical near-
oscillatory behaviour of the variables D, G and m, as shown on fig. 3,
is not apparent and the dependence of G and m on D cannot be represented by quadratic trinominals.

In order to enable this quadratic representation to be of use
for calculating G and m from synoptic data, it is necessary to find
out any kind of dependence of the coefficients A_i, B_i, U_i and V_i
upon one or more independent parameters.

The period of investigation extends from August 1975 to
December 1978 and it involves about thousand days; this amount can
be considered as sufficiently large to allow a statistical analysis
of the coefficients computed above. A careful examination of these
shows a dependence of the coeffients on two important parameters :
the time of the year and the width of the daily NGT variations (or,
in other words, two times the amplitude of the NGT daily wave), which
is defined here by

$$W = \frac{T_8(t_1) + T_8(t_2)}{2} - T_8(t_3). \qquad\qquad (12)$$

The dependence of the coefficients A_i, B_i, U_i and V_i upon these
two parameters has been expressed on a discrete basis as follows :
the time of the year is categorized by the relevant month; the
categories of W are bounded by successive integer values, such that,
within a given category, one has

$$W_{cat} - 1 < W \leqslant W_{cat}. \qquad\qquad (13)$$

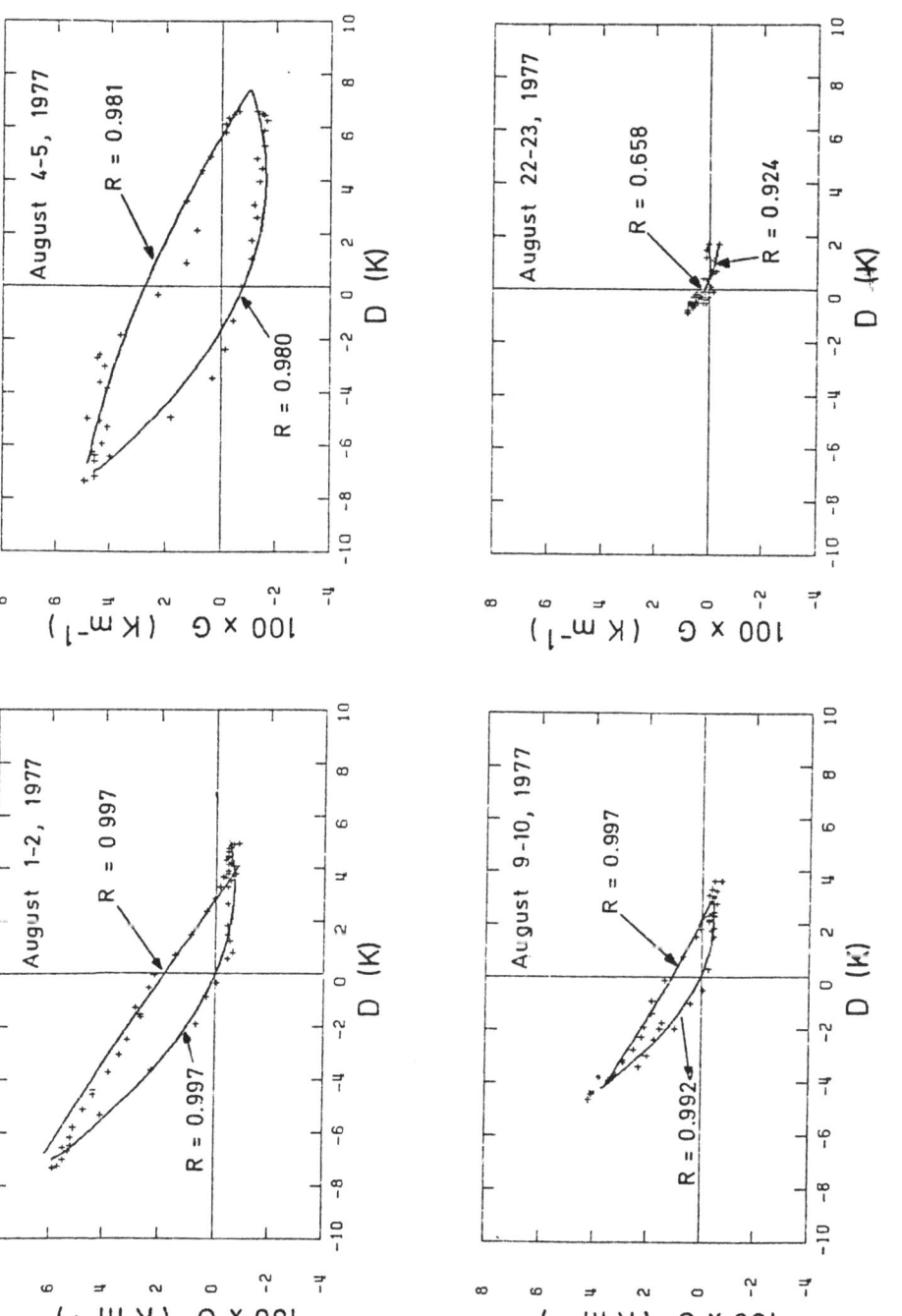

Fig. 4. Dependence between G and D (see text) for 4 selected days of August 1977, deduced from Mol tower data.

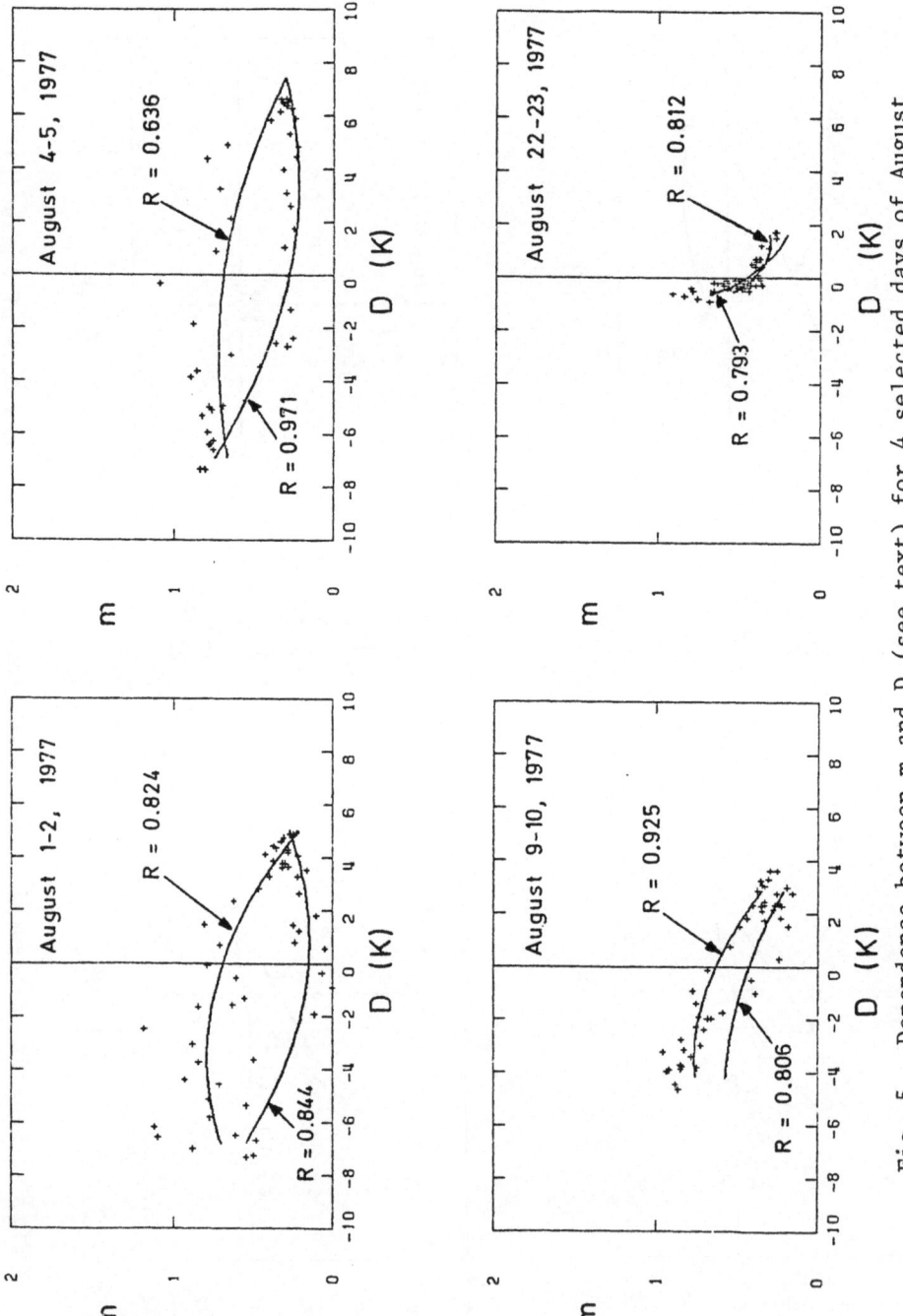

Fig. 5. Dependence between m and D (see text) for 4 selected days of August 1977, deduced from Mol tower data.

To each category belongs a certain amount of days, with their corresponding A_i, B_i, U_i and V_i coefficients. These will be replaced by a unique set of new coefficients A_i', B_i', U_i' and V_i', characteristic for the category considered. The method of deriving the latter coefficients is described below.

Let us consider a category of days, defined by the month to which they belong and by the value of W_{cat}. We then define 21 values of the variable D, ranging equidistantly from $- 0.45 \, W_{cat}$ to $+ 0.45 \, W_{cat}$. The values are called D_j, where the subscript j runs from 1 to 21. If N is the number of days in the category consi- dered, the use of relation (8) yields N values of the variable G for each value of D_j; these N values will be referred to as G_{ij}, where the subscript i corresponds to an individual day and runs from 1 to N. For each set of N values of G_{ij} where j remains fixed, an optimum value $G_{j,op}$ equal to the mode (the most probable value) of this set of values, is searched.

This procedure is then repeated for all j's and yields 21 points with the coordinates D_j and $G_{j,op}$. These points are then fitted with a quadratic trinomial, which yields the "optimal" coefficients A_i' characteristic of the category considered. This search of the optimal coefficients is illustrated by fig. 6., which shows the G vs D dependence for 4 days of August 1977, for which the values of W lie between 10 and 12 K. The crosses represent the optimum points, which enter in the final fitting (the result of the latter is not shown on the figure.

The procedure for deriving the category coefficients has been described for the A_i's only, but it is exactly similar for the coefficients B_i, U_i and V_i.

The final coefficients obtained in this way can readily be used for the determination of the E_i stability classes, as explained below. The values of these coefficients can be obtained on request.

USE OF THE COEFFICIENTS AND TEST OF THE METHOD

Let us consider a complete set of NGT and NGWS data for two successive days; the corresponding E_i classes are to be determined.

The first step is to determine the NGT maxima of the two days and to check whether these maxima fall between 4 a.m. and 12 a.m. If this is not the case, there is no means of estimating the E_i classes with our method. If the required conditions are fulfilled one determines the temperature minimum which occurs at a time between the two maxima. One then calculates the quantity D from the temperature record. The width W is obtained from from eq (12); it gives us (together with the month) the category to which the day belongs and, hence, the coefficients A_i' and B_i', we calculate

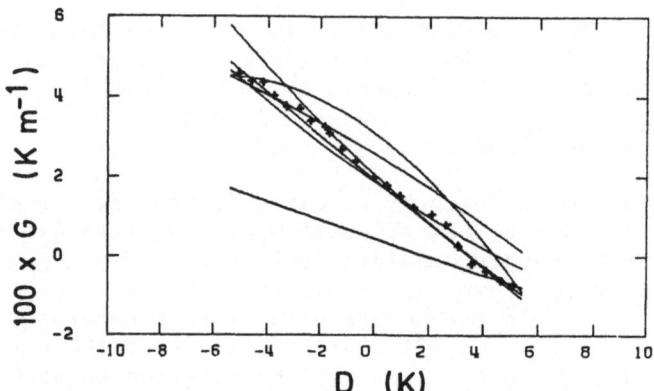

Fig. 6.   Illustration of the method by which the quadratic trinomials
          expressing the relationship between G and D, or between m
          and G, corresponding to several days belonging to a given
          category, can be reduced to a single trinomial. The full
          lines represent the individual daily fits; the crosses are
          the optimum points which, when fitted with a quadratic tri-
          nomial, give the coefficients characteristic of the category
          considered.

the gradients G.   Similarly, the use of coefficients U_i' and V_i'
and of the wind speed data yields the values of m and, hence, those
of u_{69} by using eq (3).   The knowledge of the simultaneous values
of D and u_{69} gives the value of S, which, by use of the criteria
outlined on table 1, determines the E_i class.

       In order to test the ability of the method to determine reliable
E_i classes, we first applied it to the E_i classes observed at the
Mol tower itself. In a first step, the test was made using the data
of the period which was used for the derivation of the coefficients
(August 1975 - December 1978).   The results of this test are expressed
in form of a histogram where the abscissa is the difference between
the indices of the "synthetic" and the observed E_i class, and the
ordinate is the percentage of the cases (fig. 7).   The results show
that, in 67,0 % of the cases, the synthetic E_i class corresponds to
the observed one. Further, the histogram is quite symmetrical, which
means that the overestimations and the underestimations of the quantity S are equivalent.

       We than tested the method with the Mol tower data relevant to
the year 1972, exterior to the period which has been used for the
derivation of the coefficients (fig. 8).   The same. test was also
made with the tower data of Tihange (fig. 9), Zwijndrecht(fig. 10)
and Ertvelde (fig. 11). These places are shown on the map of fig. 2.
Finally, table 3 shows, for these various tests, (a) the percentages
of the cases where the synthetic E_i class is identical to the observed
one, and (b) the percentage of the cases where the synthetic E_i class
is identical or contiguous to the observed one.

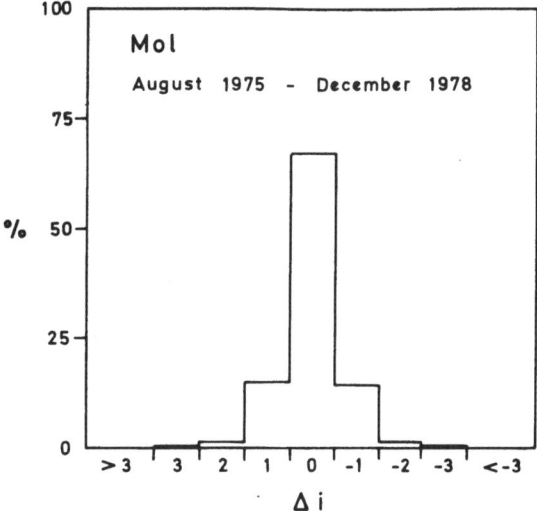

Fig. 7. Distribution of the differences between the indices i of
         the synthetic and observed stability classes, for the
         Mol data (August 1975 - December 1978 period).

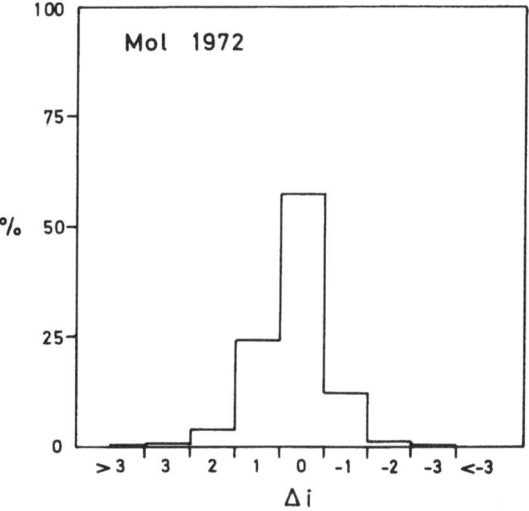

Fig. 8. Distribution of the differences between the indices i of
         the synthetic and observed stability classes, for the
         Mol data (Year 1972)

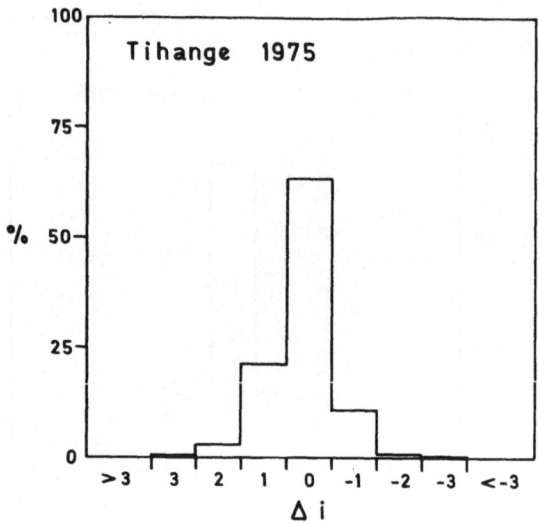

Fig. 9. Distribution of the differences between the indices i of the synthetic and observed stability classes, for the Tihange data (Year 1972)

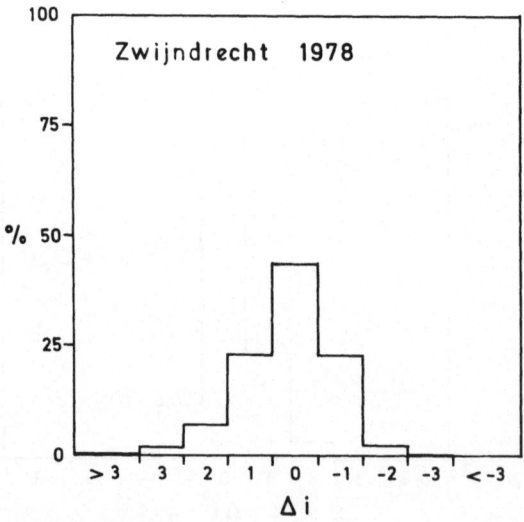

Fig. 10. Distribution of the difference between the indices i of the synthetic and observed stability classes, for the Zwijndrecht data (Year 1978)

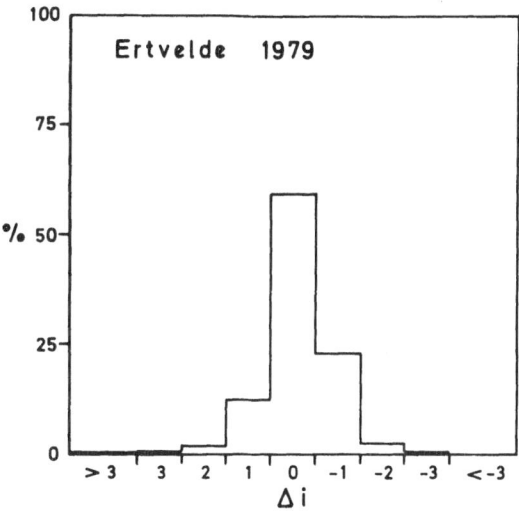

Fig. 11. Distribution of the differences between the indices i of
the synthetic and observed stability classes, for the
Ertvelde data (Year 1979)

Table 3. Percentages of the cases where (column a) the synthetic
and the observed E_i are identical, and where (column b)
the synthetic and the observed E_i are identical or
contiguous

Place and period	a	b
Mol, August 1975 – December 1978	67.0	96.4
Mol, 1972	57.3	93.8
Tihange, 1975	63.2	95.2
Zwijdrecht, 1978	43.4	88.5
Ertvelde, 1979	59.0	94.0

CONCLUSIONS

The method described in the present work shows a good applicability for a reasonably extended area (e.g. Belgium) of uniform
climatology. It has to be noted that the results for Tihange are
comparable to those for the other sites although the Tihange site
has a particular topography (Meuse river valley). That means that
the dependence between stability and NGT remains relatively unaffected
in a disturbed topography.

The synthetic E_i stability classes derived by means of the
coefficients computed in this work and the temperatures observed
at a given synoptic station are of direct use in Gaussian models
such as the IFDM code in use at the Nuclear Energy Centre at Mol
(Kretzschmar et al., 1977, 1978)

ACKNOWLEDGEMENT

This work was granted by the Belgian Ministery of Science Policy, within the framework of the National R & D Programme "Environnement-Air".

REFERENCES

Bultynck,H. and Malet,L.M., 1972, Evaluation of atmospheric dilution factors for effluents diffused from an elevated continuous point source, Tellus, 24 : 455

Gifford,F.A., 1976, Turbulent diffusion-typing schemes : a review, Nucl. Saf. 17 : 68

Klug,W., 1969, Ein Verfahren zur Bestimmung der Ausbreitungsbeding- ungen aus Synoptischen Beobachtungen, Staub, 29 : 143

Kretzschmar,J.G., Cosemans,G., De Baere,G., Mertens,I. and Vandervee,J., Some practical examples of the impact of individual sources upon the cumulative frequency distributions of the daily SO_2-concentrations in an urban and industrial area, in : "Proceedings of the Eighth International Technical Meeting on Air Pollution Modeling and its Application" (1977)

Kretzschmar,J.G., De Baere,G. and Vandervee,J., The immission frequency distribution model of the S.C.K./C.E.N., Mol, in : "Modeling, Identification and Control in Environmental Systems", Vansteenkiste, ed., North-Holland Publ. Co., 1978

Kretzschmar,J.G. and Mertens,I., 1980, Influence of the turbulence typing schemes upon the yearly average ground-level concentra- tions calculated by means of a mean wind direction model, Atm. Environ., 14 : 947

Pasquill,F., 1961, The estimation of the dispersion of windborne material, Met. Mag., 90 : 33

Singer,I.A. and Smith,M.E., 1966, Atmospheric dispersion at Brookhaven National Laboratory, Int. J. Air Wat. Pollut., 10 : 125

U.S. Nuclear Regulatory Commission, Regulatory Guide 1.23, Onsite Meteorological Programs (1972)

NET RADIATION ESTIMATED FROM STANDARD METEOROLOGICAL DATA

L. B. Nielsen[+], L. P. Prahm, R. Berkowicz[+] and
K. Conradsen[+]

Danish Air Pollution Laboratory
National Agency for Environment Protection
Rise National Laboratory, 4000 Roskilde, Denmark

INTRODUCTION

The turbulent state of the Planetary Boundary Layer (PBL) is of importance in many fields related to mankind, for example, weather forecasting, dispersion of gases and particulate matter in the atmosphere and wind power prospecting.

The 2 predominating factors determining the turbulence characteristics of the PBL are the lateral pressure pattern as driving force for the mean wind, and the sensible heat flux at the ground. Both factors are fed by the radiation energy reaching the earth from the sun and are drained by radiative energy losses to outer space. Most of the incoming and outgoing energy contributes to the energy balance at the earth's surface. Neglecting physical processes of minor importance for the surface energy balance, one obtains the well-known relation

$$R = H + LE + G, \qquad (1)$$

where the net incoming radiation R during daytime hours contributes to the ground heat flux G, to the latent heat flux LE and to the sensible heat flux H. During nighttime, the net incoming radiation is negative and is fed by the other three terms.

+ Affiliation : Department of Mathematical Statistics and Operations Research, Technical University of Denmark, 2800 Lyngby, Denmark.

Table 1. Nomenclature

ANG	Julian day angle
CEST	Central European Standard Time
D	Julian day number
F_o	solar constant
L_d	downward long-wave radiation (terrestrial)
L_{net}	net long-wave radiation at the ground
L_{no}	estimated constant L_{net} for given cloud layer
L_u	upward long-wave radiation (terrestrial)
LAT	latitude
LON	longitude
N	fractional cloud cover
N_m	modified fractional cloud cover
n	number of observations
R	net incoming radiation
r^2	squared correlation coefficient
s	standard error
S_d	downward short-wave radiation (terrestrial)
S_u	upward short-wave radiation (terrestrial)
T_g	absolute ground temperature
T_{2m}	absolute air temperature at 2 m
α_g	surface albedo
β	heating coefficient
γ	solar time correction factor
σ	solar declination
ε_g	ground emissivity
θ	solar elevation

The sensible heat flux, which is the key to the parametrisation of the PBL, can be measured by correlating temperature and
vertical eddy motion. A less sophisticated procedure is to
measure temperature and wind profiles and combine these with empirical and theoretical knowledge of momentum and heat fluxes in
the surface boundary layer. Often, none of these measurements
are available, and it is necessary to give rough estimates by
use of standard meteorological observations of clouds, temperature, humidity, precipitation and wind. At some locations, global radiation measurements, i.e. incoming short-wave radiation
measurements, are also available.

The present study concerns estimation of the net incoming
radiation R on an hourly basis by use of the standard meteorological observations. A subsequent study will treat the latent
heat flux and the soil heat flux, which finally will result in
a procedure for estimation of the sensible heat flux H. We follow to a great extent the approach used by Smith (1979), both
for the estimation of R, and for the subsequent study of the remaining terms in the energy balance equation (1). Here, we are
able to quantify R by using hourly measurements from a 10-year
period of observation at two locations in Denmark. A more complete description of the present study is given elsewhere
(Nielsen et al., 1980)

The present results are obtained over a grass surface. By use
of known albedos for other surfaces, as summarized by Kondratyev
(1969), extrapolation of the radiation relations can be made to
surfaces where the surface energy budget is of interest, e.g.
sugar fields, potatoe fields, corn fields and forests. Such an
extrapolation is based on the assumption that the net long-wave
radiation can be assumed nonvariable from one surface to another.

PARAMETERISATION OF THE COMPONENTS OF NET RADIATION

The net radiation at the surface is divided into its downwards and upwards, short-wave and long-wave components

$$R = S_d + L_d - S_u - L_u \qquad (2)$$

S_d, the global radiation, is the radiation from the sun that
reaches the surface of the earth. The irradiance from the sun
that reaches the atmosphere is the sun constant 1353 W/m^2. In
this study, the sunearth distance is assumed constant, due to
the minor importance of the variation for the derived relations.
The global radiation at the surface is a reduced part of the sun
constant due to reflection back to space and absorption by clouds,
water vapor, ozone, aerosoles and air molecules. Of these effects,
the ones due to clouds are the primary, and considering the un-

certainty in cloud observations, variations in water vapor, aerosoles and ozone can be disregarded in simple models. S_u is the reflected short-wave radiation from the ground. Using the reflection coefficient, the albedo α_g, eq. (2) can be written

$$R = (1 - \alpha_g) S_d + L_d - L_u, \qquad (3)$$

where α_g is a representative surface albedo. In simple models, α_g can be taken as a constant for a given surface (e.g. Paltridge and Platt, 1976), though it is dependent on solar elevation and spectrum of the radiation.

L_d is the downward heat flux mainly from clouds, water vapor, CO_2 and ozone to the ground. For clear skies, Swinbank (Arnfield, 1979) has developed an empirical formula

$$L_d = dT_{2m}^6, \qquad (4)$$

where d equals $5.31 \cdot 10^{-13} K^{-6} W/m^2$. This formula is derived for clear skies at night.

L_u, the upward heat flux from the ground, is given by

$$L_u = (1 - \varepsilon_g) L_d + \varepsilon_g \sigma T_g^4, \qquad (5)$$

where ε_g accounts for the fact that the surface does not absorb and emit long-wave radiation exactly like a black body.

RESULTS

From 5 years of data, empirical relations are derived for net radiation from global radiation and cloud observations. Diverse relations are found at different sites based on a rather limited amount of data. Here, we have hourly values measured on a routine basis for 14 years, out of which we have chosen data from the latest 5 years which is thought to be the most representative, and which is sufficient to give the studied relations a statistical significance.

R Estimated from S_d and Cloud Observations in the Daytime

It is found that with a good approximation, R can be written

$$R = (1 - \alpha_g)/(1 + \beta) S_d + L_{no} (N), \qquad (6)$$

where $L_{no}(N)$ is a constant for given total cloud cover N. β is a heating coefficient equal to $-dL_{net}/dR$ (Monteith and Szeicz, 1961). Estimated values of $(1 - \alpha_g)/(1 + \beta)$ and $L_{no}(N)$ from linear regressions are shown in Table 2 and the regression lines

Table 2. Linear regression between net radiation and global
radiation for different total cloud cover N. Intercept is L_{no}, $(1 - \alpha_g)/(1 + \beta)$ is the slope and s
is the standard error of estimated R.

N	L_{no} W/m^2	$(1 - \alpha_g)/(1 + \beta)$	s W/m^2	r^2	n
0	−95.0	.73	31.5	.98	759
1	−89.2	.72	30.7	.99	2874
2	−78.2	.72	31.6	.99	1734
3	−67.4	.72	35.2	.98	1869
4	−57.1	.72	33.6	.98	1570
5	−45.7	.70	32.5	.98	1791
6	−33.2	.70	28.2	.98	2043
7	−16.5	.69	21.3	.98	3494
8	− 4.3	.69	11.6	.97	4332
ALL	−28.4	.65	37.3	.94	20466

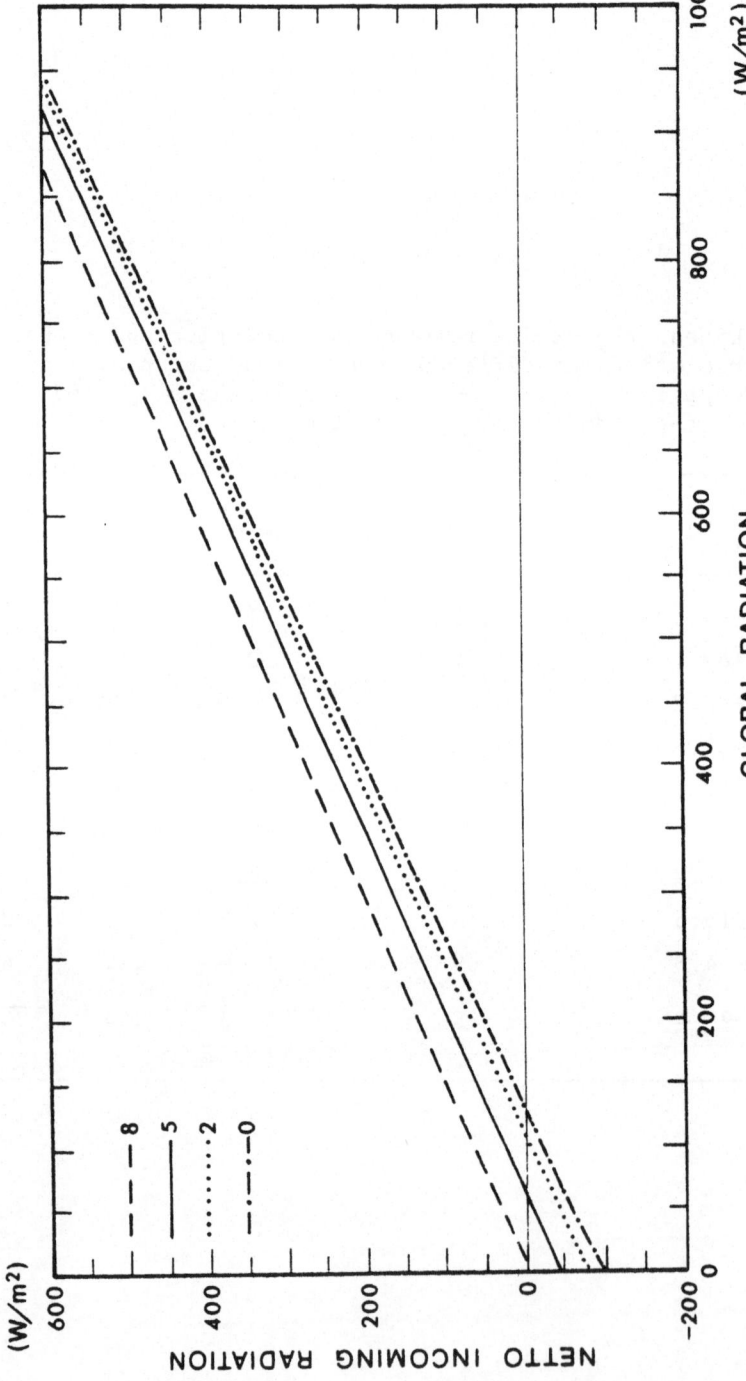

Figure 1: Regression lines for the relation between R and S_d, for different total cloud covers in oktas. Højbakkegård data.

are shown in Fig. 1. $(1- \alpha_g)/(1 +\beta_g)$ is only slightly variable with cloud cover, while $L_{no}(N)$ varies with cloud cover and is nearly zero for total cloud cover. Estimation of β from Table 2 can not be made with great statistical significance, but never-theless, extrapolation can be made to other surfaces using the estimated $(1 - \alpha_g)/(1 + \beta_g)$ values and assuming constant β. The inferred values are in agreement with other investigations (Nkemdirim, 1972; Stanhill, 1966; Gay, 1971). Several authors have reported that the heating coefficient is as much dependent on atmospheric properties as upon the surface conditions (Idso et al., 1969; Moore, 1976). However, the values generally found are small compared with 1.

R Estimated from Cloud Observations at Night

During nighttime, R is equal to L_{net}. With eqs. (4) and (5) in mind, a regression analysis was performed for separate cloud conditions at night with

$$L = a_o + a_1 T_{2m}^4 + a_2 T_{2m}^6 \tag{7}$$

The term with T_{2m}^4 was found insignificant on a 5 % level for all cloud conditions, while the term with T_{2m}^6 was found significant for clear skies and 1 oktas total cloud cover. The results from the regression are tabulated in Tables 3, 4 and 5. Table 4 shows the simple average values found at night for cloud cover greater than 1 oktas. The most significant feature from this table is that the long-wave net radiation at night is greater when the cloud base is high, compared to a medium high or low cloud base. As a simple system for predicting L_{net} at night, information on total cloud cover and height of dominating clouds are used, and a modified cloud cover N_m is defined from the al-gorithm :

$$N < 3 \to N_m = N$$

If dominating cloud cover is high, then $N=3 \to N_m = 2$

$$N > 3 \to N_m = N-2 \tag{8}$$

R Estimated from Cloud Observations in the Daytime

An empirical relationship of the form

$$R = a_o + a_1 \sin\theta + a_2 \sin^2\theta + a_3 \sin^3\theta \tag{9}$$

where θ is the solar elevation angle, was used for each N. The non-linear terms account for physical effects like refraction for low solar elevations, different radiation pathways through the

Table 3. Regression coefficients for $L_{net} = R = a_o + a_2 T_{2m}^6$ during nighttime. s is the standard error of estimated L_{net}.

N	a_o W/m^2	$a_2 \cdot 10^{15}$ W/m^{2}K^{-6}	r^2	s W/m^2	n
0	−126.7	112.0	.13	20.4	1673
1	−131.1	122.5	.17	18.0	3164

Table 4. Estimated constant L_{no} values of $L_{net} = R$ and standard error s during nighttime for different cloud heights.

N	High clouds			Medium high clouds			Low clouds		
	L_{no} W/m^2	s W/m^2	n	L_{no} W/m^2	s W/m^2	n	L_{no} W/m^2	s W/m^2	n
2	−66.4	20.7	338	−60.1	17.6	255	−68.5	20.1	812
3	−67.8	18.7	236	−58.0	19.1	267	−61.8	20.7	737
4	−65.4	17.2	117	−51.5	17.7	198	−52.5	21.7	627
5	−60.2	20.0	206	−43.5	20.8	232	−44.9	21.6	608
6	−56.9	18.9	265	−36.2	15.6	426	−35.3	20.8	773
7	−47.7	18.6	230	−28.5	14.4	732	−21.1	16.6	2188
8	−33.6	18.1	73	−20.7	12.4	695	− 8.6	9.4	5886

Table 5.   Estimated constant L_{no} values of L_{net} = R and standard
error s during nighttime for modified cloud cover
classification N_m.

N	L_{no} W/m^2	s W/m^2	n
2	−66.6	19.7	1776
3	−60.6	20.3	1222
4	−53.5	20.5	1104
5	−45.3	20.9	1088
6	−35.5	19.1	1283
7	−23.0	16.5	2929
8	− 9.9	10.5	6638

Table 6.   Regression coefficients for $R = a_o + a_1 \sin\hat\theta + a_3 \sin^3\theta$,
s is the standard error on estimated values of R.

N_m	a_o W/m^2	a_1 W/m^2	a_3 W/m^2	s W/m^2	r^2	n
0	−112.6	653.2	174.0	37.5	.95	759
1	−112.6	686.5	120.9	40.1	.94	2874
2	−107.3	650.2	127.1	57.4	.87	2042
3	− 97.8	608.3	110.6	72.0	.78	3103
4	− 85.1	552.0	106.3	75.6	.72	1944
5	− 77.1	511.3	58.5	80.1	.65	1660
6	− 71.2	495.4	−37.9	78.3	.60	1645
7	− 31.8	287.5	94.0	74.2	.54	3193
8	− 13.7	154.2	64.9	52.9	.43	4246
ALL						21466

atmosphere and the albedo dependency of solar elevation. For a
given measuring equipment, the higher order terms will also take
into account non-linearity in the equipment. The coefficients
of eq. (9) were found from a weighted regression analysis, where
the weights were the empirical variances for given $\sin\theta$. The
regressions were first performed with a classification according
to total cloud cover and height of dominating cloud layer.
A detailed investigation for all N revealed that a classification
could be made in the same way as during nighttime, with a modified N called N_m, when a simple procedure is sought.

With a classification on N_m, a regression analysis was performed to determine the constants in eq.(9). The term with $\sin^2\theta$
was insignificant on a 5 % level and was excluded. The results
are shown in Table 6 and the resulting regression curves are
shown in Fig. 2. Parts of these curves, for low solar elevations,
are below the R estimates for nighttime due to the estimation
procedure, and this feature is thought to be non-physical, because R does not decrease when the sun is rising, and does not
increase when the sun is setting. Therefore, the nighttime R
estimates are used during daytime hours if they are larger than
the daytime estimates of R for the same N.

Cumulative Distribution Function of R

As a control of the derived model for R, the cumulative distributions of measured and estimated net radiation are shown in
Fig. 3. From this figure, it is seen that there is an overall
agreement between the measured and estimated cumulative distributions of net radiation. This appears because the regressions
for the different classes are unbiased. The most severe limitations of the model formulation are the nonability for the model to predict the extreme net radiations, and the discretization
of the net values at night. Nevertheless, these limitations are
not critical for the heat budget estimation.

CONCLUSION

A procedure for estimation of the net radiation is derived
for use in heat budget calculations at the surface of the earth.
The relations for the estimation require the following information :

     a) geographical position and time
     b) global radiation measurements and/or cloud observations
     c) ground surface albedo and heating coefficient.

The net radiation varies from about -125 W/m^2 during nighttime
to a maximum of 550 W/m^2 during daytime in June. During night-

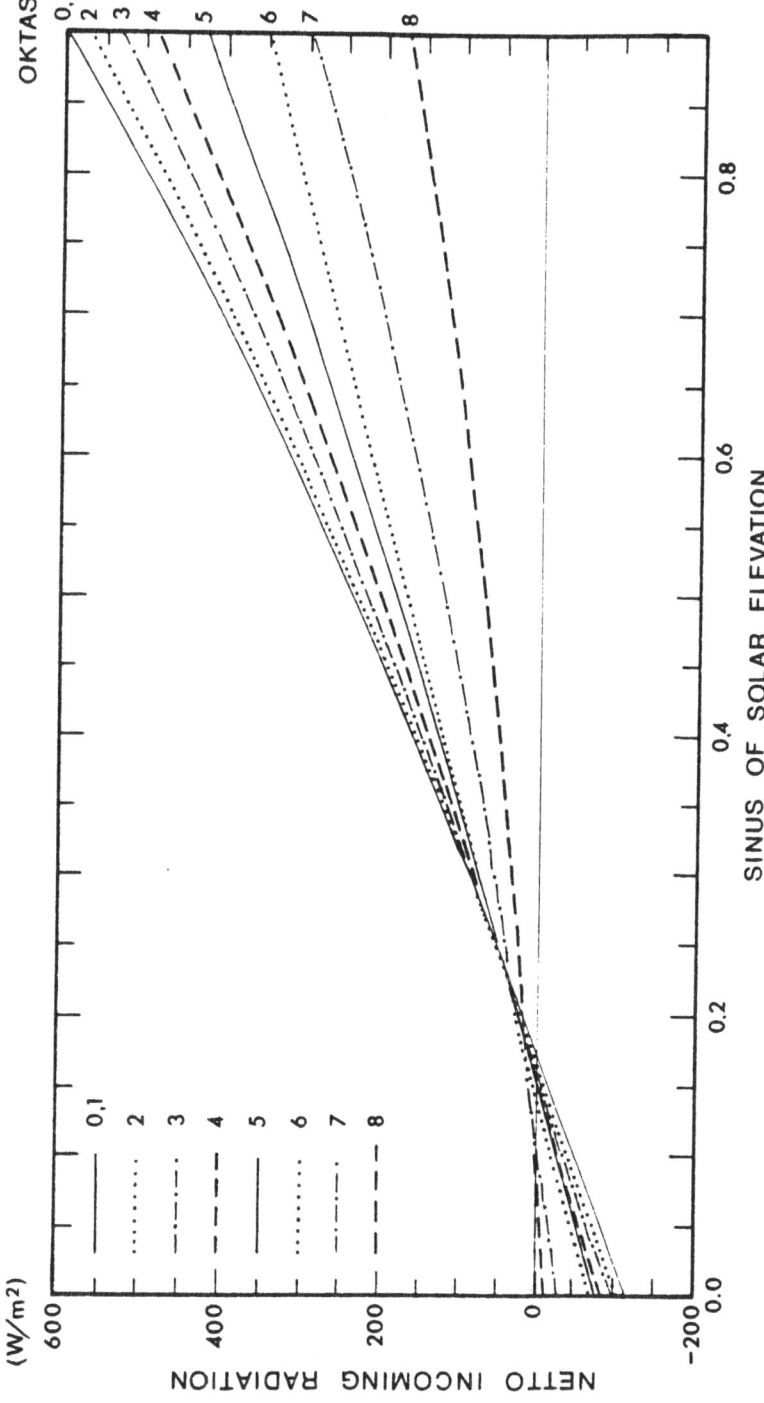

Figure 2: Regression lines for total modifed cloud cover N_m.
Højbakkegård data.

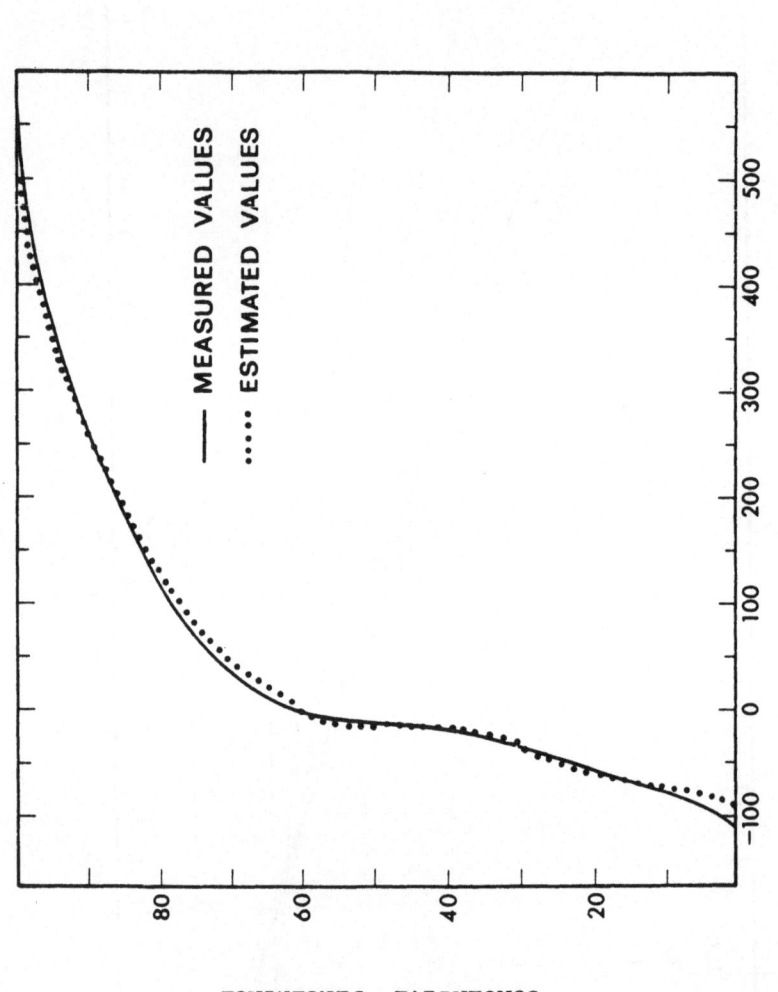

Figure 3: Accumulated distribution of estimated and measured net radiation

time the standard error on the estimated net radiations is between 10 W/m^2 and 20 W/m^2. From global radiation measurements
and cloud observations, net radiation during the daytime can be
estimated with a standard error between 10 W/m^2 for cloud cover
8 oktas, and 35 W/m^2 for cloud cover 3 oktas. If only cloud
observations are used, the uncertainty of estimated net radiations is larger than if global radiation measurements are used
and the uncertainty is about proportional with R. On the average, the standard error on estimated net radiations is smallest
for clear skies, about 37 W/m^2, and largest for 5 oktas, about
80 W/m^2.

The Danish measurements used are taken over dense irrigated
grass, but extrapolation can be done to other natural surfaces
using representative albedo values and assuming the same heating
coefficient. With greater uncertainty, extrapolation can be
done to e.g. urban areas from representative albedo values and
heating coefficients inferred from other studies. Comparison of
the Danish data with data from the Wangara experiment shows good
agreement for clear skies, but the results under cloudy conditions must be considered to be dependent upon the climate of the
measuring station.

ACKNOWLEDGEMENTS

We wish to thank Sv. E. Jensen and S. Hansen from the Hydro-
technical Laboratory, the Royal Veterinary and Agricultural College, Copenhagen, Denmark for kindly making their data available
for us, and for our many valuable discussions. Sv. E. Jensen
has been in charge of the radiation measurement program and S.
Hansen has conducted the quality control of the gathered radiation data. The synoptic data is delivered from the Danish Meteorological Institute, and we appreciate their collaboration.
We also wish to thank F. B. Smith, British Meteorological Office
for inspiring discussions.

REFERENCES

Arnfield,A. J., 1979, Evaluation of empirical expressions for
    the estimation of hourly and daily totals of atmospheric long-
    wave emission under all sky conditions, Quart.J.R.Met.Soc.,
    105 : 1041.

Clarke, R. H. with Dyer,A. J., Brook, R. R., Reid, D. G. and
    Troup, A. J., 1971, "The Wangara Experiment : Boundary Layer
    Data", Div. of Met. Physics, Technical Paper No. 19, Common-
    wealth Scientific and Industrial Research Organization, Au-
    stralia.

Gay, L. N., 1971, The regression of net radiation upon solar radiation, Arch.Met.Geoph.Biokl., B 19 : 1.

Idso, S. B., Baker, D. G. and Blad, B. L., 1969, Relations of radiation fluxes over natural surfaces, Quart.J.R.Met.Soc., 95 : 244.

Kondratyev, K. Y., 1969, "Radiation in the Atmosphere", Academic Press, New York.

Monteith, J. L. and Szeicz, G., 1961, The radiation balance of bare soil and vegetation, Quart.J.R.Met.Soc., 87 : 159.

Moore, C. J., 1976, A comparative study of radiation balance above forest and grassland, Ibid., 102 :889.

Nielsen, L. B., Prahm, L. P., Berkowicz, R. and Conradsen, K., 1980, Net incoming radiation estimated from hourly global radiation and/or cloud observations, Ibid., (submitted).

Nkemdirim, L. C., 1972, Relation of radiation plumes over prairie grass, Arch.Met.Geoph.Biokl., B 20 : 23.

Paltridge, C. W. and Platt, C. M. R., 1976, "Radiative Processes in Meteorology and Climatology", Elsevier Publishing Co., Amsterdam.

Smith, F. B. and Blackall, R. M., 1979, The application of field experiment data to the parametrisation of the dispersion of plumes from ground level and elevated sources, in "Mathematical Modelling of Turbulent Diffusion in the Environment", J. Harris, ed., Academic Press, London.

Stanhill, G., Hofstede, G. J., and Kalma, J. O., 1966, Radiation balance of natural and agricultural vegetation, Quart.J.R.Met.Soc., 92 :128.

DISCUSSION

F. B. SMITH   In our investigations using
                    data collected at CARDINGTON in ENGLAND we
                    found the largest values of net incoming radia-
                    tion when some scattered cloud were present
                    (rather than in cloud-free skies). One possi-
                    ble explanation is that clear skies are often
                    associated with rather slow-moving anticyclonic
                    situations in which visibility is often re-
                    duced by photo-chemical pollutants etc., whereas
                    scattered cloud we often linked to more mobile
                    cleaner atmospheres. Did you find a similar
                    variation and if so would you like to comment
                    on the cause ?

L. NIELSEN

In our analysis the net radiation was a maximum when the cloud cover was around 1 okta. However, on an average the net radiation for cloud cover of 1 okta was not different for that associated with clear skies. The reason for these observations is that in occasion the slight reduction in the direction beam radiation caused by light broken clouds is more than compensated by the increase of radiation associated with multiple reflections from the clouds. These effects cancel each other on an average.

ESTIMATION OF THE SENSIBLE HEAT FLUX FROM STANDARD METEOROLOGICAL DATA FOR STABILITY CALCULATIONS DURING DAYTIME

A.A.M. Holtslag, H.A.R. de Bruin, A.P. van Ulden

Royal Netherlands Meteorological Institute
De Bilt

INTRODUCTION

The stability of the atmospheric surface layer is an important feature in models of the boundary layer and air pollution dispersion. A measure of stability is the Obukhov-length. For practical applications of the above mentioned models often only standard meteorological data are available.

The Obukhov-length can be calculated from air temperature, 10 m wind speed the surface-roughness length and the sensible heat flux applying Monin-Obukhov similarity theory (e.g. Businger, 1973). The airtemperature and the 10 m wind speed are routine meteorological quantities. The surface-roughness length can be obtained with a method by Wieringa (1976).

In this paper we are concerned with the sensible heat flux. We will obtain hourly estimates of this quantity from surface energy budget considerations. In the following we will estimate the terms in the energy budget from routine meteorological data. We restrict ourselves to daytime hours with positive net radiation.

MODEL OF THE SENSIBLE HEAT FLUX

The sensible heat flux (H) is related to the flux of water vapour (E), the net radiation (Q^*) and the soil heat flux (G) by the surface energy budget:

$$H + LE = Q^* - G \tag{1}$$

where L is the latent heat of vaporization, and $(Q^* - G)$ is the total available energy flux for H and the latent heat flux LE.

Priestly and Taylor (1972) found for saturated surfaces that:

$$LE = \alpha \frac{s}{s+\gamma} (Q^* - G) \qquad (2)$$

where α is an empirical constant of order 1, s is the slope of the saturation specific humidity-temperature curve, $\gamma = c_p/L$ and c_p is the specific heat of air at constant pressure.
For daily averages of LE in summer an average value of $\alpha = 1.26$ was obtained both for water surfaces and saturated land surfaces with a short vegetation (e.g. De Bruin and Keijman, 1979).

We consider hourly values of H and LE during daytime for not necessarily saturated surface conditions. Therefore we modify (2) by:

$$LE = \alpha \frac{s}{s+\gamma} (Q^* - G) + \beta \qquad (3)$$

where α and β are empirical constants to be determined. With (1) it follows from (3):

$$H = \frac{(1-\alpha) s+\gamma}{s+\gamma} (Q^* - G) - \beta \qquad (4)$$

The constants α and β depend on the surface-moisture conditions. In our experiments we found $\alpha = 0.95$ and $\beta = 20$ Wm^{-2} for normal periods in the Netherlands. For a period after more than five days with no rain in summer we obtained $\alpha = 0.65$ and $\beta = 20$ Wm^{-2}.

THE AVAILABLE ENERGY FLUX AT THE SURFACE

The available energy flux at the surface $(Q^* - G)$ is not a routine meteorological quantity and therefore it has to be estimated from other known quantities. Burridge and Gadd (1977) use:

$$G = 0.1 Q^* \qquad (5)$$

This is confirmed by our experiments.

Now remains the problem of the estimation of the net radiation (Q^*). Herefore we use the following scheme:

$$Q^* = (1-r) K^+ + L^+ - L^- \qquad (6)$$

where r is albedo, K^+ is incoming shortwave radiation, L^+ is downward atmospheric longwave radiation and L^- is the outgoing longwave radiation from the surface. We fixed the albedo (r) on a constant value r=0.25, commonly used for short grass vegetations (e.g. Sellers, 1965). This value is in agreement with our measurements performed at our experimental site.

The shortwave radiation (K^+) is measured on a routine base at seven stations in the Netherlands. If K^+ is not measured directly, it can be estimated from (Holtslag and Van Ulden, 1980):

$$K^+ = K^+_o (1 - 0.7 N_t^2) \qquad (7)$$

where N_t is total cloud cover and K^+_o is the clear sky value of K^+:

$$K^+_o = S \sin \theta (0.48 + 0.29 \sin \theta) \qquad (8)$$

S is the solar constant taken at S = 1353 Wm^{-2}, and θ is the solar elevation.

For the downward longwave radiation we use an empirical function which is a slight modification of that of Paltridge and Platt (1976):

$$L^+ = 5.31 10^{-13} T^6 + 60 N_t \qquad (9)$$

where T is the absolute air temperature at screenheight. The outgoing longwave radiation (L^-) is parameterized by:

$$L^- = \sigma T^4 + c (1-r) K^+ \qquad (10)$$

where σ is the Stefan-Boltzmann constant, and c is a constant. The second term of the right hand side of (10) accounts crudely for the difference between the surface temperature and the airtemperature. From Holtslag and Van Ulden (1980) we obtain c = 0.07 if r = 0.25.

THE DATA SET

For this study we used experimental data which were collected at Cabauw in the Netherlands from May till September 1977. The measurements were done at a micro-meteorological field of 100 x 100 m. This field is covered with short grass of approximately 8 cm.

The net radiation (Q^*) was measured with a Funk net pyrradiometer and the incoming shortwave radiation (K^+) with a Moll-Gorczynski pyranometer. The sensible heat flux (H) and the latent heat flux (LE) were obtained with the energy budget method using

Bowen's ratio (Sellers, 1965). The latter yields also the air
temperature (T). The soil heat flux (G) was determined with the
aid of heat flux plates and thermometers at the surface and at
2 cm depth.
The total cloud cover (N_t) was interpolated from four synoptic
stations surrounding Cabauw.

We excluded hours with rain or fog because in these cases the
measurements are unreliable. This is not a serious problem because
in these conditions the sensible heat flux is small and there is
about neutral stability. Furthermore we used only hours with
positive net radiation (Q^*).

RESULTS

Figure 1 shows a comparison of calculated values against
measured values of the sensible heat flux (H), for a random sample
of our hourly data. A distinction is made between cases in normal
periods and in dry periods.
By the calculation of H we used the measured incoming shortwave
radiation (K^+) and our estimates (4), (5), (6), (9) and (10) with
the mentioned constants.

Since the parameters α and β in (4) are obtained from the
same data set, this result is certainly not a definitive test of
our model. It is our intension to verify our model in the future
with independent data. Nevertheless, we may conclude that our
approach yields satisfactory results, since the agreement is relatively good (correlation r = 0.92 and root mean square error
SE = 27 Wm^{-2}).
The correlation decreases somewhat when we use the calculated
value of the incoming shortwave radiation (K^+) with (7) and (8).
Then we find r = 0.86 and SE = 35 Wm^{-2}.

CONCLUSIONS

From this study it follows that realistic hourly averages of
the sensible heat flux can be obtained from a modified Priestly
and Taylor model with two parameters. It is found that one of the
parameters, notably α, is a function of the rainfall in the preceding period.
The parameter α varies from 0.95 under normal conditions to 0.65
in dry periods (no rain in the preceding five days).

With the modified Priestly and Taylor model and the presented
estimates of net radiation, soil heat flux and incoming shortwave
radiation we can obtain the sensible heat flux from standard
meteorological data.

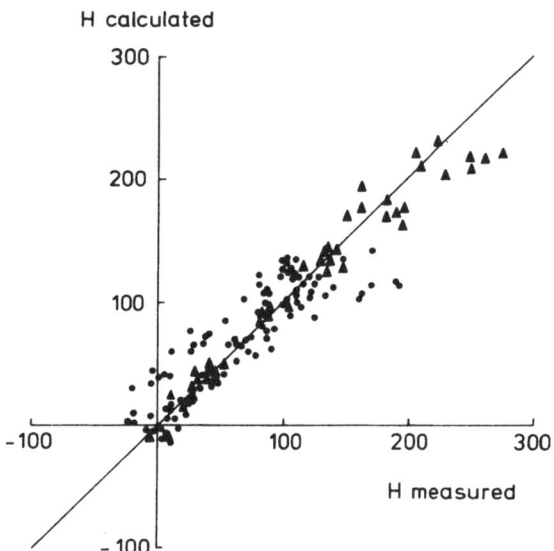

Figure 1: Comparison of calculated and measured hourly averages
          of the sensible heat flux (H). Dots represent hours
          in normal periods and triangles represent hours in
          dry periods (no rain in the preceding 5 days). The
          figure contains a random sample of 152 hours
          (correlation r = 0.92, root mean square error
          SE = 27 Wm^{-2}).

The results presented are preliminary since no independent
test of the model has been carried out. Nevertheless, we trust our
results are satisfactory for stability calculations.

ACKNOWLEDGEMENTS

This study was supported by the Dutch Ministry of Health and
Environmental Protection.
For the experiments at the micro-meteorological field we acknow-
ledge the support of the Instrumental Branch of our Institute.

REFERENCES

Burridge, D.M. and Gadd, A.J. (1977): The Meteorological Office
      Operational 10-level Numerical Weather Prediction Model,
      Meteor. Office, London, Scientific Paper 34.
Businger, J.A. (1973): Turbulent Transfer in the Atmospheric
      Surface Layer in Workshop on Micrometeorology, D.A. Haugen
      (editor), American Meteor. Soc., Boston.

De Bruin, H.A.R. and Keijman, J.Q. (1979): The Priestly-Taylor
    Evaporation Model Applied to a Large Shallow Lake in the
    Netherlands. Journal of Appl. Meteor., 18, pp. 898-903.
Holtslag, A.A.M. and Van Ulden, A.P. (1980): Estimates of incoming
    shortwave radiation and net radiation from standard meteoro-
    logical data. KNMI, De Bilt. Scientific Report, WR 80-6.
Paltridge, G.W. and Platt, C.M.R. (1976): Radiative processes in
    Meteorology and Climatology. Developments in Atm. Science 5,
    Elsevier Amsterdam.
Priestly, C.H.B. and Taylor, R.J. (1972): On the assessment of
    surface heat flux and evaporation using large scale para-
    meters. Mon. Wea. Rev., 100, pp. 81-92.
Sellers, W.D. (1965): Physical Climatology, Chicago.
Wieringa, J. (1976): An objective exposure correction method for
    average wind speeds measured at a sheltered location.
    Quart. J. Roy. Met. Soc. 102, pp. 241-253.

DISCUSSION

G. SCHAYES   The latent heat flux
               or evaporation rate is frequently considered as
               depending not only on the energy available but
               also of wind speed, as was stated originally
               by Penman. Can you comment this point ?

A.A.M. HOLTSLAG   In fact, the Priestly
    '   and Taylor model is proportional to the first
               term in Penman's equation. The second term in
               Penman's equation depends on windspeed. From
               our hourly data it followed that the windspeed
               term did not contribute to explain the varia-
               tion of the latent heat flux or the sensible
               heat flux. But, using Penman's equation we ob-
               tained under-estimates of the sensible heat flux
               at sunny days. Therefore we used the Priestly
               and Taylor model, extended with a parameter β .

A. VENKATRAM   Your data set does not
                    include direct measurements of sensible heat
                    flux. Since you have inferred the heat flux
                    from the measured net radiation the good corre-.
                    lation between heat flux estimates and "measure-
                    ments" might be a result of good estimates of
                    net radiation rather than those of the sensible
                    heat flux. Please comment.

A.A.M. HOLTSLAG   At our experimental
                    site in the Netherlands we also obtained the
                    sensible heat flux from temperature and wind
                    profiles. This heat flux compares well with
                    the heat flux from the micro-meteorological
                    field, which we have used. Also we applied our
                    scheme to "Prairiegrass" - data (Barad,1958).
                    We found a satisfactory agreement with the
                    heat flux obtained with flux profile relations
                    from the temperature and wind profiles measured
                    in the "Prairiegrass" - experiments. Thus our
                    scheme to determine the sensible heat flux
                    agrees also with the sensible heat flux from
                    profiles.

Reference :
BARAD, M.L. (1958, editor) :
Project prairiegrass, a field program in Diffusion Volume I and
II, Geophysics Research Paper n° 59 Bedford, Mass, USA.

3: ADVANCED TECHNIQUES IN AIR POLLUTION MODELLING, TO TAKE INTO ACCOUNT COMPLEX TERRAIN, HEAVY GASES, LIGHT WIND CONDITIONS

Chairman: F. B. Smith   Rapporteur: J. D. Reid

A COMPARISON OF FINITE DIFFERENCE SCHEMES, DESCRIBING
THE TWO-DIMENSIONAL ADVECTION EQUATION

B.J. de Haan

Royal Netherlands Meteorological Institute
De Bilt
The Netherlands

INTRODUCTION

In recent years much attention has been paid to research of an accurate numerical integration technique of the advection equation. As the advection equation plays an important role in a wide variety of fluid dynamical problems, all with different requirements, it is unlike that there will be found an overall best and suitable method. Therefore we shall restrict ourselves to methods which are or have been applied in the field of air pollution modelling.

The accuracy of the earlier schemes was disappointing (Molenkamp 1968). Nowadays the introduction of large and fast computers enables most modellers to use higher order finite difference schemes, which are more accurate but more time-consuming as well.

In this study we compare several recently proposed methods regarding accuracy and efficiency. The test problem is the commonly used advection of a cone in a rotating velocity field (Molenkamp 1968, Crowley 1968, Orszag 1971 and others).

THE ADVECTION EQUATION

In a two dimensional rectangular coordinate system, the differential equation discribing advection reads

$$\frac{\partial}{\partial t} c + u \frac{\partial}{\partial x} c + v \frac{\partial}{\partial y} c = 0 \qquad (1)$$

where c is the quantity advected, t time, x and y position
coordinates and u and v the velocities in the x and y directions,
respectively.
The integration domain is a square. {-1, 1; -1, 1}.
The boundary conditions depend on the flowdirection at the
boundary.

$$\frac{\partial}{\partial x} c = 0 \text{ or } \frac{\partial c}{\partial y} = 0 \qquad \text{for outflow conditions}$$

$$c = 0 \qquad \text{for inflow conditions}$$

In the test the initial condition for the quantity c is a
cone with a basis of .5, centered at {-.5, 0}.
The velocity field is that of a rotation of a solid body. The integration time is chosen such that the exact solution rotates
exactly one time. The conditions used are the same as the ones
used by previous studies (for a detailed description see Orszag
1971). Under certain conditions the velocity field of the at-
mosphere is approximately divergence free $\frac{\partial}{\partial x} u + \frac{\partial}{\partial y} v = 0$.
However the velocity field used in the test is even more con-
strained for not only $\frac{\partial}{\partial x} u + \frac{\partial}{\partial y} v = 0$ but also $\frac{\partial}{\partial x} u = \frac{\partial}{\partial y} v = 0$.
The convenience of the availability of an exact solution justifies
the use of this particular velocity field.

SCHEMES

One of the most simple schemes is the so called Euler explicit
scheme which is based on finite difference approximations of first
and second order;

$$\frac{\partial}{\partial t} c = \frac{c^{n+1} - c^n}{\Delta t} + O(\Delta t), \quad \frac{\partial}{\partial x} c = \frac{c_{i+1} - c_{i-1}}{2\Delta x} + O(\Delta x^2).$$

A scheme is called explicit if the value of the quantity c at
time (n+1)Δt is computed explicitly from values of c at (a) previous time (s). The Euler explicit scheme reads

$$c_i^{n+1} = c_i^n - \lambda/2 \, (c_{i+1}^n - c_{i-1}^n) \text{ with Courant number } \lambda = u\Delta t/\Delta x.$$

There are two ways to improve the accuracy of this
scheme. One way is to choose higher order approximations for either
of the derivatives, the other way is to start from a Taylor series
in t, the time derivatives are then substituted with (1) by expressions containing only space derivative terms (Richtmeyer and

Morton 1967). Schemes designed this way are called "characteristic interpolation" or "quasi-Lagrangian" schemes, an example is the second order Lax Wendroff scheme. This scheme reads

$$c_i^{n+1} = c_i^n - \lambda/2 \; (c_{i+1}^n - c_{i-1}^n) + \lambda^2/2 \; (c_{i+1}^n - 2c_i^n + c_{i-1}^n).$$

A straight forward extension of the Euler explicit scheme is a second order centered difference approximation for the time derivative to be combined with second (leap-frog method), fourth or even sixth order centered difference approximations for the space derivatives. A fourth order approximation for the time derivative -trapezium rule or Runge Kutta 4 method- in combination with the fourth order centered difference approximation for the space derivatives has been used for the solution of the shallow water equations (Praagman 1979).

Two efficient higher order approximations for the space derivatives have been found. The first, the compact differencing technique, has been used in combination with both explicit and implicit first order approximations for the time derivative (Adam 1977). The second is the fast Fourier transform which inspired the pseudo spectral method. This method has excellent resolution and accuracy properties (Fox, Orszag 1973).

Some "characteristic interpolation" schemes are truly extended to two dimensional advection, while others are a result of solving the one dimensional advection equation in alternating directions. A fourth order extension of the two dimensional Lax Wendroff scheme has been proposed (Gadd 1978, 1979). Most attention, however, has been paid to one dimensional schemes (see p.e. Wesseling 1972). The fourth order schemes (Crowley 1967, and others) show considerable improvement over the Lax Wendroff scheme.

The introduction of the compact differencing and the fast Fourier transform techniques led also for the "characteristic interpolation" schemes to very accurate schemes (Forester 1977) resp (Gazdag 1973), the computation costs however, increased sharply as well. A scheme of lower order using the compact differencing technique (Purnell 1976) appears to be more efficient.

From the variety of schemes discussed above a few have been selected to make a comparison. The schemes chosen are
the two step Lax Wendroff scheme (to be referred to as LW)
the fourth order extension of the scheme LW (G)
the compact differencing upstream interpolation scheme (P)
the both fourth order in time and space scheme (RK4)
the pseudospectral method (FFT).

TEST

   The effiency of a method may be expressed in terms of
accuracy and computation costs. The results of the test will be
summarized in table 1. MAX stands for the maximum value at any
grid point with exact values of 100,LAG stands for the angle the
computed maximum lags behind the exact maximum. The computational
costs will be expressed in processtime (CPU).

   The processtime is dependent on the number of timesteps (N)
and the mesh size (GRID) chosen for the schemes. The mesh size was
32x32 for all schemes but the pseudospectral method, as the good
resolution of this method (Orszag 1976) and the expense of the
fast Fourier transform make a computation on a 16x16 mesh a fair
comparison with the others. The cone width is four gridpoints on
the 32x32 mesh and two on the 16x16 mesh. None of the schemes have
been filtered. All computations have been carried out on a Burroughs
6700 computer, the programs have not been optimized.

Table 1.

method	reference	GRID	N	MAX	LAG	CPU
LW .	Richmyer Morton	32x32	200	49	27	109
G	Gadd	32x32	200	80	-4	213
P	Purnell	32x32	100	82	0	173
RK4	Praagman	32x32	100	81	5	296
FFT	Fox Orszag	16x16	200	96	1	108

COMMENTS

   The properties of the schemes become clear when we see the
images of the distribution of the quantity c as a series in time.
During the first part of the rotation most schemes round off the
sharp discontinuities of the cone, leaving the cone more bell-
shaped. In the last part this type of diffusion of the cone is
small.

   The schemes G, P and RK4 show a maximum of 80% of the original value, considering the time lag and the computation time,
method P is the best. However method P makes use of the property
of the velocity field $\frac{\partial}{\partial x} u = \frac{\partial}{\partial y} v = 0$. This makes it doubtful
whether the scheme will perform as well with more general velocity
fields.

   The pseudo spectral method behaves also better than we may
aspect under more general circumstances, as it requires that the
quantity c is periodic. The negative impact of this requirement is
the reappearance of the quantity c on the one boundary when being

advected out over the opposite boundary. Two remedies to avoid this have been proposed. One is to introduce a artificial drain term of the form $-c(x,y,t)/\tau(x,y)$ in the advection equation. The decay constant $\tau(x,y)$ should assume realistic values appropriate to the physical problems (Christensen Prahm 1976). The second proposal is to decompose the distribution of quantity c in a periodic and a polynomial part. In this way it is possible to avoid the discontinuity at the boundary (Wengle Seinfeld 1978).

CONCLUSION

We have shown that the pseudo spectral method is more efficient than any other of the tested methods with only one fourth of the number of gridpoints in the model. A drawback in real problems, however, is the requirement of periodicy.

The reduction of the number of gridpoints is certainly attractive for three dimensional models, because the number of gridpoints and consequently the data requirements and the number of computations in the vertical direction is reduced by a factor of four.

REFERENCES

Adam, J., 1977, Highly accurate compact implicit methods and boundary conditions, Journal of Computational Physics, 24, 10-22.

Christensen, O., and Prahm, L., 1976, A pseudospectral model for dispersion of atmospheric pollutants, Journal of Applied Meteorology, 15, 1284-1294.

Forester, C.K., 1977, Higher order monotonic convective difference schemes, Journal of Computational Physics, 23, 1-22.

Fox, D.G., and Orszag, S.H., 1973, Pseudospectral approximation to two dimensional turbulence, Journal of Computational Physics, 11, 612-619.

Gadd, A.J., 1978, A numerical advection scheme with small phase speed errors, Quarterly Journal of the Royal Meteorological Society, 104, 583-594.

Gadd, A.J., 1980, Two refinements of the split explicit integration scheme, Quarterly Journal of the Royal Meteorological Society, 108, 215-220.

Gazdag, J., 1973, Numerical convective schemes based on accurate computation of space derivatives, Journal of Computational Physics, 13, 100-113.

Kreis, H.O., and Oliger, J., 1972, Comparison of accurate methods for the integration of hyperbolic equations, Tellus, XXIV, 199-215.

Molenkamp, C.R., 1968, Accuracy of finite difference methods applied to the advection equation, Journal of Applied Meteorology, 7, 160-167.

Orszag, S.A., 1971, Numerical simulation of incompressible flows
    within simple boundaries accuracy, Journal of Fluid Mechanics,
    49, 75-112.
Praagman, N., 1979, Numerical solution of the shallow water
    equations by a finite element method, Dissertation at the
    University of Delft, The Netherlands.
Purnell, D.K., 1976, Solution of the advective equation by upstream
    interpolation with a cubic spline, Monthly Weather Review,
    104, 42-48.
Richtmyer, R.D., and Morton, K.D., 1967, Difference methods for
    critial value problems, New York Interscience Publishers,
    sec. ed.
Wengle, H., and Seinfeld, J.H., 1978, Pseudospectral solution of
    atmospheric diffusion problems, Journal of computational
    Physics, 26, 87-106.
Wesseling. P, 1973, On the construction of accurate difference
    schemes for hyperbolic partial differential equations, Journal
    of Engineering Mathematics, 7, 19-31.

DISCUSSION

J.L. WOODWARD   Did you try to ensure
                    mass conservation ?

B.J. DE HAAN   Earlier methods were
                    so unaccurate that mass conservation was a pro-
                    blem however for the newer schemes this is not
                    any longer the case.

R.N. MERONEY   Can you relate your
                    work to the long + Pepper paper ?

B.J. DE HAAN   The test is the same,
                    they selected some schemes that are different
                    from the ones I selected. Indeed the chapeau
                    method and the splines method were found to be
                    the best by them. The pseudospectral method is
                    in the same class (Galerkin approach) as the
                    chapeau method (finite elements method), and it
                    has been proved that the pseudospectral method
                    is the most efficient in this class.

A COMPARISON OF SOME PLUME DISPERSION PREDICTIONS WITH FIELD MEASUREMENTS

G.A. Davidson and P.R. Slawson

Department of Mechanical Engineering
University of Waterloo
Waterloo, Ont., Canada   N2L 3G1

INTRODUCTION

During four seasonal periods in 1977, measurements of the rise and spread of the powerhouse plume of the Suncor plant in the tar sands area of Alberta, Canada were collected, using both ground-based photography and an instrumented aircraft. Supporting source and meteorological data were also monitored. In addition, some limited records of ground level SO_2 concentrations during segments of the measurement period were obtained from low-level aircraft flights. Using this extensive data base, which encompasses a variety of atmospheric conditions, it has been possible to examine critically several facets of the usual approach to modelling plume rise and dispersion in the atmosphere. The aim of this paper is to present an overview of some of our results, without attempting to include all calculation details. For more detail, and for a complete description of the measurement program, the methods of data abstraction, and the models themselves, the reader is referred to Refs. 1, 2, and 3.

Our plume dispersion models, like the majority of those currently in use, follow the physical picture shown schematically in Fig. 1. Initially, due mainly to its buoyancy, the plume rises and spreads at a rate controlled by self-generated turbulence. In this buoyant rise phase, the conservation equations of fluid mechanics can be applied to predict plume behaviour. Once a final rise height has been attained, subsequent downwind diffusion of the plume is governed by the level of atmospheric turbulence. A Gaussian plume model with empirical sigma curves linked to atmospheric turbulence levels is applied to predict the concentration field in this atmospheric diffusion phase.

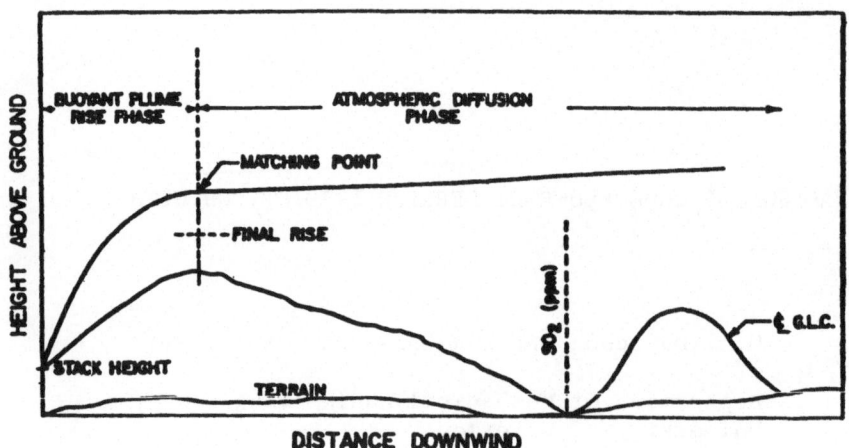

Fig. 1   Schematic of plume behaviour in stable lapse.

BUOYANT PLUME RISE PHASE

Under the assumptions that the plume is round in cross-section; that plume variables follow a top-hat profile normal to the plume direction; that the only interactions between a plume element and its surroundings arise from a buoyant force and a turbulent growth, which can be characterized by an entrainment velocity v_e; and that the turbulent growth process is isentropic, the conservation equations of mass, horizontal momentum, vertical momentum, and heat reduce to[3]

$$\frac{d}{dt}\,(\rho_p R^2 V) = 2\rho_a RV v_e \tag{1}$$

$$\frac{d}{dt}\,(\rho_p R^2 V v_x) = 2\rho_a RVU v_e \tag{2}$$

$$\frac{d}{dt}\,(\rho_p R^2 Vw) = gR^2 V(\rho_a - \rho_p) \tag{3}$$

$$\frac{d}{dt}\,[gR^2 V(\rho_a - \rho_p)] = -\rho_a N^2 R^2 Vw \tag{4}$$

where each symbol is defined in the Nomenclature. With the addition of the hydrostatic pressure law, the ideal gas law, kinematic relations between plume velocity and position, and one of the numerous empirical relations[4] for v_e, such as

$$v_e = \beta |w|, \tag{5}$$

a set of nine equations is obtained for the nine plume variables ρ_p, T_p, V, v_x, w, R, s, z, and x as functions of downwind travel time t. Solution is thus possible once source properties and atmospheric parameters ρ_a, U, and N^2 are specified. Arbitrary profiles of these atmospheric parameters can be substituted into these equations; however, in this one-dimensional, or integral formulation of plume rise, atmospheric properties, like plume properties, must be assigned one average value for each plume cross-section.

Analytical expressions for plume centerline altitude $z(x)$ and plume radius $R(x)$ can be derived from the conservation equations by introducing the additional assumptions that U and N^2 are constants, that $\rho_p = \rho_a$ = constant except in density difference terms (the Boussinesq approximation), and that $V \tilde{=} U$ (the bent-over plume assumption). Under neutral lapse conditions ($N^2=0$), the solutions become

$$R(x) = R_o + \beta z(x) \tag{6}$$

$$z(x) = \{\frac{3}{2\beta^2}[\frac{F_o}{U^3} x^2 + 2\frac{M_o}{U} x] + (\frac{R_o}{\beta})^3\}^{1/3} - \frac{R_o}{\beta} \tag{7}$$

For a point source with negligible M_o, Eq. (7) reduces to the well-known two-thirds law

$$z(x) = (\frac{3F_o}{2\beta^2 U^3})^{1/3} x^{2/3} \tag{8}$$

Under stable conditions ($N^2>0$), Eq. (7) is replaced by

$$z(x) = \{\frac{3}{U\beta^2 N^2} [F_o(1-\cos\frac{Nx}{U}) + NM_o \sin\frac{Nx}{U}] + (\frac{R_o}{\beta})^3\}^{1/3} - \frac{R_o}{\beta} \tag{9}$$

It should be noted that a corrected initial radius R_o must be used in these solutions to match the actual source mass flux to that of a "bent-over, Boussinesq" plume, according to

$$R_o = R_s \sqrt{\frac{T_{ao}}{T_{po}} \frac{w_s}{U}} \tag{10}$$

Corresponding initial flux parameters thus become

$$M_o = w_s^2 R_s^2 T_{ao}/T_{po} \tag{11}$$

$$F_o = w_s R_s^2 g(T_{po}-T_{ao})/T_{po} \tag{12}$$

During the field study, twelve plume data sets were recorded
which satisfied the requirements of the analytical formulas (constant U and N^2, and moderate to high windspeeds such that soon
after exit, V ≃ U). A comparison of plume trajectory measurements
with Eqs. (7) to (9) led to an optimum entrainment parameter of
β = 0.63, which agrees well with values reported elsewhere[4]. As
the corresponding root-mean-square trajectory error was about 13%,
it can be concluded that Eqs. (7) to (9) yield good trajectory predictions when the simplifications of the theory are realized in
the field.

The performance of the analytical models deteriorated, however,
when less ideal ambient conditions were present. When stack-top
windspeed was employed to make trajectory predictions under conditions of wind speed shear, for example, an average root-mean-square
error of about 27% resulted (for a set of 17 time mean plume data
sets). This error could be reduced to about 17% if an average
windspeed was substituted, following the averaging procedure outlined by Slawson[5]. Similar deterioration in performance was evident for low windspeed predictions, due to the failure of the
V ≃ U assumption: using an average windspeed, an average trajectory
error of about 25% resulted (for a set of 8 time mean plumes).

Improved trajectory predictions under these non-ideal ambient
conditions can be obtained, however, by returning to the original
nine conservation equations and seeking exact numerical solutions
without introducing further assumptions. Arbitrary U and N^2 profiles can be handled without difficulty, and neither the Boussinesq
nor the bent-over plume assumption is required. Numerical solution
using a Runge-Kutta technique is straightforward and involves very
little computer expense.

Under constant U and N^2 conditions, the trajectory predictions
of the numerical model, which still contains the entrainment
hypothesis of Eq. (5), are comparable with those of the analytical
models. For the numerical model, however, predictions continue to
be good under variable U and N^2 conditions, when the performance
of the analytical models begins to deteriorate. It was found that
the numerical model still performed poorly under low-windspeed
conditions if the entrainment hypothesis (5), which actually
applies to a bent-over plume, was retained. A more complete entrainment hypothesis, such as that suggested by Ooms[6],

$$v_e = \alpha \left| V - \frac{Uv_x}{V} \right| + \beta \frac{v_x}{V} |Uw| \tag{13}$$

was thus substituted. Near the source, Eq. (13) reduces to the
jet hypothesis $v_e = \alpha |V|$ while further downwind, the bent-over
hypothesis $v_e = \beta |w|$ is recovered. Using Eq. (13) with optimum
constants α = 0.15 and β = 0.68, trajectory predictions of the

numerical model agreed with measurements to within an average root-mean-square error of about 16% under all atmospheric conditions. Several other entrainment hypothesis were tested, but none performed any better than Eq. (13).

It seems clear, therefore, that average plume trajectories under essentially all atmospheric conditions can be well described by an integral plume rise model with a relatively simple entrainment hypothesis, provided that profiles of windspeed and temperature are available and that a numerical solution procedure is followed. As is illustrated in Fig. 2, such predictions appear to be reasonably accurate even under such a combination of conditions as low windspeed with windspeed shear and an elevated inversion. It is suggested that this straightforward numerical approach has much to offer, and should be employed if sufficient input data is available.

While the above discussion has focussed on plume trajectory predictions, similar comments can be made about predictions of final rise height. Under stable conditions, the first maximum reached by the analytical solution, Eq. (7), correlated well with the

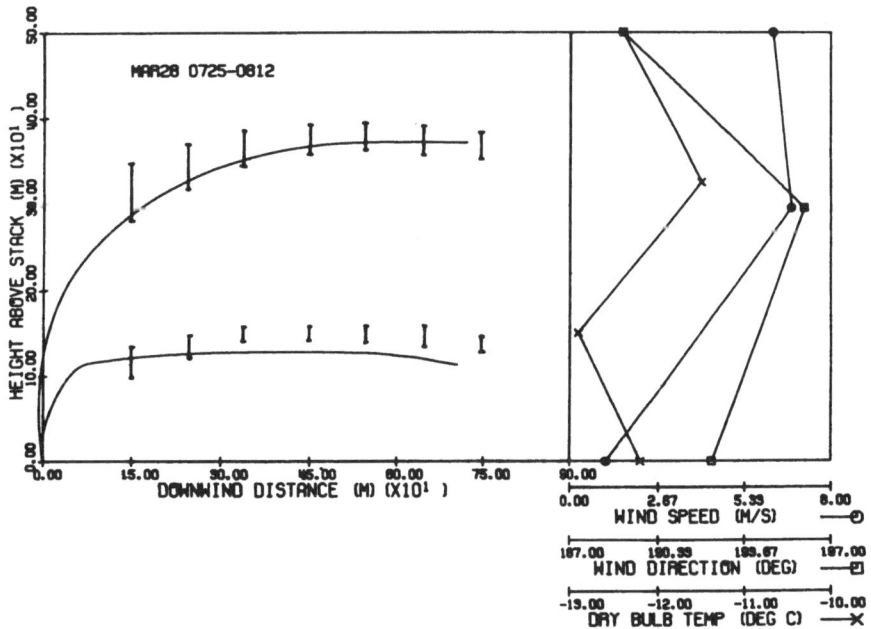

Fig. 2   Numerical plume rise prediction example.

center of SO_2 contours charted by the aircraft as long as the
restrictions of the "bent-over, Boussinesq" plume were satisifed.
Otherwise, predictions were less accurate. The numerical model
prediction of final rise height, on the other hand, fixed by the
point where the plume vertical velocity first goes negative, were
consistently in good agreement with aircraft data under all stable
atmospheric conditions. An unresolved difficulty is the determination of final rise height under neutral lapse conditions. Our
data indicates that stopping plume rise following Briggs' x^*
correlation[4] results in an underestimate of final rise height by
as much as 100%. An alternate empirical relationship developed
from Suncor data showed less scatter and closer agreement with
aircraft measurements:

$$z_f = 3.85 \ F_o/U_o \tag{14}$$

Eq. (14) is regarded as a site-specific correlation, and further
work is required to establish a more general neutral plume rise
cut off.

ATMOSPHERIC DIFFUSION PHASE

The usual basis for concentration calculations in the atmospheric diffusion phase is the solution of the steady convective-
diffusion equation with constant eddy diffusivities:

$$c(x,y,z) = \frac{Q}{2\pi U \sigma_y \sigma_z} \exp[-\frac{1}{2}(\frac{y}{\sigma_y})^2]\{\exp[-\frac{1}{2}(\frac{z+H}{\sigma_z})^2] +$$
$$\exp[-\frac{1}{2}(\frac{z-H}{\sigma_z})^2]\} \tag{15}$$

The Gaussian σ parameters, which contain the constant atmospheric
diffusivities, are then regarded as functions of x to be specified
from empirical curves related to atmospheric conditions. Eq. (15)
applies to diffusion limited only by the ground plane. A similar
equation has been derived for use when the diffusion layer is limited by an elevated inversion (see, for example, Ref. 7). Assuming that the source strength Q and average windspeed U are known
from measurements, the crucial parameters in Eq. (15) are the
plume centerline altitude above the ground, H, and the sigma parameters, $\sigma_y(x)$ and $\sigma_z(x)$.

Plume Centerline Altitude

The success of a numerical procedure for predicting plume rise
has been outlined in the previous section. The remaining problem
in the calculation of H as the plume is carried downwind is the
effect of terrain. At the Suncor site, this effect was expected

to be small, as the most significant terrain change in the area is the 80m deep, 3 km wide Athabasca River Valley, which runs alongside the plant complex. The surrounding area is otherwise relatively flat muskeg at an elevation of 360 to 365 m above sea level.

In order to estimate the importance of terrain changes on plume trajectory, the potential flow solution for flow over a step was applied to predict the streamline pattern for flow normal to the Athabasca River valley[2]. The results of this analysis, with the actual valley contour superimposed on the streamline chosen to represent the ground, are illustrated in Fig. 3. In order to avoid underestimating the effect of the valley, the chosen ground streamline actually has an elevation drop of about 20 m more than the real valley. Easterly winds, which approach the stack from the valley, will tend to lift the plume higher, while westerly winds, which carry the plume across the valley, will tend to pull the plume downwards. A quantitative analysis of these effects indicated that easterly winds would have a significant effect on plume rise, increasing H typically about 20% over the rise calculated assuming flat terrain. Unfortunately, this effect could not be verified with our data base as easterly winds are rare at the Suncor location: no case of easterly winds was recorded during our measurement program. Similar calculations indicated that the effect on plume rise due to westerly winds entering the valley was small,

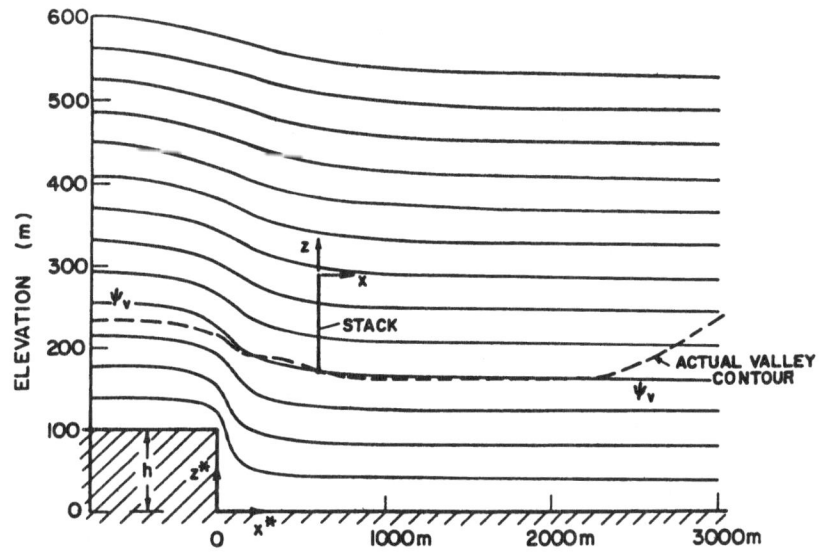

Fig. 3 Potential flow streamlines over valley.

typically reducing H only by two or three percent below the flat terrain rise: for the westerly wind situation, the plume enters the streamline pattern downwind of the step, when most of the induced vertical velocity has already died out. This prediction was borne out by our measurements of plume rise, which showed no significant correlation with wind direction for the range of wind directions encountered during the study.

An attempt was also made to predict the streamline pattern over the relatively gentle terrain changes of the surrounding plateau. Using a technique based on potential flow over a series of circular cylinders, it was predicted that the plume trajectory would remain essentially parallel to the ground. Aircraft measurements of plume cross-section concentration contours, although limited usually to two cross-sections for each plume data set, supported this conclusion.

Based on these potential flow analysis, it was concluded that, at least under neutral atmospheric conditions, the effects of terrain on H at the Suncor site could be neglected. It should be noted, however, that during very stable lapse conditions, terrain could play a more significant role in the actual plume trajectory, as valley flows essentially decoupled from the wind field over the surrounding plateau were evident in some of our measurements. A planetary boundary layer model would be required to enable prediction of these types of flows.

The Sigma Parameters

The most widely used sigma curves in plume dispersion models are probably the Pasquill-Gifford curves, although other curves such as those due to Briggs, BNL, and TVA, have also been presented[8]. The limited data base on which these sigma curves are based, and the shortcomings of the entire empirical sigma approach are well-known. An attempt was therefore made to assess this approach by abstracting sigmas from aircraft concentration measurements, and by correlating abstracted sigmas with atmospheric stability. A typical aircraft flight path for sigma measurement is illustrated in Fig. 4.

Initial attempts to classify sigmas were frustrated by considerable scatter[2]. It was found, however, that this scatter could be reduced by abstracting sigmas from entire plume cross-sections, as opposed to calculating them from SO_2 records obtained during individual transects of the plume. By constructing the contour where the concentration was 10% of the maximum level for that cross-section, (see Fig. 5), anomalous data, such as broken or disjointed plumes, were removed from the data set. Sigmas were then calculated from the remaining measurements by fitting an ellipse with

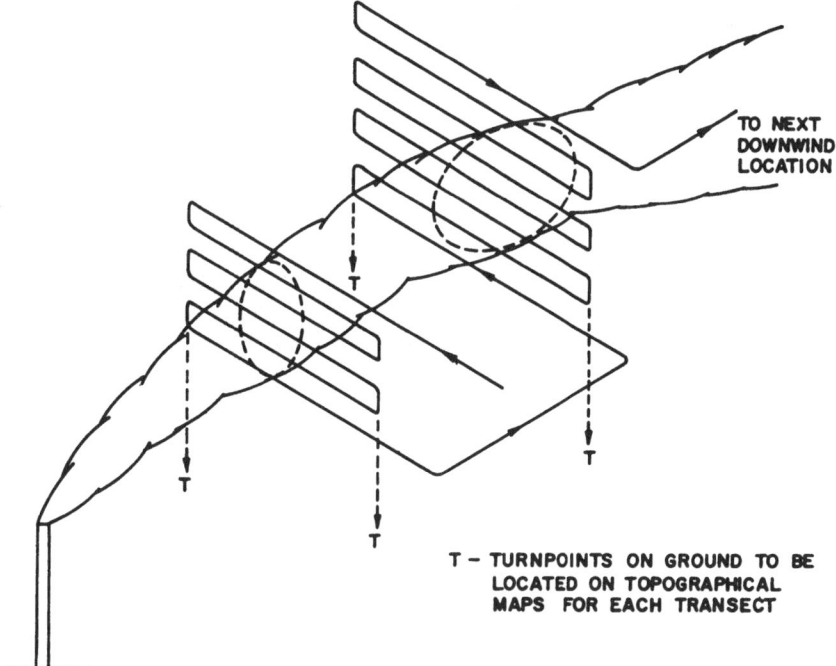

Fig. 4   Aircraft path for plume cross-section measurements.

area A equal to the measured cross-sectional area, and with minor
and major axes in the same ratio r as those measured from the
contour, such that

$$\sigma_z = \frac{1}{2.15} \left(\frac{A}{\pi r}\right)^{1/2} \tag{16}$$

$$\sigma_y = r \, \sigma_z \tag{17}$$

Aircraft contours were often sheared in the cross-wind direction.
By aligning the centers of gravity of each aircraft transect, this
shear was removed to aid in fitting a Gaussian ellipse.

Typical comparisons of sigma measurements with the common typing schemes are illustrated in Figs. 6, 7, and 8. In Fig. 6, a
σ_z versus x curve is constructed from two successive aircraft contours combined with ground-based photographic data, under the
assumption that the visible edges of plume corresponded to the 10%
concentration level, such that

$$\sigma_z = \frac{R}{2.15} \tag{18}$$

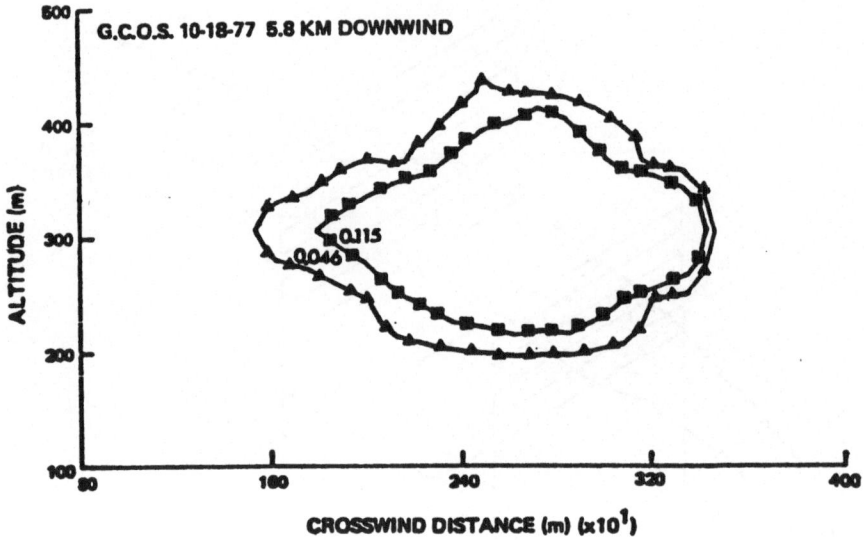

Fig. 5   A typical plume cross-section contour plot.

In Figs. 7 and 8, all aircraft data collected under near-ideal neutral lapse conditions when the wind profile showed negligible directional shear are shown. From Fig. 7, it appears that most of the data does follow Briggs class D curve fairly well. Considerable scatter remains in the σ_z curve of Fig. 8, however. This scatter is attributed to the coarse resolution of concentration data in the vertical: typically a plume cross-section would be constructed from about six transects of the plume at different discrete altitudes, whereas a continuous record of SO_2 levels in the horizontal was obtained.

Further attempts to classify sigma measurements with atmospheric stability were inconclusive due to scatter. It did appear, however, that one significant factor in the scatter was the amount of directional shear in the wind field, and a tentative correlation was proposed between the increase of σ_y above a Briggs curve prediction, and the degrees ϕ of cross-wind shear per 100 m:

$$\Delta\sigma_y = 0.19 \ \phi \ \sigma_z \tag{19}$$

From these aircraft measurements, it is concluded that the conventional sigma curves do give some indication of plume spread

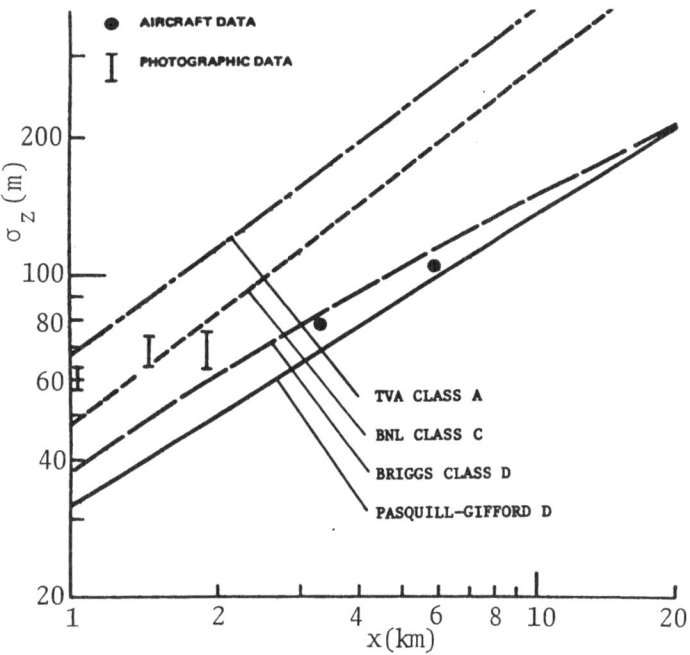

Fig. 6 σ_z measurements for Oct. 19, 1537–1650.

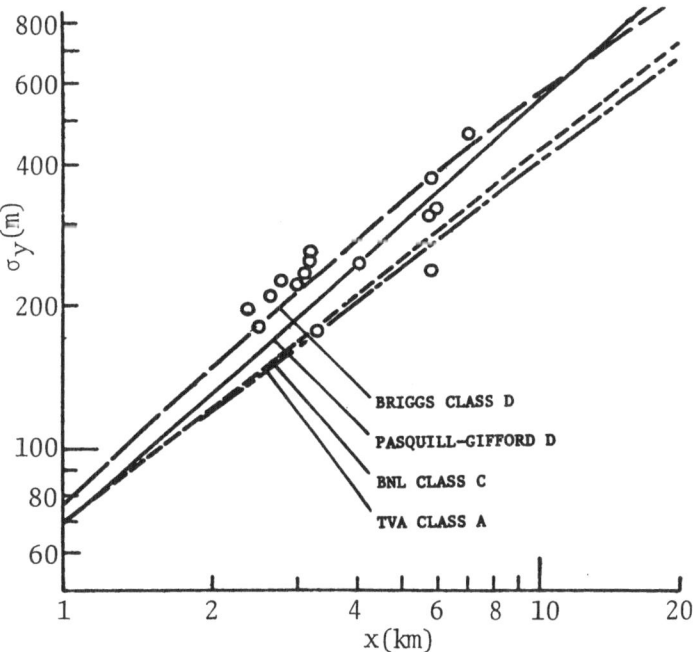

Fig. 7 σ_y data under neutral lapse with negligible cross-wind shear

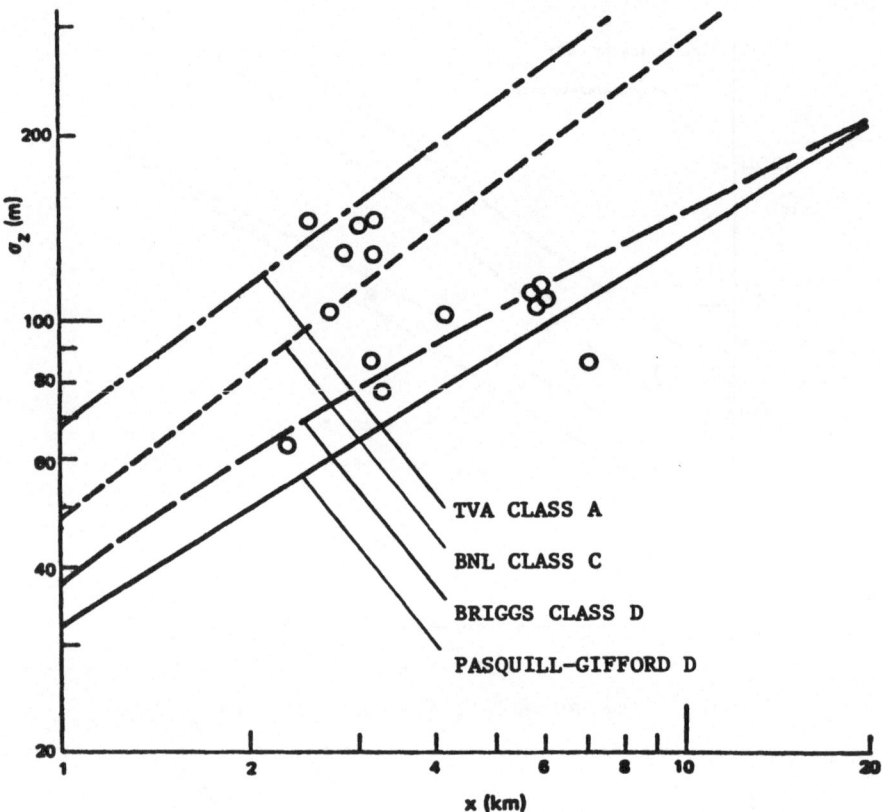

Fig. 8 Aircraft measurements of σ_z under neutral lapse with negli-
        gible cross-wind shear.

at least under the ideal atmospheric conditions of constant lapse
rate and negligible wind directional shear. Under other conditions,
however, the deviation of plume behaviour from the predictions of
the sigma curves can be significant.

GLC PREDICTIONS

The elements discussed above were combined into a plume dispersion code whose predictions could be compared with some low-level
aircraft measurements of SO_2 concentration. The main features of
the dispersion code are summarized below.

1. Plume rise was computed using the numerical solution of Eqs. (1) to (4) outlined previously.

2. Final rise during stable conditions was fixed at the point where plume vertical velocity first became negative. During neutral conditions, the correlation of Eq. (14) was applied.

3. Terrain effects were modelled according to potential flow predictions: no effect during the plume rise phase, and flow essentially parallel to the ground in the atmospheric diffusion phase.

4. Briggs' sigma curves and the σ_y enhancement of Eq. (19) were applied in the atmospheric diffusion phase.

5. Gaussian formulas appropriate to limited mixing, unlimited mixing, and fumigation were included in the code.

6. Plume rise and atmospheric diffusion phases were matched at the final rise point; that is, the plume was assumed to have radius R ($\sigma_y = \sigma_z = R/2.15$) in the plume rise phase and to grow at the rate of the appropriate Briggs' sigma curve in the atmospheric diffusion phase. In this way, an estimate of buoyancy-enhanced diffusion was included in the model.

7. Multiple sources were treated independently and the concentration fields superimposed. There are two major SO_2 sources at the Suncor site, the powerhouse and the incinerator, with source strength typically 10 to 20% of that of the powerhouse.

Typical comparisons of measured and predicted SO_2 levels are shown in Figs. 9 and 10. Conditions for the Fig. 9 case study were particularly favorable: a moderate wind (5 to 7 m/s at stack top) and constant neutral lapse remained steady during the measurement period. Agreement between aircraft data and predicted levels is good. While several cases did show agreement comparable to that of Fig. 9, several other cases were less satisfactory. Typical of these is Fig. 10, which represents a period of very light, variable winds and a stable temperature profile capped by an isothermal layer at 600 m altitude. With the measurement uncertainties in input data to the model, and the uncertainties in the model itself, this level of agreement is probably to be expected under such conditions.

It was also noted that generally the model tended to overpredict concentrations in the near field (0-6 km), while performing much better at larger distances from the source. This behaviour suggested that vertical diffusion in the near-field was greater than that included in the model, an effect that is attributed to turbulence generated by the local terrain.

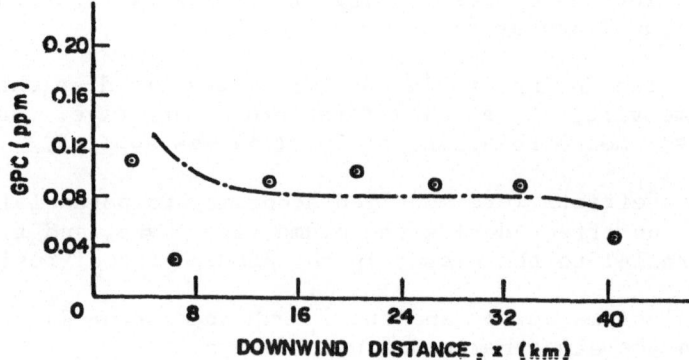

Fig. 9   Ground plane predictions and observations for Mar. 30,
         0955-1130

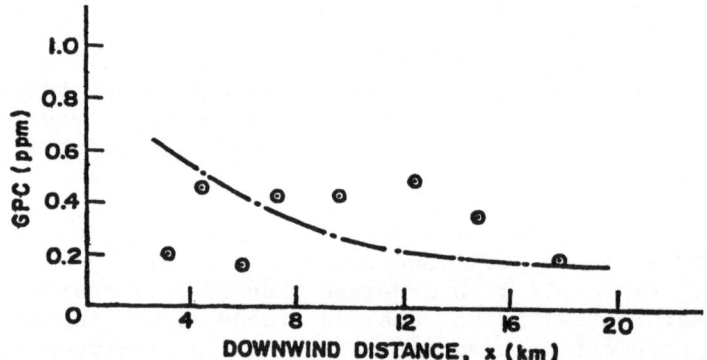

Fig. 10   Ground plane predictions and observations for Jan. 28,
          1315-1415.

CONCLUSIONS

   A well tuned site-specific plume rise and dispersion model of
the type described in this article can lead to reasonable concentration predictions. In view of this fact, and in view of the nature
of the data input to a plume dispersion model, more complex models
seem unjustified.

While it was demonstrated that terrain effects on plume rise can be negligible at sites such as those considered in this study, it was also noted that terrain generated turbulence may play a significant role in near-field diffusion. Through such effects as valley channeling, terrain can also play an important role in fixing the location of the plume trajectory in space.

REFERENCES

1. G.A. Davidson, P.R. Slawson, and S.G. Djurfors, A plume dispersion study at Mildred Lake, Alberta, Proc. 9th International Technical Meeting on Air Pollution Modeling and Its Application, Toronto, 1978, 441-454.
2. P.R. Slawson, G.A. Davidson, W. McCormick, and G. Raithby, A study of the dispersion characteristics of the GCOS plume, Syncrude Canada Ltd., Limited Circulation Report, 1978.
3. P.R. Slawson, G.A. Davidson, and C.S. Maddukuri, Dispersion modeling of a plume in the Tar Sands area, Syncrude Canada Ltd., Limited Circulation Report, 1980.
4. G.A. Briggs, Plume rise predictions, AMS Workshop in Meteorology and Environmental Assessment, Boston, 1975.
5. P.R. Slawson, Observations and predictions of natural draft cooling tower plumes at Paradise Steam Plant, Atmos. Environ. 72: 1713 (1978).
6. G. Ooms, A new method for the calculation of the plume path of gases emitted by a stack, Atmos. Environ. 6: 899 (1972).
7. D.B. Turner, Workbook of atmospheric dispersion estimates, U.S. Environmental Protection Agency (1970).
8. F.A. Gifford, Turbulent diffusion typing schemes - a review. Nuclear Safety, 17:68 (1976).

APPENDIX: NOMENCLATURE

c	pollutant concentration
g	gravitational acceleration
H	plume centerline altitude above ground
N	Brunt-Vaisala frequency
Q	source strength
R	plume radius
s	plume trajectory length
T	temperature
t	downwind travel time
U	windspeed
V	plume velocity
v_e	entrainment velocity
v_x	downwind component of V
w	vertical component of V
x	downwind plume centerline coordinate

z   vertical plume centerline coordinate
α   entrainment constant
β   entrainment constant
φ   cross-wind shear angle (deg per 100 m)

Subscripts

a   atmospheric property
o   plume property at the source
p   plume property
s   stack property

DISCUSSION

D.P. EPPEL   Can you give an experi-
            mental estimate of the range below which you
            can neglect the influence of the coriolis force ;
            perhaps on estimate of the critical Rossby num-
            ber ?

P.R. SLAWSON   We found that signi-
            ficant plume trajectory and lateral wind shear
            effects occurred beyond three to four kilometres.
            We assume these are related to coriolis effects.

J. KRETZSCHMAR   Both figures 9 and 10
            in your paper show a rather constant SO_2 level
            as a function of distance. Did you have any
            information on SO_2 levels upwind of the source ?

P.R. SLAWSON   We measured or attemp-
            ted to measure background concentrations but
            they were undetectable or close to the noise
            level at our SO_2 instrument. The measured con-
            centrations are the result of three known
            sources which were included in the model.

MODEL INVESTIGATIONS OF THE SPREADING OF HEAVY GASES
RELEASED FROM AN INSTANTANEOUS VOLUME SOURCE AT THE
GROUND

A. Lohmeyer, R. N. Meroney, and E. J. Plate

Institut Wasserbau III
University of Karlsruhe
7500 Karlsruhe 1, FRG

INTRODUCTION

Sudden release of a dense gas near the ground is accompanied by horizontal spreading caused by gravitational forces. Such clouds will drift downwind from the source location at ground level, providing an opportunity for ignition if the gas is flammable or perhaps for acute toxic effects to life in its path. An initially hemispherical volume slumps rapidly toward the ground after a sudden release. The diameter increases rapidly with an associated decrease in vertical dimension until such time as entrainment is significant. The ratio of vertical height to diameter remains quite small over most times of interest. The initial potential energy of the dense gas is converted rapidly to kinetic energy; however, this energy is also transmitted to the surrounding ambient fluid and is dissipated by turbulence at the head of the spreading plume.

The mixing of such plumes is still poorly understood despite a significant research effort of many years. The relative influence of gravity forces, viscous forces, entrainment at the plume front, entrainment at the upper surface, and modification of the background turbulent field due to stratification effects have been active subjects of discussion.

§von Humboldt Senior Scientist from Fluid Mechanics and Wind Engineering Program, Colorado State University, Fort Collins, Colorado 80521, USA.

Fay (1980) reviewed the various analytical models which have been proposed for describing the processes, and it is evident that further refinement awaits experimental verification. Hartwig and Flothmann (1980) also identify the need for new time dependent, three dimensional source experiments independent of initial gas generation or release mechanisms.

Restricting attention to instantaneous volume source behavior one finds field experiments performed by van Ulden (1974) on the sudden release of Freon-12 with an initial mixed specific gravity of 1.25, and spills of liquid natural gas (LNG) on land (AGA, 1974) or water (Feldbauer et al., 1972) with initial specific gravities near 1.5. Most recently, Picknett (1978) describes the release of air/ Freon gas mixtures with initial specific gravities ranging from 1.03 to 4.17. The LNG experiments are complicated by release mechanisms, and the recent Freon experiments suffer from instrument placement problems (Fay, 1980). Equivalent laboratory experience is limited to various lock-exchange experiments in water (Maxworthy, 1980; Huppert and Simpson, 1980), where the initial depth ratio of current to intruded fluid is often significant (Benjamin, 1968), or to finite time releases of heavy gases from area sources (Hall, 1979; Meroney and Neff, 1980).

This paper considers the results of experiments performed to examine the behavior of dense plumes during periods of gravity spread/air entrainment dominance. Havens (1977) discerned that these periods determind the lower flammability limit for LNG hazard analysis. A modified box model is presented to provide a framework of interpretation for the experiments. The experimental equipment and procedures are described. Finally, the data are evaluated and the order of magnitude of entrainment constants specified.

MODIFIED BOX MODEL FOR DENSE GAS CLOUDS

Fay (1980 has presented a generalized box model for the spread of a dense gas cloud in the form of a simple expanding cylindrical volume. The model assumes the Boussinesq assumption (ie. small density perturbations), entrainment proportional to frontal velocity, frontal velocity proportional to excess hydrostatic head, and atmospheric turbulence entrainment proportional to friction velocity. This model has been modified to retain initial inertial effects by removing the Boussinesq assumption. Hence in Fay's notation the governing relations are :

$$\frac{dR}{dt} = \alpha_1 \ (g'H)^{1/2} \tag{1}$$

$$\frac{dV}{dt} = \pi R^2 U_z + 2 \pi RH \; U_r \tag{2}$$

$$U_z = c_z \, (g'H)^{1/2}; \; U_r = c_r \, (g'H)^{1/2} \tag{3}$$

where $g' = g(1 - \rho_a / \rho)$, H = plume height, R = Radius, t = time and V = plume volume. Various proposals have been made for the values of constants c_r and c_z. Fay observes most recent plume entrainment models can be stated in terms of equivalent expressions. A survey of such models suggests the magnitude for coefficients shown in Table 1. An analytical and empirical consensus seems to exist that $\alpha_1 \simeq 1.0$, and $c_r \simeq c_z \simeq O(0.05)$.

Tabel 1   Entrainment Coefficients

Author	α_1	c_r	c_z
van Ulden (1974)	1.0	0.05	0.0
Germeles & Drake (1975)	$\sqrt{2}$	0.0	0.1
Picknett (1978)	0.94	0.82	~ 0.008[*] (Ri > 7)
Simpson & Britter (1979)	1.4	0.053	–
Cox & Carpenter (1980)	1.0	0.6	~ 0.00013[*]
Eidsvik (1980)	1.3	1.0 - 0.0	0.0005 - 0.015[*]
Fay (1980)	1.0	~ 0.01	~ 0.01
Current Experiment	1.0	0.05 - 0.1	0.05 - 0.1

[*]Under conditions modeled here

Specifying the conservation of buoyancy and an initial (index i) cloud volume where $R_i = H_i$ results in the following dimensionless (superscript $*$) expressions :

$$t^* = \frac{1}{\alpha_1 \pi^{1/6}} \int_1^{R/R_i} \frac{\left[\frac{\Delta \rho_i}{\rho a} + A \xi^a + B \xi^3\right]^{1/2} \xi}{\left[A \xi^a + B \xi^3\right]^{1/2}} d\xi \quad (4)$$

$$\frac{V}{V_i} = A\left(\frac{R}{R_i}\right)^a + B\left(\frac{R}{R_i}\right)^3 \quad (5)$$

$$\frac{H}{H_i} = A\left(\frac{R}{R_i}\right)^{a-2} + B\left(\frac{R}{R_i}\right) \quad (6)$$

where

$$a = 2\, c_r/\alpha_1$$
$$A = 1 - B$$
$$B = c_z/(3\alpha_1 - 2c_r) \quad (7)$$

When the wind speed is finite, drift distance must be adjusted for wind speed at cloud height, H. One may relate drift speed to undisturbed wind speed by

$$dX^*/dt^* = U^*(\beta H) + dR^*/dt^* \quad (8)$$

and

$$X^* = (\alpha_1 \pi^{1/6})^{-1} \int_1^{R/R} \bar{U}^*(\beta H)(\rho_a/\rho)^{1/2} d\xi + R^* \quad (9)$$

where β is an empirical constant near unity. Since $\bar{U}^*$ may be expressed in terms of a logarithmic law as

$$\bar{U}^* = \frac{U^*}{k} \ln\left\{\beta \frac{H_i}{z_o}\left(\frac{H}{H_i}\right)\right\} \quad (10)$$

and H/H_i as a function of R/R_i the relation is directly integrable. The final functional form for drift distance over a given time is

$$X^* = \frac{\bar{U}^*_{H_i} (C_f/2)}{k\,\alpha_1 \pi^{1/6}} \psi\, (R^*, \frac{H_i}{z_o}, \frac{\rho_i}{\rho_a}) + R^* \quad (11)$$

Since V_i/V is a function of R one perceives that when selfgenerated entrainment is dominant the travel distance to a given maximum varies as $\bar{U}_{\star}$.

A sudden release of volume V_i of vapor is characterized by a lenght scale $L = V_i^{1/3}$ and a time scale $T = V_i^{1/6}/(g_i')^{1/2}$ where $g_i' = g(\rho_i/\rho_a - 1)$. Huppert and Simpson (1980) suggest the initial-slump phase only exists when the depth ratio of intruding to ambient fluid is greater than 0.075. In a fluid of inifinite depth this phase is essentially instantaneous. Initial inertial/buoyancy phenomena will be present from $t^{\star} = t/T = 1.0$. Buoyancy/viscous effects are significant from $t^{\star} = [(g_i'V_i^{1/2}/\nu]^{1/3}$ if significant entrainment does not begin first from $t^{\star} = c_z^{-2/3}$. Finally background turbulence enhances mixing from $t^{\star} = g_i'V_i^{1/3}/\bar{U}^2$, until it completely dominates beyond $t^{\star} = g_i'V_i^{1/3}/U_{\star}^2$. These dominance regimes are displayed in Table 2 for the experiments presented. Only limited buoyancy/viscous effects or shear turbulence effects are expected. Hence the experiments should clearly specify the values of c_r, c_z and α_1. Data are plotted in terms of $t^{\star}$, $R^{\star} = R/V_i^{1/3}$, and V_i/V.

EXPERIMENTAL CONFIGURATION

An experiment was designed to examine in the laboratory dispersion of instantaneous volumes of dense gas released near the ground in a shear flow. The gases were released as bubbles beneath water in a boundary-layer wind tunnel, burst at the surface, and were monitored by an aspirated-hot-film katherometer.

Wind Tunnel and Bubble-Generation Equipment

A 20 cm diameter by 50 cm deep cylinder of water was mounted flush to the test section floor of the 180 cm wide by 75 cm high boundary-layer wind tunnel of the Institut Wasserbau III (Figure 1). A 50 cm^3 cup was located inverted within the cylinder. This cup was filled to the brim with the dense test gas. When the cup was rotated the gas volume formed a gas bubble which rose through the water to burst as an instantaneous volume source at the cylinder center on the water surface $\star$. Visualization experiments and concentration measurements at the source suggest the volume release occurs with limited initial dilution and release generated turbulence.

$\star$ A similar system for neutrally buoyant instantaneous volume release was used by Yang and Meroney (1972).

Table 2   Dominance Regions

Regime	Definition	t	
Initial Slump $H_i = .03$, $H = 1m$	$t^* < A$	~ 0	
Inertial/Buoyancy	$t^* > 1$	1.0	
Buoyancy/Viscous	$t^* > Re^{1/3}$	13.8	
Inertial/Buoyancy Entrainment $c_z = 0.05$	$t^* > c_z^{-2/3}$	7.4	
Entrainment/Shear Turbulence $c_z = 0.05$ $\psi = 0.045$ $U^* = .15\,\bar{U}_L^*$	$Ri(\frac{c_z}{\psi})^2 > t^* > Ri\frac{c_f}{2}$		
	$\dfrac{\bar{U}^*}{\bar{U}_L}$	FROM	TO
	0		
	0.05	460	7×10^4
	0.10	115	2×10^4
	0.20	29	5×10^3
	0.30	13	2×10^3

where

$$A = \frac{3}{2}^{-1/6}\left(\frac{H_i}{H}\right)\left(5.62\left(\frac{H_i}{H}\right)^{2/3} - 1\right), \quad \frac{c_f}{2} = \left(\frac{U_*}{U_L}\right)^2$$

$$Re = \frac{(g_i')^{1/2}V_i^{1/2}}{\nu}, \quad Ri = \frac{g_i'V_i^{1/3}}{U_*^2}, \quad t^* = \frac{t(g_i')^{1/2}}{(V_i)^{1/6}}$$

$$U^* = \frac{U}{V_i^{1/6}(g_i')^{1/2}}$$

It is possible that initial values of V_i/V lie between one and
0.5; however, it will be assumed in all data reduction that V_i/V
initially equals one. The moment of volume release was monitored
by an electricial conductivity device at the water surface; thus
a timing mark registered release during each experiment on a chart
recorder. Visualization indicated some releases occured off
coordinate origin or with finite initial lateral velocity in a
random direction.

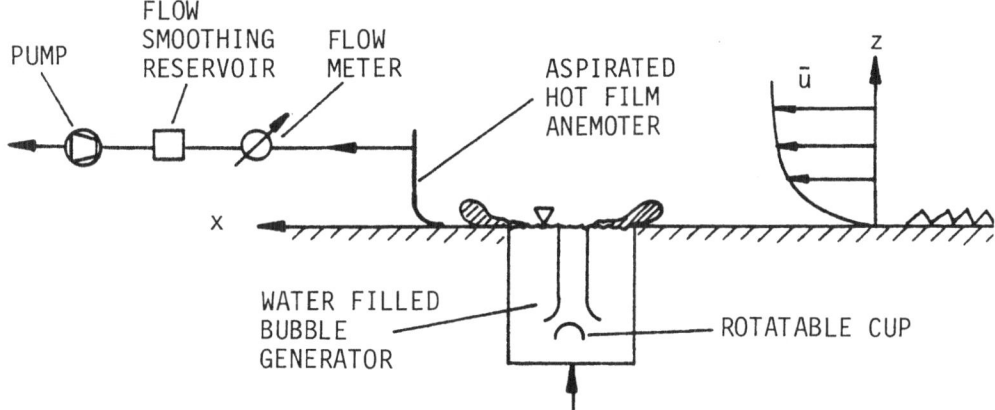

Figure 1   Experimental configuration

Shear Flow Measurements

The extremely low speeds (0.0 to 0.4 m/sec) that where required to simulate the dense cloud drift necessitated the use of a
large pressure drop constriction in the wind tunnel and special
calibration procedures for the hot wire anemometer used to measure
velocities and turbulence. DISA 55A22 hot wires monitored by a
DISA 55D01 anemometer were calibrated in a lowspeed nozzle whose
speed was seth with low vloume flowrators. Velocity and turbulence measurements were made over the test section to detect the
presence of any secondary cross currents. Velocities are reliable
to $U^* = 25$ %.

Concentration Measurements

Dense gas concentrations were measured with an aspirated hot
film anemometer (katherometer) constructed from a TSI 1441 detector in a tube. The aspiration velocity at the 1 mm diameter probe
tip was set at 0.1 m/sec to assure approximately isokinetic sampling of the plume. A fiber filter was present at the probe tip
to reduce system sensitivity to pressure perturbations during
shear flow measurements. All tests were corrected for a slight
time lag required for the sample to travel through the probe to
the detection film. Extensive tests by Meroney et al. (1978)
and again Wilson and Netterville (1980) indicate such a probe has
a flat frequency response to 150 Herz, concentration sensitivity
to 0.10 percent, and resolution within $\pm$ 5 percent of a measurement.

Since the probe is subject to drift and temperature effects it was recalibrated frequently. No significant deviations were detected.

During each realization of a volume release the katherometer response was registered on a Rikadenki DB6 chart recorder. Each sample point was recorded a minimum of three times. Time response was displayed within a resolution of $t = \pm 0.1$ sec ($t^* = \pm 3$)

RESULTS AND DISCUSSION

All experiments were performed with Freon-12 (Specific Gravity, $= 4.17$) and a 50 cm^3 initial volume; hence, there were a length scale, $L = 0.037$ m, and a time scale, $T = 0.034$ sec. Velocities were varied from zero to 0.4 m/ sec at a reference height L.

Shear Flow Characteristics

Equilibrium boundary layers were developed over 8 m of upwind channel fetch. Velocity profiles were found to fit power law relations with the exponent $p = 0.45$, and surface characteristics $U_*/\bar{U}_L = 0.170$ and $z_0/L = 0.09$. Characteristic Richardson numbers, $Ri = g_i'V_i^{1/3}/U_*^2$, varried form ∞ to 250.

Local turbulence intensity profiles were characterized by magnitudes near the ground of 0.19; thus $\bar{U}/U_*$ values were 1.12, which is below the value 2.5 quoted for classical neutral turbulent boundary layers.

Dense Puff Dispersion During Calms

The time response of the katherometer during typical source realizations is displayed in Figure 2. Initially the arrival time is short and the concentration front is sharp, subsequently the arrival time lengthens and the front becomes diffuse.

The radial growth of a dense plume in terms of dimensionless coordinates R^* and t^* is shown in Figure 3. One notes the strong inertial correction required for such a dense gas. A Boussinesq assumption would not be justified. Radial growth does not appear to be strongly sensitive to the entrainment coefficients assumed. Figure 4 and 5 describes plume dilution V_i/V versus R^* and t^* respectively. Plume concentrations decay assymptotically as $(R^*)^{-3}$ and $(t^*)^{-3/2}$. Data agrees with the modified box model when entrainment coefficientes, $c_r = c_z = 0.1$, are chosen and if no initial dilution is assumed.

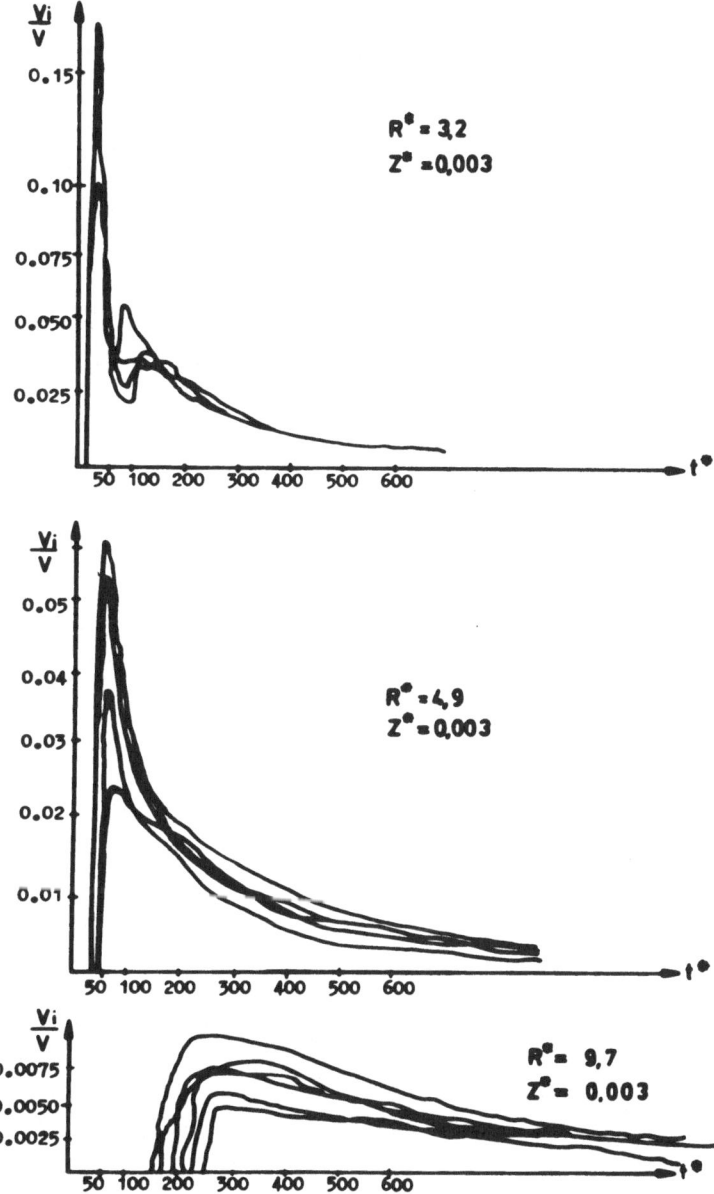

Figure 2 : Typical time response of the katherometer during
calms at various positions.

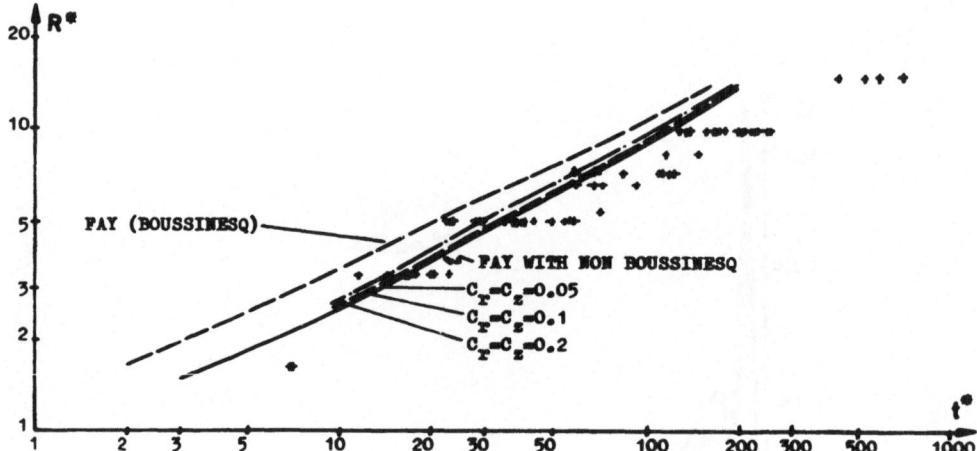

Figure 3 : Radial growth of a dense plume, $U_{x_L} = 0$.

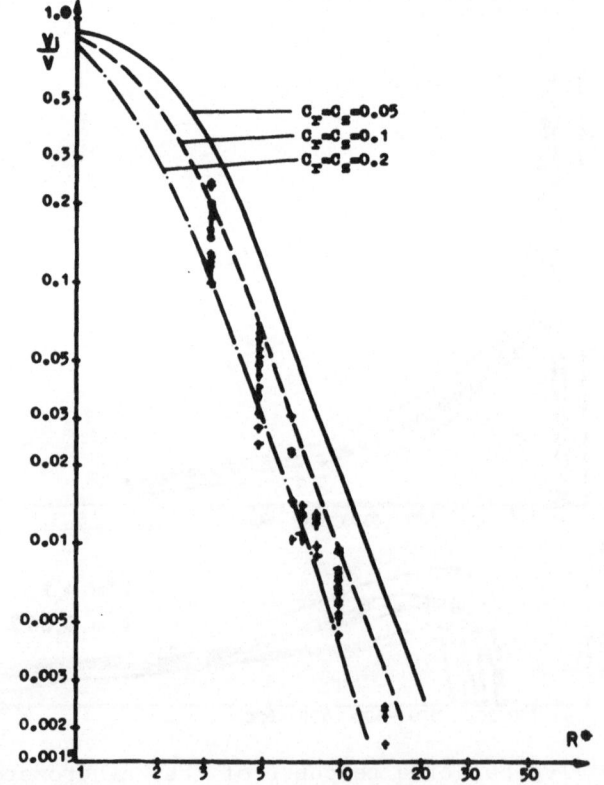

Figure 4 : Plume dilution V_i/V versus dimensionless radius
R_x, $U_{x_L} = 0$.

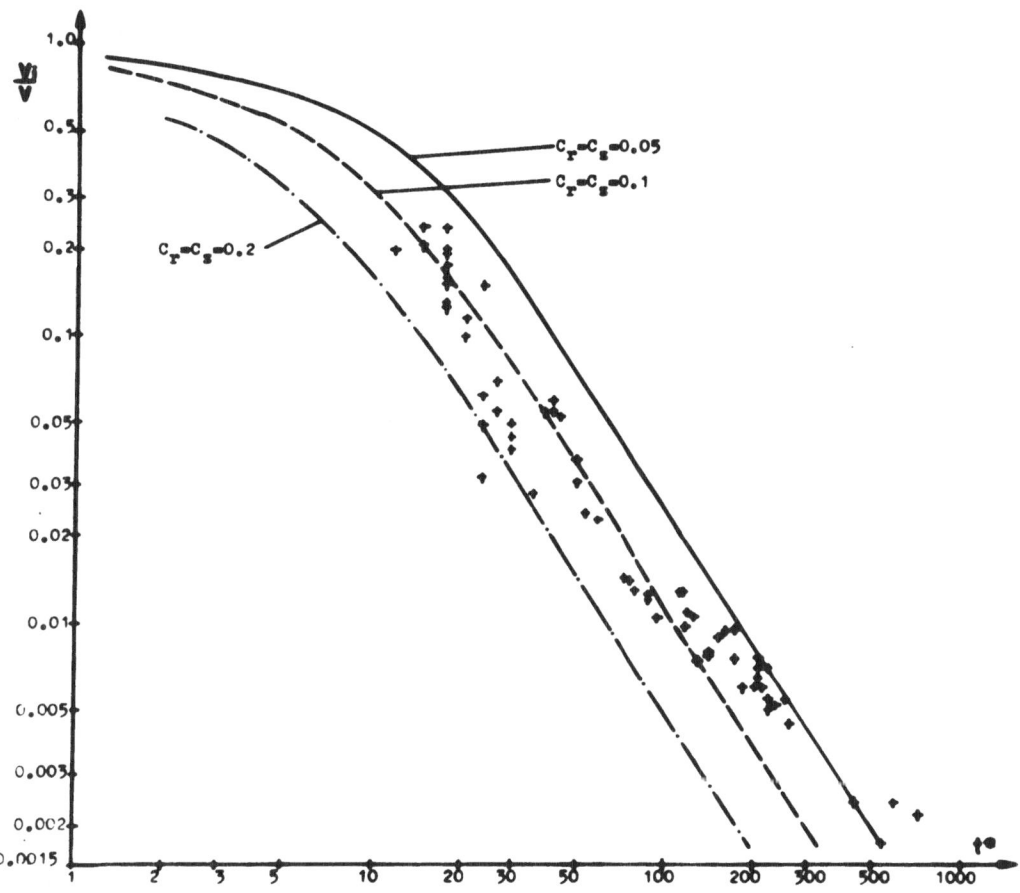

Figure 5 : Plume dilution V_i/V versus dimensionless time
t ⨯; U⨯$_L$ = 0.

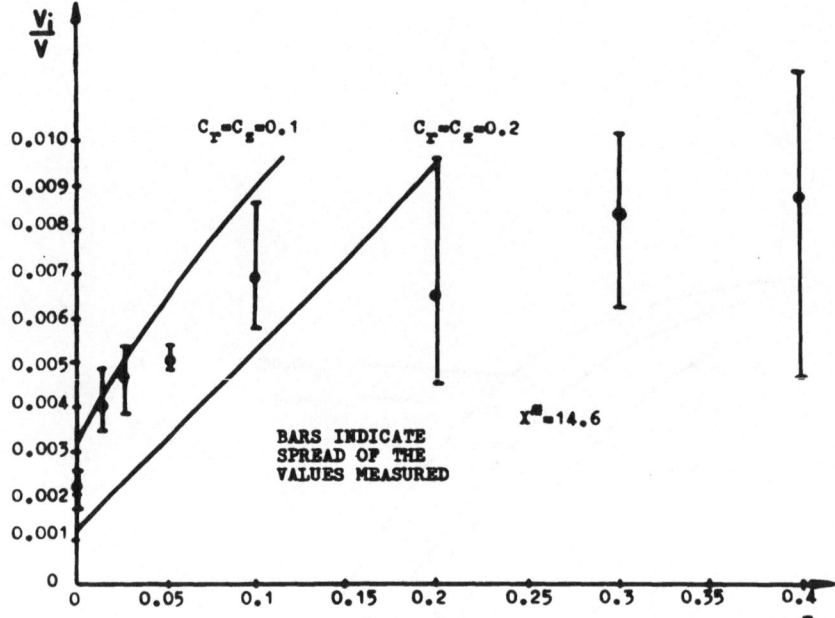

Figure 6 : Plume dilution versus dimensionless reference
           velocity U^{*}_{L} at a fixed distance X^{*} from the source.

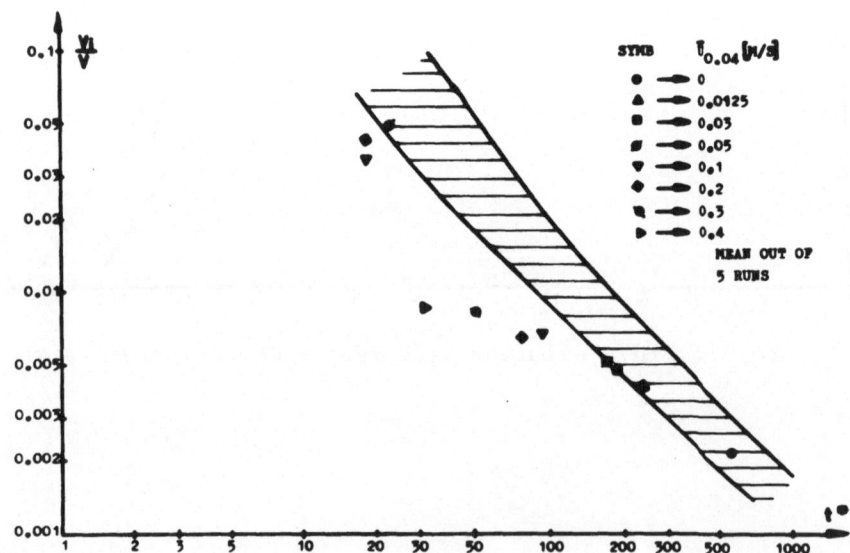

Figure 7 : Data from the wind shear experiments compared to the no
           wind data

If an initial dilution of two occurs then $c_r = c_z = 0.05$ are appropriate. No regions dominated by viscous effects are apparent before $t^x = 100$; however, subsequently spread rate slows to such low values that viscous effects result in deviations from the analysis. Error bars at long times are unexpectedly wide. Examination of replicated experiments do not seem to indicate source release mechanisms, chart recorder resulution, or katherometer response were at fault-perhaps an irregular advance of the front normally occurs.

Dense Puff Dispersion with Wind Shear
────────────────────────────────────

Results from experiments in the presence of wind shear are displayed in Figure 6 as V_i/V versus U_L^x. Note the displacement of higher concentrations to further downwind locations with increased wind speed. Only as the characteristic time scale for shear turbulence entrainment decreases to deviations from the inertial/buoyancy entrainment mechanisms occur. By use of the expression in (11) one can also examine plume behavior in terms of the viewpoint of an origin drifting with the wind. Data from the wind shear at $R^x = 4.9$ and 14.6 have been added to data of Figure 5 and replotted in Fig. 7. This permits one to note the reduced magnitudes of concentrations associated with dispersion by vertical shear not accounted for in the simple box model.

As the plume enters the mixing region dominated by shear turbulence one expects concentrations, V_i/V, to decay as $(R^x)^{-4}$ and $(t^x)^{-2}$. These data do not include distances or times suitable to discern such behavior. Fay (1980) discusses previous evidence for such dilution mechanisms in detail.

CONCLUSIONS

A series of experiments with sudden release of dense gas volumes at the ground in shear flows confirm that inertial/buoyant spreading is rapidly followed by a selfgenerated entrainment period. When Richardson numbers are sufficiently large the gas may be diluted well below 1 % before the effects of shear turbulence are evident. A modified box model for plume dilution suggests that the Boussinesq assumption is inappropriate for accurate calculation of concentration histories for very dense gases. Generalized entrainment coefficients at the plume front, c_r, and over the plume upper surface, c_z, should be between 0.1 and 0.05. Model experiments have reproduced dense plume behavior previously seen in field experiments at scales 350 times greater.

ACKNOWLEDGEMENTS

The authors wish to acknowledge support from the Institut
Wasserbau III, University of Karlsruhe, and the von Humboldt
Foundation.

REFERENCES

American Gas Association, 1974, "LNG Safety, Program Interim
    Report on Phase II Work", AGA Project IS-3-1 Battelle, Colum-
    bus Laboratories.

Benjamin, T.B., 1968, "Gravity currents and related phenomena",
    J. Fluid Mech. 31, pp. 209 - 248.

Cox, R.A. and Carpenter, R.J., 1980, "Further development of a
    dense vapour cloud dispersion model for hazard analysis",
    Heavy Gas Assessment, D. Reidel Pub. Co., pp. 55 - 87.

Eidsvik, K.J., 1980, "A model for heavy gas dispersion in the
    atmosphere", Atmospheric Environment, Vol. 14, pp. 764 - 777.

Fay, J.A., 1980, "Gravitational Spread and Dilution of Heavy
    Vapor Clouds", Second International Symposium on Stratified
    Flows, Trondheim, Norway, 24-27 June, pp. 421 - 494.

Feldbauer, G.F., Heigl, J., May, W., et al., 1972, "Spills of LNG
    on water-vaporization and downwind drift of combustible
    mixture", Report Nr. ZE6IE-72, Esso Research and Engineering
    Co., Florham Park, New Jersey.

Germeles, A.E. and Drake, E.M., 1975, "Gravity spreading and
    atmospheric dispersion of LNG vapor clouds", Proceedings of
    Fourth International Symposium on Transport of Hazardous
    Cargoes by Sea and Inland Waterways, Jacksonville, Florida,
    pp. 519 - 539.

Hall, D.J., 1974, "Further experiments on a model of an escape
    of heavy gas", Report LR (312) AP, Warren Springs Laboratory,
    Stevenage, Herts, U.K.

Hartwig, S. and Flothmann, D., 1980, "Open and controversial pro-
    blems in the development of models for the dispersion of
    heavy gas", Heavy Gas and Risk Assessment, D. Reidel Pub.
    Co., Holland, pp. 1-14.

Havens, J.A., 1977, "Predictability of LNG vapor dispersion from
    catastrophic spills into water : an assessment", Office of
    Merchant Marine Safty, U.S. Coast Guard.

Huppert, H.E. and Simpson, J.E., 1980, "The slumping of gravity currents", J. Fluid Mech., 99, pp. 785 - 799.

Maxworthy, T., 1980, "On the formation of nonlinear internal waves from the gravitational collapse of mixed regimes in two and three dimensions", J. Fluid Mech., 99, part 1, pp. 47 - 64.

Meroney, R.N., Neff, D.E. and Cermak, J.E., 1978, "Windtunnel Modeling of LNG Spills", American Gas Association Transmission Conference, 8-10 May, Montreal Canada, 26 pp.

Meroney, R.N. and Neff, D.E., 1980, "Physical Modeling of Forty Cubic Meter LNG Spills at China Lake, California", 11th International Technical Meeting on Air Pollution Modeling and Its Applications, 24-27 November 1980, Amsterdam.

Picknett, R.G., 1978, "Fluid experiments on the behavior of dense clouds, Part I, Main Report", Ptn 1 L 1154/78/1, Chemical Defense Establishment, Porton Down, U.K.

Simpson, J.E. and Britter, R.E., 1979, "The dynamics of the head of gravity current advancing over a horizontal surface", J. Fluid Mech., Vol. 94, pp. 477 - 495.

van Ulden, A.P., 1974, "On the spreading of a heavy gas released near the ground", Proceedings of First International Loss Prevention Symposium, the Hague/Delft, Elsevier, Amsterdam, pp. 431 - 439.

Wilson, D.J. and Netterville, D.D.J., 1980, "A Fast Response Heated Element Concentration Detector for Wind Tunnel Applications", To Appear in Journ. Industrial Aerodynamics, 21 pp.

Yang, B.J. and Meroney, R.N., 1972, "On diffusion from an instantaneous point source in a neutrally stratified turbulent boundary layer with a laser light scattering probe", Project THEMIS Technical Report No. 20, Colorado State University, CER72-73 BTY-RNM 17.

DISCUSSION

A.P. VAN ULDEN   In table 1 you give a list of
α values. I would like to note that Simpson
and Britter used a different heigt scale than
the one used in your paper. Thus the value is
not directly comparable.

A. LOHMEYER   The box model is very simple
and doesn't take into account a different
height between the head and tail of a density
front. In the box model the height H is just
V/ π R^2.

D. ANFOSSI   What evidence do you have that
atmospheric turbulence was well simulated your
wind tunnel.

A. LOHMEYER   The main interest of our experiments was in the no-wind cases. In the
few experiments done with wind we just attempted to simulate first order effects to see in
which direction the wind changed the situation.

PHYSICAL MODELING OF FORTY CUBIC METER LNG SPILLS

AT CHINA LAKE, CALIFORNIA

Robert N. Meroney [x] and David E. Neff [xx]

Institut Wasserbau III
Sonderforschungsbereich 80
University of Karlsruhe
D 7500 KARLSRUHE 1

INTRODUCTION

In many parts of the world there is the perception that current and planned liquid natural gas(LNG) operations and facilities present an unacceptable risk to the public. Hence, the Division of Environmental Control Technology, Department of Energy, and the Gas Research Institute, USA have supported a series of tests on liquid natural gas spilled in amounts of five and forty cubic meters onto a pond at China Lake Naval Weapons Center, California. A parallel wind tunnel model program has been performed in the meteorological wind tunnels of Colorado State University to provide field test planning information, to extend the value of a limited set of field measurements, and to validate the concept of physical modeling of LNG plume dispersion as a predictive hazard analysis tool. The measurement results described herein provided a foundation for instrument placement and interpretation of terrain effects during the 40 m^3 field experiments. Wind tunnel laboratory measurements permits a degree of control of safety, meteorological, source and site variables not often feasible or economic at full scale.

Past studies have demonstrated that the cold LNG vapor plume will remain negatively buoyant for most of its lifetime (Meroney and Neff, 1977). A hazardous mixture will extend downwind at ground level until the atmosphere has diluted the LNG vapor below the lower flammability limit,LFL, (a local concentration for

[x] 1980-1981 Von Humboldt Foundation Senior Scientist
[xx] Fluid Mechanics and Wind Engineering Program, Colorado State University, Fort Collins, Colorado 80521, U.S.A.

methane of 5 % by volume). During the fall of 1978 four experiments were performed where 5 m^3 of LNG were released through a 20 cm pipe onto a pond of water at a rate of about 5 m^3/min. (Koopman, et.al., 1979). Wind tunnel simulations were also performed before and after the field tests by Meroney and Neff (1980). Comparison of the measured concentration data with model predictions was significantly hindered by fluctuations in the wind speed and direction during the field experiments. Even when averaged over 10 second intervals, the wind speed varied by as much as a factor of three and the wind direction by as much as 60° during a single test. Nonetheless for the one case where conditions were reasonably stationary the laboratory concentrations were within 4 to 25 % of field measurements.

During the fall of 1979 a series of model tests were performed at Coloradao State University to simulate 40 cubic meter spills at the China Lake site. A new set of instrumentation, data handling procedures, and model were prepared (Meroney et.al., 1980). The results are the subject of this paper. During the summer of 1980 field spills have been performed at the China Lake facility. This data is currently being tabulated for distribution by the Lawrence Livermore Laboratory, California.

LABORATORY SIMULATION OF DENSE PLUMES RESULTING FROM CRYOGENIC SPILLS

Physical modeling in wind tunnels requires consideration of the physics of the atmospheric surface layer as well as the dynamics of the plume motion. Reliable criteria for simulating the pertinant physical properties of the atmospheric boundary layer have been demonstrated by several investigators (Cermak, 1975,1979). Frequently partial simulation suffices when the test domain is limited in time and space (i.e.Coriolis accelerations are neglected). Specific problems associated with the dispersion of cold natural gas plumes have been previously discussed by Meroney et. al. (1978, 1979). Prior experience with dense gas plume simulation has also been summarized in Meroney et.al., (1980).

Partial Simulation Criteria

When one considers the dynamics of gaseous plume behaviour, exact similitude requires the simultaneous equivalence of mass, momentum and volume flux ratios, densimetric Froude number, Reynolds number, and specific gravity. Consideration of variable property, non-ideal gas, and thermal behaviour of the plume mixture introduces additional constraints on specific heat capacity variations (Neff and Meroney, 1979).

For a plume whose temperature, molecular weight, and specific heat are all different from that of the ambient air, i.e., a cold natural gas plume, equality in the variation of the specific

gravity upon mixing must be relaxed slightly if one is to model utilizing a gas different from that of the prototype. In most situations this deviation from exact similarity is very small.

A reasonable complete simulation may be obtained in some
situations even when a modified initial specific gas ratio is stipulated. By increasing the specific gravity of the model gas compared to the prototype gas, one increases the reference velocity
over the model. It is difficult to generate a flow which is similar to that of the atmospheric boundary layer in a wind-tunnel run
at very low wind speeds. Thus the effect of modifying the models
specific gravity extends the range of flow situations which can be
modeled accurately. Meroney et al. (1974) and Isyumov and Tanaka
(1979) found that Froude number and volume flux equality provided
conservative ground-level concentrations for buoyant plumes.
Skinner and Ludwig (1978) and Kothari and Meroney (1980) obtained
similar plume trajectories when Flux Froude number and momentum
ratio equivalence are specified.

Scaling of the effects of heat transfer by conduction,
convection, radiation, or latent heat release from entrained water
vapor cannot be reproduced when the model source gas and environment are isothermal. Fortunately, in a large majority of industrial
plumes the effects of heat transfer by conduction, convection, and
radiation from the enviroment are small enough that the plume buoyancy essentially remains unchanged. The influence of latent heat
release by moisture upon the plume's buoyancy is a function of the
quantity of water vapor present in the plume and the humidity of
the ambient atmosphere. Such phase change effects on plume buoyancy can be very pronounced in some prototype situations. Fortunately the China Lake site humidity is extremely low.

The modeling of the plume Reynolds number is relaxed
in all physical model studies. This parameter is thought to be of
small importance since the plume's character is normally dominated
by background atmospheric turbulence soon after its emission. But,
if one was interested in plume behaviour near the source, then
steps should be taken to assume that the model's plume is fully
turbulent.

Simulation of the China Lake LNG Spill Plume

The buoyancy of a plume resulting from an LNG spill is
a function of both the mole fraction of methane and temperature.
If the plume entrains air adiabatically, then the plume would remain negatively buoyant for its entire lifetime. A release of an
isothermal high molecular weight gas will behave in a similar manner
to a cold plume entraining adiabatically within small corrections
for differences between specific heat capacity of source gas and air.
Hence, to simplify laboratory procedures the equality of model and

prototype specific gravity was relaxed so that pure Argon could be
used for the source gas. The equivalence of momentum flux ratio is
not physically significant for a ground source released at low flow
rates over a large area (LNG boiling on China Lake test pond);
hence, model conditions were stipulated on the basis of equivalence
between densimetric Froude number and volume flux ratio.

Argon provides almost eight times the detection sensitivity
for instantaneous concentration measurements as the carbon dioxide
used in previous studies. Over the concentration range where the
buoyancy forces are dominant the variation of the Froude number is
properly simulated. Undistored scaling of velocity components was
maintained, which implies the undistorted scaling of source strength.

Since the thermally variable prototype gas was simulated by
an isothermal simulation gas, the concentration measurements observed in the model must be ajusted to equivalent concentrations that
would be measured in the field. This relationship which is derived
in Neff and Meroney (1979) is :

$$\chi_p = \frac{\chi_m}{\chi_m + (1 - \chi_m)\dfrac{T_s}{T_a}}$$

where

χ_m = volume or mole fraction measured during the model tests

T_s = source temperature of LNG during field conditions

and T_a = ambient air temperature during field conditions.

The actual source condition, boiloff rate per unit area over
the time duration of the spill, for a spill of LNG on water is
highly unpredictable. As there was no data on the variable area
and variable volume of the different LNG tests conducted at China
Lake, the source conditions were approximated by assuming a steady
boiloff rate for the duration of the spill over a constant area.

LABORATORY METHODOLOGY

Simulation methods required to produce a model atmospheric
boundary layer have been described in some detail by Cermak (1975).
Special procedures and equipment required for dense plume measurements are considered by Meroney et al. (1978, 1979).

Wind-tunnel Facility

The Environmental Wind Tunnel (EWT) at Colorado State University was used for the LNG spill test series. This wind tunnel,
designed to study atmospheric flow phenomena, incorporates special
features such as adjustable ceiling, rotating turntables, transpa-
rent boundary walls, and a long test section (3.6 wide x 2.1. m tall

x 17.4 m long) to permit reproduction of micrometeorological behaviour at large scales. Mean wind speeds of 0.15 to 12 m/s can
be obtained in the EWT. Boundary-layer depths one meter thick over
the downstream six meters can be obtained with the use of vortex
generators at the test-section entrance and surface roughness on
the floor. The flexible test-section roof of the EWT is adjustable
in height to permit the longitudinal pressure gradient to be set
at zero. The vortex generators at the tunnel's entrance were followed by 10 m of open cardboard corrugation which produced a physical height variation of 3 mm, and a 3 m approach ramp to the 1:240
scale topography at the China Lake site.

Model

A 1:240 scale model of the China Lake topography was constructed for the use in the Environmental Wind Tunnel. The topography of the China Lake site was simulated by a layered model, each
layer (1.3 mm thick) was equivalent to 0.3 m in the field. The
model and concentration probes are shown in Figure 1. A cylindrical plenum manufactured with perforated upper plate was centered
in the middle of the test site pond. The source gas, Argon, stored
in a high-pressure cylinder was directed through a solenoid valve,
a flowmeter, and onto the circular area source mounted in the model
pond. All source release conditions were step functions; thus,
their profiles can be recreated from the data in Table 1.

Wind Profiles and Turbulence Measurements

Velocity profile measurements and reference wind speed
conditions were obtained with a Thermo-Systems, Inc. (TSI) 1050 anemometer and a TSI model 1210 hot-film probe. Turbulence measurements were made with this system for the longitudinal velocity
component and with a TSI split-film probe connected to two TSI 1050
anemometers for both longituninal and vertical component measurements. Since the voltage response of these anemometers is nonlinear
with respect to velocity, a multi-point calibration of system response versus velocity was utilized for data reduction.

Concentration Measurements

The concentrations of methane produced during a LNG spill
are inherently time dependent. It was necessary to have a frequency response to concentration fluctuations of at least 50 Hz to isolate peaks of methane concentrations above 5 % (the lower flammability limit of methane in air, LFL); hence, a set of eight aspirating hot-film probes were used for this study.

The basic principles governing the behaviour of such probes
have been discussed by Meroney et al. (1980, 1978). The hot-wire
aspirating probes were constructed with 0.1 mm diameter platinum

Figure 1 : China Lake Naval Weapons Center Spill Site Model;
           Scale 1:240

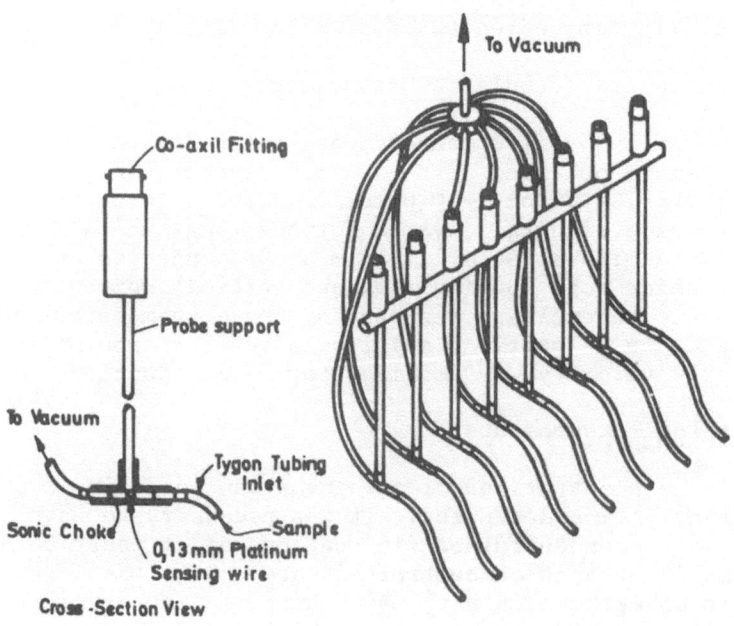

Figure 2 : Hot-wire Katharometer Probes

wire sensors monitored by an eight-channel Thermal System Inc. hot wire anemometer. The signals were conditioned for input to a Preston analog to digital converter operated by a Hewlett Packard System 1000 computer. The effective sampling area of the probe inlet is a function of the probe"s aspiration rate and the distribution of approach velocities of the gas sampled. The effective sampling area over the model was 0.5 cm^2, or resolution at full scale would be 2.9 m^2. The errors caused by a linearity assumption during data reduction, temperature drift, and calibration uncertainities is estimated to be 5 to 15 % of the measured methane concentrations.

TEST PROGRAM RESULTS

A summary of the equivalent prototype test conditions selected for the pre-field test series performed in the EWT are presented in Table 1. All dimensions reported for measured results are equivalent full scale values. The coordinate origin for all figures is the pond center of LNG spill point. The positive x axis is in the direction of the prevailing wind.

Characteristics of the Modeled Boundary Layer

Measurements of the approach flow characteristics were obtained for the modeled flow over the China Lake scale topography. In the absence of field data the characteristic velocity and length scales have been compared to values recommended by Counihan (1975) for a site of equivalent surface roughness. Table 2 compares such values as cited by Counihan and values scaled up from the model tests.

Test Series Results

The China Lake boiloff rate, duration, and wind speed for LNG spills were simulated as noted in Table 1. Concentrations were measured versus time at specified distances and heights. A digitized record of concentration time history was recorded on computer system disk files. Each test configuration was replicated several times to examine the statistics expected due to turbulence variability. Three such time history replications are presented in Figure 3 for three different ground level points downwind of spill configuration Number 7 as described in Table 1. Frome time history records such as these peak concentrations, times of 5 % arrival, 15 % arrival, 15 % departure, 5 % departure and the total dosage were evaluated.

Table 3 summarizes data for the maximum longitudinal downwind distance over which concentrations of 15, 10, and 5 % persisted for each test. The data suggest that for similar orientations and wind speeds a higher spill rate may result in slightly longer transport distance. On the other hand an increase in wind

speed for constant orientation and spill rate will result in longer
transport distances, yet for higher windspeeds the turbulence increases and the distance to LFL decreases.

        Figures 4, 5 and 6 display maximum concentration iso-
pleths when the plume moves up over a hill, along the side of a
hill, and over flat ground respectively. As might be expected the
hill delays and spreads the plume laterally resulting in shortest
downwind distances to LFL. The plume moving over flat ground tra-
vels further with smaller lateral spreading. The longest distance
to LFL appears to occur for test Number 6 when the plume moves along
the hill edge. Apparently the gravitational effect on the hill
slope decreases lateral spreading permitting the plume to exist
undilluted for longer times. The same effects are apparent in the
data in Table 3.

        For health safety purposes the data may be plotted in
terms of the maximum limits of the flammable·zone as a function of
distance and time. Figure 7 displays plume flammable zones for
test Number 1. Apparently flammable gases may persist near the
source even 10 minutes after the spill for almost all cases studied. Finally Figure 8 suggests the typical time progression of
the LFL contour over the terrain. Until 300 seconds the plume
progresses downwind; subsequently the LFL retreats toward the
source. For an idealized instantaneous release over flat terrain
one might expect the gases to retreat inward, but not necessarily
toward the source.

Table 1 : Summary of Tests : China Lake Spills

Test Number	Wind Speed 10 m (m/sec)	Spill Size (liquid) (m^3)	Spill Rate (m^3/min)	Wind Direction (Angle)
1	3	40	15	225°
2	3	40	30	225°
3	5	40	15	225°
4	5	40	30	225°
5	5	40	30	195°
6	5	40	30	255°
7	5	40	30	285°
8	7	40	15	225°
9	7	40	30	225°

Table 2 : Summary of Approach Flow Characteristics

Description	Atmospheric Data[x]	Modeled Values
z_o (m)	(0.01 - 0.15) 0.045	0.043
$1/n$	(0.14 - 0.17) 0.15	0.180
u_x/u_{10}	(0.04 - 0.05) 0.045	0.078
Λ_x (m)	(12.0 - 30.0)	31-62
Λ_y (m)	(1 - 2)	-
$(u'/u)_{30\ m}$	(0.11 - 0.18) 0.14	0.13

[x] Counihan, 1975

Table 3 : Maximum Longitudinal Distances to ULF and LFL

Test Number	Wind Speed (m/sec)	Wind Direction (angle °)	Spill Rate (m^3/min)	Longitudinal Distances (m) to 15 % UFL	10 %	5 % LFL
1	3	225	15	80	120	320
2	3	225	30	90	150	270
3	5	225	15	100	190	350
4	5	225	30	120	180	350
5	5	195	30	115	225	400
6	5	255	30	180	225	400
7	5	285	30	145	245	400
8	7	225	15	100	150	320
9	7	225	30	140	235	355

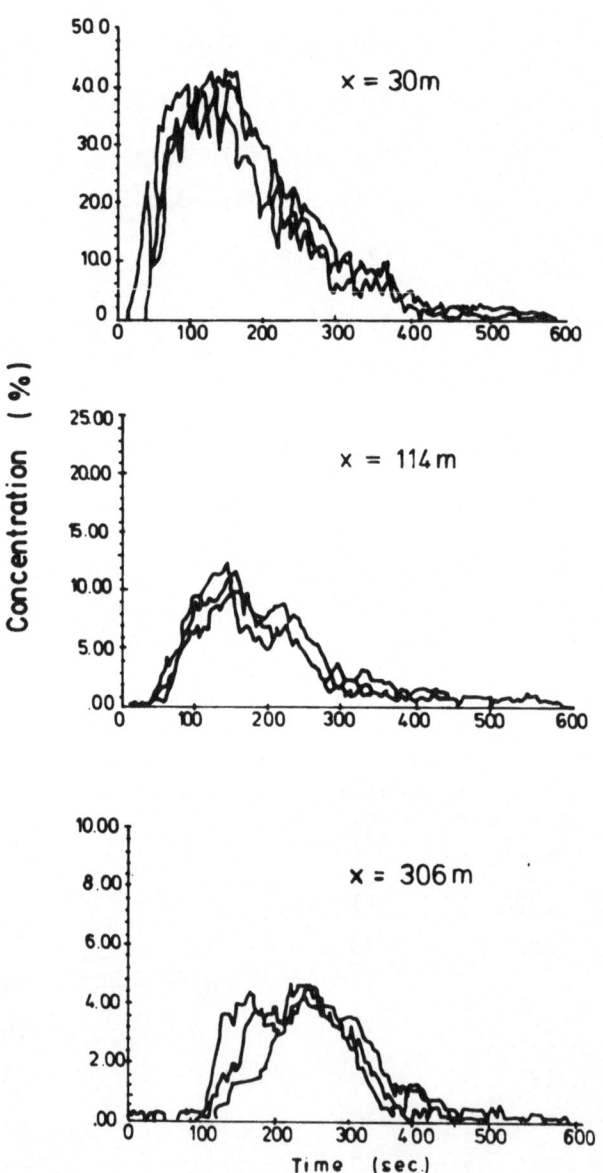

Figure 3 : Concentration Time Histories (Run Number 7)

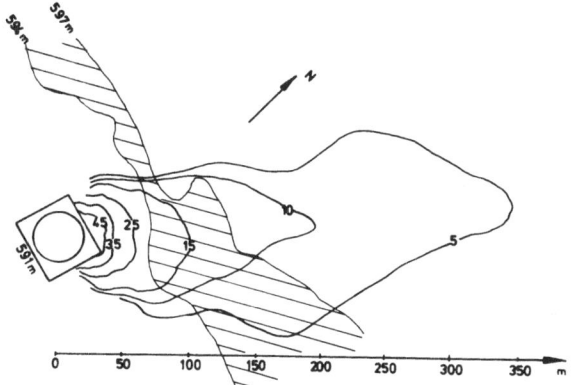

Figure 4 : Ground Level Peak Concentration Contours (Run Number 3)

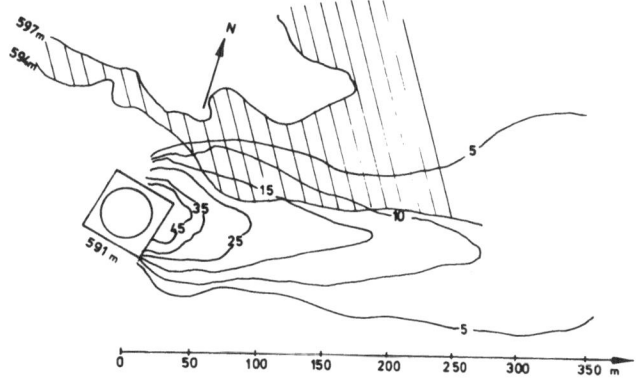

Figure 5 : Ground Level Peak Concentration Contours (Run Number 6)

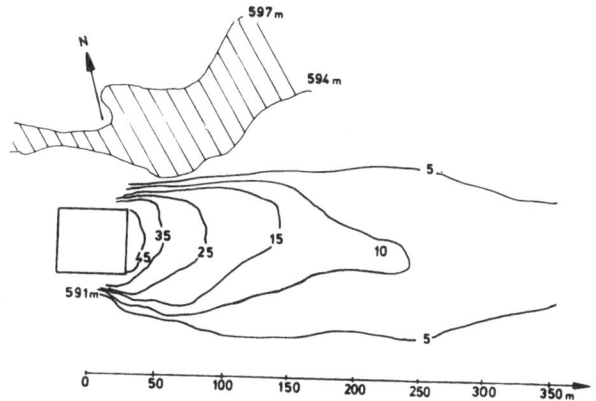

Figure 6 : Ground Level Peak Concentration Contours (Run Number 7)

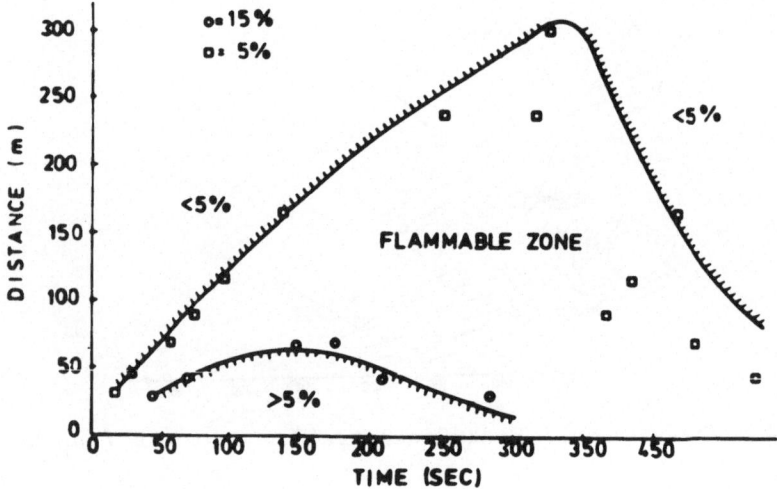

Figure 7 : Maximum Limits of Flammable Zone as a Function of
           Distance and Time (Run Number 1)

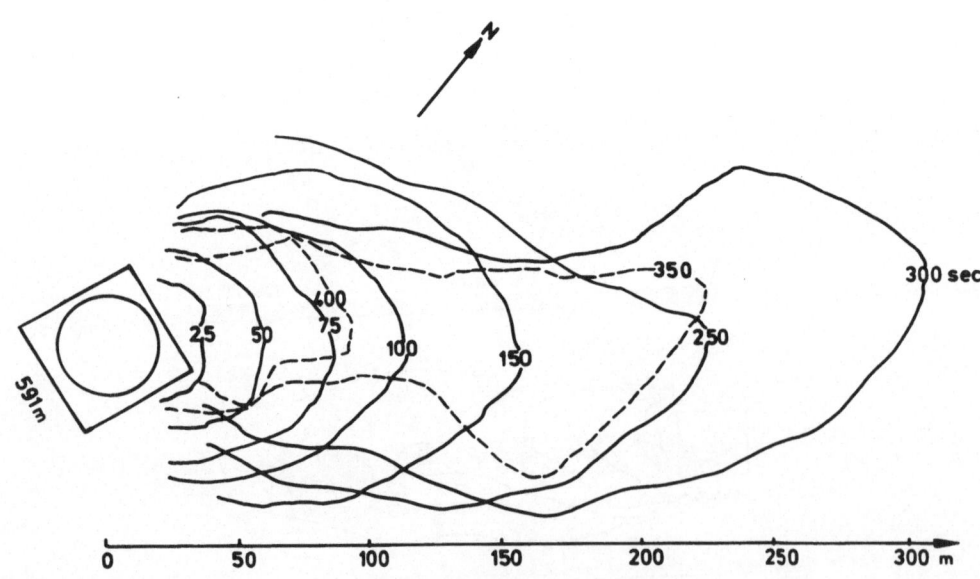

Figure 8 : Time Origression of Ground Level LFL (Run Number 1)

CONCLUSIONS

        The laboratory data await correlation and comparai-
son with equivalent field tests ; nonethless a number of interesting phenomena are discernable in the test results. Topographical effects are significant. Modest hill slopes of 1:10 can
detain dense plumes and reduce longitudinal distances to LFL.
Shallow valleys or gorges may channel the plume and sustain high
concentrations. Accelerated boiloff rates of a finite amount of
gas may result in slightly modified LFL distances ; however, the
effect of a factor of 2 variations in boiloff rate is barely
discernable in these results. An increased travel distance to a
given concentration with increased wind speed was clearly apparent for winds between 3 and 5 m/sec. As Fay(1980) notes, this
is in marked contrast to the passive dispersion of clouds where
there is an inverse dependence on wind speed.

ACKNOWLEDGEMENT

        The authors gratefully acknowledge the support of the
Gas Research Institute provided through Contract N° 5014-352-
0203.

REFERENCES

Cermak, J.E. (1975) "Applications of Fluid Mechanics of Wind Engi-
        neering, A Freeman Scholar-Lecture", J. of Fluid Engineering,
        Vol. 97, Ser. 1, N° 1, pp. 9-38
Cermak, J.E. (1979) "Application of Wind Tunnels to Investigation
        of Wind Engineering Problems", AIAA Journal, Vol. 10, N° 17,
        pp. 679-690
Counihan, J. (1975) "Adiabatic Atmospheric Boundary Layers : A
        Review and Analysis of Data from the Period 1880-1972",
        Atmospheric Environment, Vol. 9, pp. 871-905
Fay, J.A. (1980) "Gravitational Spread and Dilution of Heavy Vapor
        Clouds", Stratified Flows, Second International Symposium
        on, Vol. 1, Trondheim, Norway, 24-27 June 1980, Tapir Pu-
        blications, pp. 471-494
Isyumov, N. and Tanaka, H. (1979) "Wind Tunnel Modeling of Stack Gas
        Dispersion-Difficulties and Approximations", Proceedings
        Fifth International Conference on Wind Engineering, Cermak,
        J.E. (.) (Fort Collins, Colorado, U.S.A., July 9-14, 1979)
        Pergamon Press, 2 vols., 1400 p.
Kothari, K.M. and Meroney, R.N. (1979) "Building Effects on National
        Transonic Facility Exhaust Plume", Colorado State University,
        Fluid Mechanics and Wind Engineering Program Report CER79-
        80KMK-RNM35, Fort Collins, Colorado, December, 43 p.
Koopman, R.P. Bowman, B.R. and Ermak, D.L. (1979) "Data and calcula-
        tions of dispersion on 5 m^3 LNG spill tests, Lawrence Liver-
        more Laboratory Report UCRL-52876, 31 pp.

Meroney, R.N., Cermak, ·J.E., Garrison, J.A., Yang, B.T. and Nayak,S.
    (1974) "Wind Tunnel Study of Stack Gas Dispersal at the
    Avon Lake Power Plant", Fluid Mechanics and Wind Engineering
    Report CER 73-74-RNM-JEC-BTY-SKN35, Colorado State Universi-
    ty, Fort Collins, Colorado, April.
Meroney, R.N. and Neff, D.E. (1977) "Behaviour of Negatively Buoyant
    Gas Plumes Emitted from an LNG Spill", Proceedings of the
    6th Australasian Hydraulics and Fluid Mechanics Conference,
    Adelaide, Australia, December 5-9, 1977, pp. 472-475.
Meroney, R.N., Neff, D.E. and Cermak, J.E. (1978) "Wind Tunnel Si-
    mulation of LNG Spills", Proceedings of American Gas Asso-
    ciation Transmission Conference, Montreal, Canada, May 8-10,
    1978, pp. T217-T223.
Meroney, R.N. (1979) "Physical Modeling of Atmospheric Dispersion of
    Heavy Gases Released at the Ground or from Short Stacks",
    Proceedings of 10th International Technical Meeting on
    Air Pollution Modeling and its Application, October 23-26,
    1979, Rome, Italy, 10 p.
Meroney R.N. and Neff, D.E. (1980) "Dispersion of Vapor from LNG
    Spills Simulation in a Meteorological Wind Tunnel : Six
    Cubic Meter China Lake Spill Series", Proceedings of 4th
    Colloquim on Industrial Aerodynamics, Aachen, BDR,
    18-20 June 1980, pp. 303-320.
Meroney, R.N., Neff, D.E. and Kothari, K.M. (1980) "Behaviour of
    LNG Vapor Clouds : Tests to Define the Size, Shape, and
    Structure of LNG Vapor Clouds", Annual Report for 1979-1980,
    Gas Research Institute Contract N° 5014-352-0203, Colorado
    State University Report CER80-81 RNM-DEN-KMK.8, 80 pp.
Neff, D.E. and Meroney, R.N. (1979) "Dispersion of Vapor from LNG
    Spills - Simulation in a Meteorological Wind Tunnel of
    Spills at China Lake Naval Weapons Center, California",
    Fluid Mechanics and Wind Engineering Report CER78-79DEN-RNM41,
    Colorado State University, Fort Collins, Colorado, 77 p.
Skinner, G.T. and Ludwig, G.R. (1978) "Physical Modeling of Disper-
    sion in the atmospheric Boundary Layer", Calspan Advances
    Technology Center, Calspan Report N° 201, May.

DISCUSSION

F.B. SMITH   This is a comment rather than a question.
The value quoted in table 2 for U_x/U_{10} and attributed to Counihan must surely be a misprint. A
value nearer yours, say 0.08, is more reasonable.

R. N. MERONEY   The value quoted is correct. It is based
on data tabulated in Counihan (1975).

THE ACCIDENTAL RELEASE OF DENSE FLAMMABLE AND TOXIC GASES FROM
PRESSURIZED CONTAINMENT - TRANSITION FROM PRESSURE DRIVEN TO
GRAVITY DRIVEN PHASE

S.F.Jagger and G.D.Kaiser*

Safety and Reliability Directorate
Wigshaw Lane
Culcheth
Warrington   WA3 4LG

INTRODUCTION

Many of the toxic or explosive vapours that have been
released in accidents that have caused severe damage or loss of
life[1-5], or the need for large scale evacuation[6], are denser-
than-air. Considerable attention has, therefore, been devoted to
the modelling of the atmospheric dispersion of heavy vapours and
the current literature contains many examples[7-12].

Less attention has been devoted to the 'source terms' for the
atmospheric dispersion - that is, the expansion, entrainment
and/or evaporation mechanisms that take place, usually in a short
time, before the well-known gravitational slumping effect takes
over. A recent study[13] has shown that the way in which heavy
vapours are released from their containment can have a profound
effect on the subsequent dispersion. In the case of ammonia, the
circumstances of the escape can make the difference between a
buoyant or a non-buoyant mixture.

These source terms are treated in an ad hoc way by most
authors. This paper focuses attention on one of the modes of
release that has been responsible for some of the more severe
incidents of recent years[2,3], namely the catastrophic failure of
a pressurized container. Attention will be focused on two
existing pieces of work. The first describes experiments on

*Address from January 1, 1981. NUS Corporation, 4 Research Place,
Rockville, Maryland 20850, USA

bursting cylinders of propylene containing between 0.12 and 450kg of propylene[14], and the development of a phenomenological model for the rate of growth of the consequent vapour cloud. This model is applicable while the turbulent entrainment of air is dominated by effects driven by the initial stored energy, a regime which applies for a small fraction of a second (for a release of a fraction of a kilogram) up to the first few seconds (for a release of several tonnes).

The second piece of work is a 'box' model of the slumping and dispersion of a heavy vapour[11,15]. The purpose of this paper is to discuss the 'interfacing' of the pressure-energy and box models and to try to answer such questions as, in the case of flammable vapours, is it to be expected that the initial pressure-energy effects will dilute the cloud below the Lower Flammable Limit (LFL) or is a treatment of the slumping and atmospheric dispersion of a heavy vapour also required in order to obtain sufficient dilution? Are there circumstances in which the initial dilution can be so rapid that the gravitational slumping phase is missed altogether and there is a rapid transition to a passive puff? The paper, being necessarily a short one, concentrates on the elucidation of the physics of the expansion and slumping processes rather than on the development of detailed calculational procedures: that is, it is felt that the paper contains sufficient physical insight to make it worth while even though the process of model development is far from complete.

REVIEW OF EXISTING WORK

Phenomenological Model of Maurer et al[14]

The picture that emerges from a reading of Maurer's paper is as follows.

Once the cylinder containing the propylene, liquefied under pressure, has burst, there is vigorous bulk boiling and between 50 and 65% of the material vapourizes in a flash. The remainder is thrown into the air as fine liquid droplets. This flash expansion process is over extremely quickly. It takes place with a constant expansion velocity for a time $0.001 V_g^{1/3}$s, where V_g* is twice the volume of propylene released, measured at ambient temperature and pressure, and there is little or no entrainment of air. It leaves a hemispherical vapour/droplet mixture in a state of vigorous turbulence, which causes the entrainment of ambient air. As a consequence, the residual droplets evaporate.

---

*V_g, and the mass released, M, are introduced for scaling purposes.

While the rate of growth of the cloud is dominated by the action of its own turbulence, two regions can be distinguished. There is a central, well-mixed core of uniform concentration C_G containing about 50% of the initially released propylene vapour, where

$$C_g = \frac{0.0478 \ V_g}{(4\bar{E}t)^{1.5}} \qquad (1)$$

Here t is the time and $\bar{E}$ is a mean turbulent diffusion coefficient which has been parametrized empirically. The radius of the central core is given by

$$r_g = 1.36 \ (4\bar{E}t)^{\frac{1}{2}} \qquad (2)$$

Outside r_g, there is a concentration 'tail' C, where

$$C = C_g \ \exp \ 1.85 \ [1-\left\{\frac{r}{r_g}\right\}^2] \qquad (3)$$

and r is the radius. The form for $\bar{E}$ is

$$\bar{E} = 0.75 \ V_g^{\frac{1}{3}} \left(\frac{t}{V_g^{\frac{1}{3}}}\right)^{-\frac{1}{4}} \qquad (4)$$

Equations (1) (4) are deduced from experimental observations on tanks containing 0.124, 0.420, 1.95, 6.55, 15.6 and 454 kg of propylene. The data appear to agree well with the empirical expressions except that, for the largest release, the values of $\bar{E}/V_g^{\frac{1}{3}}$ appear to be about a factor of two smaller than the right hand side of eq (4).

The relationship $\bar{E} \ \alpha \ t^{-\frac{1}{4}}$ is universal. It stems from the use of free turbulence theory (ie turbulence theory in the absence of walls) and may be deduced from a consideration of isotropic turbulence decay and the application of the mixing length hypothesis. The coefficient of proportionality, however, is a complicated function of at least the density, the kinematic viscosity and the mass of vapour in the central core and of the initial pressure. To attempt to generalize the results of Maurer to materials other than propylene will be a difficult task and has not been attempted in this paper.

A Simple 'Box' Model for the Dispersion of Heavy Vapours[11,15]

The behaviour of a puff of heavy vapour in the atmosphere can be described by taking account of four basic processes.

These processes are:

(i)   Gravitational slumping.

(ii)   How the entrainment of air is affected by the presence of a stabilising density gradient.

(iii)   The presence of sources or sinks of heat.

(iv)   The specification of initial conditions of radius, temperature, dilution etc - that is, the definition of a source for the atmospheric dispersion model.

All treatments of heavy vapour dispersion, whether simple or sophisticated, must incorporate a description of these processes.

Gravitational slumping   Assuming that the puff is in the form of a cylinder of height h and radius r, a liquid column analogy suggests that

$$\frac{dr}{dt} = K(gh \, \Delta\rho/\rho_a) \qquad\qquad (5)$$

where g is the acceleration due to gravity, ρ is the density within the cylinder and $\Delta\rho = \rho - \rho_a$, ρ_a being the density of the ambient air; K is a constant which is often taken to be unity.

Entrainment   A second necessary element in a simple model is a prescription for the rate of entrainment of air. As far as the atmospheric turbulence is concerned, this rate of entrainment or mixing is suppressed in the presence of a stabilising density gradient. It is convenient at this point to introduce the Richardson number:

$$Ri = \frac{gl \; \Delta\rho}{\rho_a \, U_t{}^2} \qquad\qquad (6)$$

where l is a length scale and U_t is a velocity characterising the intensity of turbulence in the atmosphere. l might be the same as h, the height of the cloud, or it could be some more sophisticated measure such as a turbulence length scale in the atmosphere. The velocity U_t is usually taken to be the friction velocity or some related quantity. The Richardson number may then be interpreted as the square of the ratio of two velocities, a gravity induced one $(gl \, \Delta\rho /\rho_a)^{1/2}$ - see eq (5) - and one typical of neutrally buoyant plume growth in the atmosphere. It is reasonable to expect that the rate of entrainment of air into the top surface of the slumping puff should be a function of Ri and, indeed, the simplest prescriptions for an entrainment velocity U_e are given by[16]

$$U_e = \alpha' \, U_t \, Ri^{-1} \qquad\qquad (7)$$

which illustrates the elementary point made above, that the rate of entrainment of air is suppressed in the presence of a stabilising density. α' is an empirical constant and its value depends on the choice of l and U_t as well as on the weather category. The rate of entrainment of air through the top surface is then given by

$$\frac{dM_a}{dt} = \rho_a \ \pi r^2 \ U_e \tag{8}$$

where M_a is the mass of air in the puff. If edge entrainment is also considered, an additional term appears on the rhs of eq (8), namely

$$\frac{dM_a}{dt} = 2 \ \rho_a \ \pi r h \ U_e' \tag{9}$$

where U_e' is the edge entrainment velocity, sometimes taken to be[10]

$$U_e' = \alpha'' \ \frac{dr}{dt} \tag{10}$$

where α'' is another empirical constant.

Cloud heating   It is necessary to introduce some mechanism for heating the cloud, should this be cold, as is often the case. If the difference between the temperature of the cloud and the ground is ΔT_g, heating occurs by turbulent natural convection at a rate[17]

$$Q_c = \alpha \ (\Delta T_g)^{4/3} \tag{11}$$

where the constant α depends on such quantities as the specific heat and thermal conductivity of the cloud. Heating is also taking place because air at the ambient temperature is being entrained at a rate given by eqs (8) and (9). If the temperature difference between the cloud and the air is ΔT_a, the rate of change of temperature is given by

$$\frac{dT}{dt} = \frac{dM_a/dt \ C_{pa} \ \Delta T_a + \alpha(\pi r^2) \ \Delta T_g^{4/3}}{M_a \ C_{pa} + M_g \ C_{pg}} \tag{12}$$

where the meaning of the symbols not already defined is as follows: C_{pa} and C_{pg} are the specific heats of air and propylene (or other gas) respectively, and M_g is the mass of propylene in the cloud.

Zeman[19] argues that the turbulent convection produced is sufficient to keep the slumping puff well-mixed in the vertical

and that it can contribute significantly to the entrainment
velocity U_e.

   Source term   In order to solve the three coupled first order
differential equations in the independent variables r, M_a and T,
it is necessary to specify starting values for the variables - ie,
a source term.   For puff releases, it is usual to assume that the
puff is in the form of a cylinder, with height equal to the
radius.   In the present case, this cylinder may be assumed to
develop as the turbulent entrainment processes, driven by the
energy stored in the containment, decay.   The subject matter of
this paper is concerned with the nature of the transition from the
pressure-energy phase to the gravitational slumping phase.

A Simple Picture

   It seems that the sequence of events following the bursting
of a pressurized vessel can conveniently be divided into four
phases.

Phase 1   An initial flash expansion process, leading to the
          formation of a fine aerosol and of considerable turbu-
          lence.

Phase 2   Entrainment of air due to the action of this turbulence,
          together with the evaporation of the droplets, while the
          turbulence gradually decays.

Phase 3   Gravitational effects predominate and slumping, suppres-
          sion of the rate of entrainment of air in the presence of
          a stabilising density gradient and the heating of the
          cloud by the ground must all be considered.

Phase 4   A transition to passive diffusion.

   It will be helpful for the reader to bear this picture in
mind during the remainder of this paper.

THE TRANSITION PROCESS

When Does the Transition to Gravitational Slumping Take Place?

   An appropriate dimensionless combination* to consider is

─────────
*It is recognized that this is not the only possible dimensionless
combination

$$L = \frac{gh \, \Delta\rho}{\rho_a \, (\frac{dr_g}{dt})^2} \tag{13}$$

which is a Richardson number and is the ratio of the square of the gravitational slumping velocity and the rate of growth of the radius during Phase 2. It may be shown, using eqs (1) - (4) and $M_g = V_g/2\rho_g$, that

$$\frac{dr_g}{dt} = 1.02 \, M_g^{5/24} \, \rho_g^{-5/24} \, t^{-5/8} \tag{14}$$

where ρ_g is the density of propylene at normal temperature and pressure. Assuming that h ~ r during the pressure energy phase,

$$gh \, \Delta\rho \simeq 0.63 \, M_g^{7/12} \, \rho_g^{5/12} \, (1-\rho_a/\rho_g) t^{-3/4} \tag{15}$$

and

$$L \sim 0.6 \, M_g^{1/6} \, \rho_g^{5/6} \, (1-\rho_a/\rho_g) \, t^{1/2} \, \rho_a^{-1} \tag{16}$$

The assumption that temperature effects may be neglected has been made. It is to be expected that gravitational slumping will be important if L >> 1 and that it will be negligible if L << 1. Taking ρ_g = 1.88 kg m^{-3} (the density of propylene at NTP) and ρ_a = 1.2 kg m^{-3} (the density of air at atmospheric pressure and 20°C), L = 1 corresponds to

$$t_g = 7.4 \, M_g^{-1/3} \tag{17}$$

clearly, the value of the coefficient of t_g must be regarded as uncertain. The neglect of temperature effects, for example, is one cause of this uncertainty.

Values of t_g are shown in Table 1 for the smallest and largest releases considered by Maurer (0.12 kg and 450 kg) together with Maurer's estimate of the period for which at least part of the cloud contains an air/propylene mixture which lies between the LFL and the UFL. An extrapolated calculation for a 50Te rail tank car is also shown. Points of physical significance that emerge as as follows.

(i)   For releases of mass of the order of 0.1 kg, gravitational effects are not important, at least for predicting the duration of a flammable cloud. The dilution to the LFL is entirely governed by turbulence generated during Phase 1.

(ii)   For releases of mass of the order of tens of tonnes, gravitational slumping effets become important well before the cloud is diluted to the level of the

UFL. Indeed the time $t_g = 0.2s$ corresponds to a radius of 12.5m, which is not even large enough to contain the volume of the original flashed mass of propylene vapour, ie 50% of 50Te, so that strictly eq (2) should not be used at all. Instead, all that can be said is that gravitational slumping effects become important as soon as the initial flash expansion has taken place.

(iii)   For releases containing some hundreds of kilograms of propylene, there is clearly a transition region. This may account for the observed departure from eq (4) which suggest that, for the 450 kg release, the effective mean turbulent diffusion coefficient $\bar{E}$ is somewhat smaller than expected. If gravitational slumping effects are beginning to be important in this case, it is to be expected that the suppression of turbulent entrainment over the top surface in the presence of a stabilising density gradient cannot be neglected, and that some reduction in the value of $\bar{E}$ will be observed.

(iv)   From observations (i)-(iii) it may be deduced that there are important scaling effects. The procedure developed by Maurer et al (see eqs (1-4)) for predicting the rate of dilution of the cloud to the LFL cannot be extrapolated to 50 Te releases - indeed, the small value of t_g in Table 1 suggests that the bulk of the dilution to the LFL will occur while the cloud is in its gravitational slumping phase.

(v)   As has already been mentioned, this paper does not attempt to deal with materials other than propylene. The values of the various numerical coefficients in eqs (1)-(4) are expected to change for other gases. It is instructive to note, however, that, from eq (16)

$$t_g \; \alpha \; (1 - \rho_a/\rho_g)^{-2} \tag{18}$$

and this relationship will appear no matter what the material being studied. In the present case, $\rho_a = 1.2$ kg m-3 and $\rho_g = 1.88$ kg m^{-3} so that $(1 - \rho_a/\rho_g)^{-2} = 7.6$. For another gas with, say, $\rho_g = 1.1 \; \rho_a$, $(1 - \rho_a/\rho_g)^{-2} = 121$. It is, therefore, quite conceivable that values of t_g will be an order of magnitude or so larger for materials with densities that are nearer to those of air than is the density of propylene. If all the values of t_g in Table 1 were increased by a factor of ten or more, then the transition region could easily be moved from the few hundred kilogram region to the few tonnes or few tens of tonnes region.

(vi)   Further calculations have been carried out using equality of air entrainment velocities as a criterion for transition between Phases 2 and 3.   The values of transition times, t_g, so obtained show very little departure from those in Table 1, and indicate that the conclusions of this study remain unaltered by choice of dimensionless combinations other than that given in (13).

In ref (ii), calculations were described for a relase of 20Te of anhydrous ammonia from pressurized containment.   In order to define a source cylinder for the gravitational slumping calculations it was assumed that the initial pressure-energy driven phase entrained of the order of ten times as much air by mass as ammonia.   This rule of thumb was justified by reference to the dimensions of an observed ammonia cloud following the accidental catastrophic failure of a pressurized static storage vessel at Potchefstroom, South Africa[18].   Since ρ_a/ρ_g for cold air-ammonia mixture is considerably smaller than for propylene, it is to be expected that the effective value of t_g for ammonia-air mixtures would be larger than for propylene-air mixtures.   Thus it is not inconsistent with the foregoing to observe that, in the case of cold ammonia-air mixtures, the transition to gravitational slumping occurs at a dilution of 10:1 when several tens of tonnes of ammonia are present, while for the same mass of propylene, the transition to gravitational slumping occurs at a much smaller dilution.

The lesson to be learned from the arguments above is the following.   It appears from an analysis of accidental releases of ammonia that, for accidents involving the accidental escape of a few tens of tonnes, the rule-of-thumb that Phase 2 causes a dilution of about 10:1 is a reasonable one: however, it is dangerous to try to generalize.   If much larger quantities of ammonia are involved, or if the same mass of another material with a considerably greater density is considered, Phase 2 may cause a considerably smaller dilution.

Passive Behaviour

In this context, it is convenient to consider the ratio

$$L' = \left(\frac{dr_g}{dt}\right)\bigg/ u_*$$ (19)

where u_* is the friction velocity. If $L' \gg 1$, $dr_g/dt \gg u_*$ so that the initial turbulence is dominating the rate of growth of the cloud. If, on the other hand, $L' \ll 1$, $u_* \gg (dr_g/dt)$ and atmospheric turbulence dominates. Making use of eq (14), and taking a typical value of u_* ($= 0.5$ ms^{-1} in neutral conditions when the surface roughness length is 0.1m and the windspeed at a height of 10m is 6 ms^{-1}), $L' = 1$ corresponds to a time

$$t_a \sim 2\ M_g^{\frac{1}{3}} \rho_g^{-\frac{1}{3}}$$ (20)

Values of t_a are presented in Table 1. For the release of 0.12 kg of propylene, $t_a \ll t_g$, so that the puff makes the transition directly from the Phase 2 to the passive dispersion phase.

For both the 450 kg and 50 Te cases, $t_a \gg t_g$ and there is an intermediate phase dominated by gravitational effects. $t_a \sim t_g$ occurs for a release of 10 kg, which defines the transition mass above which gravitationally-driven effects cannot be neglected.

Table 1   Calculated values of various critical times

Mass of Propylene released (kg)	Time for dilution to between LFL and UFL according to Maurer (s)	t_g (s) see eq (18)	t_a (s)
0.12	0.05– 0.14	15	0.8
450	0.8 – 2.2	0.96[+]	12
50,000	3.8 –10.2	0.2	75

+ The computer program DENZ[15] has been used to assist in the preparation of this paper. Its use confirms the value tg $\sim$ 1s for the 450 kg release and, therefore, increases confidence in the simple arguments of Sect 3.

Here, then, is another example of an effect which prevents the simple scaling-up of procedures valid for small releases.

CONCLUSION

This qualitative study of the sequence of events that may follow the accidental failure of a vessel containing a gas lique-fied under pressure has identified four phases through which the vapour cloud evolves.

1.   An initial flash expansion process, leading to the formation of a fine aerosol and of considerable turbu-lence.

2.   Entrainment of air due to the action of this turbulence, together with evaporation of the droplets, while the turbulence gradually decays.

3.   A phase during which gravitational effects may dominate, leading to slumping and to the suppression of turbulent mixing in the vertical.

4.   Finally, a transition to passive diffusion.

In discussing these phases, a number of important physical processes have been illuminated. These have been discussed in detail in Section 3. Among the most important of the results obtained is that the time at which the transition from Phase 2 to Phase 3 occurs is proportional to the inverse of the cube root of the total mass of gas released. It follows that, there are scaling effects which imply that different modelling procedures are required when different quantities of gas are released.

The model of Maurer et al[14], which described the behaviour of the puff during Phase 2, described the dilution of a small puff containing of the order of 0.1 kg down to dilutions which are well below the LFL for the particular case of propylene. For a release containing 50 Te, however, such as might take place from a rail tanker, Phase 2 seems to be relatively unimportant and the bulk of the required dilution down to the LFL will take place during Phase 3. Hence the model of Maurer et al should not be expected correctly to predict the behaviour of such a large puff.

In the future, it is expected that effort will be devoted to making quantitative the qualitative considerations of this paper. The models to be developed will be oriented towards the prediction of quantities that are of importance in the study of the safety of chemical installations - hazard ranges for flammable clouds or toxic clouds.

REFERENCES

1. Report of the Court of Inquiry into the Flixborough Disaster,
    Her Majesty's Stationery Office, London (1975).
2. F Briscoe, 'Preliminary Analysis of the Fire and Explosion
    Incident at the Los Alfraques Camp Site on 11 July 1978',
    SRD Internal Note (1978).
3. G McMullen, 'A Review of the 11th May Ammonia Truck Acci-
    dent', City of Houston Health Department Report (1976).
    For the reasons why the Ammonia in this incident was part
    of a denser-than-air mixture, see ref (11).
4. 'Railroad Accident Report: Louisville and Nashville Railroad
    Company Freight Train Derailment and Puncture of Anhydrous
    Ammonia Tank Cars at Pensacola, Florida, 9 November 1977',
    US National Transportation Safety Board Report Number
    NTSB-RAR-78-24 (1978).
5. D H Slater, 'Vapour Clouds', Chemistry and Industry, May 1978
    Issue pp 295-302. This paper contains a tabulation of many
    incidents.   .
6. 'Mississauge Derailment - Ground Level Concentration Measure-
    ments of Chlorine and Detection of Other Compounds', A
    Report Prepared for the Ontario Ministry of the Environ-
    ment, MOE Report ARB-TDA 05-80, by SGEX INC.
7. R G Picknett, 'Field Experiments on the Behaviour of Dense
    Clouds', Report No Ptn IL 1154/78(1978), Chemical Defence
    Establishment Porton Contract Report to the Health and
    Safety Executive, Sheffield UK.
8. D J Hall, F C Barrett and M Ralph, 'Further Experiments on a
    Model of Escape of Heavy Gas', Report No LR 217 AP (1974),
    Warren Spring Laboratory, Stevenage, Hertfordshire.
9. J McQuaid, 'Dispersion of Heavier-than-Air Gases in the
    Atmosphere: Review of Research and Progress Report on HSE
    Activities', Health and Safety Laboratories Technical Paper
    8 (1979). This paper contains a reasonably comprehensive
    list of references.
10. A P Van Ulden, 'On the Spreading of a Heavy Gas Released Near
    the Ground', Proc 1st Int Symp on Loss Prevention and
    Safety Promotion in the Process Industries: Elsevier, Ed G
    H Buschmann (1974) pp 221-226.
11. G D Kaiser and B C Walker, 'Releases of Anhydrous Ammonia
    from Pressurized Containers - The Importance of Denser-
    than-Air Mixtures', Atmos Env 12 (1978) 2289-2300.
12. W G England, L H Teuscher, L E Hausman and B Freeman, 'Atmos-
    pheric Dispersion of Liquefied Natural Gas Vapour Clouds
    using SIGMET, A Three Dimensional Time-Dependent Hydro-
    dynamic Computer Model', Proc Heat Transfer and Fluid
    Mechanics Insititute (1978), Washington State University,
    Pullman, Washington, USA.
13. R. F. Griffiths and G. D. Kaiser, 'The Accidental Release
    of Anhydrous Ammonia to the Atmosphere - A Systematic

study of Factors Influencing Cloud Density and Dispersion ,
UKAEA Report SRD R154 (1979). A shortened version is to
be published in the Journal of the Air Pollution Control
Association.

14. B. Maurer, K. Hess, H. Giesbrecht and W. Leuckel, "Modelling
of Vapour Cloud, Dispersion and Deflagration After Bursting of Tanks Filled with Liquified Gas", 2nd International Symposium on Loss Prevention and Safety Promotion in
the Process Industries held at Heidelberg, West Germany,
6-9 September 1977. Preprints pp 305-319, Pub by DECHEMA
(Deutsche Gesellschaft Fur Chemisches Apparetewesen eV).

15. L. S. Fryer and G. D. Kaiser, 'DENZ - A Computer Program for
the Calculation of the Dispersion of Dense Toxic or Explosive Gases in the Atmosphere', UKAEA Report SRD R152(1979).

16. R. A. Cox and D. R. Roe, 'A Model of the Dispersion of Dense
Vapour Clouds', Heidelberg Preprints (see ref(14)), pp
359-366 (1977).

17. W. H. Mc Adams, 'Heat Transmission', McGraw-Hill, New York
(1954).

18. H. Lonsdale, 'Ammonia Tank Failure - South Africa', Ammonia
Plant Safety 17 (1975) 126-131.

19. O. Zeman, ' The Dynamics and Modelling of Heavier-than-Air
Cold Gas Releases', Lawrence Livermore Laboratory Report
UCRL-15224 (1980).

DISCUSSION

J. L. WOODWARDS   In your estimate of the time
                    duration of the gravity spreading regime did
                    not attempt to account for pool spreading and
                    evaporation of the non-flashed liquid ? This
                    will prolong the gravity spread phase conside-
                    rably.

S. JAGGER   This was not taken into account.

J. D. REID   I notice that you, along with
                    most investigators in this work, use cloud
                    height as a parameter entering for edge en-
                    trainment. The Parton films of their releases
                    don't seem to me to show entrainment across the
                    depth of the wall, but rather, significantly
                    at the top and perhaps the bottom. Do you
                    have any comment ?

S. JAGGER   The parameterization is tied
                    to the value of the entrainment velocity used.
                    I don't believe the detail of the parameteriza-
                    tion is well tested.

ENTRAINMENT THROUGH THE TOP OF A HEAVY GAS CLOUD

Niels Otto Jensen

Physics Department
Risø National Laboratory
DK-4000 Roskilde, Denmark

INTRODUCTION

After termination of the slumping phase of a heavy gas cloud, the height slowly starts to grow again. It is shown that this volume increase cannot be the result of entrainment through the side wall but must be due to entrainment through the cloud top. An entrainment relation is derived from the turbulence kinetic energy equation, and it is shown that the entrainment is driven by the turbulence in the gas cloud rather than the ambient atmospheric turbulence. As hazardous releases of heavy gases to the atmosphere is most often related to failure of pressurized vessels, the gas typically is quite cold relative to the ground. Thus, thermal convection in the cloud is the dominant turbulent mechanism, the strength of which is controlled by a surface temperature-jump Ri-number. The entrainment on the other hand is controlled by a density-jump Ri-number, and both Ri-numbers are coupled to the heat transfer at the surface and to entrainment of atmospheric air. The paper gives an approximate solution to this complex problem.

ENTRAINMENT VIA THE CLOUD'S FRONT

In the modelling of heavy gas dispersion, a useful analogy has traditionally been made to the propagation of a dense bottom current. The head of such a current advances with a velocity v, such that its Richardson number defined as

$$Ri_c \equiv \frac{g}{\rho} \frac{\Delta\rho}{v^2} h \tag{1}$$

is a constant, c^{-2}, of order unity. In the above equation g is gravity, ρ is the density of the current, $\Delta\rho$ is the density dif-

ference between the current and the surrounding fluid, and h is the
depth of the current. Hence,

$$v = c \sqrt{g \frac{\Delta\rho}{\rho} h} \quad .$$

(2)

The analogy is not perfect, though. In spreading of a heavy gas
there is not a constant supply of dense fluid, such that $\Delta\rho$ as
well as ρ is constant, but $\Delta\rho/\rho$ is rather a dependent variable.
Because of gradual entrainment of ambient air, of density ρ_a, into
the cloud the tendency is that $\rho \to \rho_a$ and $\Delta\rho \to 0$. Hence conditions are
not stationary. Despite of this it can be shown that a formula
analogous to (2) is still valid (van Ulden, 1979).

In his original treatment of the phenomenon van Ulden (1974)
considered the effect of entrainment through the side walls of the
gas cloud. In agreement with classical theory, an entrainment speed
$u_e = \alpha v$ was assumed. Here α is a constant which is about 0.2
(Simpson and Britter, 1979). The rate of change in the volume V of
the heavy gas cloud is thus $dV = A u_e dt$, where A is the frontal
area, i.e.,

$$\frac{dV}{dt} = \alpha(2\pi rh) \frac{dr}{dt} \quad ,$$

(3)

where r and h is radius and height of the cloud, respectively.
Using the substitution $2\pi rh \equiv 2V/r$, the variables in (3) can be
separated and one obtains

$$\frac{V}{V_o} = \left(\frac{r}{r_o}\right)^{2\alpha} \quad ,$$

(4)

$$\frac{h}{h_o} = \left(\frac{r}{r_o}\right)^{2\alpha-2} \quad ,$$

(5)

where subscript zero indicates initial values of the parameters.
As r increases with time due to the gravity spreading, it is seen
from (4) that the volume will be an increasing function of time.
This means that $\Delta\rho$ no longer can be regarded as a constant. A mass
balance gives the expression

$$\frac{\Delta\rho}{\rho} = \left(1 + \frac{\rho_a}{\Delta\rho_o} \frac{V}{V_o}\right)^{-1} ,$$

(6)

where ρ_a is the density of air and $\Delta\rho_o$ is the density difference
between the gas cloud and air at the beginning of the slumping
phase. Combining (4) and (6), an expression for $g\Delta\rho/\rho$ can be found,
which together with (5) and (2) gives a differential equation for
the determination of $r(t)$:

$$\frac{dr}{dt} = c \sqrt{\frac{gh_o \left(\frac{r}{r_o}\right)^{2\alpha-2}}{1 + \frac{\rho_a}{\Delta\rho_o}\left(\frac{r}{r_o}\right)^{2\alpha}}} \quad . \tag{7}$$

This expression can only be integrated for quite simple values of α as, for example, 0.5 and 1. A solution for $\alpha = 1$ (corresponding to an unusually large entrainment) and with $\rho_a/\Delta\rho_o = 1$ (corresponding to the quite large initial density difference $\Delta\rho/\rho = \frac{1}{2}$) is shown in Fig. 1. The remaining free parameters in the problem are taken as: $c = 1$, $h_o/r_o = \frac{1}{2}$ and $r_o = 10$ m. Also shown is the corresponding curve for the development without entrainment. It is seen that percent-wise there is a quite small deviation ($\sim 15\%$) between the two calculations.

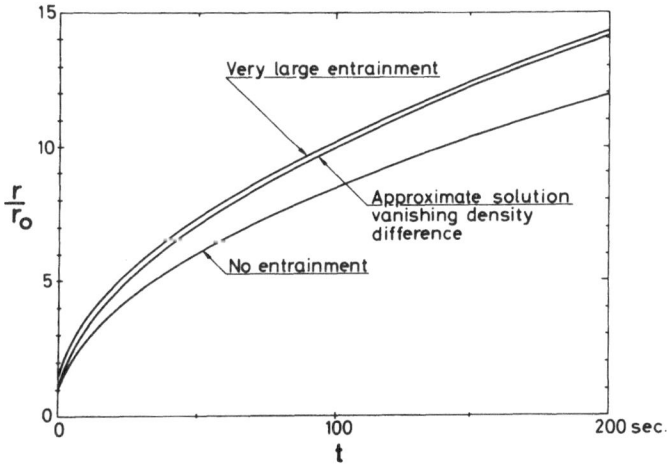

Fig. 1. The figure shows the relative growth of the radius of a heavy gas cloud in the slumping phase. The figure demonstrates that the course of development is not very sensitive to the assumptions made regarding the simultaneously occurring mixing with air (entrainment). Details regarding the calculation of the curves are given in the text.

A third curve is shown where entrainment is included, but where the relative density difference $\Delta\rho/\rho$ is assumed to be small, i.e. $\rho \simeq \rho_a$. In this situation we have from (6)

$$\frac{\Delta\rho}{\rho} \simeq \frac{V_o}{V}\frac{\Delta\rho_o}{\rho_a} . \tag{8}$$

If we use (2) and $h \equiv V/\pi r^2$ we get

$$r \frac{dr}{dt} = c\sqrt{g\frac{\Delta\rho}{\rho}V/\pi} , \tag{9}$$

which together with (9) gives

$$r \frac{dr}{dt} = c\sqrt{g\frac{\Delta\rho_o}{\rho_o}\left(\frac{\rho_o}{\rho_a}\right)V_o/\pi} . \tag{10}$$

Except for a constant factor of $\sqrt{\rho_o/\rho_a}$ it is seen that this expression is identical to that corresponding to zero entrainment. The third curve in Fig. 1 thus comes about by multiplying curve 2 with this factor, which in the actual example has a value of $\sqrt{2}$. Agreement with the exact solution is seen to be excellent.

In the case of chlorine the difference as to whether entrainment is considered or not is even smaller. If the initial mixture with air in the blow-up phase is 10:1, we have $\rho_o/\rho_a \simeq$ (10 + 70/29)/11, which only corresponds to a 6% difference in r between the entrainment/no entrainment cases.

There is thus some support for using a simplified calculation procedure where r(t) is found under the assumption that $V = V_o$.

Although entrainment has a small influence on the development of r(t), this is not true for h(t): if in eq. (5) we put $\alpha = 0$, it follows that $h \propto r^{-2}$, and thus we have h being a rapidly decreasing function of time, but if we put $\alpha = 1$, we get $h \propto r^0$, resulting in a constant value of h. The volume increase, which is a result of the process of frontal entrainment thus contributes almost exclusively to the reduction of the slumping rate and does not interfere much with the horizontal spreading of the cloud.

Regardless of how large frontal entrainment is, it can never make h a growing function of t, but at the most not decreasing. When it actually is observed that h after a certain period begins to grow instead of being reduced monotonously, it can therefore only be ascribed to entrainment occurring over the cloud's top.

ENTRAINMENT VIA THE CLOUD'S TOP

The speed with which the entrainment process develops is well known from more basic investigations: dense bottom currents (oceanography); the rate of increase of the convective atmospheric boundary layer (meteorology); and from a number of laboratory investigations (for example Kato and Philips' 1969). It is, however, in the entrainment assumptions that the various models on the market are differing.

We have already touched upon vertical entrainment. It is a difficult problem which does not become easier in the case of a heavy gas as one cannot rely on the Boussinesq approximation, but has to treat ρ and T as independent variables.

The derivation of an equation for the entrainment process can be done by using the turbulent kinetic energy budget equation. This equation can be derived from the equations of motion (see for example Busch, 1973). The equation gives the time rate of changes of the turbulent kinetic energy, e, in relation to production, transport and dissipation of that quantity (u_* is the friction velocity):

$$\frac{\partial e}{\partial t} = u_*^2 \frac{\partial \bar{u}}{\partial z} + \frac{g}{\rho} \overline{\rho' w'} + \frac{\partial}{\partial z} \left(\frac{\overline{e' w'}}{2} + \frac{\overline{p' w'}}{\rho} \right) - \varepsilon \ . \tag{11}$$

For a definition and explanation of each term, see also Busch (1973). As we consider the conditions just under the cloud's top we model each term in the following manner ($\partial e/\partial t = (\partial e/\partial z)(\partial z/\partial t)$ and $\partial z/\partial t \equiv u_{ew}$):

$$c_1 \frac{e - u_*^2}{h} u_{ew} = c_2 \frac{u_*^3}{h} - \frac{g}{\rho_o} \Delta \rho u_{ew} + c_3 \frac{e^{3/2}}{h} - c_4 \frac{e^{3/2}}{h} \tag{12}$$

where we choose h as a relevant length scale for vertical gradients. Terms are scaled in accordance with usual practice, and occur in the same order as in the previous equation. Identifying $\sqrt{e}$ as a characteristic turbulent velocity, we get by rearrangement of (12) the following equation for the determination of the entrainment velocity:

$$u_{ew} = \frac{c_2 \left(\frac{u_*}{\sqrt{e}}\right)^3 + c_3 - c_4}{c_1 - c_1 \left(\frac{u_*}{\sqrt{e}}\right)^2 + Ri} \sqrt{e} \tag{13}$$

where Ri is defined as

$$Ri = \frac{g}{\rho_o} \frac{\Delta \rho}{e} h \tag{14}$$

If the turbulence level in the gas cloud is large compared to the turbulence level in the surrounding atmosphere we get the simplified equation

$$u_{ew} = \frac{a}{b+Ri} \sqrt{e} \quad . \tag{15}$$

This is to a good approximation the case in a cold gas cloud, where a considerable portion of the turbulence is produced thermally. In the other asymptote, where the density jump Ri-number as well as convection vanishes, $\sqrt{e}$ becomes proportional to u_*, and we get

$$u_{ew} = \frac{a'}{b'} u_* \quad . \tag{16}$$

EXAMPLE OF A COLD CLOUD

Let us make an estimate of how large e typically will be in a cold gas cloud. For this purpose we will look at the convective velocity scale w_* defined as

$$w_* = \left(\frac{g}{T_0} \left(\frac{H}{\rho c_p} \right) h \right)^{1/3} \quad . \tag{17}$$

Here H is the heat flux from the ground to the gas cloud (Joule/sec m^2). As we are involved with a turbulent flow, we estimate H from an expression of the form

$$H = \rho c_p (c_H \Delta T u) \quad , \tag{18}$$

where ΔT is the temperature difference between the gas cloud and the surface, u is a typical windspeed and c_H is a heat transfer coefficient. From surface layer theory we have

$$c_H = \frac{k^2}{(\ln \frac{z}{z_0} - \psi_m)(\ln \frac{z}{z_0} - \psi_h)} \quad . \tag{19}$$

where k is a constant of approximately 0.4 (von Kármáns constant); z is the height above ground; z_0 is a parameter that describes the aerodynamic roughness of the earth surfaces; and ψ_m and ψ_h are empirical functions that corrects for the influence of temperature stratification (the heat flux H is different from zero). Empirical expressions for the ψ-functions are given by for example Businger (1973):

$$\psi_h = \ln \frac{1+y}{2}$$

with

$$y = (1 - 9 \frac{z}{h})^{1/2} \quad , \tag{20}$$

and

$$\psi_m = \ln \frac{(1+x)^2(1+x^2)}{8} - 2\text{Arctg}(x) + \frac{\pi}{2}$$

with   (21)

$$x = (1 - 16\frac{z}{L})^{1/4} \quad .$$

In these expressions, L is the so-called Monin-Obukhov length
defined by

$$L = - \frac{u_*^3}{k\frac{g}{T_o} H/\rho c_p} \quad ,$$ (22)

where T_o is a typical temperature, for example the earth surface
temperature. By using the surface layer expressions for temperature
and wind profiles, (22) can be revised to

$$\frac{z}{L} = - \frac{g}{T_o} \frac{\Delta T}{u^2} z \frac{(\ln \frac{z}{z_o} - \psi_m)^2}{\ln \frac{z}{z_o} - \psi_h} \quad .$$ (23)

The argument in the ψ-functions is thus seen to be a function of
a temperature-Richardson-number

$$Ri_{\Delta T} = \frac{g}{T_o} \frac{\Delta T}{u^2} z$$ (24)

times a correction function

$$f = \frac{(\ln \frac{z}{z_o} - \psi_m)^2}{\ln \frac{z}{z_o} - \psi_h} \quad ,$$ (25)

that is dependent upon the ψ-function itself. This means that z/L
must be found by an iterative process before c_H and in conclusion,
w_* can be found for a set of given values of ΔT and u.

        In the case of a flashing clorine spill, where the gas to
begin with is at the boiling point, ΔT is about 50°. In a typical
situation we furthermore have u = 5 m/s and h = 10 m, so that
$Ri_{\Delta T}$ becomes approximately 0.7. If we, to begin with, assign f to
the value 1, we get -z/L = 0.7. This gives inserted in (25) a new
value for f, that multiplied by $Ri_{\Delta T}$ gives a new value for -z/L
and so on. In this iteration loop z_o is considered a given parameter; a typical value is 10 cm. With the above parameter values
the cycle of iterations is illustrated in Fig. (2), where the
result is -z/L $\simeq$ 1.8

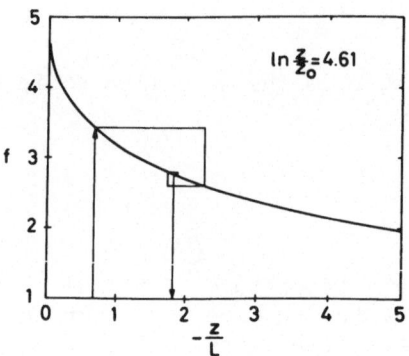

Fig. 2.   The argument in the functions that correct the normal
          logarithmic expressions for the wind and temperature
          profile is given by z/L (observation height/Monin-Obukhov
          length). L is a measure of the importance of convectively
          produced turbulence in comparison with mechanically pro-
          duced turbulence. z/L is proportional to the Richardson's
          number times a dimensionless function f, which is a weak
          function of z/L itself. The figure sketches an iterative
          course where a z/L value corresponding to Ri = 0.7 is found.

c_H can now be calculated with the help of (19). The result of
the above example is c_H = 1.37 · 10^{-2}, which is approximately a
factor of 2 larger than the result one would get if the influence
of temperature stratification would have been neglected.

By (18) one can now calculate the heat flux from the earth's
surface to the cloud, and from this result by use of (17) one can
calculate the convective velocity scale w_*. The result is w_* ≃
1.05 m/s, whereas u_* in the present case is only ≃ 0.13 m/s. From
this example it is seen that $w_* > u_*$, and with that the turbulent
kinetic energy, e, is almost exclusively determined by w_*^2. The
constant of proportionality between e and w_*^2 is of order 1.

THE EFFECT OF ENTRAINMENT

As a result of surface heating and of entrainment of air, Δρ
and ΔT gradually becomes smaller, which in turn reduces Ri and

$Ri_{\Delta T}$. As a result of the latter the surface heat transfer decreases and with that so does $\sqrt{e}$. This results first in a smaller entrainment rate. However, this is counter compensated by Ri also becoming smaller so that the entrainment process is eased. The following gives a quantitative treatment of this process.

A heat balance gives the equation

$$\rho_a c_{pa} A u_{ew}(T_a-T)dt + \rho c_p A c_H u(T_o-T)dt = (\rho+d\rho)(c_p+dc_p)VdT, \quad (26)$$

where subscript a stands for atmospheric values, A is the area of the cloud. Assuming that $T_a-T \simeq T_o-T = \Delta T$, and by dropping small terms we get

$$d(\Delta T) = -\frac{\Delta T}{h}\left(c_H u + \frac{\rho_a c_{pa}}{\rho c_p} u_{ew}\right) dt \quad (27)$$

Similarly we get from a balance of mass that

$$\rho V + \rho_a A u_{ew} dt = (\rho+d\rho)(V+dV). \quad (28)$$

Noting that dV is the combined result of entrained air, with the proper reduction in volume due to the temperature change, and the volumetric change due to surface heating of the cloud we get

$$d(\Delta\rho) - \rho\frac{d(\Delta T)}{T} = \frac{u_{ew}}{h}\left(\rho_a - \frac{T}{T_a}\rho\right) dt \quad (29)$$

where we again have dropped small terms.

The expressions give the tendency in development of $\Delta\rho$ and ΔT during a small time step, dt. With this we can, by iteration, as described above, calculate a new value for w_*, and with that a new entrainment rate that again gives the differential growth in h. A new tendency in $\Delta\rho$ and ΔT can hereafter be calculated, and so forth.

It should be clear from the above that the inclusion of vertical entrainment is not possible in connection with an exact analytical treatment of the gravity spread and side-entrainment processes.

A further step towards solving the problem analytically must necessarily include approximations. The first is that $\sqrt{e} \sim w_*$ and that Ri is so large that (15) becomes of the simple form

$$u_{ew} = \beta\frac{w_*^3}{g\frac{\Delta\rho}{\rho_o}h} \quad , \quad (30)$$

which in connection with (16) and (17) gives

$$u_{ew} = \beta c_H u \frac{\rho_o}{T_o} \frac{\Delta T}{\Delta \rho} \quad . \tag{31}$$

If we use this expression in (27) along with noting that $\rho_a c_{pa} \sim \rho c_p$ we get

$$d(\Delta T) = - c_H \Delta T u (1 + \beta \frac{\rho_o}{T_o} \frac{\Delta T}{\Delta \rho}) \frac{dt}{h} \quad . \tag{32}$$

This result, together with the entrainment relation (31) reduces the rate equation (29) for $\Delta \rho$ to

$$d(\Delta \rho) = - c_H \Delta T u (1 + \beta) \frac{\rho_o}{T_o} \frac{dt}{h} \quad . \tag{33}$$

In deriving this result we have neglected terms of order $\Delta \rho / \rho_a$ or $\Delta T / T_a$ occurring in the paranthesis. Combining the two last equations results in

$$\frac{d(\Delta T)}{d(\Delta \rho)} = \frac{T_o}{\rho_o} (1 + \beta \frac{\rho_o}{T_o} \frac{\Delta T}{\Delta \rho}) / (1 + \beta) \quad , \tag{34}$$

where the only variables are ΔT and $\Delta \rho$. The equation is of the form $dy/dx = f(y/x)$ and can as such be solved by standard methods, which involves that the variables can be separated. Defining $z \equiv y/x$, where $y = \Delta T / T_o$ and $x = \Delta \rho / \rho_o$, (34) transforms to

$$\frac{1+\beta}{1-z} dz = \frac{dx}{x} \quad , \tag{35}$$

with the solution

$$\frac{z-1}{z_i - 1} = \left(\frac{x}{x_i} \right)^{-\frac{1}{1+\beta}} \quad , \tag{36}$$

where subscript i denotes initial conditions. Thus we have the following proportionality

$$\frac{\Delta T}{\Delta \rho} \propto \left(\frac{\Delta \rho}{\rho_o} \right)^{-\frac{1}{1+\beta}} \quad , \tag{37}$$

which shows that the ratio $\Delta T / \Delta \rho$ is an increasing function of time as $\Delta \rho \to 0$. The power $(1+\beta)^{-1}$ is about 0.8. The consequence of this is that the entrainment velocity u_{ew} tends to increase with time. Hence we should expect $h(t)$, the height of the cloud, to be a strongly concave function of time.

However, as we already have seen, c_H will be relatively large in the beginning, where the surface heat transfer is large. This counteracts the above mentioned curvature. Furthermore, in the limit of $t \to \infty$, the entrainment assumption (30) becomes unrealistic compared to (13). In this limit we have $u_{ew} \propto u_*$ where the proportionality constant is of order 1. Whether a linear growth in $h(t)$ is an overall fair assumption awaits numerical computations.

REFERENCES

Busch, N.E., 1973, On the mechanics of atmospheric turbulence, Workshop on Micrometeorology, D.A. Haugen, Ed., Amer. Meteor. Soc., 1-65.
Businger, J.A., 1973, Turbulent transfer in the atmospheric surface layer, Workshop on Micrometeorology, D.A. Haugen, Ed., Amer. Meteor. Soc. 67-100.
Kato, H., and Phillips, O.M., 1969, On the penetration of a turbulent layer into a stratified fluid, J. Fluid Mech., 37, 643-655.
Simpson, J.E., and Britter, R.E., 1979, The dynamics of the head of a gravity current advancing over a horizontal surface, J. Fluid Mech., 94, 477-495.
Van Ulden, A.P., 1974, On the spreading of a heavy gas released near the ground, Loss Prevention and Safety Promotion in the Process Industries, Proceedings of the 1st International Symposium, Hague/Delft, May 28-30.
Van Ulden, A.P., 1979, The unsteady gravity spread of a dense cloud in a calm environment. Proceedings of the 10th International Technical Meeting on Air Pollution Modelling and its Application, NATO/CCMS Air Pollution Pilot Study, Rome/October 23-26.

DISCUSSION

F. B. SMITH   Would it be correct to suppose that in some releases at least the heavy gas starts with zero mean velocity and only accelerates up to its final speed rather slowly. During this acceleration phase the wind shear across the top of the heavy gas "pancake" is an important factor in determining the entrainment of ambient air into the gas cloud. Is this a fair description and is it contained in your model ?

N. O. JENSEN   NO. This type of acceleration was probably true for the Porton studies, but in practice a blow-out will often cause large quantities of ambient air to be entrained almost instantly, giving rapid acceleration.

PHYSICAL SIMULATION OF DISPERSION IN COMPLEX TERRAIN AND VALLEY DRAINAGE FLOW SITUATIONS

Robert N. Meroney[&]

Institut Wasserbau III
Sonderforschungsbereich 80
University of Karlsruhe
7500 Karlsruhe 1, B.R.D.

INTRODUCTION

Meteorologists are frequently faced with problems requiring quantitative estimates of air flow patterns and turbulence charac-teristics over complex terrain. Use of the wind flow information includes air pollution zoning, prediction of smoke movement from forest fires or slash burning, mine tailing dust dispersal, esti-mation of the movement of vegetative disease vectors or pests, and the siting of fossil fuel burning industrial facilities or power plants. In view of the practical difficulties in obtaining useful results, whether by analytical, numerical, or field in-vestigation means, it is natural to explore the possibilities of simulating flow and diffusion over irregular terrain by means of physical model experiments on the laboratory scale.

Successful modeling of some of the more complex atmospheric surface layer phenomena in a wind tunnel has only been accom-plished in the last fifteen years. Although guidelines for mode-ling flow over complex terrain are essentially similar to those for modeling that around buildings a few unique features will be discussed herein. The characteristics of wind and diffusion over complex terrain which limit physical modeling and stipulate simulation criteria are reviewed only briefly. For a more ex-tensive background the reader is referred to reviews on diffusion

_____
& Von Humboldt Foundation Senior Scientist from
  Fluid Mechanics and Wind Engineering Program
  Colorado State University
  Fort Collins, Colorado 80521, USA

in complex terrain prepared by Eagan (1975), discussions of flow
over terrain by Meroney et al. (1978), or recent symposium proceedings devoted exclusively to the subject Johnson (1979).

WIND CHARACTERISTICS IN COMPLEX TERRAIN

Mountains may alter atmospheric airflow characteristics in a
number of different ways. These effects can generally be grouped into those due to inertial-viscous interactions associated
with a thick neutrally stratified shear layer and to thermally
induced interactions associated with stratification or surface
heating.

Neutral Airflow Over Hilly Terrain

Near-neutral or adiabatic atmospheric boundary layers will
exist over mountains during situations when winds are high due
to intense synoptic pressure fields, when continous cloud banks
impede surface heating, and when sharp terrain features produce
separation eddies which mix the flow field vigorously in the vertical.

When the static stability is neutral, airflow over mountains
creates pressure gradients in the flow direction, which together
with surface friction, may produce separation, flow reversal,
and reattachment. Separation eddies at the windward or leeward
side of a mountain can alter the effective shape of the mountain
resulting in a modified wind profile at the crest. Scorer (1978)
describes eight different variations of the separation phenomenon. He notes that separation may be changed in character by
insolation, blocking, diabatic changes, and three-dimensional
effects.

Meroney et al. (1978) summarized experimental data available
from field and laboratory structures over hills, ridges, and es-
carpments. Orgill (1977) surveyed wind measurements programs
that have used wind networks (2 to 60 measurements sites) to
identify details at atmospheric surface layer behavior. Out of
the 139 field programs 3 relate to an isolated mountain, 7 to
mountain-plain flows, and 20 to complex topography.

Arya and Shipman (1978) review past measurements of diffusion
near shelter belts and simple barriers. They report that when
an elevated source is located upwind of an isolated ridge or at
the ridge top, the effect of ridge is generally to reduce the
ground-level concentrations. For low-level sources, however,
the maximum ground level concentration (g l c) in the cavity
region may become much larger than the maximum g l c without the
ridge, if a large part of the plum impinges on the separation
streamline and becomes trapped in the cavity region. Thus the

ground-level concentrations downwind of the ridge are very sensitive to source height and position. When the source is located in the lee of the ridge, increased turbulence in the wake results in much lower concentrations aloft, but higher concentrations at the ground level. Maximum g l c occurs when the source is located within the cavity region near the base of the ridge. The peak concentration decreases and its position shifts farther downwind from the source as the source height and distance from the ridge increase. The ridge also influences considerably the vertical concentration profiles. Recent field measurements of dispersion in complex terrain may be typified by those of Start et al. (1975) who report the following differences for Huntington Canyon and flat terrain diffusion : "Neutral stability tests showed five times greater dilution for canyon axial concentrations ; strong inversion tests resulted in canyon plume centerline dilutions fifteen times greater than calculations using parameters derived for flat terrain." It was concluded that plume dilutions in Huntington Canyon were affected by such physical mechanisms as enhanced mechanical turbulence associated with gradient windflows near the mountain tops, density flows originating in side canyons, and turbulent wakes from pronounced terrain irregularities.

Stratified Airflow Over Hilly Terrain

Stratification has a strong effect upon flow over and around hills. The intensity of an inversion may influence both wind velocity and direction. If inversions are frequent, acceleration of air around the hillside rather than over the hilltop may lead to larger annual average velocities away from the hill crest. Stratified flow over mountain ridges may also lead to lee waves or helm winds. There is very extensive literature dealing with orographic induced waves, but since most of it deals with cloud systems and upper atmosphere character far from the surface, it is not very helpful for dispersion analysis. Slope and valley airflows when dominated by surface heating or cooling often result in quite complicated secondary flows. According to Munn (1966) the important properties of valley airflows are :

1. Orientation of the geostrophic wind,
2. Orientation of the valley (a north-south valley has a different radiation balance from one lying in an east-west direction), and
3. The geometric dimension of the valley, such as length, width, depth, side slopes, slope of the floor and the number of bends or constrictions.

When a strong geostrophic wind blows parallel to a valley a funneling effect occurs. If the valley narrows, the wind is

speeded up as it passes through the valley. If the geostrophic
wind is blowing at right angles to the valley a large cross
valley circulation or "valley eddy" may be formed. The wind is
a combination of up and downslope winds and the background synoptic flow modified by the mountains. During high wind speed conditions (U_{10} > 5 m/sec) slope heating plays a minor role and
the valley behaves as another perturbation on a natural adiabatic shear flow approximation.

Dispersion data and detailed wind information for valley
drainage flows is lacking. Draxler's (1979) survey of recent
atmospheric dispersion experiments includes 5 data sets for rough
terrain, and only 1 for valley drainage situations.

CRITERIA FOR LABORATORY SIMULATION OF FLOW CHARACTERISTICS OVER
IRREGULAR TERRAIN

In order for the flow in any laboratory model to be of value
in interpreting or predicting the observed flow in the atmosphere, it is essential that the two flow systems should be dynamically, thermally and kinematically similar. This means that
the flow in the two systems must be described by the same equations after appropriate adjustments of the units of length, time
and other variables.

A number of authors have derived the governing parameters for
atmospheric heat, mass, or momentum transport by dimensional
analysis, similarity theory, and inspectional analysis. Another
group justify similitude by considerations of turbulence theory
and recent reviews of full scale wind data which present the
characteristics of the prototype atmospheric wind on a parametric
basis. Although all investigators do not agree concerning details,
most would concur that the dominant mechanisms can now be identified and are understandable. The following sections review
similitude criteria as they relate to adiabatic and stratified
flows over irregular terrain. Restrictive assumptions are discussed and a typical performance envelope for a large wind tunnel facility is provided.

Neutral Airflow Models

The pertinant parameters for steady turbulent, near neutral
airflows are undistorted scaling of topographical features ;
Reynolds number ($U L/ \nu$) independence ; Rossby number ($U/L \Omega$)
independence; and kinematic similarity of approach flow, i.e.,
distributions of mean velocity and turbulence.

Over irregular terrain the flow characteristics are only
weakly dependent on Reynolds number when the surfaces are aero-
dynamically rough, i.e., the flow structure will be similar if

the scaled roughness has a sufficient size to prevent the formation of a laminar sublayer. Generally the requirement for
fully rough flow is satisfied if one stipulates model conditions
such that

$$\frac{Uh}{\nu} > 10000$$

where h is average hill or mountain height, and either

$$\frac{k\,u_{\ast}}{\nu} > 20 \text{ or } \frac{k\,U}{\nu} > 400$$

where k is roughness height and $u_{\ast}$ is friction velocity
(Snyder, 1979). When terraced models are constructed the individual terrace height should be O(k). In addition an upstream
model fetch should include any ridge whose height exceeds one-
hundredth of the fetch distance, i.e., h > x/100, or any hill or
mountain whose height exceeds one-twentieth of the fetch distance ,i. e., h > x/20. The fetch indicated is of course upwind
of any proposed gas release, and an additional fetch required
for equilibrium boundary layer development must be added.

Over irregular terrain the flow field may be assumed Rossby
number independent for distances from 10 to 50 km. Hoxit (1973)
and Scorer (1978) observe that most of the time atmospheric
flows are not in equilibrium with Coriolis forces because of
thermal winds; hence the simulation of the Ekman spiral is normally only of academic interest. For flow over flat terrain an
upper limit to model length due to shear effects on dispersion
may be 5 to 10 km.

Stratified Airflow Models

The pertinant additional parameters for steady stratified
airflows are bulk Richardson numer ($g\,\Delta\theta\delta\,/\,\theta\Delta\,U^2$) equality over
the surface layer, Froude number ($\Delta\,U^2\theta\,/g\Delta\,\theta\,h$) equality over
the mountain or hill height, and sufficient lateral and longitu-
dinal model extent to account for effects of surface flow blockage or channeling. In the case of the model airflow the potential temperature, θ, may be replaced by temperature or density.

Equal gross or local Richardson numbers can be obtained in
specially designed wind tunnels. Batchelor (1953) has established that if the flowfields are such that the pressure and
density everywhere depart by only small fractional amounts from
the values for an equivalent atmosphere in adiabatic equilibrium
and if the vertical scale of the velocity distribution is small
compared to the scale height of the atmosphere, the Richardson

number distribution governs dynamical similarity. Unfortunately, these conditions are only normally satisfied in the first 100 m of the planetary layer. Batchelor has emphasized that no single local, variable quantity can be used as a similarity parameter. Similarity parameters can have meaning only when they characterize the gross features of the flow.

Snyder (1979) found that for plumes transported toward three dimensional hills in stratified flows simulation of the stability above the thin surface layer is most important to reproduce the essential features of the flow, i.e., whether plumes will impinge on the hill surface or travel over the hill top. Unfortunately in the process of meeting the Froude number criteria based on mountain height it is possible the wind tunnel duct Froude number may fall too low, i.e., $Fr_D < 1/\pi$. In this case undesirable standing wave systems tend to appear just downstream of the tunnel heat exchanger at the entrance (Yamada and Meroney, 1974).

Valley Drainage Airflow Models

Local heating and cooling of hill surfaces are the driving mechanisms for anabatic and katabatic winds which inhibit or enhance separation over hill crests. Laboratory models to simulate such conditions must still be considered largely exploratory in nature at this time. Early work includes simulations of urban heat islands by Yamada and Meroney (1971) and Sethuraman and Cermak (1974), simulation of flow and dispersion at shoreline sites by Meroney et al. (1975), and simulation of dispersion effects of heat rejected from large industrial complexes by Meroney et al. (1975).

Equality of a Monin-Obukhov type length scale ratio is appropriate for surface heating situations with a finite cross flow wind field, i.e. L/L_{mo} where L_{mo} is the Monin-Obukhov stability length. In terms of a bulk parameter this would be

$$\frac{U_m^3 \, \theta_m \, A_m \, L_p \, Q_p}{U_p^3 \, \theta_p \, A_p \, L_m \, Q_m} = 1.0$$

where Q is the total heat flux to the atmosphere, L is a vertical reference scale, A is the surface area heated or cooled, θ is the average ambient temperature, and U is the ambient reference air speed (m and p refer to model and prototype conditions). Unfortunately momentum and heat flux information are often not con-

veniently available for the field or model. In addition calculations indicate that the modeled heat flux is very sensitive to
velocity changes with background stability.

   Similarity between the flow generating mechanisms of sea
breezes and flow over certain heat islands suggests alternative
parameters. Linear numerical analysis of Olfe and Lee (1971)
and experimental and numerical studies by Yamada and Meroney
(1971) suggest the intensity of heating or cooling by the land
surface may be characterized by a heating ratio

$$HR = \frac{\Delta\theta_z L_x}{\Delta\theta_x L_z}$$

where θ_z are characteristic changes in temperature over a vertical scale L_z and θ_x are similar changes over some horizontal
scale L_x. Since the vertical-to-horizontal modeling scale is
undistorted, the parameter reduces to a single temperature ratio.

   In those situations where the gradient wind is extremely small
and where the lateral variations in surface temperature are negligible, another parameter called the Rayleigh number (g β ($\theta_w - \theta_H$)
$h^3 / \nu\alpha$) is appropriate. One notes that the scaled temperature
potential increases as the model scale cubed !

   As one can imagine there can be no firm and fixed rules for
determing the appropriate terrain area or simulation constants
to model until the flow situation is well defined. Nonetheless
a fluid model study does employ a "real" fluid; hence a well designed and carefully executed fluid model study will yield valid
and useful information.

PERFORMANCE ENVELOPE FOR WIND TUNNEL MODELING OF AIRFLOW OVER
COMPLEX TERRAIN :

   The viability of a given simulation scenario is not only a
function of the governing flow physics but the availability of
a suitable simulation facility and the measurement instrumentation to be employed. It would seem appropriate, therefore, to
suggest bounds for the range of fields situations which can
reasonably be treated by physical modeling.

   A number of boundary layer wind tunnels exist at various
laboratories. Generally these tunnels range in size from facilities with cross-sections of 0.5 m x 0.5 m to 3 m x 4 m. Several
of these facilities are equipped with movable side walls or
ceilings to adjust for model blockage. By utilizing a variety
of devices such as vortex generators, fences, roughness, grids,
screens, or jets a fairly wide range of turbulence integral

scales can be introduced into the shear layer. Varying surface
roughness permits control of surface turbulence intensity, dimensionless wall shear, and velocity profile shape. Density
stratification can be induced by means of heat exchangers, use
of different molecular weight gases, or latent heat absorption
or release during phase changes. A comparison between field and
laboratory parameter ratios are summarized in Table 1.

Neutral Airflow Models

When one combines various operational constraints into a performance envelope, a clear picture appears of the performance region
for windtunnel facilities. Figure 1 is such a performance envelope prepared for a large facility (typically 3 m x 4 m x 25 m
test section dimensions). The criteria selected to specify operations ranges are :

       Maximum model height (h $\leq$ 0.5 m)
       Minimum convenient model height (h = 0.02 m)
       Minimum Reynolds number (Re_h = U h/ ν = 10,000)
       Maximum model integral scale (L_{u_x} $\leq$ 0.5 m)

       Minimum model integral scale (L_{u_x} $\geq$ 0.05 m)

       Minimum model measurement resolution ($\Delta_z \geq$ 0.1 mm)

       Maximum model boundary depth ($\delta \leq$ 2 m)
       Minimum model boundary depth ($\delta \geq$ 0.1 m)

Since field values for some parameters are uncertain the prototype value of δ and L_{u_x} are assumed to range over complex terrain as follows

       300 m < δ < 1000 m, and
       100 m < L_{u_x} < 1000 m.

Not all previous laboratory studies meet such similitude restrictions, some experiments were performed to meet objectives other
than similitude of turbulence or mean velocity profiles; nevertheless, a number of such studies are indicated on Figure 1.
In almost all cases noted values fall within the indicated operational envelope or just outside the predicted region.

Assuming an upper value of length scale ratio of 10,000 and
a tunnel length of 25 m a distance of 50 km is well within the
capacity of existing facilities to contain in the windward direction. Assuming a lateral width restriction of 4 m suggests
a 40 km lateral maximum for the field area modeled.

Table 1 : Typical Wind Tunnel and Field Parameter Range

DIMENSIONLESS PARAMETER	WIND TUNNEL RANGE	FIELD RANGE
Reynolds number $Re_h = U h / \nu$	$0 \rightarrow 7.0 \times 10^5$	$0 \rightarrow 6.0 \times 10^6$
Richardson number $Ri_B = \dfrac{g(d\,\theta/dz)h}{U^2}$	$-1.0 \rightarrow 1.0$	$-1.0 \rightarrow 1.0$
Prandtl number $Pr = \mu c_p / k$	0.72	0.72
Skin friction coefficient or Stanton number $$\frac{C_f}{2} = \frac{\tau w}{\rho U^2} = \left(\frac{u_x}{U}\right)^2$$ $$St = \frac{q}{\rho c_p (\Delta T)}$$	$0.005 \rightarrow 0.004$	$0.001 \rightarrow 0.004$
Scaling lengths		
z_o/h	$2 \times 10^{-5} \rightarrow 0.25$	$3.0 \times 10^{-7} \rightarrow 0.15$
δ/h	$0.20 \rightarrow 50.0$	$0.10 \rightarrow 100$
$(L_{u_x})/L$	$0.06 \rightarrow \infty$	$0.001 \rightarrow \infty$
h/L_{MO}	$0 \rightarrow 5.0$	$-360 \rightarrow 240$
Power law coefficient α	$0.05 \rightarrow 0.8$	$0.05 \rightarrow 0.7$

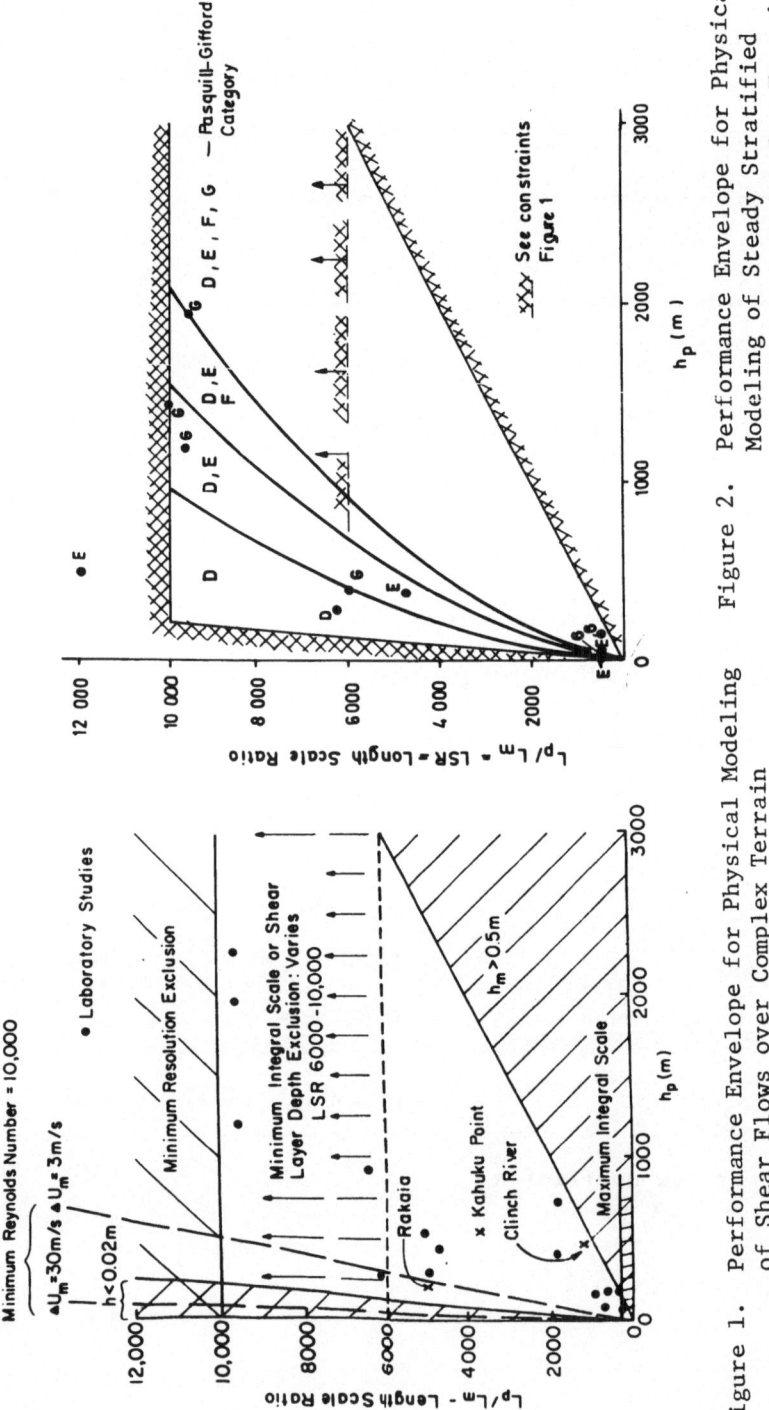

Figure 1. Performance Envelope for Physical Modeling
of Shear Flows over Complex Terrain

Figure 2. Performance Envelope for Physical
Modeling of Steady Stratified
Shear Flows over Complex Terrain

Stratified Airflow Models

One must add the additional constraints of stratification to the envelope produced in Figure 1. In this case reduced model wind speeds imply a interactive relationship between the surface roughness, surface layer Richardson number, and hill Froude number parameters. If one assumes that roughness height to hill height ratio $k/h \leq 1/10$, then the aerodynamic roughness ratio is $hU/v \geq 4000$. If one further assumes that the thermal gradients on the surface layer extrapolate to the terms required in the Froude number through power law relationships then the length scale ratio, LSR, must be

$$ LSR \leq \left\{ \frac{h^2}{4000 \ \alpha v} \left(\frac{10}{h} \right)^{1-\alpha} \left(\frac{(\delta T / \delta z) \ m \ g}{T \cdot Ri_{10}} \right)^{1/2} \right\}^{1/2} $$

where $(\delta T / \delta z)_m$ = maximum thermal gradients in model (1 °C/cm)

h = prototype characteristic hill height

Ri_{10} = characteristic Richardson number at a reference height of 10 meters

α = $\alpha (Ri, z_o)$ = velocity power law exponent

= (0.1 to 0.5)

Here Ri_{10} and α may be stipulated emperically for the various Pasquill-Gifford stratification categories. Figure 2 displays the result of stratification limitations on a performance envelope.

Valley Drainage Flows

Stipulation of strict Rayleigh number similarity would essentially eliminate the possibility of modeling valley drainage flows; however gravity current movement down steep valley walls will result in high local Reynolds numbers and turbulent mixing. In such circumstances the appropriate viscous parameter is the eddy viscosity, $\varepsilon = UL$, not v. Substitution of this variable transforms the Rayleigh number into the Richardson number introduced earlier. Aluminium shell models (LSR = 1920 and 2560) cooled by dry ice have been used by Petersen and Cermak (1980) to study drainage flows in the mountains near Crested Butte, Colorado. Richardson numbers based on the maximum surface jet

velocity and jet height were found to vary from 0.5 to 2.0. These
produced surface velocities equivalent to 1.0 to 3.5 m/s between
50 to 150 m above the ground. Specific field conditions were
not chosen in advance; rather model conditions were permitted to
develop, and the equivalent field conditions were evaluated from
the measured Richardson numbers.

VERIFICATION EVIDENCE

Not all of the cases identified on Figures 1 and 2 were
successful. Many early studies had model approach flow velocity
exponents near zero, were modeled as neutral flows when the
field observed strong stratification effects, or simulated unrealistic boundary layer depths, integral scales, or turbulence
intensities which did not match their atmospheric counterpart.
Several of the investigations on Figure 2 seem only marginally
realistic. Since plumes tend to travel undiluted for long distances in stable flow there is a tendency for the investigator
to select as large a scale ratio as possible. Since a plume's
lateral displacement over irregular terrain is determined primarily by gorge or valley meandering in stable flow even these
studies may be similar to field experience.

In seventeen of some forty physical model studies examined
by the author, which were performed over idealized or specific
irregular terrain, there are some counterpart field data available (Meroney et al., 1978). Few studies claim unreasonable correlation, and some are strongly selfcritical. Nonetheless most
studies accomplished their prestated limited objectives. Several
recent cases are examined briefly in this paper to represent the
degree of similarity possible.

Time and Space Scales Appropriate for Physical Modeling

The joint consideration of time and space scales modeled in
atmospheric boundary-layer wind tunnels may be overlaid upon the
characteristic time and horizontal scales of atmospheric phenomena. Such an overlay upon the atmospheric scales defined by
Orlanski (1975) has been proposed in Meroney et al. (1978). The
wind tunnel may study phenomena up to the mesoscale in time and
into the mesoscale in length with reasonable credibility.

When the performance envelope of physical modeling by wind
tunnels is superimposed upon the scale effects produced by terrain on atmospheric motions Figure 3 is obtained. Only the time
scales associated with turbulence or advection over terrain are
realistically included in current physical modeling repertoire.
The fluctuations in a field of motion associated with diurnal
variations in the atmosphere or with weather fronts are not developed in physical models considered to date.

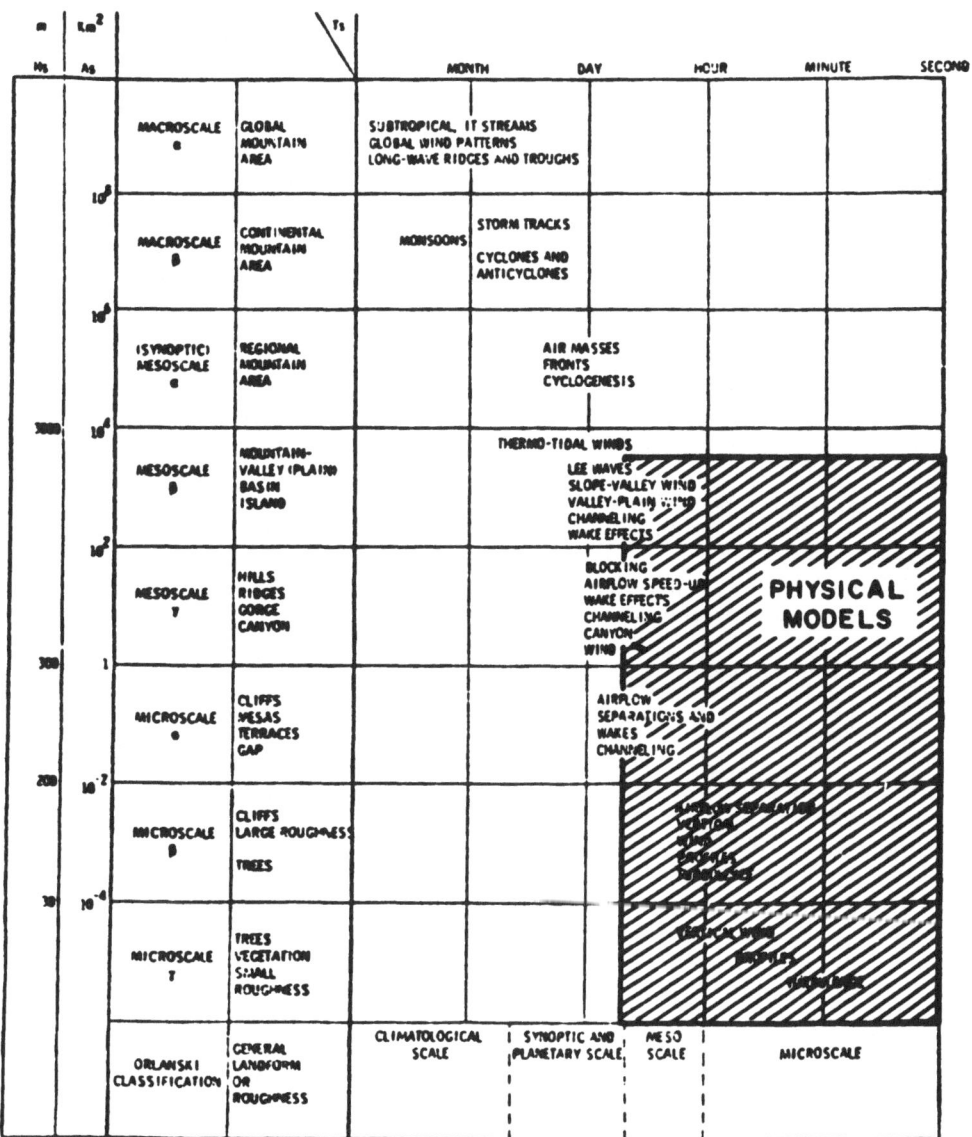

Figure 3 : A Classification of the Effects of Terrain
on Atmospheric Motions. After Orlanski
(1975)

One can, however, synthesize the average statistics of a flow field over longer time periods by associating a given measurement set with a recurring meteorological situation for which climatological probability distribution information is available. Hence it is possible to expand the physical model domain to longer time scales.

The inherent uncertainties associated with any atmospheric phenomena set bounds on any physical or numerical model verification exercise. One might say that the laboratory results simulate the average of many realizations of atmospheric flows whose boundary and initial conditions are "near" those of the laboratory model. Any single realization of atmospheric flow or diffusion over irregular terrain thus has a finite probability of varying from the mean behavior.

The prediction and model validation for scalar transport includes other uncertainties than those associated with errors in mean velocity, direction or turbulence. These include photomechanical effects, sinks, sources, depostion, etc. which are not generally modeled. Indeed scatter diagrams of numerical model to field results such as are provided by Duerver et al. (1978) suggest correlations much less than 0.5 frequently occur; hence the cases reviewed herein will primarily examine flow field variables rather than diffusion.

Correlation of Laboratory and Field Experiments

The field/model studies discussed are three familiar to the author. One is the Rakaia River Gorge region on the eastern slope of the Southern Alps in New Zealand (Meroney et al., 1980), the second is the Kahuku Point mountain-coastal region of Oahu, Hawaii, USA, (Chien et al., 1980), and finally the Clinch River valley in east-central Tennessee, USA, (Graham et al., 1979).

The Rakaia Gorge field study documented ten-minute wind speed and direction information at 27 sites during stationary conditions on two spring days selected for strong adiabatic down-valley wind flow. The Kahuku Point field study documented 24 hours average wind speed, turbulence, and directions at 32 sites during strong fall trade wind conditions. The Clinch River field study utilized a meteorological research aircraft and a 110 m meteorological tower to monitor velocity, turbulence, temperature, humidity, and direction. The aircraft flights were made during unfortunately light wind conditions and averaged over 15 second periods; nonetheless the tower data were available for strong wind situations comparable to those simulated over the model.

In the Rakaia Gorge and the Kahuku Point cases measurements were compared point to point, scatter diagrams constructed, and

sample correlation coefficients calculated. If simulation is
good there must be a high linear casual relationship between
measurements in the field and laboratory. The best estimation of
the population correlation is the sample correlation coefficients
which have been added to Table 2. Apparently wind velocities
and directions may be expected to correlate with field data at
a level of $r \geq 0.7$. Recently Holmes et al. (1979) also reported
a comparison field and laboratory measurements of maximum gust
velocities over Castle Hill (286 m) near Townsville, Australia.
Linear correlation coefficients ranging from 0.68 to 0.78 were
obtained.

Specific features of flow over irregular terrain were looked
for in both field and model data as noted in Table 3. Similar
behavior was observed in each case within the limitations of the
descisions to model or not model stratification effects.

Recent valley drainage model experiments still await counterpart field data (Petersen and Cermak, 1980). It is hoped that
the current DOE-ASCOT (Dept. of Energy - Atmospheric Studies in
Complex Terrain)studies over the Geyser area in California will
fill this gap. Nonetheless qualitative features of valley drainage flows observed by meteorologists elsewhere were observed
over the model. These include secondary cross currents caused
by valley wall drainage, down valley jets, etc.

CONCLUSIONS

The objective this paper addressed was to identify pertinent
similarity criteria for physical modeling of flow over complex
terrain. Verification required a comparison between specific
laboratory full scale experiments. A review of previous physical modeling experience provided an index of some forty case
studies relevant to terrain modeling techniques. Seventeen investigations included some field comparables. The three studies
discussed were designed to specifically test various model alternatives. Criteria for laboratory simulation of wind characteristics over irregular terrain in general have been summarized.

It would appear that the simulation wisdom developed in the
past few years is appropriate for physical modeling of flow over
complex terrain. Since the flow region of interest is usually
in the lower surface layer (z <250 m), great care must be taken
that horizontal nonhomogenuities in roughness and terrain are
faithfully reproduced. Specific conclusions from the cited
studies suggest that :

1. Un undistorted model at scales as large as 1:5000 permits
   resolution of velocity and turbulence details adequate to reproduce scalar dispersion.

Table 2. : Field to Model Data Correlations

Model or Flow Characteristic	Rakaia Gorge New Zealand		Kahuku Point Hawaii	
	Field[•]	Model	Field[•]	Model
Area of Interest (km^2)	112	112	155	155
Hill Height(m)	240	240	500	500
Model Scale	–	5000	–	3840
Power law – α	0.14±.02	0.13-0.14	0.14±.02	0.13-0.15
Roughness $-z_o$ (m)	0.05	0.045	0.11	0.11
Turbulence – u'/u_{10}	0.14±.05	0.14-0.15	0.15±.03	0.12-0.18
Shear $- \sqrt{\overline{u'w'}}/u_{10}$	0.0020	0.0019	0.0064	0.0074
Boundary depth(m) δ	750±250	500	600±300	600
Integral Scale L_{u_x} (m)	900-1000	600	167	100-120
Δ Correlation Velocities	0.68-0.78		0.71	
Δ Rank Correlation Velocities	0.78-0.95		0.84	
Δ Correlation Wind Directions	0.65-0.67		–	

[•] Counihan (1975)

Δ Compared point by point to field program

Table 3. : Features of Flow Over Irregular Terrain

Model or Flow Characteristic	Rakaia Gorge New Zealand		Kahuka Point Hawaii		Clinch River Tennessee	
	Model	Field	Model	Field	Model	Field
Approach Flow Characteristics (i.e. α, z_o, u'/u etc.)	S	S	S	S	S	S
Velocity over ridge crests	O	O	O	O	O	O
Decreased turbulence intensity over ridge crests	O	O	O	O	O	O
Increased turbulence intensity downwind of hills	O	O	O	O	O	O
Wind veering in gorges	O	O	–	–	–	–
Wind veering around hill sides	O	O	O	O	–	–
Increased turbulence downwind of coastal beach	–	–	O	O	–	–
Irregular wind and turbulence in sheltered areas	O	O	O	O	–	–
Wind veering during sea breeze and valley drainage situations	NM	–	NO NM	O	NM	–
Increased Dispersion downward ridges	NM	–	NM	–	O	O

S = simulated
O = observed during field program
NM = not modeled
NO = not observed

2. A wide range of scales and meteorological conditions may be simulated in existing boundary layer wind tunnel facilities; (see Performance Envelopes Figures 1 and 2).

3. To produce equivalent wind speeds near ground level requires accurate reproduction of surface roughness, shape, and vegetation effect. Terraced models, adequate for certain dispersion simulations where the plume is aloft or of large size compared to terrace depths, are inappropriate when the plume is at ground level and of a scale comparable to those of terrace depth.

4. Current meteorological data in complex terrain is not yet adequate to stipulate inflow conditions to physical models with confidence. Hence an adequate approach flow length must be provided to allow the surface layer to come to an equilibrium with underlying terrain undulations.

ACKNOWLEDGEMENTS

The author wishes to gratefully acknowledge support provided by the Sonderforschungsbereich 80, University of Karlsruhe, and the von Humboldt Foundation while preparing this paper.

REFERENCES

Arya, S. P. S. and Shipman, M. S. (1978) "A Model Study of Boundary Layer Flow and Diffusion Over a Ridge", 4th Symposium on Turbulence, Diffusion and Air Pollutions, Reno, Nevada, 15-18 January, pp. 584-591.

Batchelor, G. K. (1953), "The Conditions for Dynamic Similarity of Motions of a Frictionless Perfect-Gas Atmosphere", Quarterly Journal of Royal Meteorological Society, Vol. 79, pp. 224-235.

Counihan, J. (1975) "Review Paper : Adiabatic Atmosphere Boundary Layers : A Review and Analysis of Data from the Period 1880-1972", Atmospheric Environment, Vol. 9, pp. 891-905.

Chien, H. C., Meroney, R. N. and Sandborn, V. A. (1978) "Sites for Windpowers Installations : Physical Modelling of the Wind Flow Over Kahuka Point, Oahu, Hawaii", Third Int. Symposium on Wind Energy Systems, Lyngby, Denmark, August 26-29, 1980, pp. 75-90.

Draxler, R. (1979) "A Summary of Recent Atmospheric Diffusion Experiments", NOAA Technical Measurements ERL ARL-78, 38 pp.

Duerver, W. H., McCrocken, M. C. and Watton, J. J. (1978) "The Livermore Regional Air Quality Model : II Verification and

Single Application in the San Francisco Bay Area", Journal of Applied Meteorology, Vol. 17, 3, pp. 223-311.

Eagan, B. A. (1975) "Diffusion in Complex Terrain. Workshop on Air Pollution Meteorology and Environmental Assessment, American Meteorological Society, Sept. 30-Oct. 3, 1975, Boston MA.

Grahm, N. E., Taylor, G. H., Peterson, R. L., Sinclair, P. C., Frey, I. W. and Kass, T. C. (1979) "An Analysis of Terrain-induced Aerodynamic Effects Near the Kingston Steam Plant, Kingston, Tennessee", Fifth Symp. on Turbulence, Diffusion, and Air Pollution, Reno, Nevada, 15-18 January 1979, pp.113-116.

Holmes, J. D., Walkes, G. R., and Steen, W. E. (1979) "The Effect of an Isolated Hill on Wind Velocities Near the Ground Level - Initial Measurements", Wind Engineering Report 3/29, James Cook University of North Queensland, Queensland, Australia.

Hoxit, L. R. (1973) "Variability of Planetary Boundary Layer Winds", Report of Atmospheric Sciences, Colorado State University, Paper No. 199, 157 pp.

Johnson, W. B., editor (1979) Fourth Symposium on Turbulence, Diffusion and Air Pollution, American Meteorological Society, Reno, Nevada, 15-18 January 1979, 676 pp.

Meroney, R. N., Cermak, J. E. and Yang, B. T. (1975) "Modeling of Atmospheric Transport and Fumigation at Shoreline Sites", Boundary Layer Meteorology, Vol 9, pp. 69-90.

Meroney, R. N. and Sandborn, V. A. (1978) "Sites for Wind Power Installation : Physical Modeling of the Influence of Hills, Ridges and Complex Terrain on Wind Speed and Turbulence", Parts 1, II and III, D.O.E. Contract No. EY-76-5-S-06-2438, A00I, June.

Munn, R. E. (1966) Descriptive Micrometeorology Advances in Geophysics, Supplement 1, Academic Press, London, 245 pp.

Olfe, D. B. and Lee, R. L. (1971) "Linearized Calculations of Urban Heat Island Correction Effects", AIAA Paper No. 71-13, AIAA 9th Aerospace Sciences Meeting, New York, 14 pp.

Orgill, M. M. (1977) "Survey of Wind Measurements Field Programs", Batelle, Pacific North-west Laboratories, Report BNWL-2220, Wind 3, 53 pp.

Orlanski, I. (1975) "A Rationale Subdivision of Scales for Atmospheric Processes", Bull. of Amer. Meteor. Soc., Vol. 56,5, pp. 527-530.

Petersen, R. L. and Cermak, J. E. (1977) "Atmospheric Transport of Hydrogen Sulfide from Proposed Geothermal Power Plant (unit 18) : Predictions by Physical Modeling in a Wind Tunnel",

Colorado State University, Report CER-RLP-JEC- 3, 87 pp.

Scorer, R. S. (1978) Environmental Aerodynamics, Ellis Harwood Limited, London, 488 pp.

Sethu Raman, S. and Cermak, J. E. (1974) "Physical Modeling of Flow and Diffusion Over an Urban Heat Island", Turbulent Diffusion in Environmental Pollution, Advances in Geophysics, Vol. 18B, Academic Press, New York, pp. 223-240.

Snyder, W. H. , Britter, R. E., and Hunt, J. C. R. (1979) A Fluid Modeling Study of the Flow Structure and Plume Impingement on a three-Dimensional Hill in Stably, Stratified Flow, Fifth Int. Conf. on Wind Engineering, Fort Collins, CO, 8-14 July, 1979, Paper No. 132.

Start, G. E., Dickson, C. R. and Wendell, L. L. (1975) "Diffusion in a Canyon within Mountainous Terrain", J. Appl. Meteorology Vol. 14, pp. 333-346.

Yamada, T. and Meroney, R. N. (1971) "Numerical and Wind Tunnel Simulation of Response of Stratified Shear Layers to Nonhomogenous Surface Features", Project THEMIS Technical Report No. 9, Colorado State University, Fort Collins, CO.

Yamada, T. and Meroney, R. N. (1974) "A Wind-Tunnel Facility for Simulating Mountain and Heated Island Gravity Waves", Boundary Layer Meteorology, Vol. 1, pp. 65-80.

Petersen, R. L. and Cermak, J. E., (1980), Physical Modeling of Plume Dispersion at Alkali and Coal Creek, Near Crested Butte, Colorado, Colorado State University Civil Engineering Report 79-80RLP-JEC43, 283 pp.

ANALYSIS AND SIMULATION OF LOCAL CIRCULATIONS AND AIR POLLUTION OVER A COASTAL, COMPLEX SITE

E. Runca (1,3), G. Bonino (2,3), L. Briatore (2,3),
G. Elisei (4), and A. Longhetto (2,4)

(1) IIASA, Schloss Laxenburg, 2361 Laxenburg, Austria
(2) Istituto di Fisica Generale, Università di Torino, Italy
(3) Istituto di Cosmogeofisica del CNR, Torino, Italy
(4) Centro Ricerca Termica e Nucleare, ENEL, Milano, Italy

ABSTRACT

Field surveys have been done in the La Spezia complex site and a scale model of the area in a hydrodynamic tank has been built in order to both analyze the dynamic and thermodynamic structure of the air flow and to study air pollutants dispersion. Results from the field experiments made it possible to describe relevant meteorological processes such as formation of sea and land breezes, evolution of the diurnal and nocturnal inversion and others. The physical model proved to simulate the air flow of the site in satisfactory agreement with the observations. Hence, it was also operated to simulate dispersion of air pollutants under different air flow regimes. In spite of the high scale factors, the hydraulic model gave realistic simulations of the hourly average ground level concentration caused by the local thermal power plant. The physical model was then complemented by a Gaussian model in order to simulate the seasonal air pollution trend. The parameterization of the Gaussian climatological model was done by taking into account the results from the field experiments. Validation of the model through comparison with measured values over eight seasons showed that the adopted parameterization was relevant to the achievement of a satisfactory agreement with the observations and that the Gaussian model could integrate the physical model in the establishment of air quality strategies in the considered area.

INTRODUCTION

The difficulty of assessing phenomenology of local air circulations over particular sites and of investigating air pollutant
concentrations mainly derives from the complexity of the situations which characterize the conditions of equilibrium or instability of air. A description of these conditions are difficult
even in relatively flat homogeneous sites; it is therefore easy
to realize what complications arise when the site's topography
becomes irregular and etherogeneous. This is the case of mountainous and coastal environments, where the morphological and
structural variety of surfaces deeply influence the air characteristics. Thus the need arises of investigations to single out
and classify those processes whose knowledge can be used both for
describing the local meteorology and for developing and implementing air pollution mathematical models.

The site of La Spezia, due to its complicated topography,
is particularly suitable for that kind of study. It is a
coastal site, irregularly shaped and with contours a few hundred
meters high surrounding the gulf (Figure 1). A large valley
(Valle Magra) extends from W to E, north of the hilly circle. At
the western end of the hills, a headland breaks into a multitude
of little rocky islands. The presence of irregulary distributed
reliefs surrounding the gulf gives rise to land and sea breezes
which are particularly well defined and frequent. Their study,
which represents the first part of this research, was devoted
to a twofold aim: on one hand, one has tried to identify those
particularly important situations originated, or not, by the
local topography; on the other hand, to evidence the fundamental
aspects of the local flows and their interactions with the
general circulation, looking for cause-effect relations among
topography of the site, synoptical atmospheric flows and local
physical conditions of the air.

For this reason, the local aerology, particularly stationary
breezes and transient situations of breeze inversions, has been
studied by two independent and complementary ways: (1) a series
of field surveys; (2) by means of physical simulation performed
on a scale model in a hydrodynamic tank.

Field experiments and simulation by the hydraulic model have
been also done in view of their application to the analysis of
diffusion and transport of airborne stack effluents. In the case
of La Spezia, the urban and suburban areas include the main emissions: the ENEL thermal power plant close to Limone, the oil
refinery at Valdellora and the very city with its traffic and
heating pollution. In addition there are other sources as the
harbor, the shipyards, and a few small local factories.

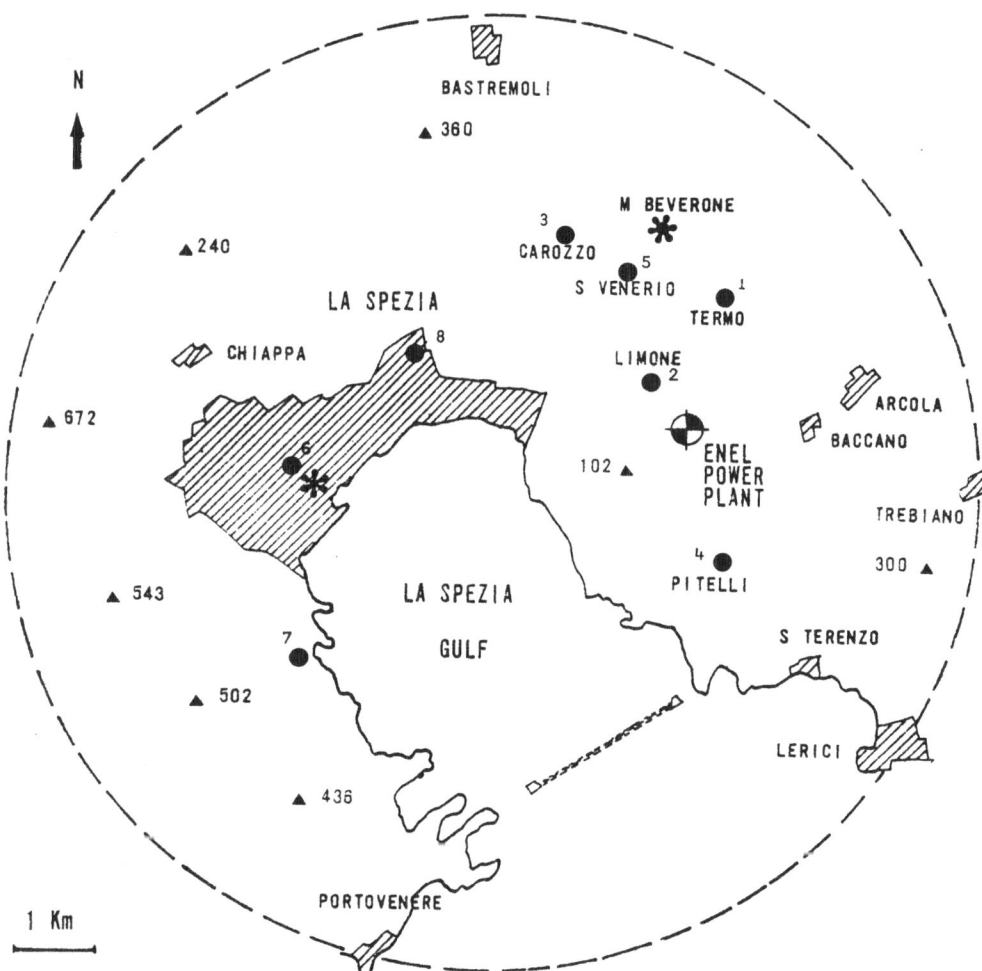

Figure 1.   La Spezia site and the circular area reproduced in
            the hydraulic model (• indicate the location of the
            SO$_2$ stations; * the meteo stations).

The realization of the hydraulic model has been supported
mainly by ENEL, also in an attempt to define previsional criteria
for operating conditions of its plant (1800 MW) on the basis of
meteorological forecasts. In order to study the seasonal air
pollution distribution the hydraulic model has been then complemented by a Gaussian type model.

FIELD EXPERIMENTS

Seven field surveys have been performed from 1974 to 1978
during significant periods of the year. They have been managed
by the C.R.T.N. of ENEL, Milano, and by the Istituto di Cosmogeofisica of C.N.R., Torino, in the frame of their respective con-
tracts "Second Environmental Research Programme" of C.E.C. and
"Progetto Finalizzato Ambiente" of C.N.R.

The surveys have been preceeded by a careful analysis of
meteorological and chemical data, recorded by the air quality
network located in the area by ENEL*. It supplied a statistical
collection of data extended over a period of several years, and
described some SO_2 pollution situations which many times could
not be related to the operating conditions of the power plant.

The detailed presentation of this analysis, as well as the
description of the instrumentation and methodologies used, will
be the subject of a specific paper. Here we restrict ourselves
to mention that the following measurements have been performed:
1) Tracking of balanced and pilot ballons;
2) Detection of thermal profiles with captive ballons;
3) Wind directions and velocities with meteo-anemometric
    control network;
4) Temperature and humidity at ground level with shelters;
5) Temperature profile above 500 m with radiosondes;
6) Lidar tracking of plumes.

Some results and findings from the field experiments have
also been discussed in different papers (Longhetto et al. 1976,
Bacci et al. 1975, Anfossi et al. 1976, Piano et al. 1976). Here
we summarize those particular aspects, relevant to physical and/or
mathematical modeling.

In the layer which extends up to the height of 1 km, within
which the diffusion phenomena we are interested in take place,
the dynamics of air masses is determined by the interaction of
synoptic flows with local ones.

---

*The network consists of 8 stations recording SO_2 concentrations
 and two meteorological stations (see Figure 1).

Two kinds of characteristic situations emerge: pseudo-adiabatic and breeze conditions. The first situation occurrs mainly from October to March, with no significant stratification. The second condition, particularly well-defined and frequent especially during the summer, involves a layer whose thickness extends, in extreme cases, from 100 to 800 m.

In short, the land breeze comes from the North with mean velocity 3 m/s and thickness 300 ÷ 400 m, while the sea breeze comes from the South with velocity 4 ÷ 5 m/s and thickness 400 ÷ 800 m; the transition between them is relatively fast (Figure 2). Each breeze is continuously growing in thickness, even if its velocity is almost constant at ground level.

The atmospheric stability conditions are connected to the alternating breezes; generally speaking, the nocturnal phase is stable (Pasquill E and F categories), while the diurnal one tends toward instability (B and C categories) (Figure 3).

During clear nights a ground based inversion develops over the land. The height of the nocturnal inversion was found to grow proportionally to the square root of time elapsing from one and a half hours before sunset (Anfossi et al. 1979).

In the morning the growth of the convective layer was proved to follow (Anfossi et al. 1976) the model described by Carson (1973, 1974).

The sea-breeze circulation which establishes after the erosion of the nocturnal inversion was proved to maintain (Anfossi et al. 1976) an unstable layer (Van der Hoven layer (Van der Hoven 1976)) which was generally capped, by an elevated inversion at approximately 500 m. The above findings on the thermodynamics behavior of the boundary layer were included in the Gaussian climatological model, described later, by imposing that during unstable situations the upward motion of pollutant matter was inhibited at 500 m, while during stable situations, the downward diffusion of pollutants was inhibited at the height of the nocturnal inversion. For the situations recorded as neutral or weakly unstable, neither nocturnal nor diurnal inversion were assumed to take place.

In view of applications of air pollution mathematical models, the field experiments were also organized to determine the dispersion parameters of pollutants released from elevated sources. Ballon trajectories were tracked for each of the six Pasquill stability conditions. The trajectories (Santomauro et al. 1978) were then used to estimate the vertical and horizontal standard deviations of the diffusing pollutants. These

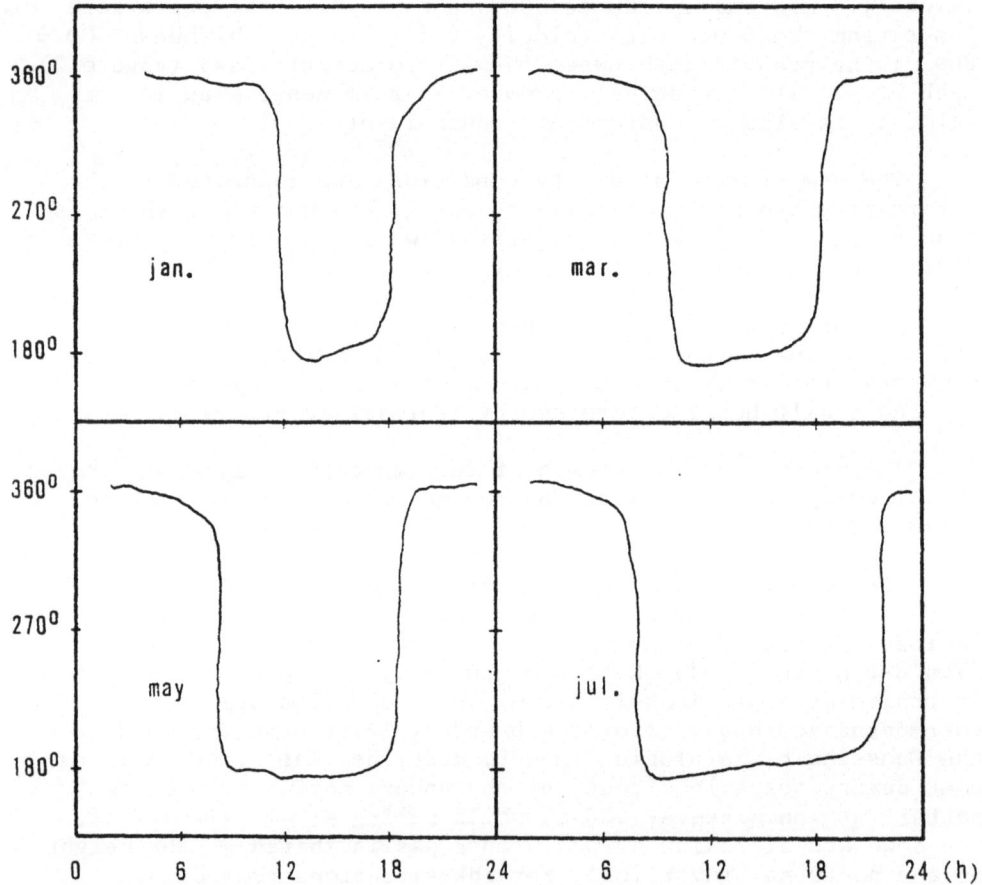

Figure 2.   Examples of the daily breeze regime observed
             at the site.

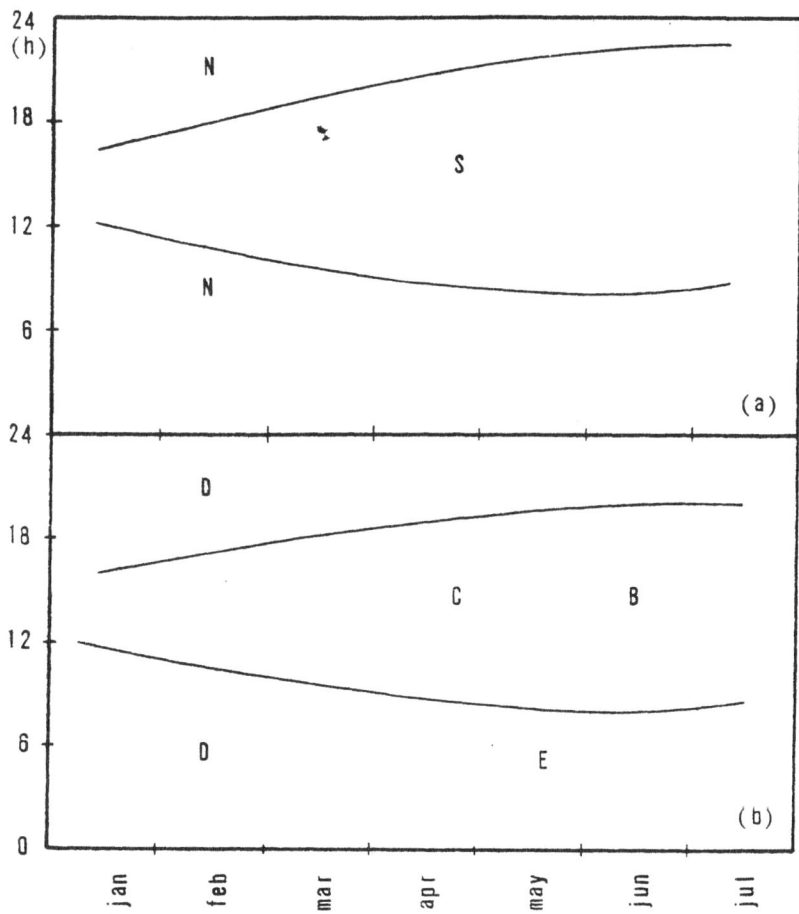

Figure 3.   Seasonal evolution of the wind direction, (a), and
the corresponding evolution of the atmospheric
stability categories, (b).

dispersion parameters were used in the Gaussian climatological model for the elevated chimneys.

Also measurements were taken to estimate the rise of the plumes of the thermal power plant. Comparisons have been made between the actual data and those expected on the basis of various models of current use. The best fit has been obtained with Briggs' formulae for the conditions of adiabatic gradient and of stability with wind u ≥ 1.5 m/s at the height of the sources.

Some typical rises calculated with Briggs' formulae were found to be in good agreement with experimental data (see Figure 4).

THE HYDRAULIC MODEL

Characteristics and Scales of the Model

A tridimensional plastic model of the La Spezia site (see Figure 1) has been built at the Alsthom Laboratories, Techniques des Fluides, Grenoble, in a hydrodynamic tank 10 m long, 1.5 m wide, and 1 m high, equipped for many types of water motion. Two cross-diffusers covered by porous walls have been placed on the channel bottom for the emission of solutions with density different from that of the main fluid (salt water for breezes simulation), distributed upstream and downstream with respect to the model; in particular, with the downstream diffuser it was possible to generate emissions against the mean stream.

When choosing the scales of significant physical quantities, two considerations were taken into account. The first one is related to the fact that the area considered, though it is very complicated, presents natural boundaries rather well defined. Therefore it has not been difficult to identify the contour of the model (Figure 1): to the N, the hilly barrier which separates La Spezia from the valleys joining Val Magra; to the E, a part of Val Magra and the reliefs between Arcola and Lerici; to the S, the mouth of the gulf, from Lerici to Palmaria Island; finally, to the W, the relief of Mount Parodi.

The second consideration is connected both to the average anemometric conditions over the site and to the emission of the power plant, since one has to scale in the same way the respective characteristic velocities.

The plant has four stacks, each one placed on a generator: two 300 MW and two 600 MW power groups, which can work independently with different power loading. Some working conditions, significant for the modeling, are listed in Table 1.

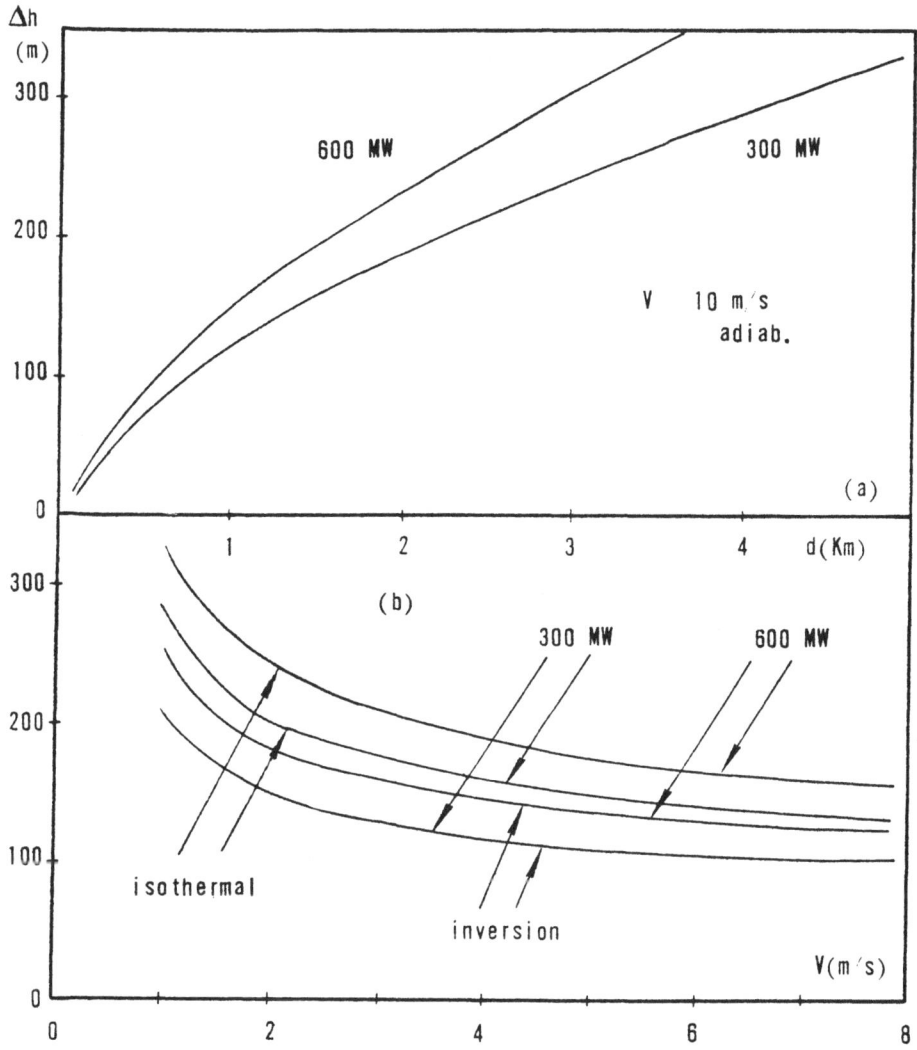

Figure 4. (a): plume rise in adiabatic conditions with 10 m/s
wind at stack's height;
(b): id. vs. wind velocity in isothermal atmosphere
and with an inversion of 1°C/100 m. Full
power loading.

Table 1. Characteristics of the power plant emission in
conditions of full loading.

Characteristics	300 MW power groups	600 MW power groups
Stack height, m	120	180 and 220
Stack diameter, m	5.50	8.50 and 6.20
Plume velocity, m/s	15.5	19 and 21
Emission temperature, °C	120	120
Gas flux at 120° C	370	630
Thermal flux, Kcal/s	10^4	$1.7 \cdot 10^4$
Buoyancy parameter, m^4/s^3	290	500
SO$_2$ flux for 3 % S fuel, g/s	1160	2320
SO$_2$ concentration at the emission (120° C, g/m^3)	3.15	3.15

The optimal scales have chosen on the basis of these
considerations and taking into account the tank dimensions and
its hydraulic characteristics. Denoting with subscript 1 the
physical quantities pertinent to the model and with subscript
2 the omologous ones in the prototype, they turned out to be
$d_1/d_2 = 1/8.10^3$ for lengths and $v_1/v_2 = 1/200$ for velocities; it
follows $t_1/t_2 = 1/40$ for times and $q_1/q_2 = 7.8/10^{11}$ for plume
fluxes. The plastic model, whose diameter is 1.5 m, thus reproduces a circular area 12 km in diameter centered on the inner
part of the harbor. The water depth in the channel was on
average 45 cm, corresponding to about 3.5 km of air.

The density stratifications of the atmosphere, corresponding
to the various temperature profiles, and the buoyancy effects of
warm plumes have been reproduced by means of similar density conditions in the channel flow and the emission (Vadot 1965; Caudron
et al. 1975). The physical similitude is based on the densimetric
Froude number conservation, i.e. by setting:

$$(\Delta\rho/\rho)dg/u^2 = const.$$

Since g is the same in the two systems, setting $\Delta\rho/\rho = \Delta$,
the similitude condition $\Delta_1 d_1/u_1^2 = \Delta_2 d_2/u_2^2$ yields $\Delta_1/\Delta_2 = 1/5$.

The emissions in the model have been simulated with a mixture alcohol-fluorescine. The sampling for measuring ground level concentrations (g.l.c.) was made by using small diameter pipes, continuously drawing liquid. Each sample for spectrofluorimetric analysis was of $50cm^3$ and the sampling time was of 1 min. The measurement of the density vertical profile was done in a section of the channel centered on the model by vertically sampling the saltness with a conductivity probe. Alcohol-fluorescine and salt water fluxes were measured by flow meters.

Possibilities of Model Simulations

From a general point of view, it must be pointed out that the method of density stratification allows to reproduce a wide range of environmental conditions and that the site phenomenology turns out to be accurately modeled. For example, in conditions of sea breeze v_2 = 6 m/s, h_2 = 800 m, the vertical velocity profiles above two points of the model have been measured, and the following results have been obtained: the profiles have a peculiar shape over the area comprising S. Venerio, Limone and Termo (Figure 1), where two well-defined maxima are present around 300 and 700 m, separated by a relative minimum around 500 m, while in the inner part of the gulf close to the outer dike, where a maximum of velocity is reached in the 300 ÷ 500 m layer, the profile is of the usual kind. This situation has been actually observed during the field surveys.

Also the use of different dyes for different emission and/or air movements allows immediate visualization of microcirculations, whose observation in the prototype would be rather long and difficult. An example is that of certain situations of stationary flow that occur during the inversion from sea to land breeze and that lead to the formation of an air layer, lenticularly shaped, which settles over the W and NW parts of the city where it remains, slowly rotating counterclockwise.

Simulated situations

The following simulations have been performed:
a)   the first 800 m completely adiabatic;
b)   steady-state synoptic winds with the breezes;
c)   formation of sea and land breezes.

In each of these the power plant emissions, in the hypothesis of the use of 3% sulphur fuel, have been studied for half and total power loading conditions.

The adiabatic situations, (a) above, have been reproduced in the case of 7 m/s winds at stack height, coming from N, NE, E, SE, and S.

The simulation of stationary breeze conditions, (b) above, with the corresponding thermal stratifications and with simultaneous presence of superimposed winds coming from N and S, have been carried out for two mixing heights to account for the evolution of the diurnal inversion. Their mean velocity was of 4.5 m/s for sea breezes and 3 m/s for land breezes and their thickness was measured over the central part of the gulf. For the case of sea breezes, various thermal profiles have been considered, because of the complexity of their structure: isothermal-adiabatic, adiabatic with inversion, and superadiabatic-adiabatic-isothermal.

The thermal profile for land breeze is normally simpler than that of the sea breeze. We have limited our experiments to those of an inversion-adiabatic type, the ground-based inversion height varying from 200 to 500 m. Some velocity profiles in the hydraulic channel and some g.l.c. distributions of SO_2, which illustrate typical results of the simulations of type (b), are shown in Figure 5 and 6 respectively.

Finally we have studied the transient situations of breeze inversion, (c) above. The visualization of water flows by means of various dyes is particularly interesting in these complicated cases. Photographic recording allowed an immediate and comprehensive picture of motions that would be very difficult to observe at the site. As an example, we show in Figure 7 what happens during the earlier phases of an inversion from land to sea breeze (Figures 1-7 are taken from Briatore et al. in press).

Comparison Among Field Measurements and Physical Model

Comparison of the physical model with the prototype has been done for those situations having maximum similarity of boundary conditions.

The physical model was able to reproduce the counterclockwise rotation of the wind during the transition from land to sea breeze, as it was observed in reality. Also in agreement with the field observations, the model did not show a definite tendency of wind direction during the opposite process, i.e. the transition from sea to land breeze.

Simulations performed by the physical model of dispersion of air pollutants from the low sources (urban sources), as well as from the chimneys of the thermal power plant, gave a substantial agreement with the observations.

In the case of urban sources, the model showed that during weak flows the pollutants stagnate in localized air pockets over

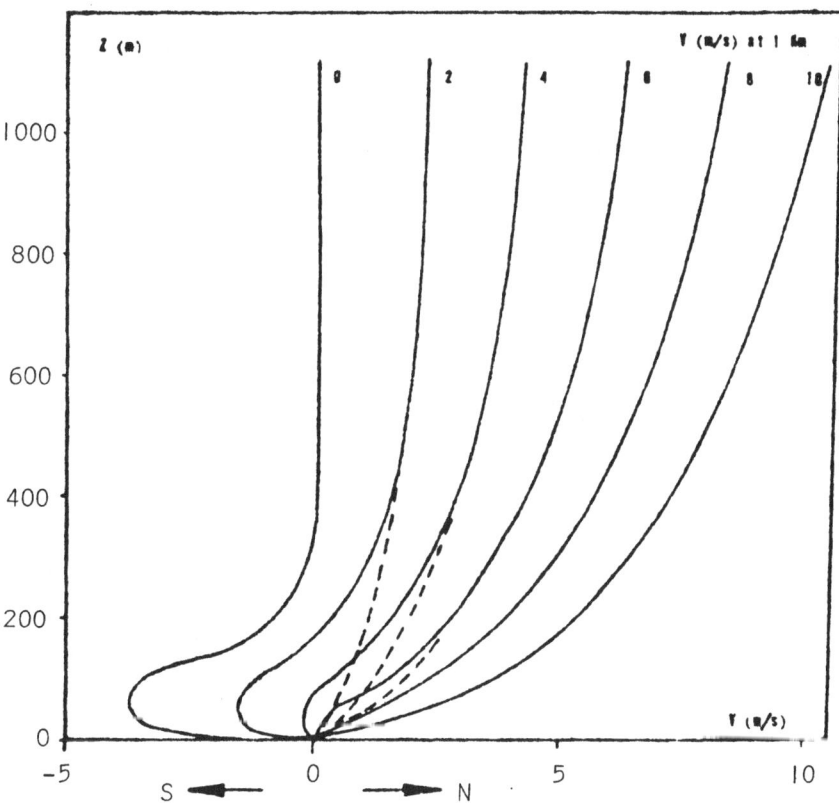

Figure 5. Example of vertical velocity profiles simulated
in the hydraulic channel: initial progression
of a sea breeze 150 m thick.

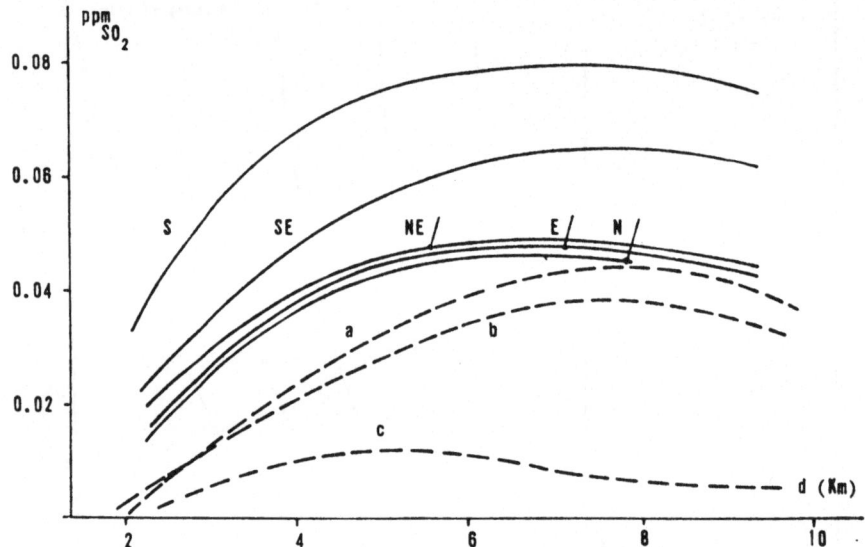

Figure 6. Examples of g.l.c. measured in the hydraulic model.
Continuous lines: adiabatic conditions, wind velocity
7 m/s, various directions of the incoming flow, full
power loading; dashed lines: sea breeze; (a) and (b)
refer to an adiabatic breeze 700 m thick with a strong
inversion in the 700-900 m layer, for full and half
loading; (c) refers to a stable layer extending from
300 to 700 m, comprised between two adiabatic zones;
half power loading.

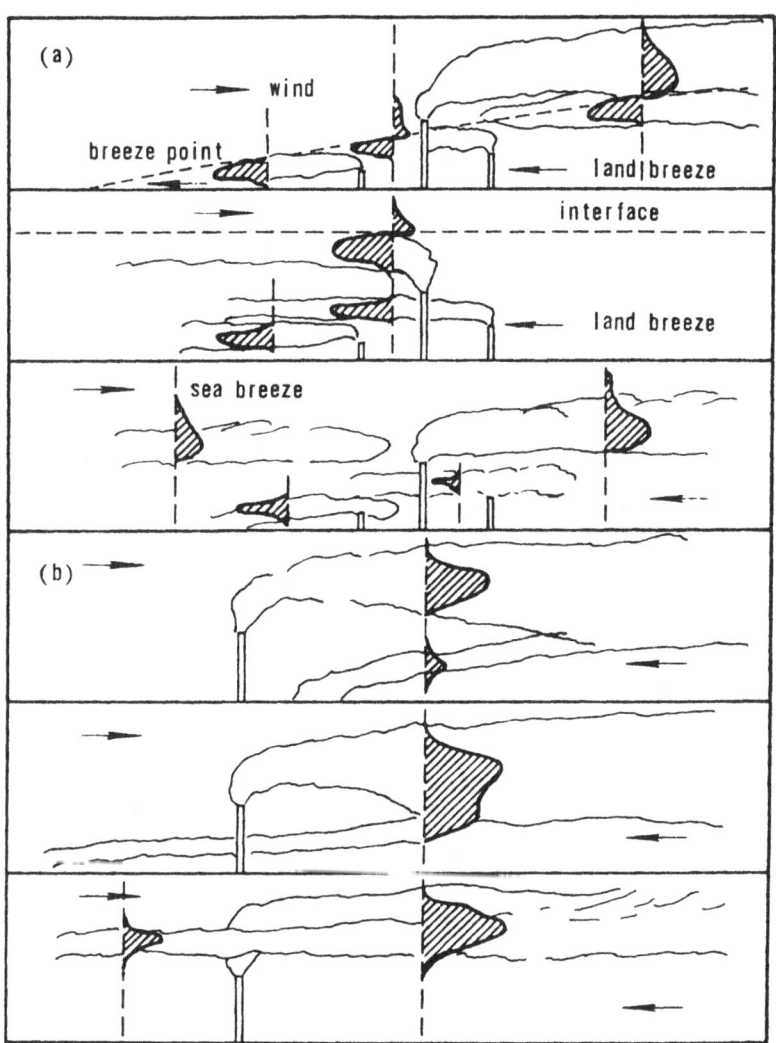

Figure 7.   (a): inversion from land to sea breeze;
             (b): details of evolution of a land breeze.

         = concentration distribution

the western urban part of the gulf. High pollutants concentrations
were measured as it occurs in reality, due to the low height of
urban emissions.

Several simulations were done in order to analyze the impact
of the thermal power plant on the air quality of the area. In all
the analyzed circumstances the plant operating at full loading
conditions gave the lowest ground level concentrations. In fact,
the plumes attained in this case such altitudes as to be no longer
influenced by the locally induced circulations; their contributions
to air pollution in the area of interest were very low and never
exceeded the hourly average value of 0.05 ppm. These results were in
perfect agreement with the SO_2 concentration measurements recorded
by the monitoring stations of the air quality network. In the
stations of the network downwind of the thermal power plant (not
affected by other sources), the recorded pollution level was
generally quite low and often close to the instrumental detection
threshold.

In conclusion, for all the considered cases, the hydraulic
model gave consistent results. In particular, for the simulations
concerning the analysis of air pollutant dispersion, the results
showed that for sampling time of approximately one hour, the hydraulic model was able to reproduce satisfactorily diffusion in addition
to transport. This a priori seemed uncertain because of the high
scale factor involved.

THE MATHEMATICAL MODEL

The assumption that the distribution of gaseous pollutants in
a plume is Gaussian has constituted a basis for many studies of
computation of air pollutant concentrations. Although there have
been examples of Gaussian models applied to the simulation of
short term (hourly-daily) average concentrations (see e.g. Shieh
et al. 1972) the Gaussian concept is mainly considered adequate
for long term (month-season) averaging times (see e.g. Martin 1971;
Calder 1971; Calder 1971; Prahm and Christensen 1971; Runca et al.
1976).

Air pollution models applied to the simulation of long-term
average concentrations are also named climatological models. Since
the development of an industrial and/or urban area must guarantee
a low exposure (i.e. a low long-term average concentration) to
air pollution for all the points in the area of interest, these
models provide a substantial help to planners and decision makers.
However, the credibility of Gaussian climatological models is not
well established, especially in application to topographically
complex sites. This is, to a large extent, due to the inadequacy
of model parameters to actually represent the physical situations

under which pollutant dispersion occurs. Thus, in the case of
La Spezia, the availability of very detailed information on the
local meteorology offered also the opportunity both to study the
physical factors which could be included in the model to improve
simulation results and to verify the applicability of the Gaussian
concept to such a complex site.

As shown by the following, the physical information from the
results of the field experiments was relevant to the implementation
of the model.

Description of the model application is reported by Bonino et
al. (1980). Here, first, we report briefly about the data and the
model used by Bonino et al. (1980) and the results achieved; second,
we describe a sequence of validation tests to analyze the effects
of the adopted physical parametrization on the results given by
the model. In order to check to which extent the statistical averaging of errors affects the agreement between measured and computed
concentration values, the model results are also compared with
measurements, separately for the stability situations (A-B), (C-D),
and (E,F), defined according to Pasquill (1974).

Meteorological and Concentration Data

As already mentioned the control of the local pollution
situation is performed by an air quality monitoring network, located in the area by ENEL. Of the two meteorological stations one
located in the city and the other at 226 m above sea level (Mt.
Beverone stations), only the data recorded by the latter have been
used as they were considered more adequate for description of the
transport of air pollutants over the La Spezia basin.

As already discussed, the wind in the area is prevailingly
blowing from North and South. Although the local orography creates
channeling effects and during sunny days a sea breeze flow super-
imposes the synoptic circulation, in general, the wind directions
recorded by standard instrumentation can be considered as the ones
along which, on a long-term averaging time, the transport of
pollutants occur.

Emission Data

Fundamental information required to apply the model concerns
the SO_2 emission distribution and rate. It was evaluated by
dividing the sources into the three following classes: (a) industrial; (b) urban, and (c) sources due to the harbor activities.

The industrial sources included in the model were the refinery operating in the area and the local power plant. The urban and harbor ones were represented by 62 and 12 equivalent point sources respectively (Bonino et al. 1980).

Diffusion Equation and Parameters

Assuming that: a) on long-term averaging times pollutant distribution is uniform in the crosswind direction in every sector of the wind rose (Calder 1971; b) the vertical pollutant distribution is Gaussian; and c) ground and inversion layers are impermeable barriers to air pollutants, the diffusion equation of the model can be put in the following form:

$$
C_P = \frac{2^{-3/2} N}{\pi^{3/2}} \sum_{k}^{M} \frac{Q_k}{D_{P,k}} \sum_{\substack{id,iw \\ is,it}} \frac{F\left(id,iw,is,it\right)}{U\left(id,iw,is,it\right)\sigma_z\left(D_{P,k},is\right)}
$$

$$
\sum_{n=-\infty}^{n=\infty} \left\{ exp\left[-\frac{\left(Z_P - h_k(u,T,is) + 2nL\right)^2}{2\,\sigma_z^2\left(D_{P,k},is\right)}\right] + exp\left[-\frac{\left(Z_P + h_k(u,T,is) + 2nL\right)^2}{2\,\sigma_z^2\left(D_{P,k},is\right)}\right]\right\}
$$

where: C_p = concentration at receptor point ($\mu g/Nm^3$); N = number of wind rose sectors (8 in this study); M = number of point sources; Q_k = emission rate of the Kth source ($\mu g/s$); $Dp_{,k}$ = distance of point P from the Kth source projected on the wind direction (m); F(id,iw,is,it) = frequency of wind blowing into a given 45^0 sector of the wind rose (id), for a given wind speed class (iw), atmospheric stability class (is) and temperature class (it); u(id,iw,is, it) = average wind speed corresponding to the meteorological situation having frequency F(id,iw,is,it); σ_z = vertical standard deviation of the diffusing particle (m); z_p = receptor point height (m); H_k = height of the plume axis (geometric stack height + plume rise) (m); T (id,iw,is,it) = average temperature corresponding to the meteorological situation having frequency F(id,iw,is,it)(^0C); L = inversion layer height (m).

According to this equation the seasonal average concentration at receptor point P is the sum of the contributions of the M sources due to the meteorological conditions recorded in the area, each multiplied by its frequency of occurrence in the considered season.

The joint frequency meteorological distribution was derived from the data recorded at Mt. Bevorene station by taking eight wind directions, six atmospheric stability according to Pasquill-Gifford, the following five classes for wind velocity: $1 \div 2$; $2 \div 3.5$; $3.5 \div 5.5$; $5.5 \div 8.5$ and greater than 8.5 m/s, and the following four classes for temperature: < 0, $0 \div 10$, $10 \div 20$, and greater than $20^{\circ}C$. Temperature is included in the joint frequency meteorological distribution to account for the effect of buoyancy on the plume rise.

The implemented Gaussian equation is a modification of the formulation originally proposed by Pooler (1961) and successively applied in more or less modified form, by many authors (see, e.g. Martin 1971, Calder 1971, Prahm and Christensen 1977, and Runca et al. 1976).

Physical Parametrization

Compared with the formulation used by Runca et al. (1976), the one reported here and in the previous study by Bonino et al. (1980), includes the effects on dispersion due to diurnal and nocturnal inversion.

The former is taken into account by means of an infinite number of image sources (in the application only the first eight image sources were considered); the latter by assuming that pollutants released above the height of the nocturnal inversion cannot penetrate through it .

In order to take somewhat into account the complex topography of the area the concentration is computed at the height (z_p in the equation) above sea level of the monitoring stations. In general, this choice is not the proper one as the relative position of the pollutants trajectories and the underlying surface is effected by the hilly characteristics of the terrain. However, in this study, considering the distribution of the sources and the topography of the area, collocation of the receptor points at their height above sea level appeared to be the most reasonable assumption. Also numerical tests showed that this choice gave the best agreement between simulated and measured values.

The σ_z values derived from the field experiments were used for the elevated sources; the σ_z values of Gifford and Pasquill (see e.g. Gifford 1975) were adopted for the low sources. In order to take into account the turbulent mixing introduced by the urban roughness and thermal effects the P-G σ_z values have been moved one class towards instability in stable situations (see e.g. Runca et al. 1976; Prahm and Christensen 1977).

Evaluation of the rise of the plumes of the power plant stacks
was done in accordance with the field experiments, by Briggs' formulas for power plant emissions, while the Concawe model (see e.g.
Detrie 1969) was used for the refinery plumes as no experimental
data were available for these emissions. The height of urban and
harbor emissions were taken constant (Bonino et al. 1980).

Model Validation

Seasonal SO_2 average concentrations were computed by the model
for eight seasons from March 1975 to February 1977, i.e. for the
period for which both emission and meteorological data were
available. The seasons have been defined as follows: Spring =
March, April and May; Summer = June, July and August; Autumn =
September, October and November; Winter = December, January and
February.

Simulated and measured values in the eight stations are plotted
in Figure 8 which is taken from Bonino et al. (1980). The agreement is very satisfactory. The model correctly describes both the
seasonal pollution pattern and the SO_2 spatial distribution in
every season. Station 1 and 4 being out of order respectively
during Summer 1976 and Autumn 1976 were not considered. For the
data of Figure 8 the correlation coefficient is equal to 0.9;
slope and intercept of regression line (simulated versus measured)
are 0.89 and 0.002 ppm respectively.

Although the correlation coefficient does not provide an absolute way to evaluate the model, the value obtained here is considered
high even for models applied to situations simpler then the one of
La Spezia. (The regression line shows that the model slightly
underestimates the highest concentration values and overestimates
the lowest concentration values).

The results achieved with the chosen physical parametrization
of the model could be due to statistical averaging of errors. In
order to establish the credibility of the model the analysis
described below has been carried out.

Seasonal concentration values occurring separately with the
stability conditions (A-B), (C-D), and (E-F) have been simulated
and compared with the corresponding measured values. Correlation
and regression line coefficients for the cases and for the others
considered are reported in Table 2.

The lowest correlation coefficient is found for the stable
situation, the highest for the stable situation. However, if
during stable situations the effect of the nocturnal inversion
had not been considered, the correlation coefficient would
have dropped from 0.92 to 0.71, the intercept of the

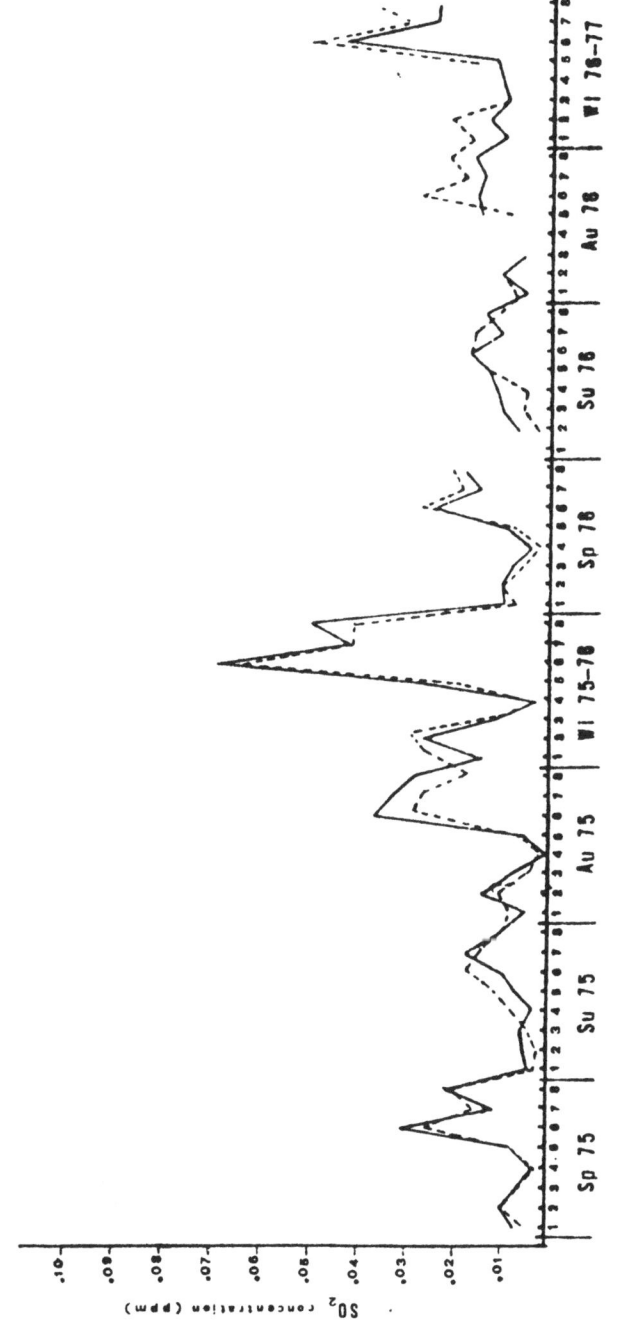

Figure 8. Measured and simulated seasonal average SO_2 concentration at monitoring stations:
—— measured values; - - - simulated values. Sp: March–April–May.
Su: June–July–August. Au: September–October–November. Wi: December–January–
February.

Table 2.   Correlation coefficient, slope, and intercept of the
regression line for the cases illustrated in the first
column. The reference case is the one in Figure 8. The
other differ from the reference case either in considering
only those situations with a given atmospheric stability
or in modifying some of the parameters used the reference
case.

Considered cases	Correlation coefficient	Reg. Line: $C_{SIM} = \alpha C_{meas} + \beta$	
		α	β (ppm)
Reference case	0.90	0.89	0.002
Stability (A-B)	0.75	0.61	0.005
Stability (C-D)	0.83	0.87	0.002
Stability (E-F)	0.92	1.03	0.003
Stability E-F (no noct. inv)	0.71	1.05	0.018
Stability E-F ($\sigma_{P.G.}$ not mod)	0.93	1.43	-0.002
Reference case (σ_{Briggs})	0.81	0.55	0.004

regression line would have risen from 0.002 to 0.018 ppm, and the slope would have became larger than 1. The model, therefore, would strongly overestimate the measured concentration values as shown in Figure 9.

The results given by the model for stability (E-F), in addition to considering the nocturnal inversion, were obtained by moving the σ_z values applied to the low sources of one class toward instability as discussed above. If this condition had not been imposed, the model as shown by the slope and intercept of the regression line (see Table 2) would have strongly overestimated the highest concentration values and slightly underestimated the lowest concentration values. Since the largest concentrations occur in the station located in or in proximity of the urban area, this result confirms that the dilution of pollutants in an urban area is greater than in an open country due to the increased turbulent mixing introduced by urban aerodynamic roughness and thermal effects. The correlation coefficient, however, is as high as for the case in which the σ_z values were moved one class towards instability. This indicates on one side that there is basic agreement between the model and the real process, on another side that the correlation coefficient as already said, cannot be taken as the only estimator of the credibility of the model.

The latter consideration can be extended to all the cases quoted in Table 2. The correlation coefficients reported in Table 2 range from 0.71 to 0.93 and thus the model appears to be an image of the real process. The focus of the image depends on the introduction into the model of some of the physical factors which are relevant to the process (such as the nocturnal inversion condition) and on the refinement of some parameters.

The model has been also run by replacing for the low sources the dispersion parameters given by Pasquill-Gifford with the ones proposed by Briggs for urban emissions (see, Gifford 1975). In agreement with the previous considerations the correlation coefficient remained satisfactory high. The low sensitivity of the correlation coefficient to dispersion parameters was also reported by Prahm and Christensen (1977). The slope of the regression line, however, dropped to 0.55, indicating a strong increase of the deviations between measured and simulated values. Due to this result the utilized dispersion parameters appeared to be a very suitable choice for this study.

The final comment on the validity of the adopted physical parametrization concern the introduction of the diurnal inversion in the model. Simulations without this condition proved that it had very little influence on the concentration field. Given that this condition mostly influences the contribution to concentrations from the elevated sources, this result was expected as a consequence

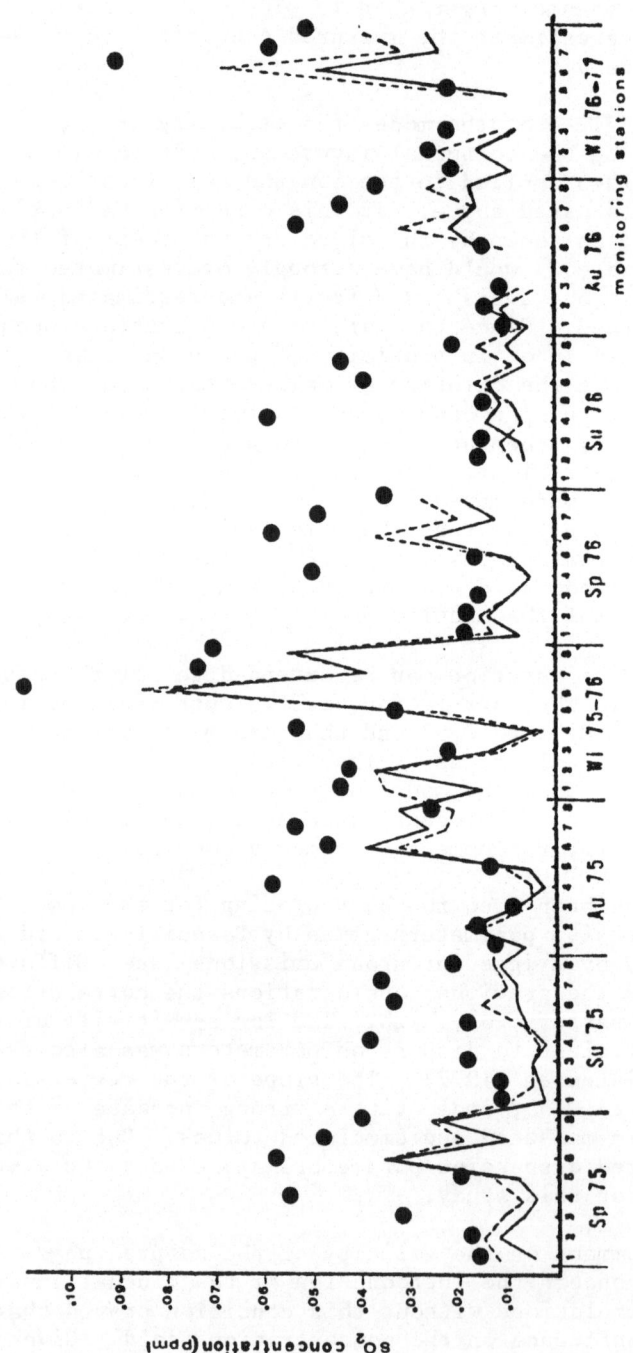

Figure 9. Measured and simulated SO$_2$ concentration at monitoring station for E and F
stability categories: ——— measured values; – – – simulated values;
● simulated values without nocturnal inversion condition.

of the relative closeness of the tall chimneys to the receptor points and to the inadequacy of the diffusion equation described in section "Diffusion Equation and Parameters" to represent the complexity of vertical mixing in unstable situations (see, e.g. Lamb 1978).

CONCLUSIONS

The micrometeorology and dispersion of air pollutants in the La Spezia complex coastal site has been studied by means of field surveys, a physical model in a hydrodynamic tank, and a Gaussian-type climatological model.

The main results obtained by this study can be summarized as follows:
1) Field experiments were fundamental to describe the main aspects of the local meteorology in La Spezia and provided the basic infor- mation both for the realization of a physical model based on the principle of hydraulic analogy and for the implementation of a Gaussian climatological model;
2) Comparison between physical model simulations and observations, being very satisfactory in reproducing both the local air flow dynamics and the dispersion of air pollutants, proved that hydraulic modeling is a reliable technique which can be applied to complex sites avoiding more expensive field observations;
3) Validation analysis of the Gaussian climatological model showed that the adopted parametrization based on the data deduced from the field experiments was fundamental to the achievement of a substantial agreement with the observed concentrations and proved that this type of mathematical model could integrate the hydraulic model to establish air quality strategies at the considered site.

REFERENCES

Anfossi, D., Bacci, P., Giraud, C., Longhetto, A., and Piano, A., 1976, Meteorology Surveys in La Spezia site, in Atmospheric Pollution Elsevier, Amsterdam.
Anfossi, D., Bacci, P., Bonino, G., Briatore, L., Elisei, G., Giraud, C., Longhetto, A., Piano, A., and Richiardone, R., 1979, Aerologia di un sito costiero complesso: il Golfo di La Spezia. Published by Consiglio Nazionale delle Ricerche, Roma, Ref. Collana del P.F. Promozione della Qualità dell'Ambiente, AQ 3/3.
Bacci, P., Longhetto, A., and Anfossi, D., 1975, Proc. 6th Int. Tech. Meet. on Air Poll. Modeling, Frankfurt.
Bonino, G., Longhetto, A., and Runca, E., 1980, A physical approach to air pollution climatological modeling in a complex site. Nuovo Cimento 3C, in press.
Briatore, L., Elisei, G., and Longhetto, A., 1980, Local air circulations over a complex coastal site: a comparison

among field surveys, hydraulic, and mathematical data,
Nuovo Cimento, in press.

Briggs, G.A., 1969, Plume rise. USAEC, Div. of Tech. Info.,
TID - 25075.

Calder, K.L., 1971. A climatological model for multiple source
urban air pollution, pp I1-I33, NATO/CCMS, Air Pollution
Rep. n° 5.

Caudron, L. and Viollet, A., 1975, Méthodes et moyens d'étude des
panaches d'effluents rejetés dans l'atmosphere (Electricité
de France; Paris).

Carson, D.J., 1973, The development of a dry inversion-capped
convectively unstable boundary layer. Quart. J. Roy. Meteor.
Soc. 99: 450-467.

Carson, D.J. and Smith, F.B., 1974, Thermodynamic model for the
development of a convectively unstable boundary layer.
Advances in Geophysics 18A: 111-124.

Detrie, J.P., 1969, La Pollution Atmospheric. Dunod, Paris.

Gifford, F.A. and Hanna, S.R., 1973, Modelling urban air pollution
Atmos. Environ. 7: 131-136.

Gifford, F.A., 1975, Atmospheric dispersion models for environmental
pollution applications. Air Pollution and Environmental
Impact Analysis A.M.S.: 35-58.

Lamb, R.G., 1978, A numerical simulation of dispersion from an
elevated point source in the convective planetary boundary
layer. Atmos. Environ. 12: 1237-1304.

Longhetto, A., Anfossi, D., Bacci, P., Giraud, C., and Piano, A.,
1976, Proc. Int. Conf. on Air Poll., Pretoria.

Martin, D.O., 1971, An urban diffusion model for estimating long-term
average values of air quality. J. Air Poll. Control Assoc.
21: 16-19.

Pasquill, F., 1974, Atmospheric Diffusion. Halsted Press, John
Wiley, New York.

Piano, A., Anfossi, D., Bacci, P., Giraud, C., and Longhetto, A.,
1976, Compt. Rend. XIV Journées de l'Hydraulique (Paris)

Pooler, F., 1971, A prediction model of mean urban pollution for
use with standard wind rose. Int. J. Air Water Pollution 4:
199-211.

Prahm, L.P. and Christensen, M., 1977, Validation of a multiple
source Gaussian air quality model. Atmos. Environ. 11:
791-795.

Runca, E., Melli, P., and Zanetti, P., 1976, Computation of long-
term average SO_2 concentration in the Venetian area. Appl.
Math. Mod. 1: 9-15.

Santomauro, L., Bacci, P., Longhetto, A., Anfossi, D., and
Richiardone, R., 1978, Experimental evaluation of diffusion
parameters on a local scale by means of no-lift balloons.
J. Appl. Meteor. 17: 1441-1449.

Shieh, L.J., Halpern, P.K., Clemens, B.A., Wang, H.H., and Abraham,
F.F., 1972, Air quality diffusion model: application to New

York City. I.B.M. J. Res. Develop. 16: 162-170.
Vadot, L., 1965, Etude de la diffusion des panaches de fumée dans
l'atmosphere (Centre Interprofessionnel Technique d'Etudes
Pollution Atmospherique; Paris).
Van der Hoven, I., 1976, Atmospheric transport and diffusion at
coastal sites. Nucl. Safety 8: 490-499.

DISCUSSION

R. STERN

How many measurements have you used to calculate the σz values for the elevated stacks ? Do you have measurements for every stability class ?

E. RUNCA

I do not remember the number of measurements which have been used to compute the σz values for the elevated stacks. The σz values have been computed for all the stability classes.

A. BERGER

What are the numbers of degrees of freedom involved in the correlation coefficient determination and what is their significance level ?

E. RUNCA

This was not checked. The agreement shown is believed to be valid on physical grounds, that is, relevant physical parameters (such as treatment of nocturnal inversions) and an accurate emission inventory were included.

A NEW GAUSSIAN PUFF ALGORITHM FOR NON-HOMOGENEOUS,

NON-STATIONARY DISPERSION IN COMPLEX TERRAIN

Paolo Zannetti

Simulation Modeling Department
AeroVironment Inc.
Pasadena, California 91107 (U.S.A.)

ABSTRACT

This paper presents a complete algorithm for applying the Gaussian method to the most general and complex dispersion conditions. The plume evolution is described by transport and diffusion of a series of independent puffs, thus allowing the treatment of non-homogeneous, non-stationary conditions which characterize both short-range dispersion in complex terrain and long-range transport of pollutants. Physical and numerical problems related to the application of the puff model are discussed, and a general computer-oriented methodology proposed. This method provides a more numerically-accurate representation of plume characteristics and can handle both transport and calm conditions.

INTRODUCTION

Regulatory problems, especially in the U.S.A., have strongly affected air pollution dispersion studies. Recently, in fact, air quality modeling has been elevated in the U.S.A. to the quantitative decision-controlling level due, essentially, to the Clean Air Act Amendments of 1970 and 1977, and to the subsequent action of the U.S. EPA (U.S. EPA, 1978). Not only are mathematical models used for obtaining emission permits for existing sources, where field experiments allow some calibration and verification of the models, but due to a relatively new doctrine -- the Prevention of Signifi cant Deterioration (PSD) -- air quality models are also one of the most important tools in industrial development and land use planning, especially in the Western United States.

Unfortunately, the accuracy of the applied modeling technique has not always been proportional to the importance of the regulatory problem. Oftentimes, with the justification that such methods provide "conservative" estimates, the physical assumptions used have been too simple for the simulation of very complex phenomena. More important, only recently has any attention been given to the problem of statistical evaluation of air quality model performance (Zannetti and Switzer, 1979; Hillyer et al., 1979; Bencala and Seinfeld, 1979).

Historically, the urgent need for simple, clear, and effective numerical methods for computing the environmental impact of an emission source has had the positive effect of boosting research and investigation in this field, but had the negative effect of forcing available methods to work beyond their physical and numerical limits.

A typical example is the extensive use of the Gaussian steady-state dispersion formula which has become -- and probably will remain for long time -- an essential regulatory tool for air pollution problems in the U.S.A. Such a methodology gives a plume formula whose validity requires the main assumptions of 1) spatial homogeneity, 2) stationary conditions, and 3) flat terrain. However, since many dispersion cases, especially in complex terrain or for intermediate ranges, do not follow these assumptions, much research has been and will be devoted to both the determination of empirical correction factors in the Gaussian approach and to the development of new algorithms for regulatory purposes.

In the past, the more complex, time-varying applications of air quality simulation modeling have made extensive use of dynamic grid model techniques (mainly finite-difference simulations following the K-theory approach). Recently, however, a growing concern has arisen regarding some important limitations of such a numerical approach and, precisely, of 1) its numerical advection errors, 2) its often incorrect representation of plume growth, and 3) the difficulty of relating the diffusion coefficient K to standard atmospheric measurements.

In addition, the whole Eulerian approach used in grid model simulations has been criticized, since a growing number of scientists agree in considering air pollution dispersion a Lagrangian phenomenon, to be simulated by specific source-receptor numerical methods, without the computation of spatially-averaged grid concentrations.

For the above reasons, some modelers have recently attempted to 1) develop new transport and diffusion techniques for the more complex applications (e.g., the particle-in-cell method;

Sklarew et al., 1971; Lange, 1978), and 2) extend the applicability of the Gaussian method by breaking up the plume into a series of independent elements, segments (Chan and Tombach, 1978), or puffs (Lamb, 1969) to treat, in particular, non-stationary, non-homogeneous dispersion conditions.

The following sections of this paper discuss the physical and numerical problems relating to the utilization of the puff method and define a new puff algorithm for both an accurate and computationally efficient simulation of plume characteristics by a series of puffs.

THE PUFF APPROACH

Puff Dispersion

In past dispersion studies (e.g., see Pasquill, 1974), puffs and plumes have shown different dispersion characteristics. In fact, while the spread of a plume (measured by the horizontal and vertical standard deviations of its spatial concentration distribution) tends initially to grow proportionally to the downwind distance and, ultimately, proportionally to the square root of such a distance, puff dispersion characteristics have shown different behaviors. In particular, Gifford (1957), in examining smoke puff data, found the existence of two predicted (Batchelor, 1952) growing regimes for the puff sigma, with an initial proportionality to the downwind distance, followed by a sigma growth proportional to the 3/2 power of such a distance for intermediate dispersion.

Puff Modeling

In spite of the above considerations, most of the practical applications of puff modeling algorithms have avoided the treatment of real puff dispersion characteristics. In fact, it has been found that a Gaussian plume can be represented by a series of "equivalent" (or "fictitious") puffs if 1) the puff sigmas grow in the same way as the plume sigmas, and 2) the distance between two contiguous puffs is sufficiently small. Then, the existence in the computational domain of many puffs, instead of a stationary plume, allows some degree of representation of non-stationary, non-homogeneous emission and dispersion conditions, otherwise impossible to handle with standard Gaussian formulae.

However, this approach must not be seen as a perfect representation of the dispersion phenomenon. In particular, if the utilization of a plume sigma function for puff modeling is correct in nearly-stationary conditions, very little is known of its use during fully non-stationary situations and, especially, in nearly-calm conditions, which are often associated with air pollution episodes.

In conclusion, the common "puff model" must be considered only a step ahead in the application of the Gaussian semi-empirical approach. However, the method is potentially very powerful and future data on tracer dispersion experiments are expected to strongly improve the applicability of the entire methodology.

The Basic Puff Algorithm

The primary computational algorithm of the puff model consist of five computational substeps at each time step Δt.

1. Emission. One puff is generated at each emission exit point (plus, the eventual additional elevation due to the plume rise contribution).

2. Advection. The center of each existing puff is moved according to the local wind.

3. Diffusion. All puffs sigmas are increased according to the local turbulence state.

4. Deposition, Precipitation Scavenging, and Chemical Decay. The mass of each puff is reduced (e.g., by exponential reduction) to take these effects into account.

5. Contribution to Receptors. The contribution to each receptor is computed by summing all single contributions of the existing puffs.

Figure 1 shows an example of puff dispersion simulation (on a horizontal plane) from two point sources, where three-dimensional, non-stationary, non-homogeneous meteorological conditions can be properly taken into account by the method.

Recent improvements of the puff model have been developed for taking into account wind shear effect, dynamic plume rise (Sheih, 1978), and determining the correct time increment Δt for puff advection and diffusion computations (Ludwig et al., 1977). The latter problem is especially important to maintain an accurate plume representation without creating serious problems of computer storage and CPU running time, caused by a too small Δt. The solutions proposed for this problem consist of a puff-merging algorithm (Ludwig et al., 1977) and the incorporation of the mean-wind advection into the streamwise dispersion coefficient (Sheih, 1978). However, these methods are practical for use only in nearly steady-state conditions, since it can be easily proved that fully non-stationary simulations can require a computationally prohibitive Δt of a few seconds to avoid concentration overestimation and/or underestimation due to sudden changes of wind direction. In fact, the basic condition for a perfect representation of a Gaussian

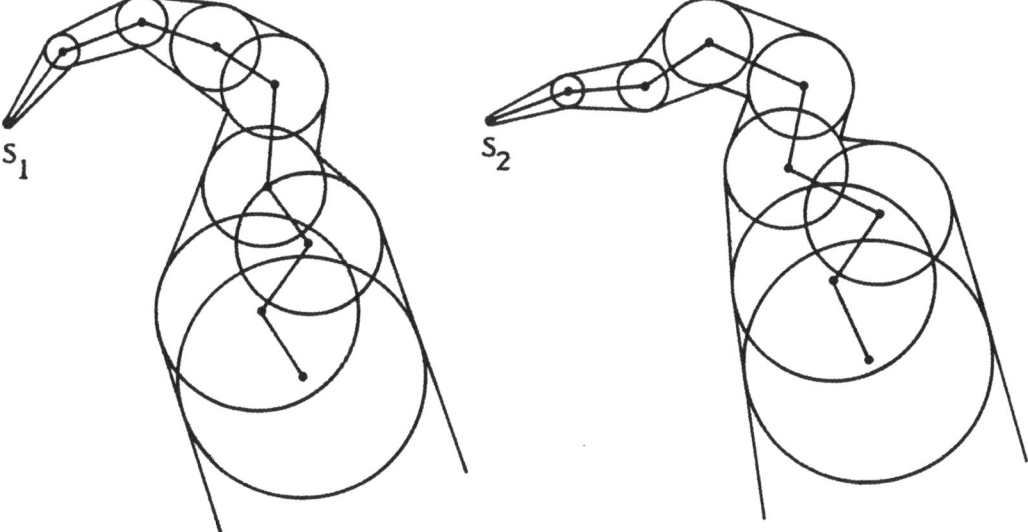

Fig. 1.   Example of puff model dispersion computation (two point
sources) in non-stationary, non-homogeneous meteorological conditions.

plume, by a series of independent puffs, requires the distance
between the centers of two contiguous puffs to be less than their
streamwise standard deviation of the concentration distribution
(Ludwig et al., 1977); a condition that requires a too small Δt for
concentration computations at receptors not too far from the source.

THE AVPPM$^{(*)}$ METHOD

A new method has been proposed (Zannetti, in press) for a more
accurate plume simulation by a series of puffs. This method handles
general non-stationary conditions without creating serious computer
storage and CPU time-consumption, as mentioned in the previous
section. Moreover, a general algorithm has been incorporated for
treating calm conditions.

The entire methodology has been codified into a computer
package (AVPPM/2A) which, if a suitable meteorological input is
available, seems particularly appropriate for the numerical simulation of non-homogeneous dispersion conditions (as in complex
terrain) or non-stationary phenomena (as in intermediate or long-
range dispersion simulations).

Representation and Handling of Puffs

The key point in the AVPPM algorithm is the selection of a Δt
large enough (e.g., 5 to 10 minutes) to avoid serious computer
storage and CPU time-consumption problems. However, such a large
Δt provides a poor resolution for the plume description, since, the
basic puff condition (separation between contiguous puffs less than
their streamwise sigmas) is met only very far downwind of the source.
In other words, the puffs give only the geometry of the plume's
evolution and the puffs' masses cannot be used directly for computing the concentration at the receptor points.

Figure 2 shows a non-stationary plume represented by puffs at
time intervals t and t + Δt. If a "chain" is kept between the
puffs, their ages and source points at each time interval
[t, t + Δt], then the area affected by the entire plume (see
Figure 2) can be represented by a certain number of quadrilaterals
with vertices in the puffs' centers at the beginning and end of the
time interval. Then, if a receptor is sufficiently close to a
quadrilateral (e.g., less than 4 or 5 puffs' horizontal sigmas), a
splitting technique is applied to generate enough puffs in the quadrilateral to meet the basic puff condition discussed above.

$^{(*)}$AeroVironment Puff Plume Model

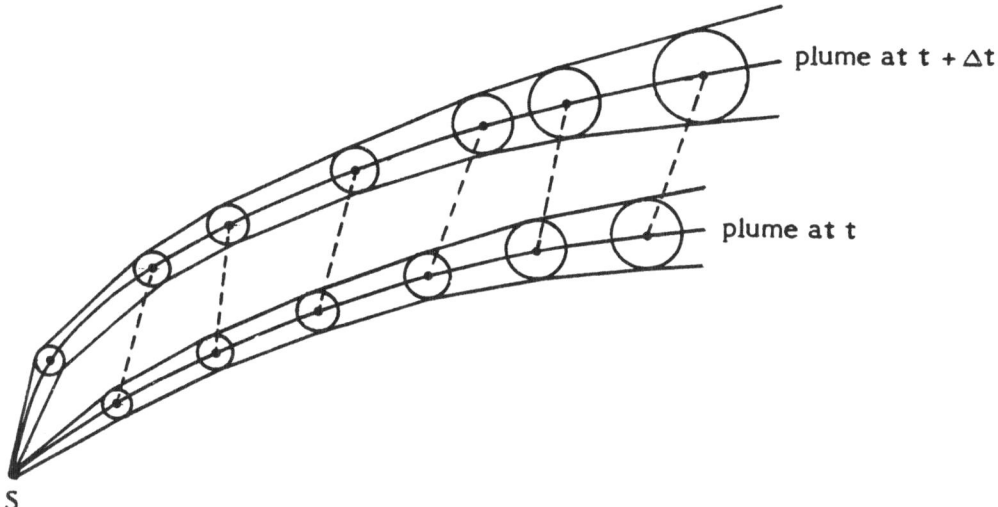

Fig. 2.   Plume horizontal representation by a series of puffs at
           different time steps during non-stationary, non-homo-
           geneous conditions.   The mass of each puff, during the
           time interval [t, t + Δt], has actually affected the area
           of the corresponding quadrilateral.

It must be noted that this splitting technique will be applied only when required for a particular receptor.· In this way, we solve our computer storage and time problems without losing the accuracy of the computation.

Treatment of Calm Conditions

The AVPPM methodology assumes that, in the study area, wind speeds greater than u_{min} define transport conditions, while wind speeds less than u_{min} are associated with calm conditions, where u_{min} must be determined from experimental data (a first guess can be 1 m/s).

We can then assume, in transport conditions, for example, the common power law:

$$\sigma = a \ d^b \ , \tag{1}$$

where the sigma evolution of each puff is a function of the downwind distance d, and the parameters a and b are dependent upon the turbulence state.

In calm conditions, statistical considerations of the turbulence characteristics (Pasquill, 1974) allow the description of a similar sigma growing law

$$\sigma = a' \ (t - t_0)^{b'} \ , \tag{2}$$

where $t - t_0$ is the age of the puff since its generation at t_0.

Then, for wind speed u equal to u_{min}, Equations (1) and (2), using $d = u \ (t - t_0)$, give

$$a' = a \ u_{min}^b, \qquad b' = b \ . \tag{3}$$

The AVPPM algorithm in the puff diffusion step uses Equation (1) for transport conditions and Equation (2) for calm conditions, where a' and b' are given by Equation (3).

The above methodology is not restricted to the sigma function of Equation (1), since similar considerations can be developed for virtually any sigma function. However, the above scheme should provide a first approximation algorithm for a consistent treatment of calm conditions, where future tracer experiments are expected, in particular, to supply a better estimation of a' and b'.

Virtual Distances and Ages

At each time step, diffusion is taken into account by increasing the sigmas of each puff. In transport conditions, for example, from Equation (1) we have

$$\Delta\sigma = \frac{\partial\sigma}{\partial d}\Delta d = abd^{b-1}\Delta d \; , \tag{4}$$

where Δd is the distance traveled by the puff during Δt.

The dependence of $\Delta\sigma$ on the total downwind traveling distance d requires particular care and special consideration. It is easy to see, in fact, that the real downwind distance d is not generally appropriate for this computation. Figure 3 shows the time evolution of a single puff where a change in the turbulence state at time t (in our case, an increase of turbulence) changes the growing rate of the puff's sigmas. It is clear that, after time t, the puff growth is affected not by the total actual downwind distance d, but by the "virtual" downwind distance d_v from the virtual source S'. This virtual distance d_v is defined as the downwind distance that the same puff, to have the same sigmas, would have had to travel from the source if the turbulence state had always been the most recent one during its entire life.

More generally, with different horizontal and vertical turbulence characterizations, virtual distances of a horizontal d_h and vertical d_z, should be defined. Moreover, when the puff grows according to its age (in calm conditions), similar considerations are brought to the definition of a virtual age t_v or, more generally, of a horizontal t_h and vertical t_z virtual ages of the puff.

According to the above considerations, the increase of the puff's sigmas at time t, assuming Equations (1) and (2), is computed using

$$d_v = \left[\sigma(t)/a*\right]^{1/b*} \; , \tag{5}$$

and

$$t_v = \left[\sigma(t)/a'*\right]^{1/b'*} \; , \tag{6}$$

where a* and b* are the coefficients of Equation (1) for the current turbulent state j* (at time t) computed at the puff's center, and a'*, b'* are the analogous coefficients of Equation (2).

As a final important remark on this method, it must be pointed out that the sigma power law of Equation (1) is not essential for the algorithm. Any general formula

$$\sigma = f\left[d\right] \tag{7}$$

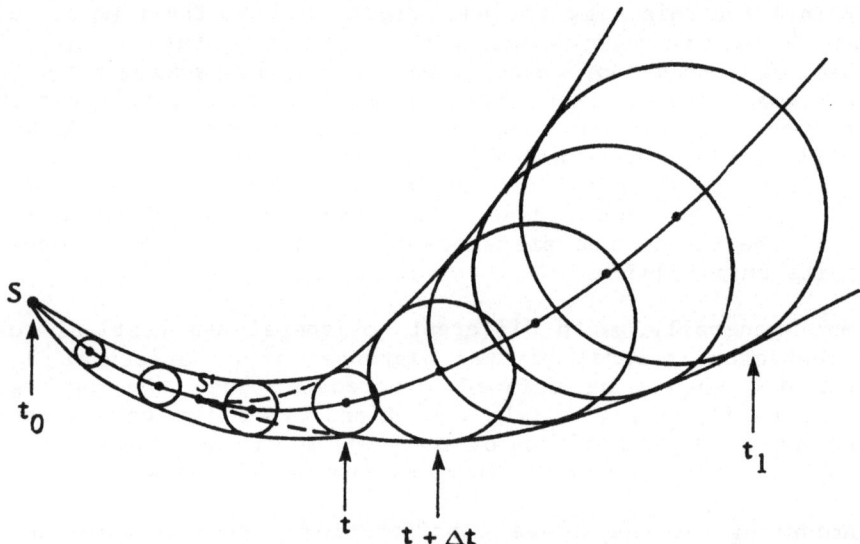

Fig. 3. Evolution of one puff during the time interval (t_0, t_1).
Turbulence is low during (t_0, t) and increases during
(t, t_1), then requiring the computation of the virtual distance of the puff from the virtual source S'.

can be used, then requiring

$$d_v = f^{-1}[\sigma(t)],\qquad(8)$$

for the computation of the virtual distance.

However, with formulae more complex than Equation (1), the computation of Equation (8) is not straightforward as in Equation (5) and, consequently, an iterative algorithm may be required. Analogous considerations hold for the virtual age.

CONCLUSIONS

A new puff dispersion algorithm for plume representation has been discussed. The utilization of this methodology should increase the applicability of the Gaussian method to treat both non-stationary, non-homogeneous conditions, and calm conditions.

The puff method discussed above has been codified into a computer package (AVPPM/2A) which, with suitable input data on wind, turbulence, and emissions, can simulate dispersion experiments using various options on sigma parameters, plume rise functions, ground and inversion reflection parameters, etc. Moreover, for complex terrain simulations, the elevation above the ground of the center of each puff is forced, from one step to the next, to remain within a pre-fixed range to better take into account terrain irregularities.

The code has been successfully compared against steady-state Gaussian computations. As an example, simulations that required Δt of the order of 10 seconds, according to the basic puff condition, have been correctly performed with a Δt of one order of magnitude greater.

However, simulation runs have pointed out the importance of using an accurate splitting technique. In fact, in some cases, the generation of artificial puffs, by a linear interpolation of the characteristics of the four puffs at the vertices of the quadri-lateral, does not provide sufficient accuracy.

Numerical tests using a correct interpolation technique have shown that a more accurate computation is particularly important for splitting inside the first quadrilateral (the one which contains the source), where the highest error is produced by simple linear inter-polation. However, if some attention is paid to the choice of Δt, the computationally faster linear interpolation can be successfully used.

Experimental data are expected soon which will validate the entire methodology, especially the semi-empirical method which has

been proposed for the treatment of calm conditions. Data are also
expected to suggest concentration distributions different from the
Gaussian one and easily incorporable into the method.

Acknowledgements

This study has been partially supported by the Western Oil and
Gas Association (WOGA).

REFERENCES

Batchelor, G.K., 1952, Diffusion in a field of homogeneous turbu-
lence, II. The relative motion of particles, Proc. Camb. Phil.
Soc., 48.
Bencala, K.E., and Seinfield, J.H., 1979, An air quality model
performance assessment package, Atmos. Environ., 13:1181.
Chan, M.W., and Tombach, I.H., 1978, "AVACTA -- air pollution model
for complex terrain applications," AeroVironment Inc. report
AV-M-8213, Pasadena, CA.
Gifford, F.A., 1957, Relative diffusion of smoke puffs, J. Met.,
14:410.
Hillyer, M.J., Reynolds, S.D., and Roth, P.M., 1979, "Procedures for
evaluating the performance of air quality simulation models,"
U.S. EPA 450/4-79-033.
Lamb, R.G., 1969, "An air pollution model of Los Angeles," Master's
Thesis, University of California at Los Angeles.
Lange, R., 1978, ADPIC a three-dimensional particle-in-cell model
for the dispersal of atmospheric pollutants and its comparison
to regional tracer studies, J. Appl. Met., 17:320.
Ludwig, F.L., Gasiorek, L.S., and Ruff, R.E., 1977, Simplification
of a Gaussian puff model for real-time minicomputer use, Atmos.
Environ., 11:431.
Pasquill, F., 1974, "Atmospheric Diffusion," Ellis Horwood Limited,
England.
Sheih, C.M., 1978, A puff pollutant dispersion model with wind shear
and dynamic plume rise, Atmos. Environ., 12:1933.
Sklarew, R.C., Fabrik, A.J., and Prager, J.E., 1971, "A particle-in-
cell method for numerical solution of the atmospheric diffusion
equation, and application to air pollution problems," Systems,
Science and Software, Inc. Report 3SR-844, LaJolla, CA.
U.S. Environmental Protection Agency, 1978, "Guideline on air
quality models," EPA-450/2-78-025, OAQPS No. 1.2-080.
Zannetti, P., in press, An improved puff algorithm for plume disper-
sion simulation, submitted to an international journal.
Zannetti, P. and Switzer, P., 1979, "Some problems of validation and
testing of numerical air pollution models," presented at the
Fourth Symposium on Turbulence, Diffusion, and Air Pollution,
January 15-18, Reno, NV.

DISCUSSION

F. AXENFELD   Your are using the Gaussian model for
an instantaneous release. What about your σ_x?
You seem to take $\sigma_x = \sigma_y$, but in most experiments
only σ_y and σ_z are determined. How do you justify
using $\sigma_x = \sigma_y$?

P. ZANNETTI   The AVPPM package can take into account
$\sigma_x \neq \sigma_y$ if this information is available. If the
experiment provides only σ_y and σ_z, probably the
assumption $\sigma_x = \sigma_y$ is the most reasonable one.

IMPACT STUDY IN COMPLEX TERRAIN

K.E. Grønskei and B. Sivertsen

Norwegian Institute for Air Research (NILU)
P.O. Box 130, N-2001 Lillestrøm
Norway

ABSTRACT

Tracer experiments, using sulphur hexafluorid, SF_6, were carried out to study the transport and diffusion of gases released from an aluminium hall in a complex valley-fjord area of western Norway. Data from these experiments were applied to study : 1) wake effects behind the aluminium hall; 2) transport and dispersion on scale of a few kilometers and 3) concentration pattern. Continuous records of meteorological parameters were used to ascertain the statistical significance of a limited number of tracer experiments. Additional radiosonde and pilot balloon data were collected during the tracer experiments to describe the 3-dimensional variation of wind and temperature.

The "local similarity" treatment, as proposed by Pasquill, and based upon wind and temperature profile data, together with cross-wind fluctuations (σ_β) were used to describe vertical and horizontal dispersion. Estimated concentrations were compared with measured SF_6 concentrations and reasonable correspondance was found for the 15-minutes average concentrations. The effect on 24-hr average concentrations of variable wind and diffusion conditions is discussed and the frequency distribution of these values in the area is adjusted by using observed SO_2 concentrations from one location.

1   INTRODUCTION

      Serious air pollution problems may occur in complex terrains.
Dispersion models for general application in such areas terrain are
not available. Impact studies therefore must be based upon relevant
measurements in the actual area. For use in Norway existing dispersion models often has to be modified for assessing impact in
terrains where their applicability may be questioned. Comprehensive
investigations to improve the description of dispersion in complex
terrain have previously been undertaken in Norway (1), (2). Recently
better and more flexible tracer techniques have been introduced (3).
Simplified models to describe the main characteristics of the
atmospheric turbulence and to evaluate dispersion parameters from
measured meteorological parameters have been proposed (4), (5).

      Proposed modernization, rebuilding and expansion of the al-
uminium smelter in Høyanger required an impact study. This study
made it possible to test, evaluate and apply the new tracer and
modelling techniques in extremely complex and mountainous terrain.

2   TOPOGRAPHY AND CLIMATOLOGY

      The aluminium smelter at Høyanger is located at the end of
Høyangerfjord, a deep northerly branch of Sognefjord.   The mountains surrounding the fjord valley system rise up to 1000 m, slop-
ing on the average 45 deg, as shown in Figure 1.

      The wind conditions in the area are dominated by channeling
of wind along the fjord-valley axis, and are further characterized
by strong diurnal variations due to land/sea breeze and mountain/
valley wind patterns. These patterns are especially dominant
during the summer season, as shown in Figure 2.

      These complex topographical and meteorological conditions
frequently lead to complicated transport and dispersion patterns,
not adequately handled by existing dispersion models.

      During the winter season pollutants emitted from the smelter
are brought out over the fjord 60-70% of the time. On the other
hand, the winds are towards land 80-90% of the time during day
time in summer. From an environmental impact point of view, the
summer situation is the more serious one and was selected for a
closer study.

      Three meteorological stations were set up to continuously
record the winds and stability of the area for one year. The data
obtained were also applied to clarify the statistical significance
of a limited number of tracer experiments carried out. Addition-

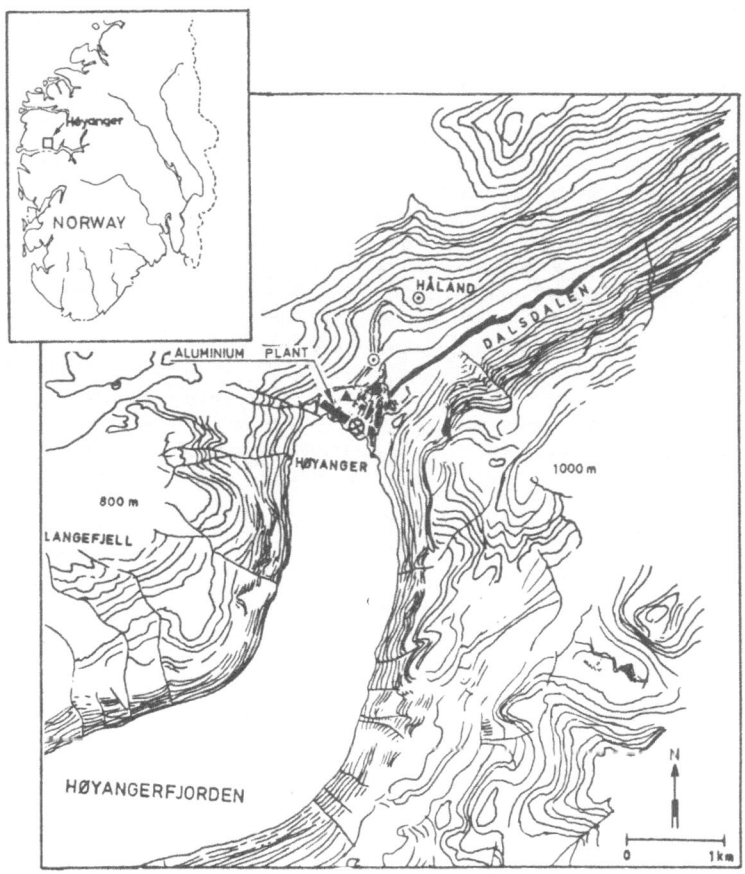

Figure 1: Topographical map of the Høyanger area.

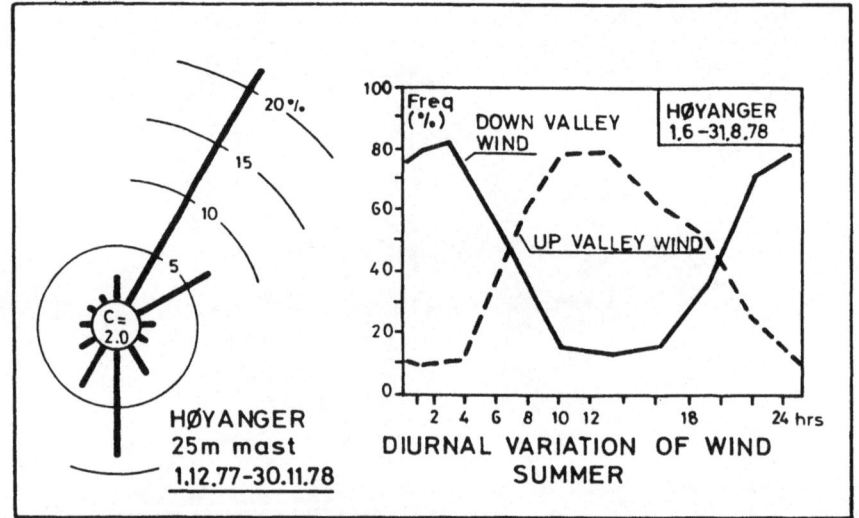

Figure 2: *Annual wind frequency distribution (windrose), and the diurnal variation of up- and down-valley wind during summer in Høyanger.*

ally radiosonde and pilot balloon data were collected during the experiment periods. The location of the stations and experimental set up are shown in Figure 3.

3   TRACER EXPERIMENTS

Twelve tracer experiments were carried out to study the transport and diffusion of gases released from the aluminium smelter hall. The results were applied to:

- study wake effects behind the smelter hall
- establish concentration patterns within the nearest few kilometers
- quantify land/seabreeze impact
- obtain data for model validation
- estimate future impact due to proposed changes in emissions

Sulphur hexafluorid (SF_6) was used as a tracer. SF_6 is an inert nontoxic, colourless, odourless gas which can be detected at concentrations as low as 10^{-12} p SF_6/p air by electron capture gas chromatography (3). Background levels of SF_6 in Scandinavia are less than 10^{-12} p SF_6/p air (6).

3.1 Experimental procedures

The methods employed included the continous release of a known amount of SF_6 simultaneously from 3 points on the roof of the aluminium hall, to simulate the emissions from the roof top vents. The representativeness of using only 3 release points had been established previously. In some tests SF_6 was released at one elevated point from a tube held aloft by a helium filled balloon, to simulate a stack release. The approximate locations of the release points are shown in Figure 3. Fifteen-minute averaged, and instantaneous air samples were collected in 20 cm^3 disposable plastic syringes. Data from Lamb (3) indicate that samples collected in the syringes do not change in concentration more than 5% over approximately one to two weeks. In this study, sample analyses were completed within a few hours of collection.

3.2 Meteorological conditions during the tests

Eight tracer experiments were carried out during 23-26 May 1978, and four experiments during 13-15 June 1978. A typical diurnal pattern of the up- and down-valley wind component, shown in Figure 4, indicates that the down-valley night time flow is shallow (~ 100 m in depth) while the day time seabreeze up-valley wind region extends up to at least 600 m.

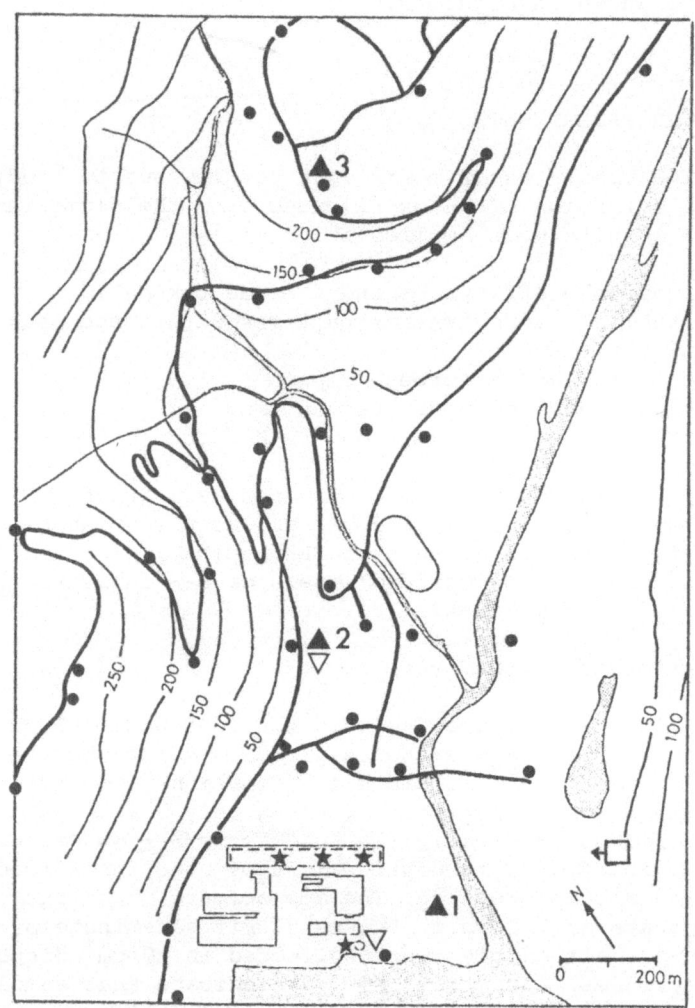

Figure 3: Location of releases and sampling stations.
▲1 : *Automatic weather station with 25 m mast*
▲2 : *Mechanical wind station, 10 m mast*
▲3 : *Temperature measurements at 2 m (220 masl)*
● : *15 min average SF$_6$ samplers*
★ : *SF$_6$-release point*
◄□ : *Time lapse, photos*
▽ : *Radiosondes, pilot balloons*

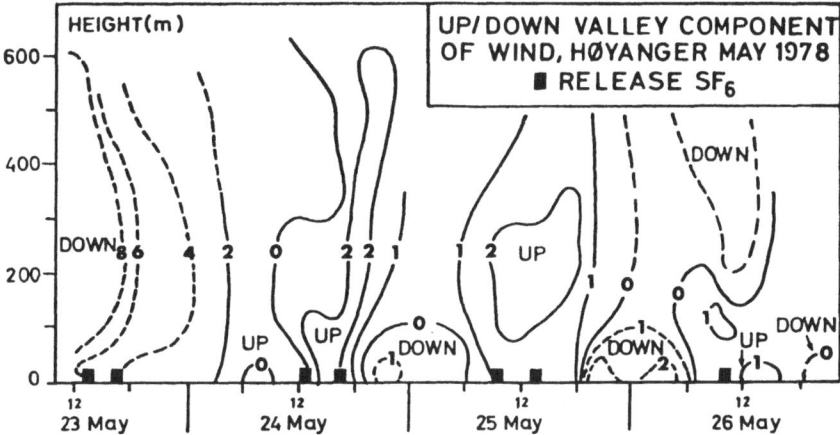

Figure 4: *The up- and down-valley wind component in Hoyanger during*
          23 May – 26 May 1978.

Superimposed on this mesoscale wind pattern are smaller scale
upslope/downslope winds due to hot air buoyancy or cold air drainage. In the morning hours with the sun heating the western slopes
of the valley and the eastern side still in shadow, the temperature
difference, as shown in Figure 5, between the two sides of the
valley (only 1-2 km apart) might be more than $5^{\circ}C$ and cause an
upslope wind.

3.3 Concentration patterns

A typical concentration pattern measured in the early morning
hours is shown in Figure 6. The upslope wind on the heated side entrained the pollutants from the smelter, and carried them to higher
elevations along the slope, where they leveled off and followed the
main valley axis at about 200-250 masl across Håland and into the
main valley of Dalsdalen.

Figure 7 presents a somewhat different pattern typical for the
midday experiments. The local thermal effect on the wind pattern is
less dominant than during the morning hours.

The vertical cross section of SF_6 concentrations ~ 400 m downwind from the source shows, however, that parts of the cloud were
still brought up to ~ 200 masl before transported into the main
valley.

SF_6 was released in the early morning landbreeze on 15 June
1978 to study the return flow of pollutants after the on set of
the seabreeze. The SF_6 release from the aluminium hall was stopped
when the landbreeze shifted to seabreeze. At this moment stratified
plumes of pollutants were visually observed over the fjord. 15-min
avearge SF_6 concentrations measured at Håland (Station 3 in
Figure 3), 220 masl and about 1 km north-northeast of the release,
are plotted in Figure 8. SF_6 was detected at Håland 40 minutes
after the wind shift (release cut off) and remained at ~ 0.5 $\mu g/m^3$
concentration for about 2 hrs.

3.4 Estimation of SO_2 and fluoride concentrations using a simple
proportionality model

Assuming that the releases of SF_6 reasonably well simulate
the emissions of pollutants, a simple proportionality concept may
be applied to estimate the concentration, C_p, of a given pollutants when its emission rate, Q_p, is known:

$$C_p = C_{SF_6} \cdot (Q_p/Q_{SF_6}) \tag{1}$$

Strictly speaking Equation (1) is valid only when the pollutant and SF_6 are dispersed in the same way and when no deposition

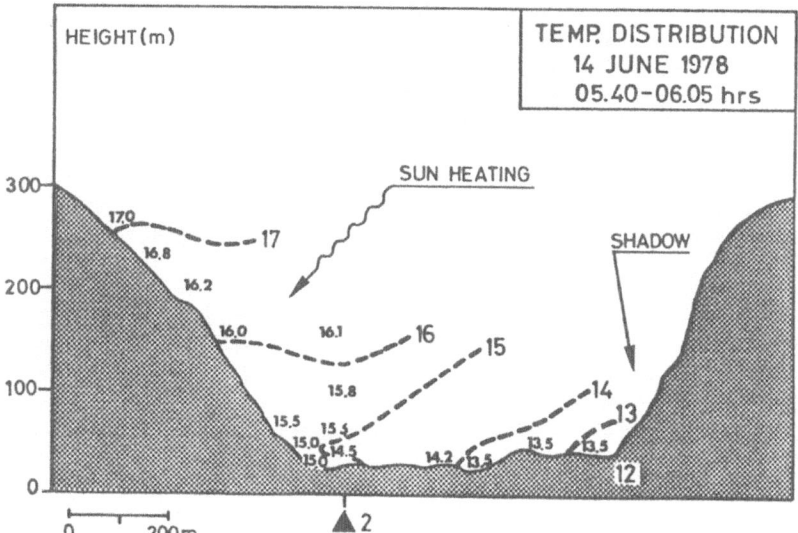

Figure 5: Temperatures (°C) through a cross section at station 2 of the Høyanger valley in the morning of 14 June 1978 at 0540-0605 hrs.

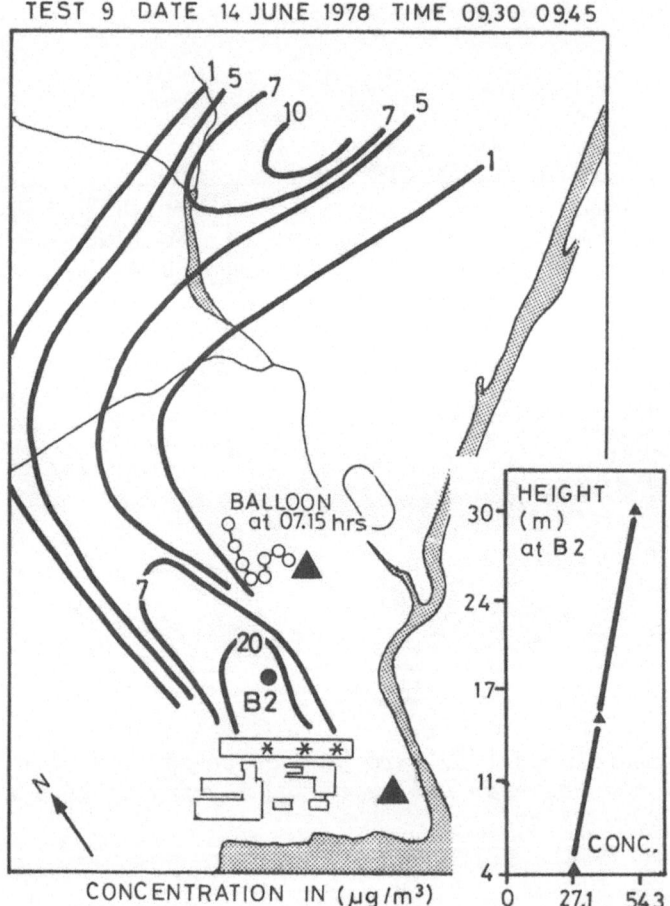

Figure 6: SF₆-concentration distrubution (μg/m³) at 0930-0945 hrs on June 1978. Releases from 3 points at the smelter hall roof

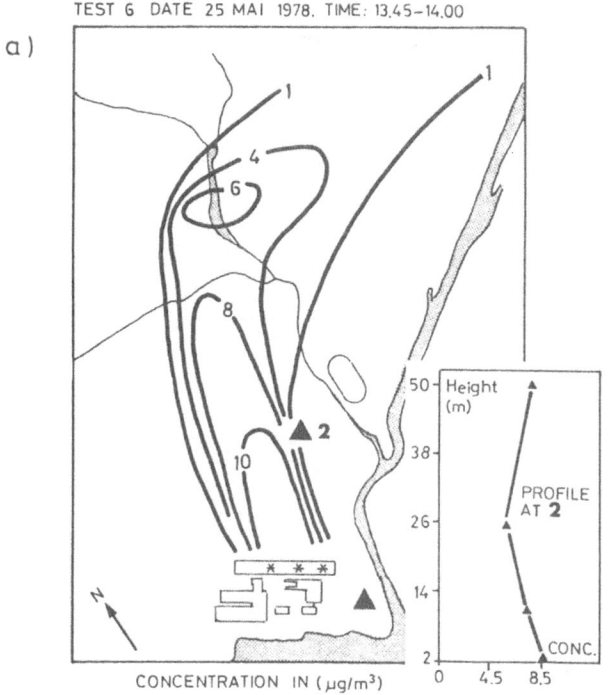

TEST 6 DATE 25 MAI 1978. TIME: 13.45–14.00

a)

CONCENTRATION IN (μg/m³)

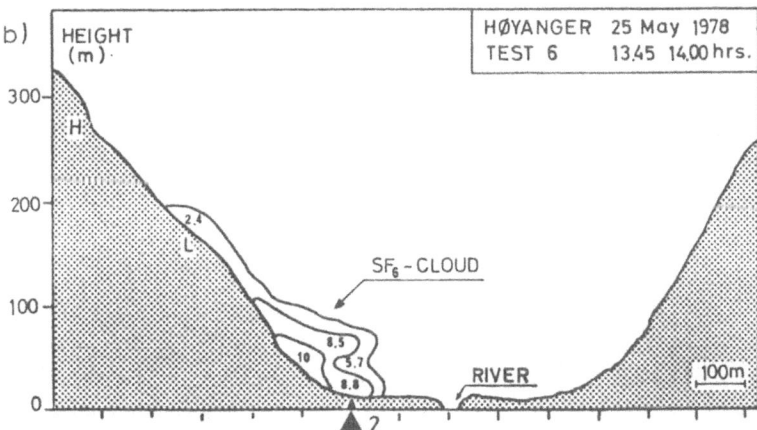

Figure 7: SF₆ concentrations (μg/m³) at 1345–1400 hrs on 25 May 1978
a) ground level concentrations and vertical profile at 2
b) vertical cross-section through 2.

or reactions are taking place during transport. The transport times
involved in the experiments in Høyanger were so short that these
assumptions should apply for most cases.

Only in the land/seabreeze experiments the concentrations of
SO_2 and fluorides might have been overestimated, since dry deposition of SF_6 is smaller than for other substances.

Figure 9 illustrates how releases of fluorides from the
existing hall, from a hypothetical 70 m high stack and from the
seabreeze return flow may be combined to estimate the morning hour
concentration pattern of fluorides. Similar estimates of maximum
ground level concentrations of fluorides and SO_2 were carried out
for 6 alternative release rates including the actual 1975-76
releases.

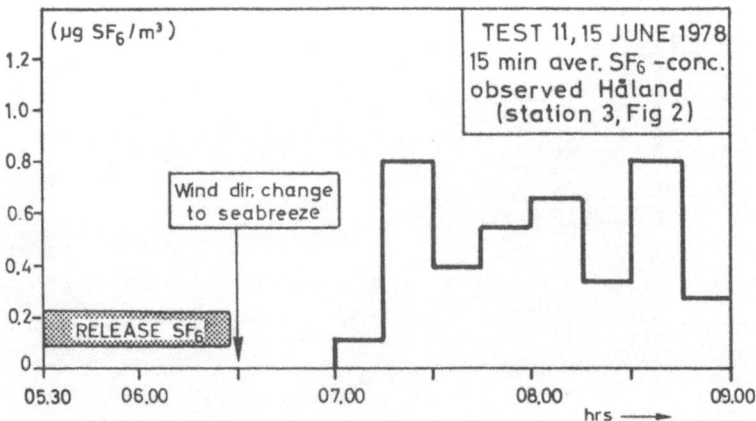

*Figure 8: SF_6-concentrations measured at Håland (station 3, Figure 2)
during a shift from land breeze to sea breeze in the Høyanger
area. SF_6 releases were stopped at the time of wind direction
shift.*

Table 1 summarizes the maximum concentrations estimated for the 12 diffusion experiments.

Table 1: Estimated maximum ground level concentrations of total fluorides ($C_F(max)$) and SO_2 ($C_{SO_2}(max)$) from observed maximum SF_6 concentrations.

Test no	Release	$C_{SF_6}(max)$ μg/m³	$C_F(max)$ (μg/m³) Emission alternatives Fluorides						$C_{SO_2}(max)$ (μg/m³) Emission alternatives SO_2					
			1975-76	1	2	3	4	5	1975-76	1	2	3	4	5
1	Hall C	14.4	73	22	33	50	71	19	63	166	213	238	229	118
4	Hall C	5.1	20	6	9	14	19	5	17	45	59	64	62	32
5	Hall C	7.6	30	9	14	21	29	8	26	68	87	97	93	49
6	Hall C	10.0	27	8	8	19	26	7	23	61	79	87	85	44
7	Hall C	2.0	8	2	3	5	8	2	7	19	22	25	24	12
8	Hall C	2.8	11	3	5	8	11	3	10	25	32	35	34	13
9	Hall C	27.3	68	22	30	46	66	19	60	156	199	224	216	112
10	40 m chimney	31.4	-	28	38	44	28	57	-	521	521	521	559	593
11	In the land-breeze	2.4	9	3	4	6	9	2	8	21	27	30	29	15
12	70 m chimney	5.9	-	5	7	8	5	11	-	98	98	98	105	112

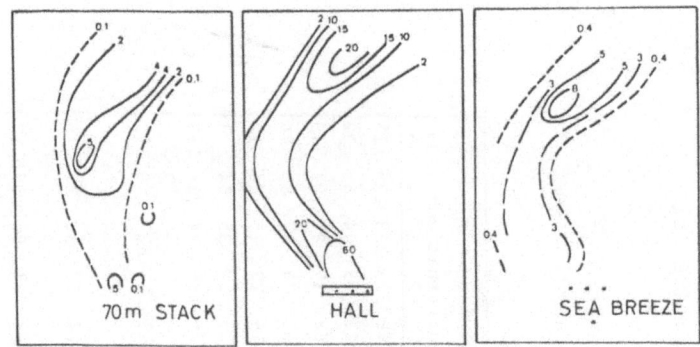

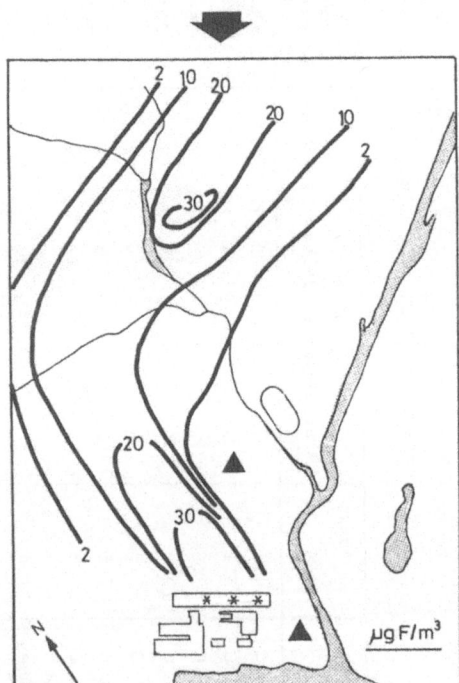

Figure 9: *Fluoride concentrations (15 min average) estimated from measured SF_6 concentrations, assuming releases from the smelter hall, from a stack and contributions from the sea breeze return flow.*

4   EVALUATION OF METEOROLOGICAL DATA

To evaluate the dynamics of the mean airflow close to the
ground, the following "forces" are important:

1. The horizontal stress (τ) has to be known for the horizontal
   equation of motion. $\tau/\rho \overset{def}{=} u_*^2$, where ρ denotes the density
   of air, and u_*^2 is the time average of the vertical
   exchange of momentum.

2. In a stratified atmosphere a buoyancy force influences the
   vertical motion. It may further be shown that this force is
   proportional to the Vaisäla-Brundt frequency (ν_s) for small
   perturbations:

$$\nu_s^2 = \frac{g}{T}\frac{\delta\theta}{\delta z}$$

g : the acceleration of gravity
T : the temperature

$\frac{\delta\theta}{\delta z}$: vertical variation of potential temperature

3. For the thermodynamic energy equation, the vertical flux of
   heat ($\overline{w'\theta'}$) has to be known:

$$\overline{w'\theta'} = u_* T_*$$

In the complex terrain of Høyanger other "forces" (than
listed above) may be important for the air stream e.q. the temperature stratification shown in Figure 5 indicates a local horizontal pressure gradient of the same order of magnitude like the
horizontal stress term. Nevertheless as a hypothesis the similarity
theory was used to describe turbulent exchange close to the ground.
A closure of the equations would then be obtained.

Monin-Obukhov's similarity theory assumes that the vertical
fluxes of heat and momentum, together with the height z and the
Vaisäla-Brunt frequency ν_s, constitute the important parameters for
the mean air stream.

A length parameter, L, and a temperature parameter θ_* are
defined:

$$L \overset{def}{=} -\frac{u_*^3 T}{\kappa g\, u_* \theta_*} , \text{ and } \theta_* \overset{def}{=} \frac{\overline{w'\theta'}}{u_*} \qquad (3)(4)$$

Two non-dimensional products may then be defined, and the π
theorem used to describe the vertical gradients in wind speed, $\frac{du}{dz}$
and potential temperature, $\frac{d\theta}{dz}$:

$$\frac{\kappa z}{u_*} \frac{du}{dz} = \phi_m\left(\frac{z}{L}\right), \quad \frac{\kappa z}{\theta_*} \frac{d\theta}{dz} = \phi_h\left(\frac{z}{L}\right) \qquad (5)\ (6)$$

It has been proposed (5) to use Businger's empirically determined universal functions ϕ_m and ϕ_h (7).

The integrated equations are used to estimate parameter describing the turbulent exchange u_*, θ_* and L. The equations were solved by iteration using as input wind and temperature observations from two levels (25 m and ground level). An average empirical value was used for the roughness length (z_o = 40 cm).

A more complete description of the equations may be found in (8). If this simplified description of turbulence may be used to estimate dispersion of pollution in the area, the results will indicate that the description of turbulence may be applied as a closure scheem in the equation describing the dynamics of the air stream.

4.1 Vertical dispersion

According to Hanna (9) there is a close relation between turbulent exchange of heat (K_h) and turbulent exchange of pollution (K). The following formulae was used:

$$K \approx K_h = 0.35\ u_* z/\phi_h\ (z/L) \qquad (7)$$

A finite difference approach to estimate vertical exchange, based on K-theory, requires numerical simulation in many atmo-spheric layers. To avoid the effort and cost of detailed numerical solutions, Pasquill (4) has proposed a "local similarity" treatment of vertical spread of pollutants from a ground source.

The basic principle of the method is to express the rate of dispersion, $\frac{d\bar{z}}{dt}$ (where $\bar{z}$ is the mean displacement of an ensemble of particles after a given travel time), in terms of basic parameters of the turbulent boundary layer. The assumption is made that the increase in vertical spread, $\bar{z}$, or alternatively z_m (the extreme vertical displacement), is always determined by to local properties:

a) σ_w : turbulence intensity, expressed as the standard deviation of vertical wind velocity,

b) $\ell(z)$: scale of turbulence prevailing at $\bar{z}$ or z_m.

If it is further assumed that $\ell(z)$ is proportional to the spectral scale, λm, dimensional analysis gives:

$$\frac{d\bar{z}}{dt} = \sigma_w \; f_1 \; (\frac{\lambda_m}{z}) \bar{z} \tag{8}$$

$$\frac{dz_m}{dt} = \sigma_w \; f_2 \; (\frac{\lambda_m}{z}) z_m \tag{9}$$

As a hypothesis, Pasquill proposed the following universal functions:

$$\frac{d\bar{z}}{dx} = f_3 \; (\frac{K}{uz}) \bar{z} \tag{10}$$

$$\frac{dz_m}{dx} = f_4 \; (\frac{K}{uz}) z_m \tag{11}$$

By using detailed numerical dispersion calculations, these functions are estimated as simple power functions that apply from unstable conditions ($L = -7$ m) to moderately stable conditions ($L = 4$ m):

$$\frac{d\bar{z}}{dx} = a \; (\frac{K}{uz})^b, \text{ where } a = 0.95; \; b = 1.06. \tag{12}$$

$$\frac{dz_m}{dx} = c \; (\frac{K}{uz})^d, \text{ where } c = 3.56; \; d = 2.7. \tag{13}$$

When $\frac{K}{u}$ is constant with height:

$$\bar{z} = (a(\frac{K}{u})^b ((1+b)x + z_o^{1+b}))^{1/(1+b)} \tag{14}$$

When the vertical distribution of pollution is gaussian $\sigma_2 = \sqrt{\pi/2} \cdot \bar{z}$, similar equations may be derived using K-theory to describe horizontal variations in the first moment of the vertical pollution distribution.

4.2 Horizontal dispersion

Based on instantaneous wind observation every 5 minutes, the average wind for each tracer experiment were calculated. To describe the horizontal variation in wind direction, twelve instantaneous observations were used to calculate the standard deviation σ_θ. The results show that σ_θ was decreasing with increasing windspeed (34-14 deg) and increasing with stability (16-32 deg). The numbers in brackets denotes typical variation in σ_θ. The average value was

about 25 deg when the wind was blowing along the valley axis and
above 50 deg when the wind was blowing in other directions.

Since the transport time of pollution from the source areas
to the receptor points was probably less than the Lagrangian time
scale in the area, and the following equation was used to calculate
the horizontal standard deviation (σ_y) of a pollution cloud.

$$\sigma_y = \sigma_\theta X \cdot S_p \tag{15}$$

Values of S_p as proposed by Pasquill (4) are given in the
table below:

X (m)	100	200	400	1000	2000
S_p	0.8	0.7	0.65	0.6	0.5

These data were verified in areas with small roughness lengths
(z_o ~5cm) in Norway (8).

4.3 Plume rise and initial dispersion

Based on photos of the plume of pollutants emitted from the
smelter hall under different meteorological situations, the following empirical plume rise formula was estimated:

$$h = h_s + \frac{a}{u}, \quad \text{where } a = 11 \text{ m}^2/\text{s}. \tag{16}$$

h_s = height of the roof emission.

From tracer studies of the downwash behind the smelter hall
(10) the following initial horizontal dispersion (σ_{yi}) and vertical dispersion (σ_{zi}) were determined as:

$$\sigma_{yi} = 30 \text{ m}, \qquad \sigma_{zi} = 34 \text{ m}.$$

5   RESULTS OF CALCULATIONS

Observed and calculated SF_6 concentrations using the method
presented in section 4 for two typical experiments are shown in
Figure 10. It is seen that the measured values may deviate considerably from the observed ones. The deviation may be explained
by:

a)   surface wind variations in the area,
b)   topographical conditions,
c)   "normal" stochastic fluctuation of concentrations,
d)   vertical motion and convection within the area.

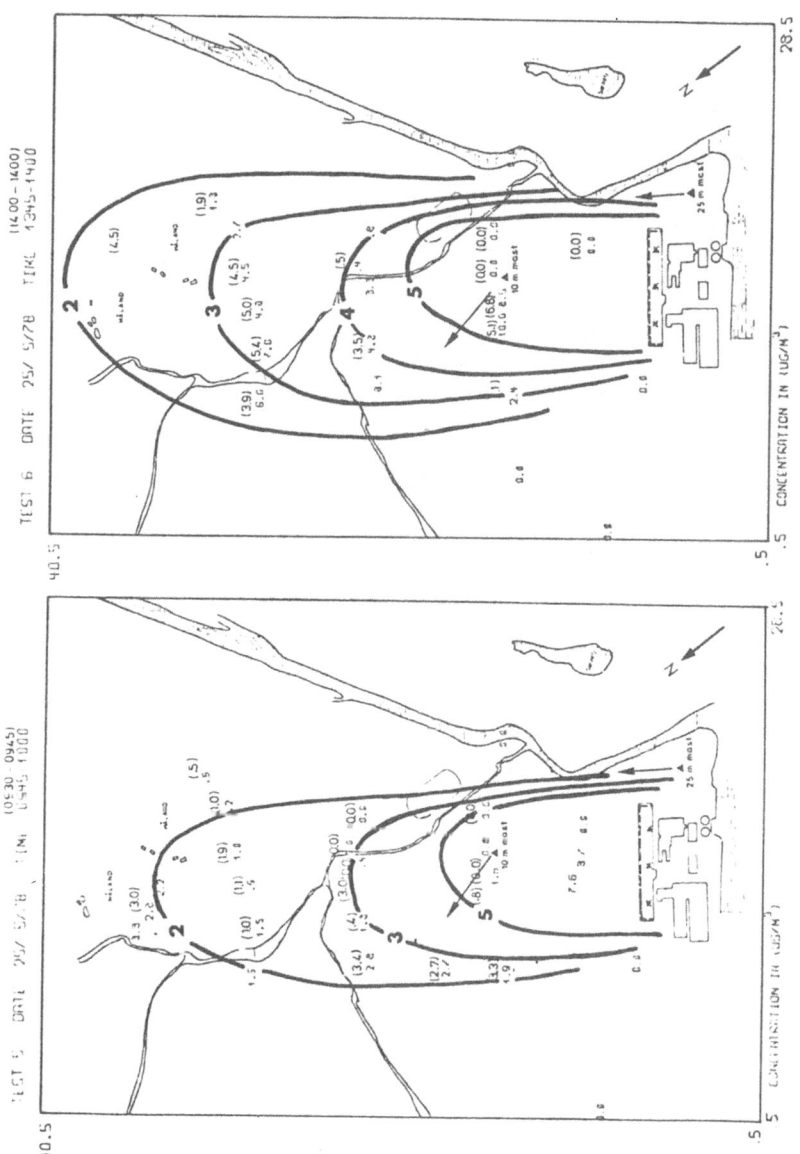

Figure 10: Calculated and observed SF₆-concentrations for tracer
experiments number 5 and 6. The windobservations are
marked as vectors at locations. Unit: µg SF₆/m³.

Generally it is seen that on a short term basis the calculation procedure overestimates the observed values.

However, when the procedure is used to estimate 24-hour mean
values of SO_2 in the area, the observed values are underestimated.
The frequency distribution of observed and calculated 24-hour mean
values are given in Figure 11. The deviation may be explained by:

a) uncertainties in emission data
b) possibility that polluted air, transported out over the
   fjord during the night, is returned to the measuring
   stations by the seabreeze during the day

When expected concentrations around the new aluminium plant,
were estimated, the calculated values were adjusted according to
the present correspondence between the observed and the measured
frequency distribution. The background SO_2 concentration
3 µg SO_2/m^3 was added. Further the adjustments were assumed to
be proportional to the total emissions from the plant (see Figure
11) and adjustment factors were normalized with respect to total
emissions. When future concentration distributions were estimated,
the calculated distributions were adjusted, adding the background
concentration and multiplying the different percentile values
with the respective normalized adjustment factor and the future
total emission.

To take into account the variable wind conditions of the area,
a puff model was used to simulate the tracer experiment in a land/
seabreeze wind system. Using wind data from Station 1 only, the
calculated movement of the SF_6 cloud is shown in Figure 12. The SF_6
measurements around the smelter show that the pollution cloud returned over the valley and that it is not removed from the area
wind measurement at Station 1 or Station 2 would imply.

6   CONCLUDING REMARKS

This study has demonstrated the effectiveness of tracer techniques for the study of air pollutant dispersion in complex terrain.

To estimate the frequency distribution of pollution concentrations for comparison with air quality standards, dispersion
models have to be applied. The applicability of available models
is still questionable. The results of this study show that:

a) The dispersion caused by small scale atmospheric turbulence,
   seem to be reasonably well described. Data on emission and
   meteorology in existing dispersion models to describe the
   downwind pollution concentrations within the maximum impact
   area around the aluminium smelter halls.

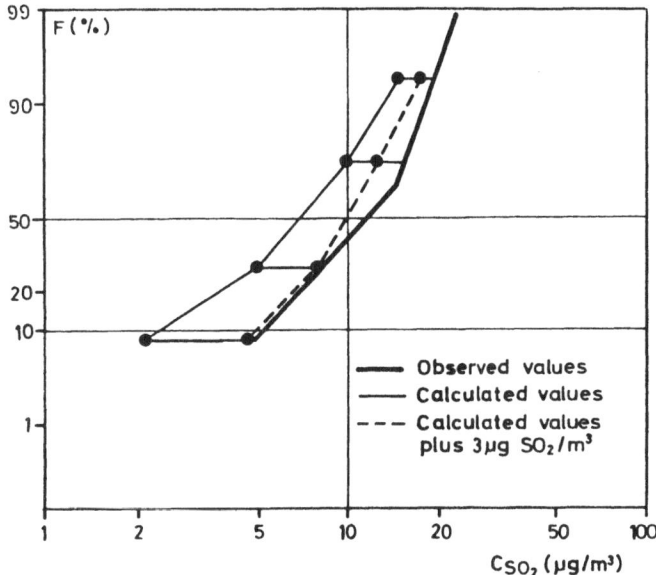

Figure 11:   *Frequency distributions of 24th mean calculated and ob-*
              served SO₂-concentrations for the period 3 May - 11 May
              1978, (μg/m³).

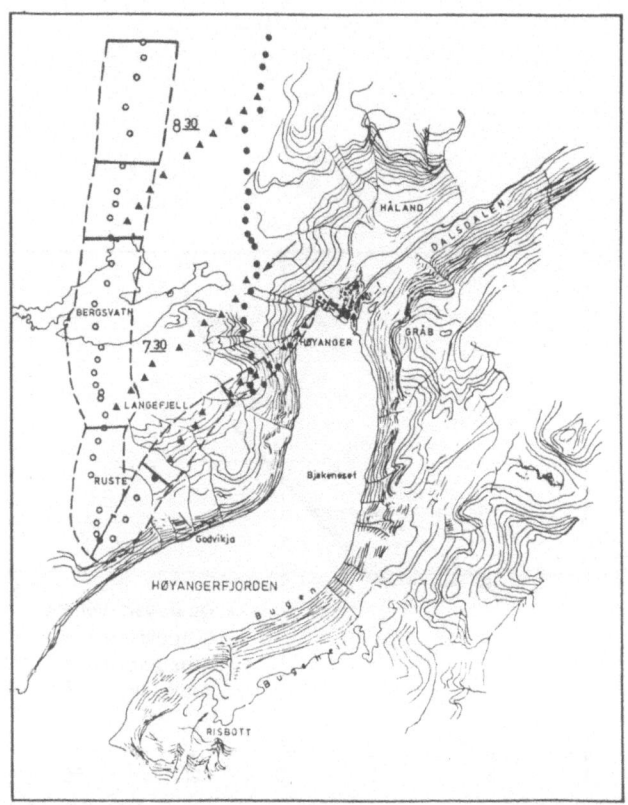

Figure 12: Calculated position of the SF₆-cloud, carried out over
the fjord with the land breeze during the morning of 15
June 1978. (SF₆-emissions: 0527-0630 hrs). Instantaneous wind measurements each 5 minutes at the 25 m level
(station 1), was used for calculating puff-trajectories.
The broken line shows the width of the plume.

○ : the position of the puff emitted at 0530
● : the position of the puff emitted at 0630
▲ : the position of the puffs emitted during the period
0535-0625.

b) The description of the mean wind field from measurements at
   only two measuring stations was insufficient to describe the
   pollution distribution in the fjord/valley area during the
   occurance of land/seabreeze and to predict the total venti-
   lation of the valley. The tracer experiments and air quality
   measurements indicate that part of the emitted pollution may
   during certain situations remain in the valley atmosphere
   for several hours.

c) Thermal upslope winds on valley sides exposed to the sun
   influence the concentration patterns considerably and lift
   pollutants above the valley flow. Even crude modelling of
   this flow pattern would improve the dispersion estimates
   considerably.

d) In accordance with previous studies (1) and (2) the results
   indicate that when the wind blows along the valley, turbulence depend on surface roughness and on vertical wind and
   temperature distributions in much the same way as in open
   terrain.

REFERENCES

1. K.J. Eidsvik and F.K. Hansen, "Turbulent diffusion in the
   surface boundary layer of near neutral stratified flows
   along four valleys". (Norwegian Defence Research Establishment FFIVM. Intern rapport VM-6.) Kjeller (1972).
2. Y. Gotaas, "Atmospheric dispersion in valleys". Part I
   (Norwegian Defence Research Establishment NDRE Report
   no 65). Kjeller (1974).
3. B.K. Lamb, "Development and application of dual tracer
   techniques for the characterization of pollutant transport
   and dispersion". Ph. D. Thesis, Calif. Inst. Technol. (1978).
4. F. Pasquill, "Some topics relating to modelling of dispersion
   layer". (EPA 650/4-75-015) Washington D.C. (1975).
5. N.E. Busch, S.W. Chang and R.A. Anthes, "A multi-level model
   of the planetary boundary layer suitable for use with mesoscale dynamics models". J. Appl. Meteor., $\underline{15}$ 909-919 (1976).
6. M. De Bortoli and E. Peechio, "Measurements of some halonogatil
   compounds in air over Europe". Atmos. Environ. $\underline{10}$, 921-923
   (1976).
7. J.A. Businger, "Turbulent transfer in the atmospheric surface
   layer". In: Workshop on Micrometeorology, D.A. Haugen, Ed.,
   Amer, Meteor. Soc., 67-98 (1973).
8. B. Sivertsen, "Dispersion parameters determined from measurements of wind fluctuations, temperature and wind profiles".
   In: Proceedings of the ninth international technical meeting
   on air pollution modelling and its application. (NATO/CCMS
   no 103). Toronto (1978).

9.   S.R. Hanna, "A review of the influence of new boundary layer
     results on diffusion prediction techniques". In: WMO
     Symposium on boundary layer physics applied to specific
     problems of air. (WMO-No 510) Geneve (1978).
10. B.K. Lamb, V. Vitols and O.F. Skogvold, "Atmospheric tracer
     techniques and gas transport in primary aluminium industry".
     J. Air. Poll. Contr. Assoc. <u>30</u>, 558-566 (1980).

ACKNOWLEDGEMENT

     The authors wish to thank

     - Årdal and Sunndal Verk A/S for sponsoring the study.
     - Dr. B.K. Lamb for his guidance in performing SF_6-tracer
       experiments and for his participation in the study.
     - Frederick Gram for developing the puff model.
     - Val Vitols for suggestions improving the manuscript.

CONVERSION RATE OF NITROGEN OXIDES IN A POLLUTED ATMOSPHERE

R. Guicherit, K.D. van den Hout, C. Huygen

TNO Research Institute for Environmental Hygiene,
Delft, The Netherlands

H. van Duuren, F.G. Römer, J.W. Viljeer

N.V. KEMA, Environmental Research Department,
Arnhem, The Netherlands

INTRODUCTION

The major part of emitted NO_x is NO. The first step in the removal of NO_x from the atmosphere is the oxidation of NO into NO_2. Then, NO_2 in a range of reactions is converted into oxidation products such as nitrates, nitric acid and PAN. Besides, because of photolysis in the daytime, NO_2 will contribute to the formation of ozone.

Knowledge of the NO_x removal rate from the atmosphere is of importance for a number of reasons:
- for long range transport problems of NO_x in general and for estimation of the impact of urban NO_x on downwind areas in particular;
- for an understanding of tropospheric chemistry;
- for modelling exercises. Especially in models of the Gaussian plume type, conversion and annihilation of pollutants are treated by the introduction of some exponential loss term which is directly related to the removal rate of the pollutant under discussion.

NO_x may be emitted at ground level or by elevated sources (plume NO_x). Because the removal rate of NO_x emitted by elevated sources needs not per se be the same as that of NO_x emitted at ground level, the two cases will be treated separately.

The residence time (τ) for NO_x in an air mass near the ground in a well-mixed layer under average meteorological conditions is given by:

$$\tau^{-1} = \tau_d^{-1} + \tau_w^{-1} + \tau_c^{-1} \tag{1}$$

τ_d is the residence time by dry deposition removal
τ_w is the residence time by wet deposition removal
τ_c is the residence time by chemical conversion

The residence time of NO_x by dry deposition is given by:

$$\tau_d = H.\nu_d^{-1} \tag{2}$$

τ_d is the residence time (s)
H is the mixing depth (m)
ν_d is the deposition velocity ($m.s^{-1}$)

The loss rate of NO_x by dry deposition on an hourly basis amounts to:

$$\Delta NO_x = 1 - \exp{(-3600.\tau_d^{-1})} \tag{3}$$

or:

$$\Delta NO_x = 1 - \exp{(-3600.H^{-1}.\nu_d)} \tag{4}$$

If H is assumed to be 1000 m and the dry deposition velocity ν_d for NO_x is assumed to be $\frac{1}{4}$ of the dry deposition velocity for SO_2 i.e. 0.2 $cm.s^{-1}$, ΔNO_x is << 0.01 h^{-1} (TNO, 1980).

Loss of NO_x by wet deposition is given by:

$$c_t = c_o \exp{(-\lambda t)} \tag{5}$$

c_t is the pollutant concentration at time t
c_o is the original pollutant concentration
λ is the "wash out" coefficient (s^{-1}), estimated to be $\leqq 10^{-5} \, s^{-1}$
(TNO, 1980)
Since $\tau_w = \lambda^{-1}$, this amounts to a residence time of NO_x by wet deposition of about 28 h. This calculation holds for a precipitation intensity of about 1 $mm.h^{-1}$. From this figure a yearly average removal rate of 0.001-0.002 h^{-1} and a residence time of 3 to 6 weeks may be calculated for precipitation duration and intensity valid for Western Europe.

The chemical conversion of NO_x is dependent on many factors. Smog chamber studies, modelling experiments and field monitoring reveal that the NO_x removal rate depends to a considerable extent on the degree of photochemical activity. From what has been discussed before, it seems that in many circumstances loss of NO_x by wet and dry deposition is small compared with chemical conversion. In the following paragraphs an estimate of the oxidation of NO and the removal rate of NO_x, mainly by chemical conversion reactions, will be given.

NO OXIDATION IN PLUMES

Unlike the oxidation of NO in the free atmosphere, the oxidation with oxygen also plays a role in the oxidation of NO in plumes. The major dark reactions in plumes are (Elshout et al., 1978; Van Duuren et al., 1979):

$$2NO + O_2 \xrightarrow{\quad k_1 \quad} 2NO_2 \tag{6}$$

$$NO + O_3 \xrightarrow{\quad k_2 \quad} NO_2 + O_2 \tag{7}$$

$$NO + RO_2{}^{\cdot} \xrightarrow{\quad k_3 \quad} NO_2 + RO^{\cdot} \tag{8}$$

Because of absorption of UV irradiation the following reactions occur during daylight hours:

$$NO_2 + h\nu \xrightarrow{\quad k_4 \quad} NO + O \;(\lambda < 400 \text{ nm}) \tag{9}$$

$$O + O_2 + M \longrightarrow O_3 + M \tag{10}$$

$$O_3 + h\nu \longrightarrow O_2 + O_1^D \;(\lambda < 320 \text{ nm}) \tag{11}$$

$$H_2O + O_1^D \longrightarrow 2OH^{\cdot} \tag{12}$$

In the daytime the so-called photo-stationary state can be reached in very well-mixed parts of the atmosphere:

$$\left[O_3\right]\left[NO\right] = \frac{k_4}{k_2}\left[NO_2\right] \tag{13}$$

In an atmosphere heavily polluted with hydrocarbons radicals will be generated. For this reason with reaction (7) competitive reactions will proceed. This results in a higher NO_2 level and consequently in the formation of ozone. This situation is especially of importance for plumes from industrial areas, such as the Rijnmond area, near Rotterdam (Van Duuren et al., 1979).

Model studies (Elshout and Steenkist, 1974; Varey et al., 1978; Cocks and Fletcher, 1979) have shown that the oxidation of NO with oxygen in plumes occurs close to the stack (transport times up to 200 s). For power station plumes with initial NO concentrations between 100 and 500 ppm, the degree of oxidation in the daytime is mostly less than 0.1 (Elshout and Steenkist, 1974; Varey et al., 1978).

NO oxidation with ozone is determined by the mixing velocity of the plume with ambient air, as was proved by Hegg et al. (1977). It was also confirmed experimentally that reaction (7) is diffusion controlled and that the NO_2/NO ratio is highest on the edges of the plume.

Varey et al. (1978) developed a numerical model for the calculation of NO oxidation in power station plumes. This model, based on reactions (6), (7), (9) and (10), also proved that the oxidation rate integrated over the plume cross-section is determined by the degree of mixing of the plume with ambient air, expressed in the eddy diffusivity D (Fig. 1.). Under stable atmospheric conditions (Pasquill stability class F, with $D \approx 0.1$ $m^2.s^{-1}$ the oxidation of NO is mainly determined by the reaction with oxygen. Under good mixing conditions with D values of > 10 $m^2.s^{-1}$, the reaction with ozone prevails. The model results of Varey et al. (1978) have been confirmed by measurements in plumes of the Maasvlakte power station (NO/NO_x emission ratio ≈ 0.98)(Elshout et al., 1978). In the period 1975-1980 measure-

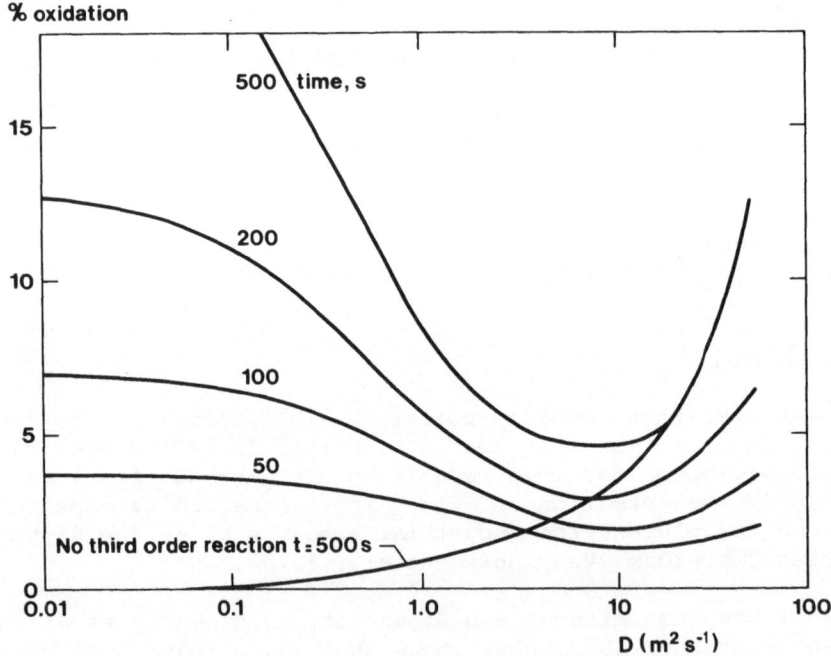

Fig. 1.   Percentage oxidation, integrated over the plume cross-section as a function of D for various times. $[NO]_o$ = 450 ppm; $[O_3]_o$ = 0.03 ppm. Results of model calculations (Varey et al., 1978).

ments were carried out on 27 days at distances between 0.2 and 9 km
from this power station (two 540 MW units). During these measurements
with background concentrations of O_3 from 20 to 65 ppb, oxidation was
found to be 50-60% (averaged over the plume cross-section). Fig. 2.
shows the NO oxidation at the Maasvlakte power station as a function
of the O_3 concentration. For Pasquill classes B and C (eddy diffusivity D > 5 $m^2.s^{-1}$) oxidation increases with increasing O_3 concentration. However, under neutral conditions (eddy diffusivity D ≈ 1 $m^2.s^{-1}$)
oxidation is independent of the O_3 concentration. In this case mixing
of ozone into the plume is limited, so for the transport time of
interest the oxidation rate is mainly controlled by the reaction with
oxygen (Fig. 1.).

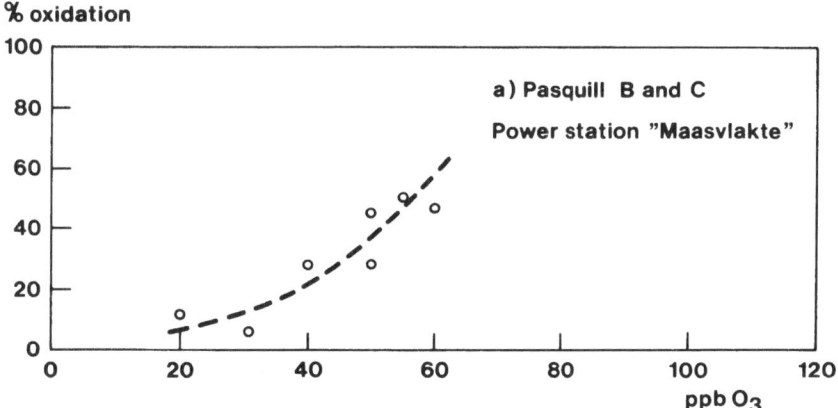

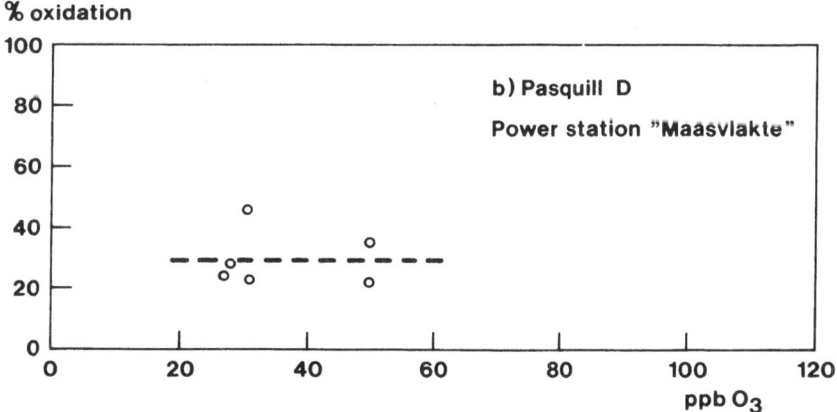

Fig. 2.   Oxidation of nitric oxide in power station plumes. Percen-
          tage oxidation, integrated over the plume cross-section, as
          a function of the ambient ozone concentration for various
          Pasquill stability classes. Travelling time 500 s.

Measurements by Melo et al. (1978), carried out under more stable conditions than those in Fig. 1., show that even at high O_3 concentrations up to 110 ppb the oxidation is independent of the O_3 concentration (approximately 0.1 for a travelling time of 500 s). This implies that under these conditions the reaction with oxygen is the main contribution to the formation of NO_2.

Cocks and Fletcher (1979) developed a reactive plume model to study the gas phase reactions in power station plumes. This model enables the oxidation of NO to be calculated. The model has been applied for five well-defined measurements of those carried out at the Maasvlakte power station: distance 2-9 km, travelling time 150-900 s, O_3 concentration 20-55 ppb, unstable to neutral atmosphere. The results of the calculated values of the NO/NO_x ratio (average value for the plume cross-section) agree with the values found. The calculated values vary between 0.41 and 0.86 and those measured between 0.60 and 0.81. The mean value of the difference between measured and calculated ratio was found to be 0.09; the largest difference amounts to 0.20 (absolute value).

NO_x REMOVAL RATE IN A POLLUTED ATMOSPHERE

The concentration of a reactive pollutant in an urban atmosphere changes with time as a result of chemical conversion reactions, changes in emissions, and by dispersion and/or dilution. On the other hand the concentration of an "inert" pollutant in an urban atmosphere will only change with time as a result of changes in emissions, and by dispersion and/or dilution. If a reactive species and an "inert" species have a common source it can be shown (Chang et al., 1979) that

$$k_r = - \frac{\Delta (\ln R.I^{-1})}{\Delta t} \tag{14}$$

k_r is the lower bound of the pseudo first-order reaction rate coefficient for the reactive species
R and I are the concentrations of the reactive and "inert" species respectively.

From equation (14) it follows that from the daily trend of the NO_x/CO ratio (CO is treated here as an inert species), obtained by long-term average aerometric data measured at a fixed ground station, the removal rate of NO_x can be estimated; on condition, however, that NO_x and CO have a common source.

The main source of CO in The Netherlands is automobile exhaust gas. Automobile exhaust gas is also a major NO_x source. Guicherit and Hoogeveen (1978) have shown that the amount of NO_x in air parcels, reaching a sampling site, which is attributable to mobile (automotive) sources can be estimated by comparing the NO_x to acetylene ratio typical of mobile source emissions for various driving conditions. Acetylene is regarded as a traffic tracer. The ratio of NO_x/C_2H_2

Table 1. NO_x removal rate, k_r (h^{-1}),
           from sunrise to sunset
           (lower bound)

winter (1975)	0.01
summer (1975)	0.05
photochemical episodes (1975) O_3 (max.) > 200 $\mu g.m^{-3}$	0.14
September (1973) average maximum O_3 concentration 150 $\mu g.m^{-3}$	0.12

Table 2.   Summary of the most important removal reactions of NO_2 in
           ambient air

NO_2 + OH $\rightarrow$ HNO_3	$k_5 = 1.1 \times 10^{-11}$ $cm^3.mol^{-1}s^{-1}$ Hamson and Garvin (1978)
NO_2 + O_3 $\xrightarrow{\text{intermediate steps}}$ HNO_3	k_6 net
NO_2 + RCO_3 $\xrightarrow{\text{intermediate steps}}$ PAN	k_7 net

depending on driving conditions varied from 3 to 8 in 1975. For higher ratios it is concluded that other major emission sources are involved. In the analysis of the daily trend of the NO_x/CO ratio, only those wind directions were considered for which the major emission source was established to be automobile traffic.

The daily trend of the NO_x/CO ratio measured at a ground station near Delft (1973 and 1975 data) is given in Figs. 3. and 4. From the graphs the removal rate of NO_x has been calculated. The results are shown in Table 1.

The most important reactions of NO_2 in ambient air are given in Table 2.

From the September removal rate of NO_x based on reaction with OH radicals only, an OH radical concentration of 3×10^6 $molec.cm^{-3}$ can be deduced. From measurements in this part of The Netherlands it follows, however, that the PAN/HNO_3 ratio is about 1. Further, it is assumed that 75% of the HNO_3 formation is due to reaction of NO_2 with OH and 25% by reaction of NO_2 via O_3. So a more realistic OH

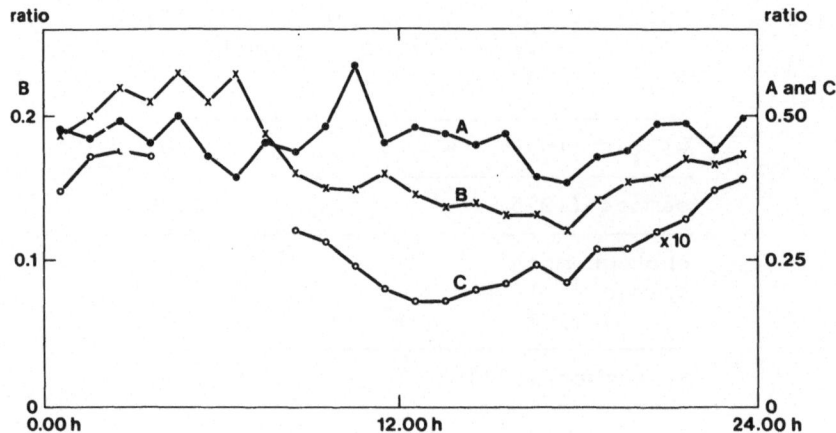

Fig. 3. Daily trend of pollutant ratios (September 1973 data)
         A: i-pentane/acetylene ratio
         B: propylene/acetylene ratio
         C: NO_x/CO ratio

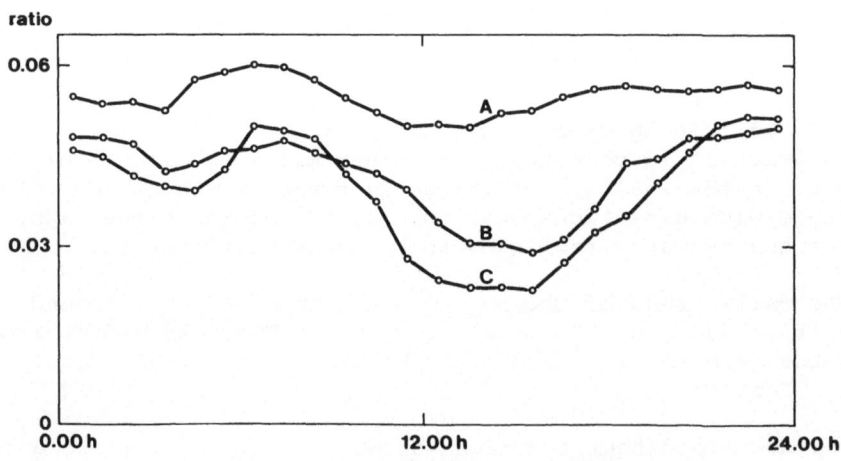

Fig. 4. Daily trend of the NO_x/CO ratio (1975 data)
         A: winter data January, February, March, October, November,
            December
         B: summer data April through September
         C: days with photochemical smog formation (O_3 concentrations
            > 200 $\mu g.m^{-3}$)

Table 3. Calculated average OH radical concentration (from sunrise to sunset)

period	NO_x removal rate from Table 1. (h^{-1})	OH concentration $(molec.cm^{-3})$	literature quotations
winter average (1975)	0.01	$< 2.5 \times 10^5$	$\approx 1 \times 10^5$ Fishman and Crutzen (1978)
summer average (1975) O_3 (max.) = $90\ \mu g.m^{-3}$	0.05	5×10^5	$\leqq 1 \times 10^6$ Ehhalt et al. (1980) $(3.5-5) \times 10^5$ Fishman and Crutzen (1978)
September (1973) O_3 (max.) = $150\ \mu g.m^{-3}$	0.12	$(5-10) \times 10^5$	5×10^5 Fishman and Crutzen (1978)
smog formation (1975) $O_3 > 200\ \mu g.m^{-3}$	0.14	1.5×10^6	2×10^6 Derwent and Hov (1979) 1.2×10^6 Guicherit et al. (1978)

radical concentration of about 10^6 molec.cm^{-3} is obtained. Table 3. summarizes the OH radical concentrations following this procedure. These OH radical concentrations are in accordance with those determined from the removal rate of i-pentane (0.01 h^{-1}) and propylene (0.07 h^{-1}). This removal rate is derived from the daily rate of the i-pentane/acetylene and propylene/acetylene ratio for September 1973 (Fig. 3.). From the equations

$$k_{r_{iC_5H_{12}}} = k_a \cdot [OH] \tag{15}$$

and

$$k_{r_{C_3H_6}} = k_b \cdot [OH] + k_c \cdot [O_3] \tag{16}$$

the OH radical concentrations of 9×10^5 molec.cm^{-3} and 5×10^5 molec.cm^{-3} can be calculated for September using $k_{(a,b,c)}$ values from Greiner (1970), Atkinson and Pitts (1975) and Hamson and Garvin (1975).

THE NO_x REMOVAL RATE IN PLUMES

The removal of NO_x from the atmosphere is to a considerable extent effected by the oxidation of NO_2 to secundary products such as nitric acid (nitrates) and PAN. Hydrocarbons as a source of reactive radicals and ozone both natural and as a product of photochemical air pollution play a role in this oxidation process.

As to the NO_2 conversion in power station plumes, it was explained before that the oxidation process in plumes highly depends on mixing with ambient air. Because of the relatively low ozone and hydrocarbon concentrations compared with those of NO_x in the plume, no oxidation products will be formed. If, however, under suitable meteorological conditions such a plume mixes with an urban-industrial plume, this will be possible. According to Cocks and Fletcher (1979) very little nitric acid was produced from NO_x in the plume after a 100 min travelling time. High concentrations of hydrocarbons, however, attribute considerably to the formation of N-containing acids (Cocks and Fletcher, 1979).

According to (provisional) measuring results by Hegg and Hobbs (1979) NO_x conversion to nitrate in the plumes of three coal-fired power stations is limited. The conversion of NO_2 to nitrate was calculated and for one of the measurements values between 0.003 h^{-1} and 0.005 h^{-1} were found for distances of 5 and 27 to 43 km respectively. Forrest et al. (1979) carried out measurements in the plume of an oil-fired power station in which they also found low values which hardly differed from the background values. During two measurements some formation of nitrate from NO_x was detected at distances of 5 and 16 km. As was the case for the measurements by Hegg and Hobbs (1979), nitric acid was not determined so that for this reason the calculated conversion rate may have been too low. In refinery plumes at Los Angeles Parungo et al. (1980) measured relatively large nitrate particulates that had not been emitted as primary components. It was assumed that heterogeneously produced nitric acid ($3NO_2 + H_2O \rightarrow 2HNO_3 + NO$) reacted with NaCl to form nitrate particulates. These nitrate particulates melt at a relative humidity as low as 60% and can grow to large particulates (1-10 μm) before equilibrium is reached. According to the authors inorganic nitrate can be produced on cloudy days and at night. One of their conclusions is that under these conditions the reaction $M + OH + NO_2 \rightarrow HNO_3 + M$ is not the most important one for the formation of nitrate, for on sunny days with expected high OH radical concentrations no increase in the amount of inorganic nitrate is found.

For the determination of the NO_x removal rate in urban areas Spicer (1980) has carried our measurements in isolated plumes of Phoenix (Arizona) and Boston (Massachusetts). The flights were arranged in such a way that the measuring data refer to the same air parcel. During these measurements, besides NO_x, oxidation products,

tracer concentrations, and also various other parameters were determined. The measurements at Phoenix showed a NO_x removal rate of
< 0.05 h^{-1}. This result has to be considered as provisional. From
the experiments at Boston the expected NO_x concentration was calculated from the initial urban tracer/NO_x ratio and the downwind tracer
concentration. Based on these experiments, the rate of the NO_x removal
from the Boston air varied from 0.14 to 0.24 h^{-1}, with an average
rate and life time of 0.18 h^{-1} and 5.8 h respectively. The reported
values are applicable only to daylight hours under photochemical
conditions (maximum O_3 concentrations in the plume 90 to 140 ppb).

    The NO_x conversion rate with regard to that of SO_2 was calculated
from the measuring results of the plume of the Rijnmond industrial
area (Van Duuren et al., 1979). The measurements were carried out by
day and focussed on that part of the industrial area occupied with
the major oil refineries and chemical industries (Fig. 5.). Each
flight consisted of three tracks flown at four different altitudes
(150, 300, 450 and 600 m). Track 1 was located at 2-14 km upwind the
industrial area. Track 2 was located immediately behind the area,
and track 3 was at 15-20 km downwind. Track lengths were 12-18 km.
The flight scheme was: measurements at 150 m on tracks 1, 2 and 3
respectively, then at 300 m in the same sequence and so on for the
other altitudes · This implies that the measurements were done in
three different air parcels. It is assumed, however, that the air
pollution pattern is uniform within the approximate 2 h duration of
the measurements.

    The flights were carried out on days when the area between tracks
2 and 3 was a rural area. It was assumed that no major NO_x and SO_2
was emitted anymore in this area. As to SO_2 the measuring results
confirmed this assumption. Because the SO_2 removal rate is relatively
low with regard to that of NO_x this component can be used as a tracer
for the determination of the NO_x removal rate. From differences in
average NO_x and SO_2 concentrations between tracks 2 and 3 (averaged
for the altitude of 150-600 m) for three measuring days 18-23% of NO_x
available on track 2 was calculated to have disappeared between the
two tracks.

    The transport times were calculated from wind velocity averaged
for the altitude interval, and wind velocity was estimated from the
surface wind velocity. The NO_x conversion rate calculated in that
way amounts to 0.25-0.40 h^{-1}. Considering any errors in wind velocity
estimated and in relatively low NO_x and SO_2 concentrations, the NO_x
conversion rate will have been at least 0.15 h^{-1}. The O_3 concentration at the time of measurements was 25-30 ppb. Under photochemical
conditions (temperature 24°C, O_3 concentration 80-100 ppb), a conversion rate of 0.19 h^{-1} was measured. During that flight, tracks 2 and
3 were located at distances of 10 and 40 km from the industrial area.
The altitude was between 250 and 900 m. The average wind velocity
in the mixing layer, mixing height 1000 m, was determined with a
weather balloon.

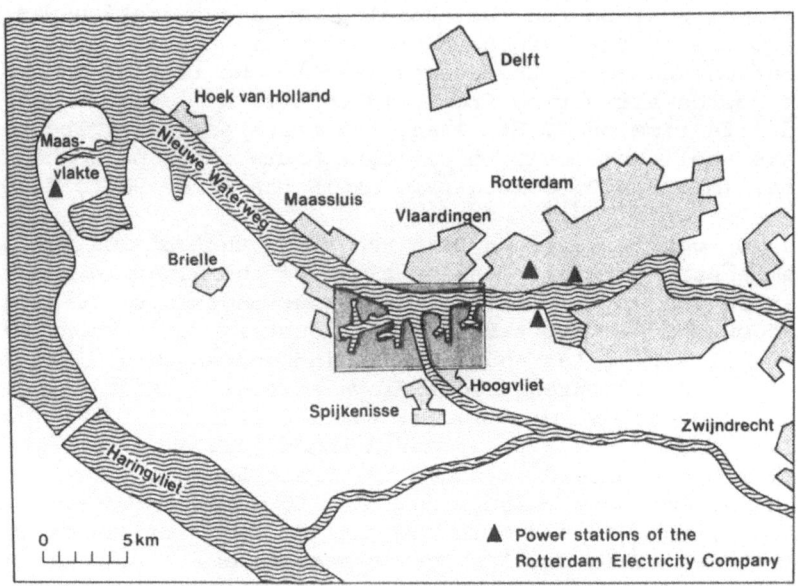

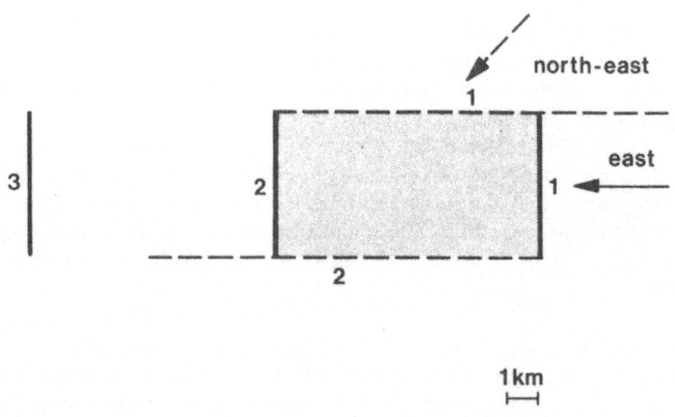

Fig. 5.   The Rijnmond measuring area; locations of flight track
          patterns at winds from east and north-east.

DISCUSSION

The oxidation of the primary emittent NO into NO_2 is the essential first step in the removal of NO_x from the atmosphere. It has been proved that the atmospheric conditions (time of the day, season, turbulence) strongly determine the oxidation rate of NO both in individual plumes and at ground level. The measured data are summarized in Table 4.

Though NO_x removal is mainly governed by chemistry, it is for reasons of not fully understood measuring techniques that we are faced with difficulties and sometimes the impossibility to determine the NO_x removal rate from the measured data of secondary products as well as gaseous components. For instance, when studying the results of chemiluminescent measurements of NO_2, concentrations too high a value (and consequently too low a removal rate) shall have to be reckoned with, depending on the converter type used.

An accurate determination of the travelling time of air parcels in plume measurements is essential for the calculation of the removal rate and the residence time. Besides, in case of k_r determination in plumes, corrections have to be made for SO_2 losses due to dry deposition.

Table 4. Conversion of NO into NO_2 under different atmospheric conditions

measurement	degree of oxidation (%) (NO_2/NO_x)	conditions
power plant plume*	5	stability class B, C; travelling time 500 s; O_3 concentration 20 ppb; 7 flights
	60	idem; O_3 concentration 60 ppb
	30	stability class D; travelling time 500 s; O_3 concentration 25-50 ppb; 6 flights
urban atmosphere**	45-55	winter, maximum increase between 6.30h-8.30h MET
	30-80	summer, maximum increase between 6.30h-10.30h MET

* – averaged over the plume cross-section
  – initial NO_2/NO_x ratio about 2%
** maximum differences for averaged values

NO OXIDATION IN A POLLUTED ATMOSPHERE

Fig. 6. shows the daylight hourly curves for the $NO_2/(NO+NO_2)$ ratio for an urban area for the winter months (January, February, November and December 1975) and for the summer months (May through August 1975). In winter, the NO_2 concentration proves to be approximately 50% of the NO_x concentration with only slight variations during the day. In summer, on the other hand, the NO_2 concentration early in the morning at the beginning of the rush-hour makes up 30% of the NO_x concentration, increasing up to nearly 80% within a few hours. Then there is a gradual increase up to 85% till 15.00h, decreasing to approximately 50% after sunset.

According to Chang et al. (1979), the conversion rate is:

$$k_{r(NO \rightarrow NO_2)} = \frac{1}{1 - NO_2/(NO + NO_2)} \left[\frac{\Delta NO_2/(NO + NO_2)}{\Delta t} \right] \qquad (17)$$

This is based on the assumption that NO_2 is converted so much more slowly, that the NO_2/NO_x ratio is hardly affected by it. So in fact, $k_{r(NO \rightarrow NO_2)}$ is larger. From Fig. 6. conversion rates of > 0.35 h^{-1} are calculated.

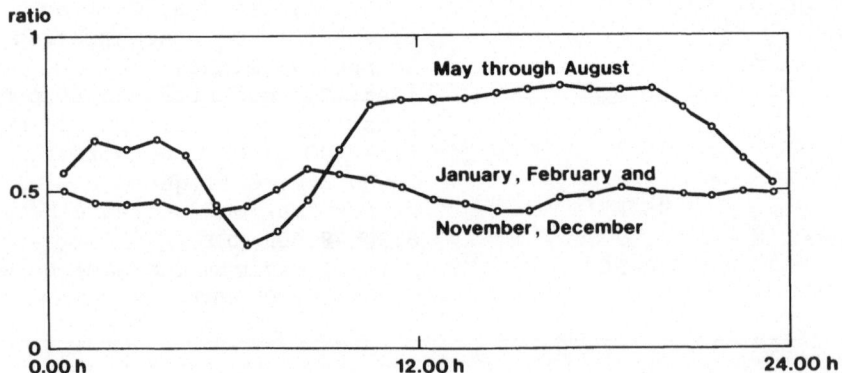

Fig. 6.   Daily trend of the $NO_2/(NO+NO_2)$ ratio at Delft in 1975.

Because of limitations in the measuring equipment, the results have to be regarded as not conclusive. Nevertheless, the results generally agree fairly well with values mentioned in the literature for $k_r < 0.15$ h^{-1} on ground level (polluted atmosphere) and $k_r > 0.15$ h^{-1} in plumes. The results have been listed in Table 5.

Table 5. Survey of data for the NO$_x$ removal rate in urban atmospheres and urban plumes

method	removal rate (h^{-1})	remarks	reference
		urban atmosphere	
NO$_x$/CO	0.01	Delft 1975, lowest value, winter daytime	this article
	0.05	Delft 1975, summer average	this article
	0.14	Delft 1975, highest value, photochemical episodes, O$_3$>200 μg.m^{-3}	this article
	0.04	Los Angeles 1970-1975, lower bound, all seasons 8.00h-16.00h value	Chang et al.(1979)
	0.11	Los Angeles 1970-1975, average 9.00h-13.00h value for May-October	Chang et al.(1979)
$\frac{PAN+HNO_3}{NO_x+PAN+HNO_3}$	0.05	Los Angeles basin 1973, 5 weeks summer period, 10.00h-14.00h	Spicer in: Chang et al. (1979)
	0.04	St. Louis 1973, 5 weeks summer period 8.00h-13.00h	Spicer in: Chang et al. (1979)
		urban plume	
NO$_x$/C$_2$H$_5$ & NO$_x$/CH$_4$	0.09	LARPP study 5 November 1973, 8.00h-13.00h, O$_3$ 100 ppb	Calvert in: Chang et al. (1979)
tracer/NO$_x$	0.18	Boston 1978, 4 experiments (0.14-0.24 h^{-1}), photochemical conditions, O$_3$ 90-140 ppb (tracers CO, F-11, C$_2$H$_2$)	Spicer (1980)
SO$_2$/NO$_x$	≥0.15	Rijnmond (Rotterdam) April 1977-August 1978, 2 experiments, O$_3$ 25-30 ppb	this article
	0.19	idem, August 1979, 1 experiment, O$_3$ 80-100 ppb	this article

An important assumption adapted in this study, is the neglect of the mixing of relatively old pollutants into the air considered in equation (14). More specifically, for the aeroplane measurements the air outside individual stack plumes as well as urban-industrial plumes, containing pollutants of a generally different composition, is gradually mixed into the plume. In the case of the ground monitoring measurements, the aged smog layer, transported at night above the nocturnal inversion layer and mixed downward in the morning, may be of importance. When this air mass contains relatively clean air, or air with a concentration ratio of reactive and inert pollutants comparable with that of the air considered in equation (14), the influence on the results is rather small. At present, however, this effect has not yet been fully evaluated. One of the difficulties is that the nocturnal chemistry of aged smog is not well established, and consequently the early morning concentrations of NO_x aloft are unknown. Insight into this might probably be gained from aeroplane measurements performed in the early morning.

Another point that requires further evaluation, is the assumption, used in the derivation of equation (14), that the sources are horizontally homogeneously distributed. Although this is clearly not the case, it is not evident how strictly this condition should be met. The averaging over many days, that has been performed, implies averaging over many wind directions and wind speeds. This tends to smooth the distribution of times of travel, which determines the time available for reaction. To clarify this it might be interesting to use results obtained from monitoring stations at various distances from important source areas.

SUMMARY

The removal of NO_x from the atmosphere is to a great extent determined by chemistry. As, in general, these chemical pathways proceed through NO_2, also a description and determination of the oxidation of NO into NO_2 in plumes and urban (polluted) atmospheres is given.

For power station plumes the degree of oxidation (NO_2/NO_x) strongly depends on the stability of the atmosphere. Model calculations for a travelling time of 500 s are qualitatively in accordance with measured values. These are 30% for stability class D and O_3 concentrations of 25-50 ppb and between 5 and 60% for classes B and C and O_3 concentrations between 20 and 60 ppb respectively. For the removal rate ($k_{r(NO \to NO_2)}$) in an urban atmosphere a value of > 0.35 h^{-1} can be calculated.

For the determination of the NO_x removal rate in an urban atmosphere the procedure followed by Chang et al. was used. Thus for the daily trend of the NO_x/CO ratio lower bounds for k_r of 0.01 h^{-1}

during the winter daytime and of 0.14 h^{-1} during photochemical
episodes (maximum O_3 concentration > 200 $\mu g.m^{-3}$) were found. From
k_r derived OH radical concentrations are in agreement with concentrations derived from the removal rate of i-pentane and propylene
in the same air parcels.

The NO_x removal rate has also been determined in an industrial
plume (Rotterdam area). The calculations are based on the assumption
that SO_2 is far more slowly removed than NO_x. The removal rate amounts
to at least 0.15 h^{-1} taking into account the deviation in the estimated wind speed.

Under photochemical conditions (O_3 concentrations 80-100 ppb)
for longer distances and a well-known transport time, a removal rate
of 0.19 h^{-1} was determined.

In a survey of the measured results and values from literature
for k_r the comparability of the two different cases appeared to be
satisfactory. Nevertheless the data must be interpreted with care,
because several assumptions had to be made.

REFERENCES

Atkinson, R. and Pitts Jr., J.N., 1975, Rate constants for the
    reaction of OH-radicals with propylene and butenes over the
    temperature range 297-425°K, J. Chem. Phys. 63:3591
Chang, T.Y., Norbeck, J.M. and Weinstock, B., 1979, An estimate of
    the NO_x removal rate in urban atmosphere, Environ. Science
    Technol. 13:1534
Cocks, A.F. and Fletcher, I.S., 1979, A model of the gas-phase
    chemical reactions of power station plume constituents, CERL
    laboratory report no. RD/L/R 1999
Derwent, R.G. and Hov, O., 1979, Computer modelling studies of
    photochemical air pollution formation in North-West Europe,
    AERE report R, 9434, HMSO, London
Duuren, H. van, Elshout, A.J., Römer, F.G., Viljeer, J.W. and Noks,
    E., 1979, Luchtverontreiniging boven de Rijnmond, Elektrotech-
    niek 57:267
Ehhalt, D., Perner, D. and Platt, U., 1980, Trace gas measurements
    by optical absorption, Communication by Platt at the COST 61A bis
    Workshop on transport and modelling (The Hague)
Elshout, A.J., Duuren, H. van, Viljeer, J.W. and Noks, E., 1978, De
    oxidatie van stikstofmonoxide in rookpluimen, Elektrotechniek
    56:429
Elshout, A.J. and Steenkist, R., 1974, A numerical approach to the
    effect of the oxidation of NO in the dispersion of these
    pollutants from a point source, N.V. KEMA technical note
Fishman, J. and Crutzen, P.J., 1978, The distribution of the hydroxyl-
    radical in the troposphere, Atmospheric Science paper no. 284,
    Colorado State University

Forrest, J., Garter, R. and Newman, L., 1979, Formation of sulfate ammonium and nitrate in oil-fired power plant plume, Atm. Environment 13:1287

Greiner, N.R., 1970, Hydroxyl radical kinetics by kinetic spectroscopy VI. Reactions with alkanes in the range 300-560°K, J. Chem. Phys. 53:1070

Guicherit, R., Blokzijl, P.J. and Plasse, C.J., 1978, Some notes on the abatement of photochemical ozone production, IMG-TNO report

Guicherit, R. and Hoogeveen, A., 1978, in:"Photochemical smog formation in The Netherlands", TNO-report, R. Guicherit, ed., Delft

Hamson, R.F. and Garvin, D., 1975, Chemical kinetic and photochemical data for modelling atmospheric chemistry, National Bureau of Standards Technical Note 866, USGPO, Washington D.C.

Hamson, R.F. and Garvin, D., 1978, Reaction rate and photochemical data for atmospheric chemistry, National Bureau of Standards, Special Publ. 513, USGPO, Washington D.C.

Hegg, D., Hobbs, P.V., Radke, L.F. and Harison, H., 1977, Reactions of ozone and nitrogen oxides in power plant plumes, Atm. Environment 11:521

Hegg, D.A. and Hobbs, P.V., 1979, Some observations of particulate nitrate concentrations in coal-fired power plant plumes, Atm. Environment 13:1715

Melo, O.T., Lusis, M.A. and Stevens, R.D.S., 1978, Mathematical modelling of dispersion and chemical reactions in a plume – oxidation of NO to NO_2 in the plume of a power plant, Atm. Environment 12:1231

Parungo, F.P., Pueschel, R.F. and Wellman, D.L., 1980, Chemical characteristics of refinery plumes in Los Angeles, Atm. Environment 14:509

Spicer, C.W., 1980, The rate of NO_x reaction in transported urban air, Proc. of the 14th International Colloquium Paris, Studies in Environmental Science Vol. 8:181

TNO report 1980, Rekensysteem Luchtverontreiniging VII. Deelonderwerp: Verwijderingsprocessen SO_2, NO_x, CO en sulfaat, L.A.M. Janssen, ed., CMP-TNO, Delft

Varey, R.H., Sutton, S. and Marsh, A.R.W., 1978, The oxidation of nitric oxide in power station plumes. A numerical model, CERL laboratory note no. RD/L/N184/78

4: FORECASTING OF POLLUTANT CONCENTRATION UNDER EPISODIC CONDITIONS

Chairman: A. Berger   Rapporteur: M. Van Dop

FORECASTING OF FUMIGATION EPISODES IN THE PO VALLEY

P. Bacci[*], A. Longhetto[*] and D. Anfossi[x]

[*] ENEL/CRTN Bastioni Porta Volta 10, Milano, Italy

[x] CNR - Istituto Cosmogeofisica C. so Fiume 4, Torino
Italy

INTRODUCTION

Air quality management in the Po Valley needs particular
attention. It is the more densely populated region of Italy and
there are various and complex pollution sources in the area. It
experiences a complex climatology. Winds, which are induced by
the surrounding Alps and Appennines, are generally weak and with
large shears.

To be able to help the responsible authorities in taking
the proper measures, it is essential to have at one's disposal
forecasting models, as simple as possible but rigorously validated.
In order to develop such models, the physical processes responsible of the more dangerous at least potentially, pollution episodes, are to be known, studied and discriminated. Three conditions,
mainly, bring about situations of potential pollution risk in the
Po Valley : fumigations, diffusion under A or A/B stability categories (both occurring during the warm season) and persistent fog
(occurring in the late autumn and winter).

In the present work, we will deal with the first two
situations. The study was done on the site of Turbigo (35 km to
the west of Milano). The Po Valley is a flat region having dimensions of about 400 km x 100 km, with an average slope less than
one in a thousand. The presence of smaller size valleys make the
air circulation complex. Turbigo is situated at the bottom of
such a valley, the Ticino valley. The region is almost flat with
the mountains about 40 km to the North and 100 km to the South.

At Turbigo there is an ENEL Power Plant having a full load capacity of 1365 MW with six sections : 3 sections of 320 MW each and 3 of 260, 75 and 70 respectively. Two of the three 320 MW units are jointly connected to one stack, while all the other units have single stacks (see tab.1). The Power Plant has a network of five SO_2 samplers positioned as shown in fig. 1. As it can be seen, they are aligned along one particular direction, that of the prevailing mountain and valley breeze. Tab. 2 shows the distances (in metres) among stacks and SO_2 samplers.

The present analysis is based on some of the data automatically recorded, as a function of time, by the Power Plant computer. The parameters are : SO_2 g.l.c. in each sampling station, unit loads, wind speed measured at 20 and 60 m, air temperature at screen level. The paper examines the summer measurements (June to October) for two years (1977 and 1978). Only the situations in which, at least one monitoring station gave a g.l.c. greater than 0.1 ppm were considered. From the detailed analysis of the data, we tested which phenomenon (dispersion in A category, fumigations or other unknown causes) could be responsible of the g.l.c. peaks. Then we tried to evaluate the possibility of forecasting the occurrence of the peaks.

MODELS USED IN THE COMPUTATIONS

For the dispersion of plumes under unstable or neutral conditions we used the gaussian model

$$\chi = \frac{Q'}{\pi \, \dot{u} \, \sigma_y \, \sigma_z} \; e^{-\frac{y^2}{2\sigma_y^2}} \; e^{-\frac{H^2}{2\sigma_z^2}} \tag{1}$$

For the σ_y and σ_z curves we used those obtained previously in the Po Valley (Santomauro et al., 1978).

The word fumigation is used for any process carrying buoyant plumes to the ground (Gifford, 1975). Hence it is necessary to specify which processes we have identified in this analysis. Considering the peculiar climatology of the site under study, we examined :

(a) High wind fumigation. g.l.c. were calculated by means of

$$\chi_{max} = \frac{Q'}{\pi \, e \, H \, \Delta H \, u} \; \frac{\sigma z}{\sigma_y} \tag{2}$$

Table 1 – Turbigo Power Plant specifications

Unit (MW)	Stack	Stack height (m)	Exit diameter (m)
70	Ponente 1	48	3.6
75	Ponente 2	48	3.6
260	Levante 1	96	5.8
320	Levante 2	150	4.2
320+320	Levante 3	150	6.0

Table 2 – The five monitors are too close to the
3 Levante stacks to measure trapping

Stack / SO_2 Sampler	$P_1 + P_2$	L_1	L_2	L_3
1	2950	3150	3200	3225
2	4800	4625	4615	4699
3	2175	1950	1925	1900
4	340	625	700	815
5	865	550	520	490

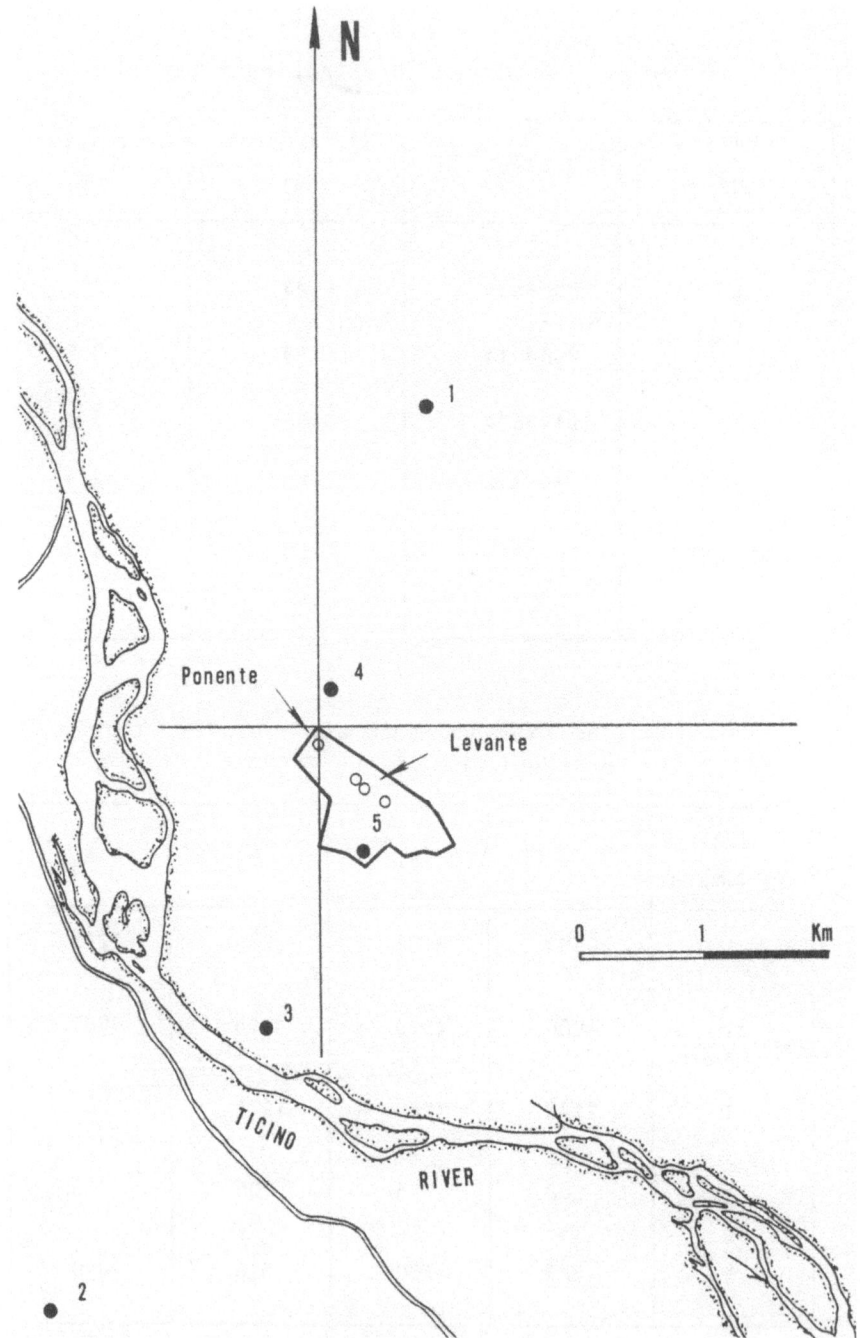

Figure 1 : The Power Plant has a network of
five SO_2-samplers

(Gifford, 1975), in which $H = Hs + \Delta H$ and σ_z/σ_y was set
equal to 0.7 as suggested both by Smith curves (1973)
and by Bacci et al. (1974). ΔH was computed (as in the
case of eq.1) by means of Briggs model (1969)·

$$\Delta H = 1.8 \ F^{1/3} \ u^{-1} \ (3 \ x_{\ddot{x}})^{2/3} \tag{3}$$

with $x_{\ddot{x}} = 2.16 \ F^{2/5} \ Hs^{3/5}$.

(b) Calm wind fumigation. It was used the model suggested by Briggs
(reference from Slade, 1968)

$$\chi_{max} = \frac{0.05 \ Q'}{F^{1/2} \ s^{-1/4}} \ H \tag{4}$$

where $H = Hs + 5.1 \ F^{1/4} \ s^{-3/8}$ $\tag{5}$
s being the stability parameter.
(c) Trapping or limited mixing fumigation. We used the formula

$$\chi = \frac{Q'}{\sqrt{2\pi} \ u \ h \ \sigma_y} \ e^{-\frac{y^2}{2\sigma_y^2}} \tag{6}$$

(Gifford, 1975) in which h is the height of the stable layer
capping the unstable or neutral layer in contact with the
ground. The exponential term allows us to calculate g.l.c.
when the sampling stations are not exactly on the plume axis
direction.

Strictly speaking, trapping occurs when aloft there is
an intense temperature inversion. In this case (Smith, 1973) one
has g.l.c. peaks lasting from 15 to 20 minutes and at downwind
distances greater than 20 Hs. In this respect, from tab. 2 it can
be seen that the five monitors are too close to the 3 Levante
stacks to measure trapping. There is instead the possibility that
the network could record some limited mixing fumigation from the
lower Ponente stacks. However, the analysis of our data did not
show any case of trapping from the lower stacks.

Later, it will be shown that peaks in g.l.c. are generally present between 11.00 and 14.00 (local time) and a preliminary
estimate gave us the impression that they were correlated to the
growth of the mixing height (M.H. from now on). Thus we tried to
verify if the following scheme of limited mixing fumigation could
be correct or not.

When the M.H. is growing from about 300 to 600 m, there
is a situation in which the plumes from Levante stacks are in the
slightly stable (or isotherm) layer. At this time the rising M.H.
starts to interact with the lower plume border. This phenomenon

limits the diffusion at the plume top while enhances that at the bottom. This could be a fumigation due to the transition between a diffusion classification of the type A + E (E aloft and A below) to A (extended to all the interested layer).

In this conditions, we put h (in eq. 6) equal to the M.H. This last quantity was evaluated by means of the Carson model (1973) in which the entrainment parameter was set to zero. This fits very well the experimental measurements of the M.H. in the Po Valley (Bacci et al., 1975 and 1979). Carson divides the M.H. growth in five phases. For the needs of the present analysis we shall consider only the first two. During the first one the nocturnal radiative inversion is destroyed. In this phase we should find trapping from the two 48m stacks. During the second one the growth rate is at its largest since the temperature profile is less stable and the solar radiation is reaching its greater values.

To compute the time at which this kind of fumigation occurs, we used

$$t_f = \frac{N}{\pi} \; \arccos \left(1 - 10.7 \times 10^{-4} \; \frac{H^2}{I_o} + 10.7 \times 10^{-4} \; \frac{\gamma 1 - \gamma 0}{\gamma 1} \; \frac{zi^2}{Io} \right) \quad (7)$$

This formula is derived for the second phase of the Carson model setting the M.H. equal to the height H of the plume and assuming a sinusoidal trend for the sensible heat flux. The meaning of the other symbols is :

t = fumigation time (hours)
N = Number of insolation hours during the examined day
I_o = total daily flux of sensible heat (cal/cm^2)
z_i = height of the top of the nocturnal inversion at dawn (m)
$\gamma 0$ = potential temperature gradient in the inversion layer (°K/m)
$\gamma 1$ = potential temperature gradient in the layer over the nocturnal inversion (°K/m).

I_o was evaluated from the relation (Bacci et al., 1979) $I_o = 0.275 \; Q$ where Q is the total incoming radiation. Q was deduced from the Smithsonian Tables, setting an atmospheric transmission coefficient equal to 0.75 (average value for the Po Valley). $\gamma 1$ was set equal to 0.0095 K/m, while $\gamma 0$ was obtained by the relationship

$$\gamma 0 = \frac{d\theta}{dz} = 0.0098 + \frac{0.63}{z_i} \; \Delta T \quad (8)$$

where ΔT is the temperature decrease, at screen level, during the night (Anfossi et al., 1976). z_i was calculated by means of

$$z_i = 70 \sqrt{t} \quad (9)$$

(t being expressed in hours and z in metres) obtained in the Po
Valley (Anfossi et al., 1974). In eq. 9, t represents the time
passed from sunset (in this case from sunset to dawn).

In all the calculations of this paragraph, we extrapolated wind speed measured at 60 m to 150 m (for the higher stacks)
by means of the power law expression

$$U_{z_2} / U_{z_1} = (z_2/z_1)^{0.3}$$ (10)

From analysis of data 0.3 was an appropriate exponent for Turbigo.
As wind direction was the more uncertain parameter in all the analyses (winds are weak and largerly sheared), we resorted to compute g.l.c. for the direction θ indicated by the anemometer and
also for the two directions θ + 20° and θ - 20°. The chosen concentration was the largest among the three calculated.

EXPERIMENTAL OBSERVATIONS

We examined all the experimental data before any computation. We found :

(a) During the examined periods 60 cases (i.e. g.l.c. greater than
    0.1 ppm) were analysed.
(b) Peaks can be present both in only one monitor and in more.
    They may last 1 or 2 hours.
(c) In a few cases peaks are contemporarely present in monitors
    which are opposite with respect to the stacks (that is : one
    upwind and the other downwind).
(d) Peaks were found in periods during which both Levante and Po-
    nente were operating and also when only Levante units were in
    operation.
(e) Wind speed shows a great variability (from calm to 15 m/s,
    recorded during foehn conditions). Prevalent are light winds.
(f) Wind direction is always scarcely defined. In fact direction
    is uncertain in light winds and it is difficult to extrapolate
    it higher up.
(g) If a few cases are excluded, the temperature decrease, at
    screen level, during the nights preceeding the examined days,
    is larger than about 13 degrees. On the other hand, it is not
    true the opposite, that is that every time we found a large
    Δ T at ground, we also found a g.l.c. peak.
(h) The peak shape is variable and the time of the recording does
    not allow any fine structure to be detected.
(i) While in the 70 % of cases the unit loads are constant in the
    hour preceeding the peak, in the remaining 30 % of cases there
    is a substantial load increase at least in one unit.
(1) The temperature sounding made at Linate (near Milano) at 00.00
    showed the presence of a stable layer aloft in about half of
    the cases. But only in 9 cases the stable layer was at the
    interested heights. In the other ones it was present above 1000 m.

(m) The distribution of occurrence of peaks in the monitors, or-
    dered from 1 to 5, is the following : 28 %, 13 %, 5 %, 37 %
    and 17 %.
    The average time of occurrence was 12.30 (local time) for the
    two stations (1 and 4) in the North and 11.20 for the three in
    the South. This fact reflects the prevailing breeze direction.

RESULTS AND DISCUSSION

    The 60 cases were divided into 3 categories according to wind
speed. In the first group (u ⩽ 1m/s) we have 6 cases, in the second one (1 < u < 5 m/s) 42 cases and in the third one (u ⩾ 5 m/s)
12 cases. We will now discuss these three groups of data.

(a) Calm conditions (u ⩽ 1 m/s)
    We tested the hypothesis of wind calm fumigation. Eq. (4) was
    used to compute g.l.c. Tab. 3 shows the data and the results.
    In the last column we indicated the stacks which seem to be
    responsible of the peaks. This tentative attribution is based
    on both the calculated g.l.c. and the distance stack-monitor.
    It is to be added that in all the cases but the first, there
    was an intense subsidence inversion at a height of 1000 ÷
    1200 m.
(b) Moderate wind (1 < u < 5 m/s).
    We tested the hypothesis of limited mixing fumigation. Time
    of occurrence of the fumigation was computed by means of eq.(7).
    The differences $\Delta t_f = |t_f^{meas} - t_f^{calc}|$ were also calculated.
    It was assumed that only those cases giving Δt_f less than
    1.5 hours could be considered belonging to the limited mixing
    fumigation.
    The limiting value of 1.5 h was chosen considering that the
    operations of changing the fuel burnt into the Power Plant
    from high to low sulphur content last about 30 min. This
    change is to be done any time meteorological conditions produ-
    cing large g.l.c. are foreseen. 24 cases out of 42 satisfied
    the limiting condition yielding an average Δt equal to 1 h
    (± 25 min).
       As regards the remaining 18 cases we had to test the hypo-
    thesis of normal diffusion under unstable conditions (A or B
    category). Unfortunately, only for 7 cases we could compute
    the g.l.c. since in the other cases it was not possible to
    establish a well defined wind direction. Tab. 4 shows the re-
    sults of such cases.
    χ = 0.0 in the last four columns means that the computed con-
    centration was less than 0.01 ppm. In the last five cases, the
    computed concentrations support the hypothesis of diffusion un-
    der unstable conditions. On the contrary, the first two cases
    seem to be due to other unknown reasons (such as, for example,
    other sources).

Date	Hour	Monitor	% meas (ppm)	% Calc. (ppm)				Distance Levante (m)	Distance Ponente (m)	Liable Stacks
				1L	2L	3L	P			
9/8/78	10.30	5	0.10	0.11	0.10	0.10	0.08	500	865	P+1L
9/8/78	10.15	4	0.14	0.21	0.10	0.11	0.08	340	700	P+1L
10/1/77	14.00	3	0.14	0.08	0.18	0.11	0.07	1925	2175	1L+2L+3L
10/1/77	14.30	5	0.13	0.09	0.10	0.11	0.07	300	865	P+1L
10/22/77	11.30	1	0.25	0.09	0.03	0.10	——	3200	2925	1L+2L+3L
10/22/77	11.30	4	0.16	0.05	0.03	0.10	——	340	700	1L (?)

Table 3   Data and results

Date	Hour	Monitor	U (m/s)	Dist. L (m)	Dist. P (m)	% meas (ppm)	Calculated (ppm)			
							L1	L2	L3	P
6/28/78	12.30	1	1.8	3200	2925	0.15	0.0	-	0.0	-
9/13/78	9.45	4	1.3	340	700	0.15	0.0	0.0	0.0	0.02
10/14/78	13.30	4	3.1	340	700	0.29	0.0	0.0	0.0	0.13
10/18/78	13.45	1	2.4	3200	2925	0.17	0.10	0.07	0.08	0.04
10/21/78	11.30	1	2.0	3200	2925	0.18	0.08	0.06	0.05	0.06
10/23/78	13.45	1	2.6	3200	2925	0.13	0.09	0.04	0.08	0.04
10/24/78	13.00	1	2.9	3200	2925	0.21	0.09	0.06	0.05	0.10

Table 4

Table 5

Date	U (m/s)	Dist. L (m)	Dist. P (m)	ϰ Meas. (ppm)	ϰ Calc. (ppm)				Case Number	Liable Stacks
					1L	2L	3L	P		
8/31/78	7.5	500	900	0.12	0.10	0.06	0.06	0.32	1	P
9/12/78	5.0	500	900	0.25	0.07	0.05	0.05	0.22	2	P
9/12/78	15.0	4800	4800	0.19	0.15	0.09	0.09	0.24	3	P+1L
7/4/77	5.5	700	450	0.11	—	0.05	0.05	—	4	
7/5/77	7.7	3150	2950	0.22	—	0.07	0.07	—	5	
7/11/77	5.5	3150	2950	0.17	—	0.05	0.05	—	6	
7/12/77	8.4	3150	2950	0.26	—	0.06	0.06	—	7	
7/13/77	5.0	3150	2950	0.22	—	0.05	0.05	—	8	
8/11/77	5.3	4800	4800	0.10	—	—	0.05	—	9	
8/29/77	14.6	3150	2950	0.26	0.13	—	0.09	—	10	1L
10/12/77	5.5	700	450	0.13	0.08	0.05	0.05	—	11	
10/25/77	7.3	3150	2950	0.32	0.10	0.06	0.07	0.31	12	P+1L

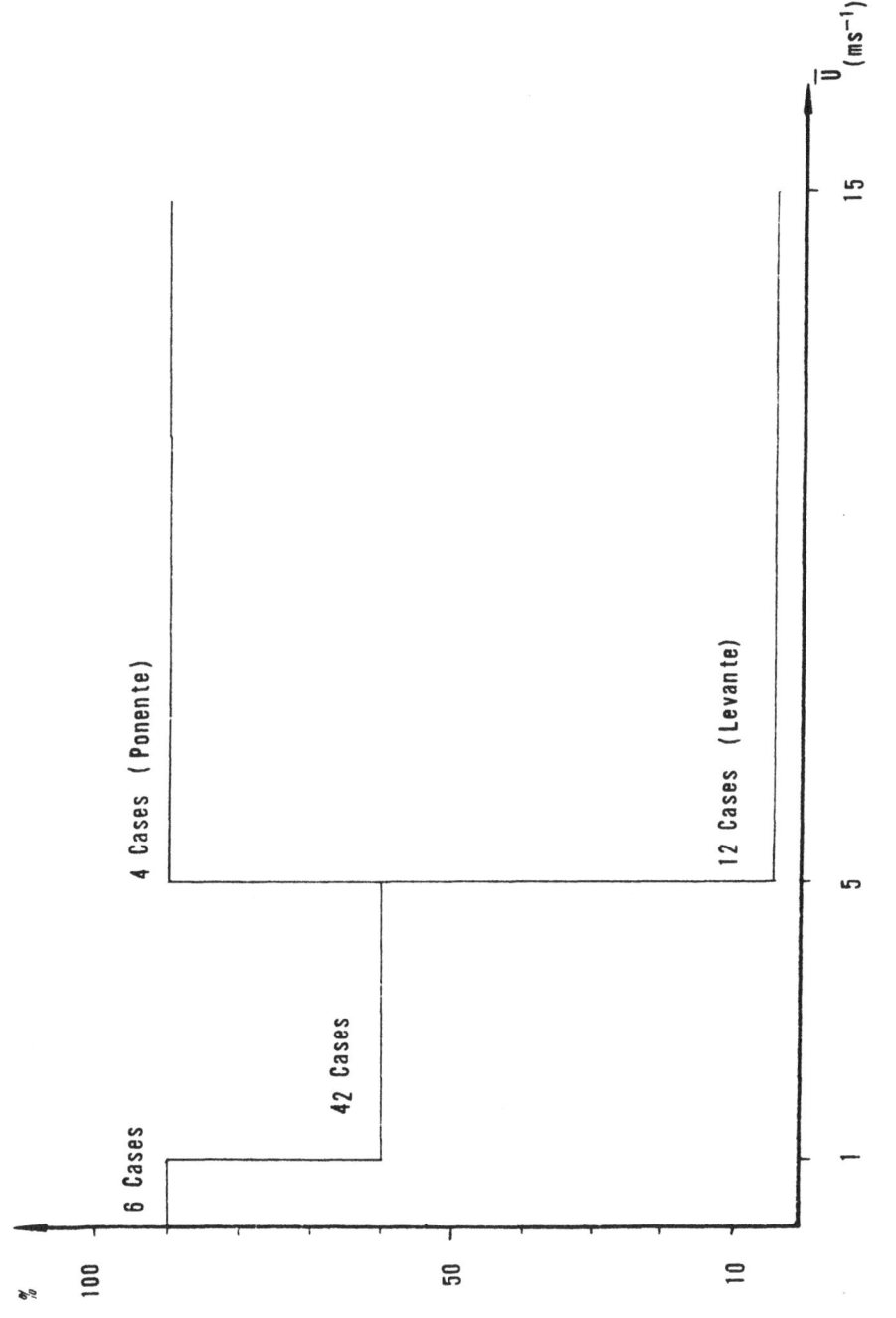

Figure 2

(c) High wind (u $\geqslant$ 5 m/s).
   We tested the hypothesis of high wind fumigation. G.l.c. were
   computed by means of eq. (2). Following Gifford (1975), we
   evaluated for the three stacks heights (48, 96 and 150 m) the
   three downwind distances 2150, 7400 and 16000 m at which the
   maximum g.l.c. is likely to occur. Keeping this information
   in mind and looking at Tab. 5, we observed that only in 5 cases
   we were able to attribute the observed peaks to the high wind
   fumigation (in the last column the supposed responsible stacks
   were indicated). In particular it appears that Ponente stacks,
   when in operation, are always considered responsible of the
   peaks. Foehn conditions, as it is nearly obvious, produce
   high wind fumigation. For the other 7 cases, not explained
   by fumigation, we calculated also the g.l.c. under neutral conditions (eq. 1). We did not find any agreement between these
   last computations and the measured concentrations. This induced us to conclude that also for these cases the cause is unknown.

CONCLUSIONS

   The major aim of this study was to test the possibility to
forecast pollutant dispersion mechanisms such as fumigations as a
function of the wind assumed as a predictor.

   Fig. 2 shows the results obtained considering all the situations in which g.l.c., at least at one sampling station, was
greater than a certain reasonable threshold value (by us set equal
to 0.1 ppm).
It can be clearly seen that the forecasting is satisfactorely for
extreme wind conditions (calm and u $\widetilde{~}$ 15 m/s). It decreases for
moderate winds. Thus, in this region, there is a competition between fumigation and usual diffusion mechanism which is described
by classical models. Hence forecasting in such situations is more
difficult. It is interesting to see that calm conditions, which
are generally considered the most difficult to be described, agree
with a simple model.

   The large variability of wind direction and of its vertical
shear introduce further complications. However, from a pollution
management point of view, the exacte angle of pollution fall out is
less important than the concentration as a function of the distance on a circle around the source.

REFERENCES

Anfossi D., Bacci P. and Longhetto A., 1974, An application of
   Lidar technique to the study of nocturnal radiation inversion.
   Atmos. Environ., 8:537
Anfossi D., Bacci P. and Longhetto A., 1976, Forecasting of ver-

tical temperature profiles in atmosphere during nocturnal radiation inversions from air temperature trend at screen height, Quart. J. R. Met. Soc., 102:173

Bacci P., Elisei G. and Longhetto A., 1974, Plume rise and dispersion at Ostiglia power station, Atmos. Environ., 8:1177

Bacci P., Longhetto A. and Anfossi D., Models of temperature profiles under convective and inversion conditions connected to air pollution, "Proceedings of 6th ITM on Air Pollution Modeling", Frankfurt, 313

Bacci P., Longhetto A. and Anfossi D., 1979, Sviluppo dello strato convettivo in Pianura Padana e previsione di inquinamento elevato, in "Atti della riunione del Subprogetto Aria, AC/3/149, 29

Briggs G. A., 1969, "Plume rise", Report TID - 25075, USAEC

Carson D. J., 1973, The development of a dry inversion capped convectively unstable boundary layer, Quart. J. R. Met. Soc., 99:450

Gifford F. A., 1975, Atmospheric dispersion models for environmental pollution applications, in "Lectures on Air Pollution and Environmental Impact Analyses", D. A. Haugen ed., A.M.S., Boston

Santomauro L., Bacci P., Longhetto A., Anfossi D., Richiardone R., 1978, Experimental evaluation of diffusion parameters at local scale by means of no-lift balloons, J. Appl. Meteor., 17:1441

Slade D., 1968, "Meteorology and Atomic Energy", USAEC ed.

Smith M.E., 1973, "Recommended guide for the predictions of the dispersion of airborne effluents", A.M.S. of Mech. Engin.

DISCUSSION

A. VENKATRAM   Could you explain what you mean by satisfactory agreement between model predictions and measurements. Do you have a quantitative measure of the degree of agreement ?

D. ANFOSSI   In validating our procedure we consider "satisfactory agreement" when the forecasted fumigation episode is actually recorded by the monitoring network. In preparing our procedure instead, we considered satisfactory the agreement when we were able to assign a fumigation episode to a precise stack and to a precise fumigation classification. The amount of agreement is then not quantitative but related to subjective evaluations.

M.L. KRAMER

1. The concentrations you report as "high"
   range up to 0.125 ppm.(in table 5, there
   are measured g.l.c. of 0.32 ppm!). How
   do these compare to your country's standards?

2. Has the forecast plan you discussed been
   implemented ?

D. ANFOSSI

1. Italian limit law is of 0.3 ppm over
   30 minutes. We started to consider
   ground level concentration of 0.1 ppm
   since this is a treshold value chosen
   to start the pre-alarm procedures. In
   table 5, the maximum measured g.l.c. is
   0.32 ppm. Furthermore, we did not report
   the data regarding all the 60 cases,
   but only a sample of 25 cases.

2. During the last part of this summer the
   Power Plant Staff started to apply our
   procedure. After the next summer we
   will have enough information to choose
   which road we have to follow in order
   to implement our procedure. New remote
   sensing instrumentation is now being in-
   stalled on this site (i.e.,R.A.S.S.,
   three-static doppler sodar and flowmeter).

NUMERICAL COMPUTATION OF HIGH AIR POLLUTION LEVELS

Cl. Demuth, G. Schayes, P. Hecq, M. Cravatte
Institute of Astronomy and Geophysics
Catholic University of Louvain
2   Chemin du Cyclotron
B-1348   Louvain-la-Neuve
Belgium

Abstract. The particular conditions during which
high air pollution levels are observed can be
simulated by an appropriate atmospheric boundary-
layer model (PBL) coupled with a diffusion model.
The atmospheric dispersion is described here by
the product of the cross-wind and vertical diffu-
sions. This is evaluated by a numerical 2-dimen-
sion model taking into account sedimentation,
deposition and/or other interactions with the
atmospheric environment. The numerical solution
of this 2D-model is obtained by finite difference
technique, the wind and vertical diffusion coefficient being provided by the PBL-model or a PBL-
parameterisation scheme.

Introduction

In this study, we wish to show that the application of a
simple diffusion model coupled to a PBL-model or PBL-parameterisation
scheme can give the concentration field when the particular case of
high atmospheric pollution levels occur in our country. In order
to achieve this goal, the following points will be developed successively :

a) the values of the meteorological parameters used in diffusion
   problems will be determined through the analysis of the observed
   pollution in the Gent region from June to December 1978.

b) the diffusion and PBL-models and PBL-parameterisation will then
   be summarized. The PBL-model is shown to simulate correctly
   the atmospheric layer for the Wangara experiment, which corresponds to a very particular atmospheric situation similar to

609

those related to high concentration levels in Gent. This success justifies the use of the PBL–model in order to simulate turbulent diffusivity profiles during these kinds of atmospheric conditions.

c) with those as an input to a diffusion model, atmospheric concentrations of pollutants are simulated in connection with a particular field experiment conducted in Belgium. Ratios between observed and computed concentrations appear to be between 0.7 and 1.1.

Characteristics of extreme pollution levels

In order to define the extremes, we have retained parameters which are commonly used in diffusion study, as windspeed, stability of the atmospheric boundary layer and mixing or inversion height. The stability is here defined in the sense of Bultynck-Malet[1] with temperature at two levels (8m, 114m) and windspeed at 69m.

Figures 1 to 3 show SO_2-concentrations greater than 400 µg/m^3 as a function of stability and windspeed. They are measured at thirteen receptors set out in the region of Gent during the period from June till December 78. These concentrations are measured every half an hour with identical instruments. The wind was measured at 30 meters heights.

Analysis of these figures indicates shows that the high concentration situations appear, in order of importance, as follows :

    i) in case of stability with a windspeed between 1 to 6 m/s,
   ii) in case of instability with a weak or zero wind,
  iii) in case of neutrality with a wind greater than 6 to 7 m/s,
   iv) in case of instability with a strong wind around 10 m/s.

High concentrations in stable situations arise with a ground based inversion which has a thickness of 200 meters. The influence of strong wind only appears for one station.

Excluding calm wind cases where pure convection is preponderant during the day, the other extreme cases can be handled with a diffusion model which can take into account the actual situation.

Modelling Procedure

This analysis shows how important it is to know accurately the vertical structure of the temperature and the vertical wind profile, if we want to simulate high concentrations by diffusion models. Modelling atmospheric diffusion of a pollutant in the atmosphere needs a planetary boundary layer model in order to take into account adequately the physics of the PBL and the diffusion processes. Owing to the fact that diffusion models and planetary

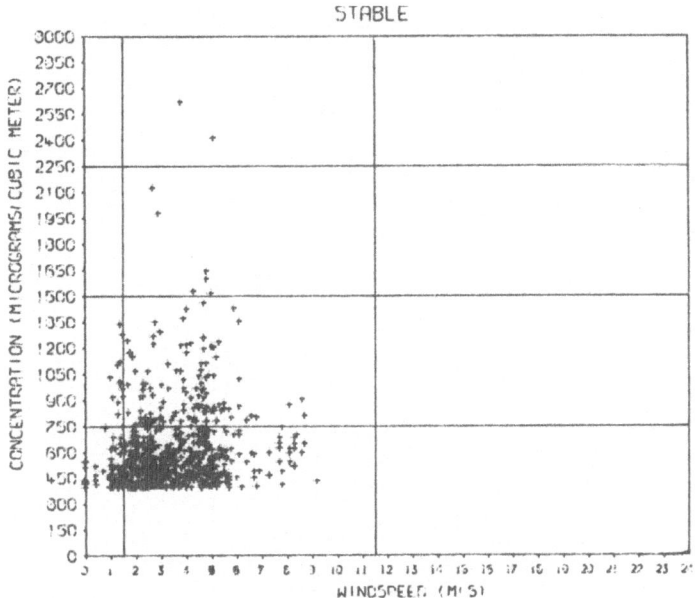

Fig. 1. Representation of SO$_2$- concentration measured by half-
hour at 13 receptors in Gent as a function of the windspeed in case of stability. (Concentrations $\geq$ 400 μ/m^3).
The measurement period is June to December '78.

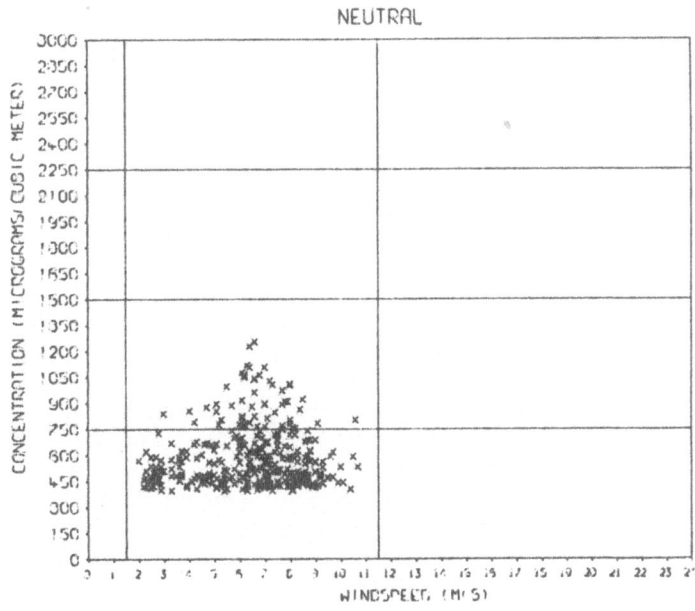

Fig. 2. Representation of the SO$_2$- concentration measured by half-
hour at 13 receptors in Gent as a function of the windspeed in case of neutral atm. (Concentrations $\geq$ 400 μg/m^3).

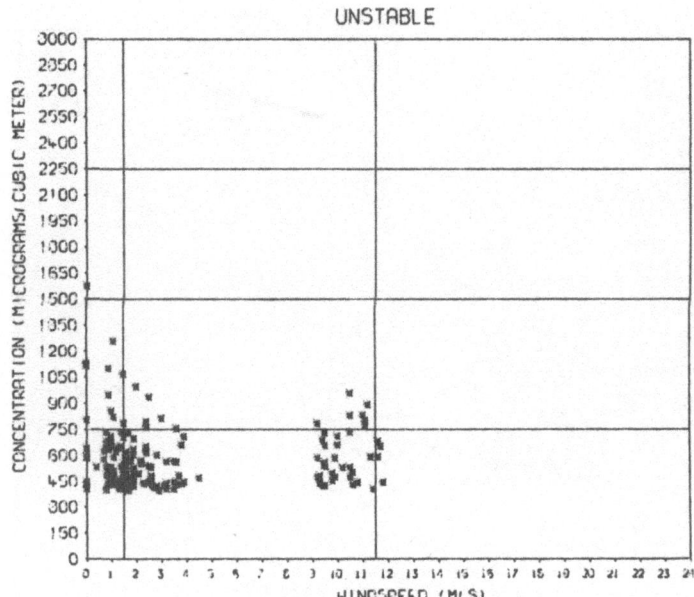

Fig. 3.   Representation of the SO_2^- concentration measured by half-
          hour at 13 receptors in Gent as a function of the wind-
          speed in case of instability. (Concentrations $\geq$ 400 $\mu g/m^3$).

boundary layer models are associated, we shall first define the
diffusion model and after that, the possibilities will be discussed
to model the planetary boundary layer simply and correctly.

     The type of diffusion model is dependent on the effective
duration of the pollution release : for continuous releases of a few
hours, stationarity hypothesis in time intervals (mean period) which
are compatible with the most commonly used meteorological scales,
for example, half an hour can be introduced. For a short release,
or in meteorological situations such as calm wind, an evolutive
model is indispensable. In addition to the stationarity and assuming the horizontal homogeneity hypothesis to be valid, we shall consider that the diffusion is the product of the gaussian transversal
diffusion, given by an analytical expression, by the diffusion in the
vertical plane obtained by numerical methods. This diffusion problem
is represented at Table 1.

Table 1. Steady-state equation for the dispersion in the vertical
          plane.

EVOLUTION EQUATION FOR THE POLLUTANT

$$U(z) \ \frac{\partial Z}{\partial x} = \frac{\partial}{\partial z} \ K_z(z) \ \frac{\partial Z}{\partial z} + w_s \ \frac{\partial Z}{\partial z} - \lambda \ Z + Q\delta(x)\delta(z-H)$$

advection = diffusion+sedimentation+interaction+source

INITIAL CONDITION : $Z(0,z) = 0$

BOUNDARY CONDITIONS: $K_z \ \dfrac{\partial Z}{\partial z} + w_s Z = aZ$ $\qquad$ at $z = 0$

$\qquad\qquad\qquad\qquad K_z \ \dfrac{\partial Z}{\partial z} + w_s Z = 0$ $\qquad$ at $z = L$

PARAMETERS

- Z : average concentration
- x : distance downwind
- z : height above ground
- L : mixing height
- U : mean wind speed, a function of height
- K_z : diffusivity along z, a function of height
- w_s : fall speed of particulate matter (constant)
- λ : proportionality constant in the interaction term, the
      resulting effect of all interactions is approximed
      by a linear destruction term
- Q : strengh of the continuous source
- a : proportionality constant for the flux into the ground,
      the ground is assumed to act as a sink
- H : effective source height

The transversal diffusion is defined on the basis of the knowledge of the dispersion parameter σ_y depending on the chosen mean period. For half-hourly periods, the σ_y parameters are those of Bultynck-Malet. The vertical diffusion is deduced from a planetary boundary layer (PBL) model or from a parameterisation of this layer. The choice of one of these methods is depending on available meteorological data.

a) Model of the planetary boundary layer

To find an approximation of the diffusivity profile, a PBL model is used to simulate the evolution of the boundary layer in specific meteorological situations. The different kinds of such models depend on the turbulence closure scheme adopted. For this closure, the turbulent energy equation (also called 1.5 order closure) gives a better physical simulation than the simple first order-closure, without having the complexity of higher order models.

Since the horizontal homogeneity is assumed, a one dimensional model (only in z) is used. This model, developped by Schayes-Cravatte[2], gives the time-evolution of vertical profiles of different meteorological variables. The initial conditions are vertical profiles of wind, temperature, humidity and turbulent energy. Boundary conditions are geostrophic wind, temperature and humidity at ground-level.

The efficiency of this model for extreme concentration levels with a non-zero wind has been demonstrated with the experiment of Wangara in Australia where the meteorological situations are close to the two first high concentration situations.[3] For the 33rd day of this experiment, the model still reproduces very well the temperature profile during day and night. The wind profile being also close to reality, the PBL-model allows to simulate correctly the diffusion coefficient. Figure 4 reproduce the turbulent diffusivity profiles for day and night. During the day, the K_z-profile presents a maximum of 300 m^2/s at 600 m above the ground. For the night, these profiles are more irregular with a maximum near the ground and a second above.

b) Parameterisation scheme of the atmospheric boundary layer

This PBL model assumes that sufficient information is available for the initialisation of the differential equations, which is usually not the case. Very often, only synoptic meteorological data are known at several points in a region. Sometimes, temperature and wind are known at a few levels along a meteorological mast. Following the nature of the available data, a parameterisation of the atmospheric boundary layer is possible which allows to determine the upward buoyancy flux Q_0 at the ground with the friction velocity u_* for a given roughness length z_0. With the knowledge of Q_0, u_*, the turbulent diffusivity profile can be

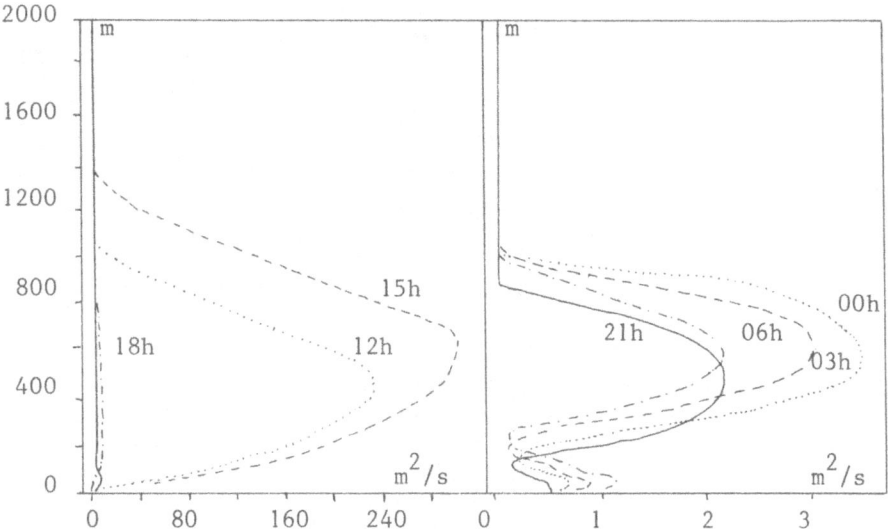

Fig. 4. Computed profiles of eddy diffusivity coefficient during
day 33 (left) and during night 33-34 (right) at Wangara.

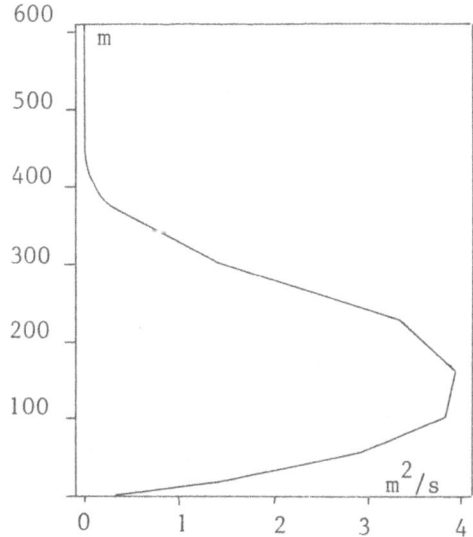

Fig. 5. Computed profile of eddy diffusivity coefficient by PBL-
model.

defined by an analytical function of the stability of the PBL. In
this study, we shall apply the Wipperman's function[4] given by the
expression

$$K_z(z) = k \, u_* \, z \, \exp\left[-c(\mu) \, \left(\frac{z \, f}{ku_*}\right)^{0.764}\right] \tag{1}$$

where k is the Von Karman constant (0.35), f is the Coriolis parameter, μ is the PBL stability parameter and $c(\mu)$ is an empirical
function of μ. Rao and Snodgrass[5] present, for stable conditions
($\mu \geq 0$), a turbulent diffusivity profile represented by

$$K_z(z) = \frac{k \, u_* \, z}{(1 + 4.7 \frac{z}{L})} \, \exp\left[-9.1 \, \frac{z}{L \, \mu^{1/2}}\right] \tag{2}$$

where L is Monin-Obukhov stability length. Other relations are
possible but, as our aim is not a sensitivity study to K_z, we have
adopted only relation (1).

When synoptic meteorological informations are only available,
Schayes-Cravatte's scheme for obtaining the PBL parameters is useful.

In our case, as we start with the knowledge of the horizontal
wind u and the potential temperature θ at some levels along a mast,
Louis's[6] scheme is adopted. He proposed the determination of parameters u_* and Q_0 as function of the bulk Richardson number which is
given by :

$$Ri_B = \frac{\beta \, z_1 \, \delta_1 \theta}{|u(z_1)|^2} \tag{3}$$

where β is the buoyancy parameter (g/T_0), T_0 is the mean temperature
between the levels z_1 and z_0, $\delta_1\theta$ is the difference between the
θ-values at z_1 and z_0 and $u(z_1)$ is the horizontal wind at the level
z_1. If the potential temperature and the wind are measured at levels
z_1 and z_0, then it is possible to obtain u_* and Q_0 following the
scheme presented into Table 2.

In our application and taking into account the definition of
the stability classes of Bultynck-Malet, we will use in (3) temperature
at levels 8m and 114m and wind at the intermediate level of 69m.

This approach allows to define a profile for K_z following a
power law. This is applied over a characteristic vertical length
by fitting the lower part of the real profile and assuming a constant value above the maximum. In this case, the application of an
analytical diffusion model can so be considered[7]

Table 2.   The Louis scheme for the representation of the vertical surface temperature flux Q_0 and the friction velocity u_*:

$$u_* = b^{1/2} u(z_1)$$
$$Q_0 = -c \; \delta_1 \theta \; u(z_1)$$

UNSTABLE CASE
$Ri_B < 0$:

$$b = a^2 \left(1 - \frac{9.4 \; Ri_B}{1 + 69.56 \; a^2 \; (-Ri_B \; z_1/z_0)^{1/2}}\right)$$

$$c = \frac{a^2}{0.74} \left(1 - \frac{9.4 \; Ri_B}{1 + 49.82 \; a^2 \; (-Ri_B \; z_1/z_0)^{1/2}}\right)$$

STABLE CASE
$0 \leq Ri_B$:

$$b = \frac{a^2}{(1 + 4.7 \; Ri_B)^2}$$

$$c = \frac{b}{0.74}$$

where $a^2 = \dfrac{k^2}{\ln^2(z_1/z_0)}$ is the drag coefficient in neutral case.

Application

The diffusion model coupled to a   PBL-model has been applied to a data base obtained from field experiment. The sounding of the atmospheric boundary layer with captive balloon allowed the initialisation of all necessary parameters for the application of the PBL-model. Without going into the details of the measurements but knowing that the meteorological situation is close to the third case of high concentration situation in Gent, the PBL-model gives the diffusivity profile represented in Fig. 5 for the first half an hour, the most important mean period for the release.

The application of the diffusion model presented above with the K_z-profile deduced from the PBL-model, gives acceptable values for the ratio between calculated and measured concentration at each receptor (between 0.7 and 1.1). This is particularly significant if you consider the limited number of receptors. However, close to the source, the observed concentrations are much larger than the computed ones, this being the result of the local mechanical turbulence induced by the nearby buildings. Table 3 presents these results obtained with the K_z-profile from the PBL-model and the power law deduced from the profile. The diffusion model with a power law for K_z overestimates the observed concentration field. In each calculation, the wind profile has been defined by a power

law where the parameters have been taken from the Bultynck-Malet
stability classes.

Table 3.   Ratios between calculated and measured concentration at
            five receptors for a K_z calculated by the PBL-model and
            by a power law deduced from this model. The release
            of the continuous source is of 1 g/s.

Receptors	ψ_{cal} in $\mu g/m^3$	
	(PBL)	(power law)
1 (1.6 km)	0.878	1.263
2 (1.5 km)	0.034	0.052
3 (1.6 km)	0.383	0.566
4 (1.7 km)	1.145	1.577
5 (3.5 km)	0.529	0.578

The K_z-profile obtained from Louis parameterisation scheme
does not give a realistic profile except if the roughness length
z_0 is taken as 0.3m, which seems a too large value for the site
under investigation.

Conclusion

Application of the steady-state diffusion equation with a
vertical profile for eddy K_z given by a PBL-model allows to obtain
the concentration values which agree with observations. The use
of a parameterisation scheme in order to define the K_z-profile from
temperature gradient seems more difficult. In both cases, the method
presented in this paper needs to be applied to other experiments before definite conclusions may be drawn but the present results are
encouraging.

Acknowledgements

This research is a part of the National R-D Programme on
Environment-Air, financed by the Services for Science Policy Programming, Belgium.

References

1. Bultynck H., Malet L., 1972 : Evaluation of atmospheric dilu-
     tion factors for effluents diffused from an elevated conti-
     nuous point source. Tellus XXIV, 5.
2. Schayes G., Cravatte M., 1980 : Diffusivity profiles deduced
     from synoptic data. 11th ITM-NATO/CCMS, Amsterdam, Nov. 1980.

3. Clarke et al., 1971 : The Wangara experiment. Boundary Layer
     Data. Techn. Paper n° 19 CSIRO, Australia.
4. Wippermann F., 1972 : Universal profiles in the barotropic
     boundary layer. Contr. Atm. Phys. $\underline{45}$, pp. 148-163.
5. Rao K.S., Snodgrass H.F., 1979 : Some parameterizations of the
     nocturnal boundary layer. Boundary Layer Meteorology $\underline{17}$,
     pp. 15-28.
6. Louis J.F., 1979 : A parametric model of vertical eddy fluxes
     in the atmosphere. Boundary Layer Meteorology $\underline{17}$, pp. 187-
     202.
7. Demuth Cl., Berger A., Jacquart Y., Schayes G., 1977 : A K-
     analytical model, including calm wind situations. in 8th
     NATO/CCMS on Air Pollution Modeling and its Applications,
     Louvain-la-Neuve. NATO/CCMS n° 80, pp. 610-631.

APPLICATION OF A PHOTOCHEMICAL DISPERSION MODEL TO THE NETHER-

LANDS AND ITS SURROUNDINGS

P.J.H. Builtjes (1), K.D. van den Hout (2), C. Veldt (1),
H.J. Huldy (1), J. Hulshoff (3), W. Basting (1), R. van
Aalst (1)
(1) MT/TNO, Postbus 342, Apeldoorn, the Netherlands
(2) IMG/TNO, Post 214, Delft, the Netherlands
(3) IWIS/TNO, Postbus 297, Den Haag, the Netherlands

ABSTRACT

The investigation deals with the application of a grid-
based Eulerian photochemical dispersion model developed by
Systems Application Inc., U.S.A.. The model calculates concentrations of O_3, PAN, NO, NO_2, SO_2, etc. with emissions,
meteorological parameters and initial and boundary values as
input data.

The modelling region covers an area of 230x310 km^2, and
includes the Netherlands and its surroundings, the Ruhr-area
and Antwerp. Concentrations are being calculated on an hourly
basis for grids of 10x10 km^2, with 5 levels in the vertical
direction.

The purpose of this investigation is to compare calculated concentrations for the episode of 7/8 June 1976 with
measured concentrations, and subsequently investigate the
usefulness of this model for the study of photochemical air
pollution problems.
Preliminary results are presented.

INTRODUCTION

Commonly used dispersion models are mostly of the
Gaussian plume model type. These models are capable of
calculating concentrations caused by emissions of an inert
or simple reacting species (such as SO_2).

However, there is often an interest in the concentration
levels of chemically reactive pollutants and in concentration

levels of inert as well as chemically reactive pollutants under
special meteorological conditions (episodes).

Especially in the United States a great deal of attention
has been given to the development of methods suitable to determine concentration levels during photochemical episodes.

An often used method at this moment is the so-called
'Empirical Kinetic Modelling Approach' (EKMA) (1). This model
is based on an isopleth diagram generated by computer simulations of the chemical transformations that occur among pollutants in a relatively large well-mixed box subjected to
dilution and time-varying light intensity.

By using measured O_3-concentrations and NO_x and NMHC-
concentrations (non methane hydrocarbons) in a certain area
it is possible to determine the necessary decrease of NO_x
and NMHC-emissions to be able to reach a desired O_3-level.
However, the treatment of emissions, pollutant transport and
their interactions with chemical processes is limited. In
addition, the model cannot predict spatial effects.

In view of these restrictions it was decided to investigate the usefulness of more sophisticated dispersion models,
next to the study of the EKMA-model. The aim of the study
was to have available a physically sound dispersion model
which is capable of calculating spatial and temporal effects.
This dispersion model should, in principle, be capable of calculating the dispersion of any arbitrary pollutant during
episodes, and not only be directed to photochemistry.

In the Netherlands, a large project is carried out directed to the study of photo-chemistry (2), which project
is partly sponsored by the government. This corresponds to
the increasing interest in Western Europe in the problems
of photochemical air pollution, especially in the United
Kingdom, Sweden, Norway, Germany and the Netherlands.
Another large project, sponsored by the Dutch government
is directed to the dispersion of mainly SO_2 and NO_x, and
has the purpose to calculate the influence of future
emission changes on airquality (3). It is intended to use
the dispersion model to be developed and considered in
the study described here also to study the impact of control strategies and future emission changes on air quality.

DESCRIPTION OF THE DISPERSION MODEL

The purpose of this study is to develop a dispersion
model, or to apply an existing model to describe and calculate

photo-chemical dispersion processes. After a literature survey
it was decided to apply the so-called SAI-airshed model, developed at Systems Applications Inc., U.S.A.. The SAI-model takes
into account all the relevant physical and chemical processes
in a sound manner. Furthermore the model calculates concentrations of pollutants in a spatial and temporal resolution.
For a general description of the model see Reynolds e.a. (4).
The SAI-model is a so-called grid model. With the emissions
of NO, NO_2, SO_2, hydro-carbons etc. and meteorological information it calculates the concentrations of O_3, PAN, NO, NO_2,
SO_2, for every grid for hourly averaged concentrations. The
size of the area studied is typically about 150x150 km^2, with
grids of 2x2 km^2 to 10 x 10 km^2 and several levels in the
vertical direction. The total simulation is based on one or
two days. The model has been used for several regions in the
United States, especially for regulation questions such as
determination of the most suitable place of a large power
station, the impact of certain car-exhaust regulations, the
air-quality changes by different economical and industrial
development scenarios, etc. (5, 6, 7). Up till now, the model
has been used only once in Europe (8). It should also be
mentioned that a modified version of the model has been used
to study the dispersion of SO_2 during wintertime eposides (9).

The basis of the model is the well-known diffusion equation for each pollutant considered:

$$\frac{\delta C_i}{\delta t} + \frac{\delta(u \; C_i)}{\delta x} + \frac{\delta(v \; C_i)}{\delta y} + \frac{\delta(w \; C_i)}{\delta z} =$$

time
dependence   advection

$$\frac{\delta}{\delta x}\left(K_H \frac{\delta C_i}{\delta x} \right) + \frac{\delta}{\delta y}\left(K_H \frac{\delta C_i}{\delta y} \right) + \frac{\delta}{\delta z}\left(K_V \frac{\delta C_i}{\delta z} \right)$$

turbulent diffusion

$+ \; R_i$ $+ \; S_i$ (1)

chemical   emission
reaction

C_i is the pollutant concentration which is a function of
space (x, y, z) and time (t). The equations of the different species are coupled by the chemical reaction term R_i.

One of the major problems in using grid models based
on the solution of equation (1) is the error introduced by the
so-called numerical diffusion. The SAI-model uses a numerical
solution scheme called SHASTA, which avoids numerical diffusion
to a sufficient degree of accuracy.

A possible treatment of the advection provided by the model
is based on the solution of the conservation of mass equation
using measured wind velocities. A complete hourly averaged wind
field is calculated in this way using certain interpolation
techniques.

The turbulent diffusion is specified by the turbulent
diffusivities K_H and K_V. Because the advection dominates the
horizontal transport and no grids smaller that $2x2$ km^2 are
applied, a constant value of $K_H = 50$ m^2/s is used. For the
vertical transport, the vertical diffusivity is dominant.
A scheme is used which calculates K_V as a function of mainly
the stability class, the ground-level wind speed and the surface roughness. The diffusion break (mixing height) is used
to determine at which height K_V assumes a small value.

The chemical mechanism used in the SAI-model is the
so-called Carbon-Bond Mechanism. Its main feature is the
splitting of all hydrocarbons in four 'effective' hydrocarbon groups: aldehydes, aromates, olefins and paraffins.
In addition to this scheme the model can also calculate the
oxidation of SO_2 and the formation of aerosols.

Ground-level and point-source emissions are considered.
The emissions are instantaneously mixed in the grid cell
in which they are emitted. For the point-source emissions
the plume-rise formula of Briggs is used. Of importance are
also the initial and boundary conditions for all species.
The deposition velocities are calculated according to species and landuse.

The model is divided in sub-models for each part of
the calculation, for example a windfield submodel, a chemical sub-model, a boundary value sub-model etc.. The output
of each sub-model can be considered separately, before
these outputs are used as input to the simulation program
itself.

In using the model it is of importance to keep the
accuracy of the different inputs in balance. The construction of the SAI-model is such that the level of sophistication of the different parts of the model is also in
balance.

THE APPLICATION STUDY

The simulation has been carried out using a CDC-171 computer with a total capacity of 128 K. The total calculation time (cpu-seconds) on this relatively slow computer was about 12 hours for a 24 hour real time simulation. The total simulation area has been divided into 6 segments to cope with the memory needed. In the following a description will be given of the modelling region, the episode and the input-data.

The modelling region and the episode considered

In general the choice of the modelling region will be highly dependent on the distribution of emission sources in the area of interest. In contrast to the situation in the U.S.A., where some cities can be considered as isolated the distance between industrialized areas in Western Europe is relatively small. It is then desirable to chose the modelling region such that the pollutant concentrations present in the area of interest are significantly determined by emissions inside the modelling region. Therefore the modelling region was taken large enough to include important up-wind source areas in Belgium (Antwerp) and Germany (Ruhr-area), see Fig. 1. Practical considerations concerning among others the availability of emission data and computer capacity prohibited further enlargement of the area.

The model is applied to the simulation of the photo-chemical episode of 7 and 8 June 1976. An extension of a high pressure area centered over the North Sea was drifting North-Eastwards, resulting in weak winds during daytime of June 7 and a South-Easterly flow during the remainder of the episode. Surface wind speeds at June 8 were around 5 m/s. The maximum temperature was around 28°C. Based on the desired resolution and computer capacity the area is divided into grids of 10x10 km^2, there are five levels in the vertical direction. The size of the vertical cells is varying. The ground cell has a height of 50 m, the total height of the modelling region is 2100 m.

The emission-input

The total emissions are divided in area sources and point sources. The point sources are tall stacks from power plants, refineries and metal industry. In total 226 point sources are used in the emission inventory. Stacks situated close together are often combined to one stack. The emissions from stacks are SO_2, CO and NO_x. The area sources have an area of 10x10 km^2, and the emissions are

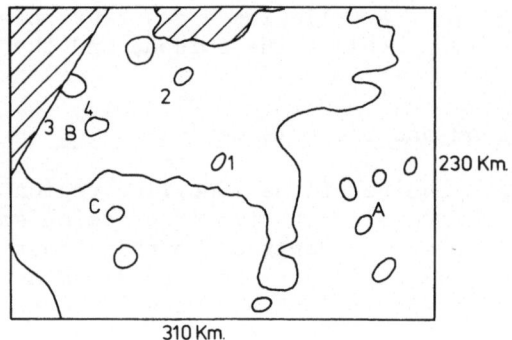

Fig. 1. The modelling region. A: the Ruhr area. B: the Rynmond area. C: Antwerp

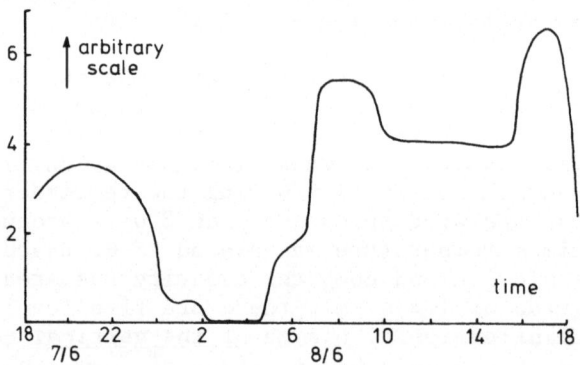

Fig. 2. Local traffic emission-changes with time

the added contributions from traffic, shipping, small industries
inhabitents-related activities such as painting and natural emissions and contain SO_2, CO, NO_x and hydrocarbons. The hydrocarbons
are divided into four categories; aldehydes, paraffins, olefins and aromates. The total NO_x emission is divided by
weight into 85% NO and 15% NO_2. Table 1 gives an overview
of the emissions used in the modelling area.

Table 1. Emission inventory for the episode June 7,8 1976

Total Emissions in kg/hour averaged over the periode	Point sources / Area sources		Germany/Belgium/The Nether-lands		
SO_2 :370.000	56%	44%	63%	26%	11%
NO_x :195.000	40%	60%	66%	13%	21%
CO :760.000	6%	94%	68%	11%	21%
Paraffins: 44.000	–	100%	38%	23%	39%
Olefins : 14.000	–	100%	36%	17%	47%
Aromates : 40.000	–	100%	28%	13%	59%
Aldehydes: 7.000	–	100%	46%	12%	42%

As an example of the time-dependency of some emission categories Fig. 2 shows the traffic emission changes with time.

The meteorological input

The meteorological data consist of wind-field data, mixing
height, ground-level temperature, vertical temperature gradients, stability class of the atmosphere, water concentration
atmospheric pressure and photolysis rate. As wind-data wind
velocities at every hour, has to be given. The preparation
of this wind field is carried out by first applying an interpolation scheme to derive from measured ground-level wind
velocities at about 60 stations in the region wind velocities for all ground-level grid cells. The vertical velocity
profile is produced by defining at the four corners of the
region velocity profiles based on an analysis of measurements at several meteorological towers and on rawinsonde
data. Using again an interpolation schema, combined with
a decreasing factor of influence between ground-level data
and upper-wind velocities with increasing distance to the
ground, windfields were created for every hour. In Fig. 3
vertical velocity profiles are given for various times.
These profiles exhibit a clear nocturnal jet. An important factor in the vertical dispersion processes is the
mixing height. The mixing height has been derived from
rawin sonde measurements at several locations in the

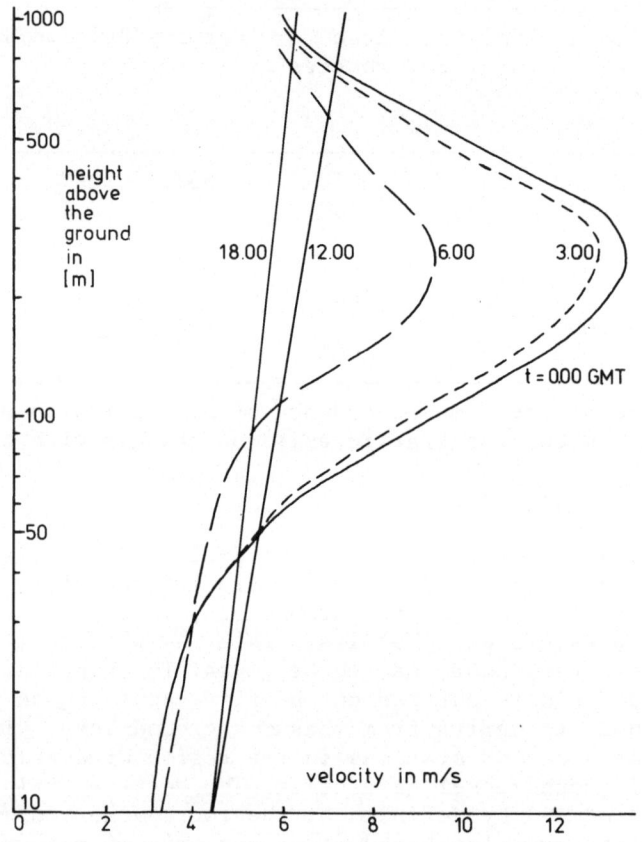

Fig. 3.   Vertical velocity profiles at June 8

area, and calculations of the mixing height using mainly the measured heat flux based on a model developed by Tennekes (10). Because no significant horizontal variations were observed, the mixing height, shown in Fig. 4, is used throughout the whole region. For the chemical reaction mechanism, the photolysis rate of NO_2 is an important input parameter. Fig. 5 shows the photolysisrate used, which is based on UV-measurements. The model takes into account the additional decrease of the photolysis rate caused by the calculated formation of aerosols.

Surface characteristics

For every ground-level grid cell a roughness and a vegetation factor is needed based on the type of terrain. The roughness factor influences the dispersion, the vegetation factor determines the removal of pollutants by the deposition (for detailed information, see (11)).

The initial and boundary conditions input

The determination of the initial and boundary conditions of the concentrations often creates difficulties, due to the lack of outdoor measurements for many species, especially at higher levels. The influence of the initial conditions will decrease with increasing simulation time, the influence of the boundary conditions will remain throughout the simulation. Due to the limited data, the uncertainty in the conditions used in the first simulation calculation is large. Although the region is chosen so as to contain the major emission-sources in the surroundings of the Netherlands, the total amount of influx of pollutants across the boundary caused by the boundary conditions based on the few outdoor measurements, is large. The total influx of SO_2 is about 20% of the emission of SO_2 in the area, for NO_x this number is about 40%. The influx of CO is about 7 times larger than the emissions, for NMHC this value is about 6. Consequently, the influence of the boundary conditions, especially of the value for NMHC, on the formation of photo-chemical pollutants will be large. In the near future very careful consideration has to be given to the values used for the boundary conditions.

THE CALCULATED RESULTS

The above-described input data have been used to perform the first simulation with the model, for the period of June 7, 18.00 GMT until June 8, 18.00 GMT. The model output consists of the concentrations of 13 pollutants, containing SO_2, O_3, NO, NO_2, PAN and hydrocarbons. The concentrations are given for every grid cell of 10x10 km^2, with 5 levels in the vertical direction up to 2100 m. The results are given at every hour, as well as

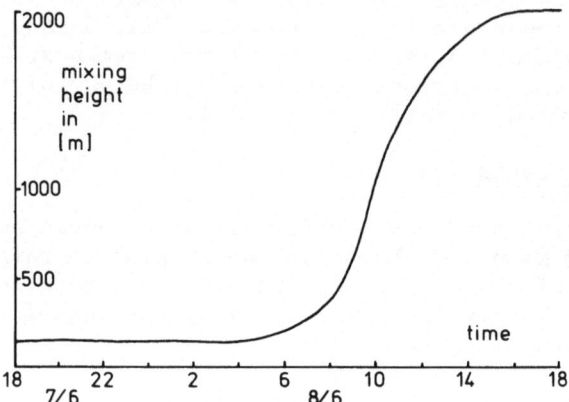

Fig. 4.   The mixing height

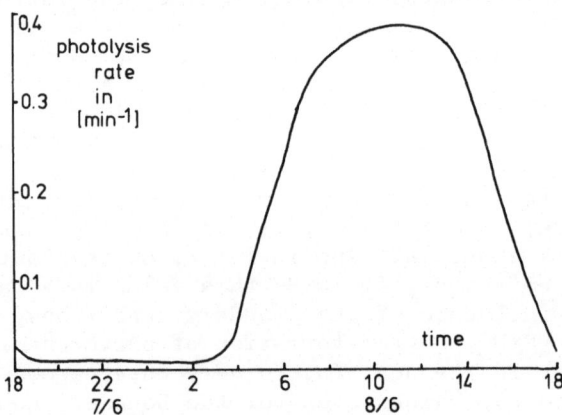

Fig. 5.   The photolysis rate

values averaged over one hour. The O_3-concentrations calculated averaged between 15.00 and 16.00 GMT show the highest values. At that time the O_3-concentration in Germany is about 185 $\mu g/m^3$, the level increases in the downwind direction and reaches a value of about 250 $\mu g/m^3$ in the most downwind area. The value averaged over the whole region is 205 $\mu g/m^3$, the standard deviation is about 10%. These results show that there are only small concentration gradients calculated for this secondary pollutant. The usefulness of a model is often judged by comparing the calculated and the observed concentrations. Before doing this, the following remarks are made (12). A distinction should be made between model verification and model validation. Model verification means the consideration of the fundamental correctness of the model formulation in view of the description of the physical and chemical processes. In this respect, the SAI model formulation is considered to be rather good compared with other models. Model validation means a comparison between measured and calculated concentrations. In the case that a perfect model is used, still the calculations can differ from the measurements. This is caused by the following facts. The measurements are pointwise, the calculated values are volume-averaged. There can be instrumental errors and there can be errors in the model input parameters.

Several statistical methods exist to compare predicted and observed values (12). At the present stage of this study, only graphical comparisons have been made. Fig. 6 shows a comparison of O_3-concentrations at three monitoring stations. The location of these stations is indicated in Fig. 1. The comparison is based on results averaged over one hour, the value given e.g. at 16.00 GMT is the averaged value between 15.00 and 16.00 GMT. In general the calculated O_3-pattern compares well with the measured pattern. The calculated pattern shows less detail, possibly caused by the averaging over 10x10 km^2. The maximum O_3-concentration calculated is comparable to the measured values, but is reached, on the average, about one hour earlier. The O_3-concentrations calculated during the night are higher than the observed results, which probably indicates a too large vertical exchange or a too low NO-emission. The results for the monitoring station at Vlaardingen of NO and NO_2 are shown in Fig. 7. The calculated NO-concentrations are clearly lower than the observed values. The measured NO-values are influenced by local sources, such as traffic, the calculated values are averaged values over a gridcell of 10x10 km^2. The calculated NO_2-concentrations are in reasonable agreement with the observations, although the pattern with time is different. These results clearly indicate the difficulties created in comparing pointwise measurements with volume-averaged predictions. However, it should be remarked that still the O_3-concentrations observed and calculated at Vlaardingen (see Fig.6) are in close agreement.

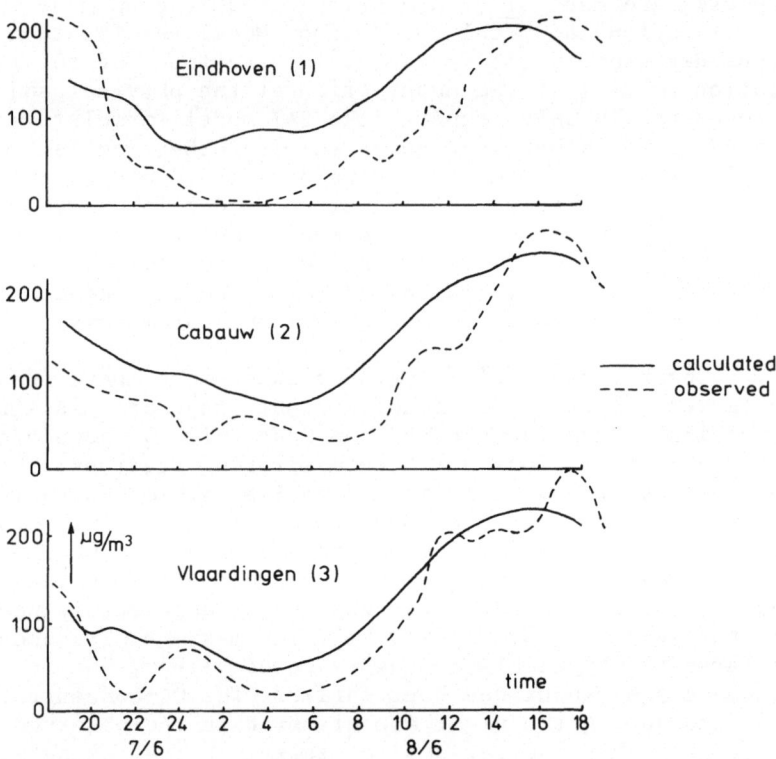

Fig. 6. Comparison of O_3-concentrations (see also Fig. 1)

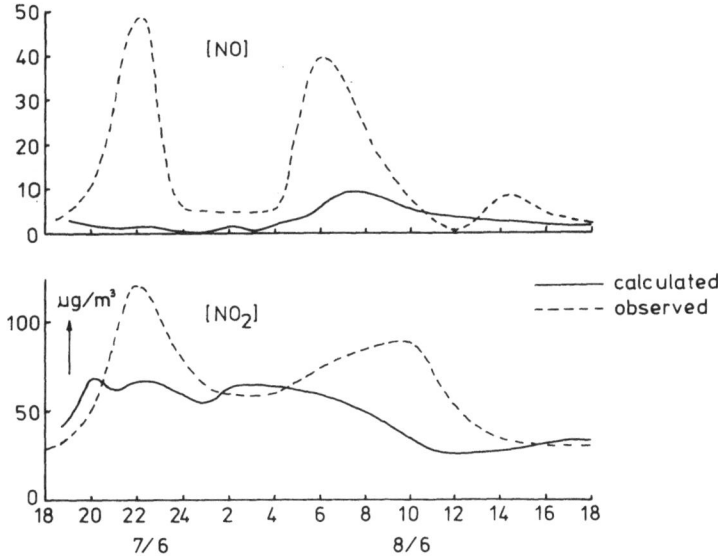

Fig. 7.   Comparison of concentrations of NO and NO$_2$ at
          Vlaardingen (3)

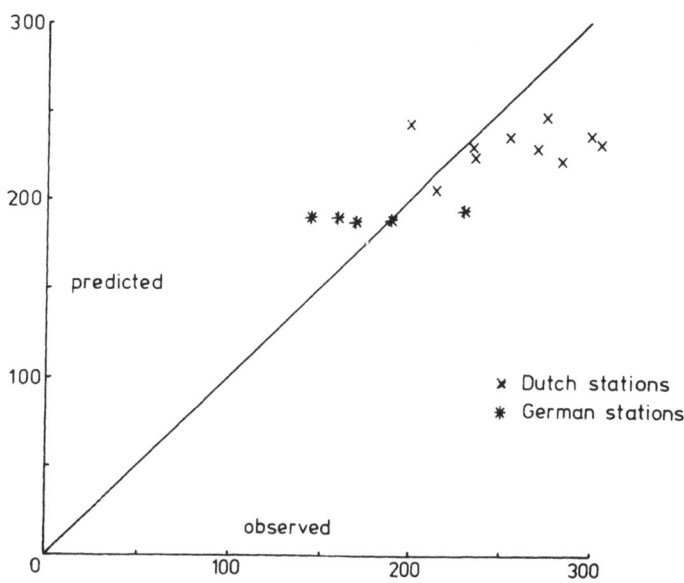

Fig. 8.   A comparison of measured and calculated maximum O$_3$-concen-
          tration in μg/m^3 at 15 measuring stations.

From the viewpoint of air quality a great deal of attention will be given to a comparison of observed and predicted maximum O_3-concentrations. In Fig. 8 such a comparison is made based on 15 monitoring stations, from which 5 are located in Germany. The agreement shown is reasonable, deviations are observed up till 20%. The general pattern shows an overprediction of the lower maxima and an underprediction of the higher maxima. The calculated downwind O_3-gradient across the area is smaller than the observed gradient.

CONCLUSIONS AND FUTURE RESEARCH

In view of the uncertainties related to the input data the calculated O_3-concentrations for this first simulation run are in reasonable agreement with the observed data. The maximum O_3-concentrations calculated are within $\pm 20\%$ of the observed maxima. The general outcome of the O_3-concentrations for this episode shows a relatively homogeneous cloud of O_3 covering the whole region during the hours of the maximum O_3-concentrations, with a distinct increase of O_3 of about 35% from Germany to the Western part of the Netherlands.

In the near future more detailed statistical comparisons will be made between observed and predicted values, also for other pollutants such as SO_2. Based on the results of these further comparisons, sensitivity runs will be carried out. These runs will be related to the influence of changing the initial and boundary conditions of hydrocarbons and O_3 and changing (the ratio of) the emissions of hydrocarbons and NO_x. Also, in parts of the area considered, calculations will be carried out with a finer grid resolution. Before real application studies will be carried out, it seems worthwhile to simulate also another episode with different weather conditions in order to be able to support results from application studies more firmly.

ACKNOWLEDGEMENT

The authors are indepted to the people from Systems Applications Inc., and especially to Dr. S.D. Reynolds, for their assistance and advice during this study.

(1)   EPA. "Uses, limitations, and Technical Basis of Procedures for Quantifying Relationships between Photochemical Oxidants and Precursors".
      EPA-450/2-77-021a (1977)
(2)   R. Guicherit e.a. "Photochemical smogformation in the Netherlands" TNO-Den Haag (1978)
(3)   L.A.M. Janssen e.a. "Calculationsystem Airquality" TNO-CMP 80/7, in Dutch (1980)

(4) S.D. Reynolds e.a. "An introduction to the SAI-airshed model
     and its usage" SAI, 950 Northgate
     Dr. SAN RAFAEL, CA 94903
(5) L.E. Reid e.a. "Evaluation of airshed model performance in
     Tulsa" SAI-no. 223-E580-164 (1980)
(6) G.E. Anderson e.a. "Airquality in the Denver metropolitan
     region 1974-2000, SAI No. EF 77-222 (1977)
(7) T.W. Tesche e.a. "Simulated impact of alternative emissions
     control strategies on photochemical oxidants in Los Angeles.
     EF 78-22R (1978)
(8) J.H. Weaving e.a. "An experimental and theoretical investi-
     gation of photochemical pollution in Turin-Committee of
     Common Market Automobile Constructors (1979)
(9) R.H. Kummler e.a. "A comparative validation of the RAM and
     modified SAI models for short term SO_2-concentrations in
     Detroit" J. of APCA 29.7.720 (1979)
(10) H. Tennekes, J. Atm. Science 30, 558 (1973)
(11) J.P. Killus e.a., Continued Research in Mesoscale air pol-
     lution simulation modeling, Vol. V. EF 77-142 (1977)
(12) K.E. Bencala, Atm. Environment 13, 1181 (1979)

DISCUSSION

H. VAN DOP   Your calculated concen-
          tration of both ozone and NO show an increase dur-
          ing the morning hours of the second day, while
          NO_2 concentrations decrease. Is this not in
          contradiction with the general observations that
          O_3 concentrations can only increase when NO has
          been eliminated.

P.J.H. BUILTJES   The calculated results
          are averaged results over a gridcell, which will
          influence these results. Also the relation be-
          tween O_3/NO/NO_2 is influenced by hydrocarbons,
          which can explain the difference between the
          calculations and the general observed behaviour.

K.E. GRONSKEI   Considering the discre-
          pancy between observed and calculated O_3-values
          during nighttime could this discrepancy be caused
          by the diurnal variation in dry deposition ?

P.J.H. BUILTJES   Dry deposition (surface
          resistance) is taken into account, but is inde-
          pendent of time.
          Transport to the surface is dependent of time.

Explanation of too high O_3-levels :

- too low NO-emission close to the ground
- too large vertical diffusivity
- too small dry deposition.

A. BERGER   I was surprised to see
the height of the mixing layer during nighttime
($\sim$ 200 m) after having seen, on the previous
slight a so strong vertical wind shear with a
jet around 500 m. Don't you think that we have
to improve our definition of mixing layer,
taking into account mechanically induces mixing ?

P.J.H. BUILTJES   The mixing height is
based only on temperature gradients. I agree
that it would be better to determine the
'mixing height' in temperature gradient as well
as shear of the mean velocity profile. However,
up till now, nobody does that.

P. MELLI   Have you compared the
results of your model with those of a no simple
model ?

P.J.H. BUILTJES   There are no simple
models available which can calculate secondary
pollutants with this space and time resolution.

B.E.A. FISHER   Have you been able to
assess the importance of sub-grid scale mixing
on the rate of chemical reactions ?

P.J.H. BUILTJES   This is a problem.
The model used in this way is only valid
rather far from a point source. There are sub-
models for calculation of the local influence
of roadways (NO/O_3) and tall stacks (NO/O_3).
Work is carried out considering reactive plume
models to incorporate the effects in a more
sound way.

E. RUNCA   Does your model take
into account that some of the pollutants you
mentioned during night time are emitted above
the height of the nocturnal inversion layer and
can travel to large distances from the emissions ?

P.J.H. BUILTJES   These pollutants will
be transported first and then entrained during

daytime in the mixing layer when this layer is growing.

J.M. QUINAULT   May I have some comments about the turbulent diffusivity profile you used ?

P.J.H. BUILTJES   The vertical diffusivity is calculated according to an algoritm based on the PBL-model of Deardorff. The vertical diffusivity is a function of stability (MONIN-OBUKHOV length) windspeed, surface roughness, (see for details the references given in the paper).

AN APPLICATION OF A POLLUTION EPISODE PREDICTOR DERIVED FROM A K-THEORY MODEL

Pietro MELLI
IBM Scientific Center - Rome

Giorgio FRONZA
Istituto Elettrotecnica ed Elettronica
Centro Teoria Sistemi, Politecnico, Milan

INTRODUCTION

The fitting of K-models (see for instance Randerson (1970), Shir and Shieh (1974), Runca et al. (1979)) to pollutant concentration measures is often unsatisfactory under very stable or unstable conditions in the lower atmosphere. Unfortunately, not only these situations are very frequent but (in presence of low and tall sources, respectively) correspond to the most severe pollution episodes at ground level.

This paper describes the application of a K-model to summer sulphur dioxide pollution from a power plant in the Po Valley. The case has been studied from a completely different viewpoint also by Finzi et al. (1978) and Bacci et al. (1981). In particular, the K-model aims et explaining summer fumigation episodes in conditions of strong instability, namely in presence of plume breakdown by an enhancing unstable connective layer (see for instance Slade (1968)).

In detail, a good performance by the K-model is looked for in the paper through the following adjustments.

- A relatively accurate calibration of the vertical profile of the diffusion coefficient. Such calibration takes into account the vertical profile of atmospheric stability, reconstructed by ground level measurements via simple mathematical models. Basically, these models respectively describe the dynamics of the boundary of the lower stable layer (during clear nights) and the dynamics of the boundary of the unstable convective layer (during sunny days).

- The use of real-time forecast instead of batch simulation of the
  K-model, in accordance with the techniques suggested by Bankoff
  and Hanzevack (1975), Desalu et al. (1974) and Fronza et al.
  (1979). This approach leads to the use of the so-called Kalman
  predictor (see for instance Kalman (1960), Jazwinski (1971)),
  which is based on a correction (filtering) of the concentration
  field at each time step, in accordance with the measurements
  available at that step. In gross terms, filtering at each time
  step avoids "the propagation of model errors" from the step to
  the following ones.

     The K-model, the assignment of its parameters and the Kalman
predictor derived from the model are described in the next section.
The last section is devoted to the presentation of the results of
the application to the case study and to the related comments.

DESCRIPTION OF THE POLLUTION PROBLEM AND K-MODEL

The pollution problem and measurement network

     The area under consideration is shown in Fig. 1, together
with the polluting source PP and the measurements points (this is
an"equivalent" simplified measurement network, the actual network
consists of a larger number of sensors).

     The area is flat, so there is no significant orographic ef-
fect to be taken into account.

     In summer 1973, the year to which the present study makes
reference, the plant had a target power of 320 Megawatt. The pollutant is released through a stack of 120 m.

     Hourly wind speed and direction have been recorded by an
anemometer located at 10 m above the ground. Hourly solar radiation (in Cal $cm^{-2}hr^{-1}$) has been recorded by a pyrheliometer.
Finally, Pasquill category at the end of the night, a datum
required for the evaluation of a model input (see next section)
has been determined by wind data and temperature vertical
structure as resulting from air temperature trend at ground level.

     More specifically, Pasquill stability class is defined via
wind speed and cloudiness. In turn, cloudiness, which is usually
not measured directly, is strongly correlated to net radiation
(see Pasquill (1976)), which is, in turn, highly correlated to
the ground level temperature profile (e.g. Anfossi et al. (1976)).

The K-model

     The model, here briefly reported for commodity, has been
extensively illustrated by Runca et al. (1979) where its main

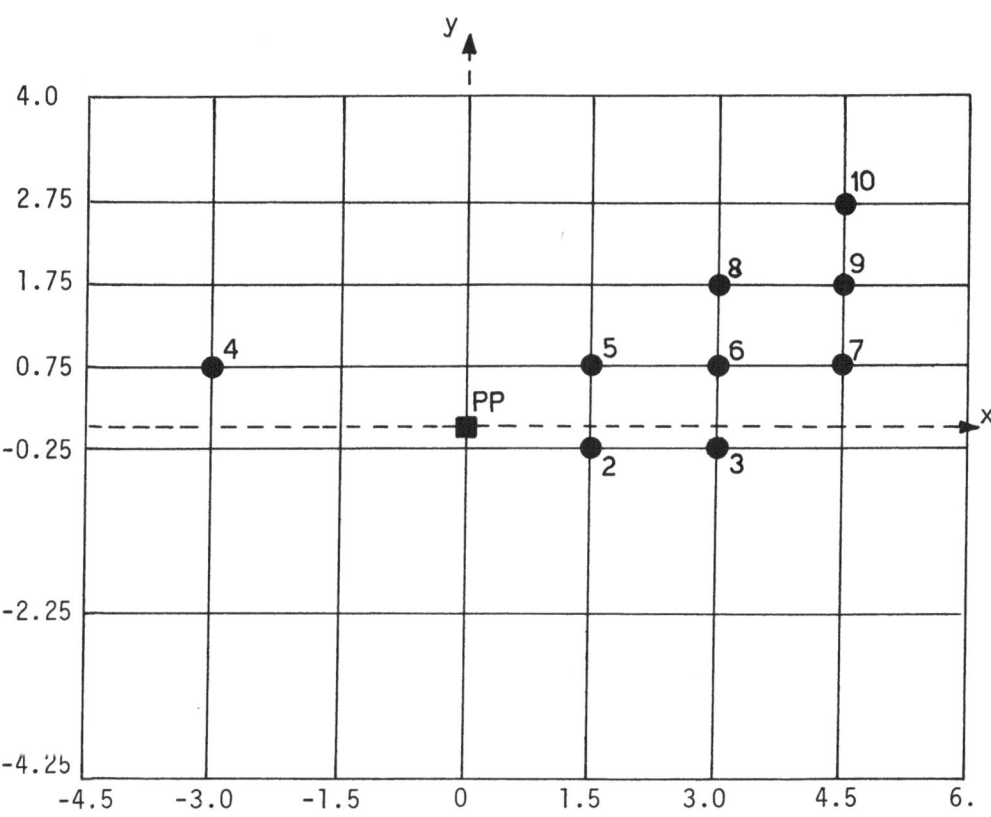

Fig. 1 - The polluting source, the measurement network (●)
and the integration grid at ground level (distances
in km).

characteristics (in terms of numerical stability, accuracy...) are
also discussed. The starting point is the three dimensional advec-
tion-diffusion equation (with the usual boundary and initial conditions)

$$\frac{\partial c}{\partial t} + v_x \frac{\partial c}{\partial x} + v_y \frac{\partial c}{\partial y} = K_x \frac{\partial^2 c}{\partial x^2} + K_y \frac{\partial^2 c}{\partial y^2} + \frac{\partial}{\partial z} \left[K_z \frac{\partial c}{\partial z} \right] + S \quad (1)$$

where

c   = pollutant concentration;
v_x, v_y = wind components in the horizontal plane;
K_x, K_y, K_z = diffusion coefficients;
S   = source term.

The scheme proposed by Runca et al. (1979) for the numerical
integration of eq. (1) is a fractional step algorithm (see for
instance Yanenko (1971)). Precisely, eq. (1) is splitted into the
six equations

$$\frac{\partial c}{\partial t} = S \tag{2a}$$

$$\frac{\partial c}{\partial t} + v_x \frac{\partial c}{\partial x} = 0 \tag{2b}$$

$$\frac{\partial c}{\partial t} + v_y \frac{\partial c}{\partial y} = 0 \tag{2c}$$

$$\frac{\partial c}{\partial t} = K_x \frac{\partial^2 c}{\partial x^2} \tag{2d}$$

$$\frac{\partial c}{\partial t} = K_y \frac{\partial^2 c}{\partial y^2} \tag{2e}$$

$$\frac{\partial c}{\partial t} = \frac{\partial}{\partial z} \left[K_z \frac{\partial c}{\partial z} \right] \tag{2f}$$

If c^k is the concentration field at the k-th time step, the
field c^{k+1} is computed as follows. A field c^* is obtained by the
numerical integration of eq. (2a) with initial condition c^k. Then,
a field c^{**} is obtained by the numerical integration of eq. (2b)
with initial condition c^* . Then a field c^{***} is obtained by the
numerical integration of eq. (2c) with initial condition c^{**}
and so on. The numerical integration of eq. (2f) provides c^{k+1}. In
more detail, in Runca et al. (1979), the integration of eqs. (2b),
(2c) (contribution of the advection terms) is supplied in accordance with Carlson's scheme (see for instance Richtmyer and
Morton (1967)), while the integration of eqs. (2d)-(2f) (contribution of the diffusion terms) is supplied by the procedure due
to Crank and Nicolson (1947).

Assignment of model inputs and parameters.

The wind field has been assumed to have zero component in the vertical plane (see eq. (1)) and to be uniform in each horizontal plane. Such uniform field has been evalutate by the power low

$$v(k,z) = v_R(k) \left(\frac{z}{z_R}\right)^{\alpha(s(k,z))}$$

where

$v(k,z)$ = average wind speed in the k-th time step, at level z;

z_R = level where wind measures are taken;

$v_R(k)$ = average wind speed in the k-th time step, at level z_R;

$s(k,z)$ = Pasquill stability class in the k-th time step, at level z;

$\alpha(s)$ = given function of stability reported in Table 1 (second column).

The vertical diffusion coefficient has been assigned through the formula

$$K_z(z,s(k,z)) = K_D(k) \; z \; \exp(-\rho(s(k,z)) z/\bar{H})$$

where

$\rho(s)$ = function reported in Table 1 (third column);

$\bar{H}$ = height of the integration region

$$K_D(k) = z_R^{-1} K_{zR}(s(k,z_R)) \exp(\rho(s(k,z_R)) z_R/\bar{H})$$

$K_{zR}(s)$ = vertical diffusion coefficient at level z_R (Table 1, fourth column)

Table 1 - Wind and diffusion parameters versus Pasquill stability classes

s	α	ρ	K_{zR} $(m^2 s^{-1})$	$K_x = K_y$ $(m^2 s^{-1})$
A	0.05	6	45	250
B	0.1	6	15	100
C	0.2	4	6	30
D	0.3	4	.2	10
E	0.4	2	0.4	3
F	0.5	2	0.2	1

As for the horizontal diffusion coefficients, it has been assumed $K_x(s(k,z)=K_y(s(k,z))$ = function reported in Table 1, fifth column.

If, for a certain k, the stability class changes with z within the integration region, the continuity of the vertical profiles of wind speed and diffusion coefficients has been maintained by interpolation. From this viewpoint, typical situations occur in a clear night (development of a stable layer) and in a clear day (development of an unstable convective layer). In the former case, the level of the boundary of the stable layer has been evaluated at each time step through the formula due to Bacci et al. (1975). In the latter case, the top of the unstable connective layer has been evaluated at each time step by the formula due to Carson (see again Bacci et al. (1975)).

The Kalman filter

In Fronza et al. (1979) it is shown in detail how the above mentioned numerical schema for the solution of eq. (1) can be transformed into a stochastic dynamical system, namely into the two vector equations :

$$\underline{X}(k+1)=\emptyset(\underline{s}(k+1),\underline{v}(k+1),d(k+1)\underline{X}(k)+\emptyset(\underline{s}(k+1),\underline{v}(k+1)),\underline{E}(k+1)+\underline{n}(k) \tag{3a}$$

$$\underline{Y}(k) = H\underline{X}(k) + \underline{w}(k) \tag{3b}$$

where, for the k-th time step

$\underline{X}(k)$ = vector of the average concentrations in the points of the integration grid;

$\underline{s}(k)$ = vector of the stability classes at the various levels of the integration region;

$\underline{v}(k)$ = vector of average wind speed of the various levels of the integration region;

$d(k)$ = average wind direction;

$\underline{E}(k)$ = (suitably defined) emission vector;

$\underline{Y}(k)$ = vector of average concentration measurements.

Furthermore

$\emptyset(s,v,d)$ = (suitably defined) transition matrix;

H = suitable zero-one matrix;

$\{\underline{n}(k)\}_k$ = purely random zero mean process (white noise), called process noise;

$\{\underline{w}(k)\}_k$ = white noise called measurement noise.

The one-step ahead Kalman predictor for system (3) is

$$\hat{\underline{X}}(k|k) = \hat{\underline{X}}(k|k-1) + G(k) \left[\underline{Y}(k) - H\hat{\underline{X}}(k|k-1) \right]$$

$$\hat{\underline{X}}(k+1|k) = \emptyset(\hat{\underline{s}}(k+1|k), \hat{\underline{v}}(k+1|k), \hat{d}(k+1|k)) \hat{\underline{X}}(k|k) +$$
$$+ \emptyset(\hat{\underline{s}}(k+1|k), \hat{\underline{v}}(k+1|k), \hat{d}(k+1|k)\underline{E}(k+1)$$

where

$\hat{\underline{X}}(k|k)$ = filtered state, namely a posteriori (i.e. at time k) estimation of $\underline{X}(k)$ on the basis of the new available datum $\underline{Y}(k)$

$\hat{\underline{X}}(k+1|k)$ = prediction of $\underline{X}(k+1)$, made at time k.

G(k)   = suitable weight matrix (Kalman gain) which drives the correction (4) of the previous forecast ($\hat{\underline{X}}(k|k-1)$ into the filtered state $\hat{\underline{X}}(k|k)$, and the entries of which also depend upon the variances of process and measurement noises·

$\hat{\underline{s}}(k+1|k)$, $\hat{\underline{v}}(k+1|k)$, $\hat{d}(k+1|k)$ = forecasts of $\underline{s}(k+1)$, $\underline{v}(k+1))$,d(k+1) made at time k.

For the details concerning the actual implementation of the meteorological forecasts and the assignment of noise covariances see again Fronza et al. (1979).

RESULTS

Up to now, the above mentioned K-model has been simulated on a few episodes in the region of Fig. 1. A typical 2h forecast performance is shown in Fig. 2-5, where, for brevity only for stations are reported (the performance has been similar in the remaining ones). This episode is due to fumigation in a clear day after a clear night, with weak winds mainly from WSW.

The main comments are the following

- The fitting is rather satisfactory, also with respect to the episode profile. Naturally, the fitting is slightly worse during the episode.
- The improvement by the Kalman predictor is very slight or questionable (see Table 2). This means that a better calibration of noise covariances of the stochastic system (3) must be searched,
- The relatively satisfactory performance can be ascribed to the calibration of model and input parameters, in particular, to taking into account the vertical profile of stability. In more detail, the use of a single layer K-model has given a really unsatisfactory result. This susggests to look for model improvements by a more accurate calibration of the K_z vertical profile. In particular such improvements will be searched by

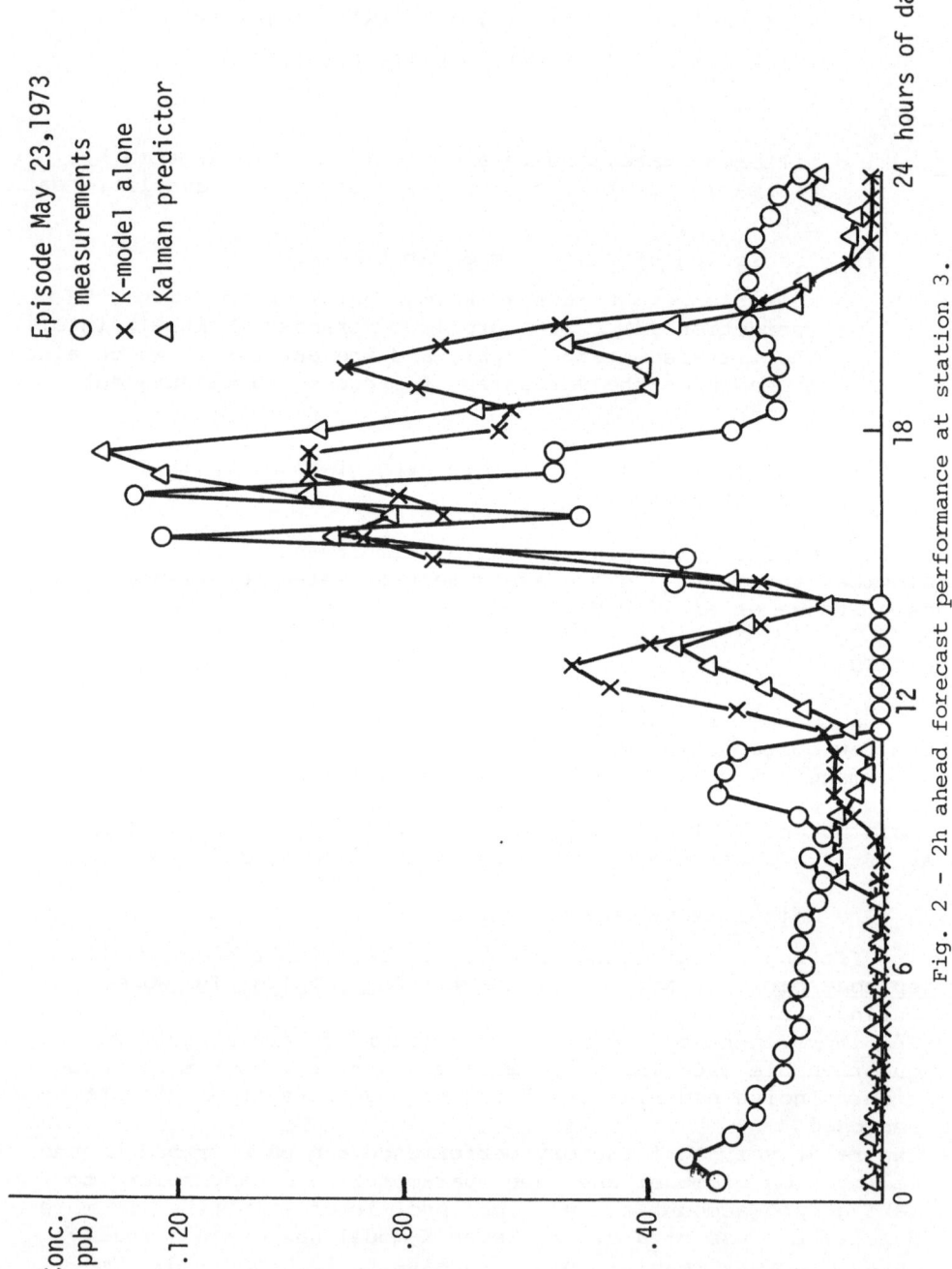

Fig. 2 - 2h ahead forecast performance at station 3.

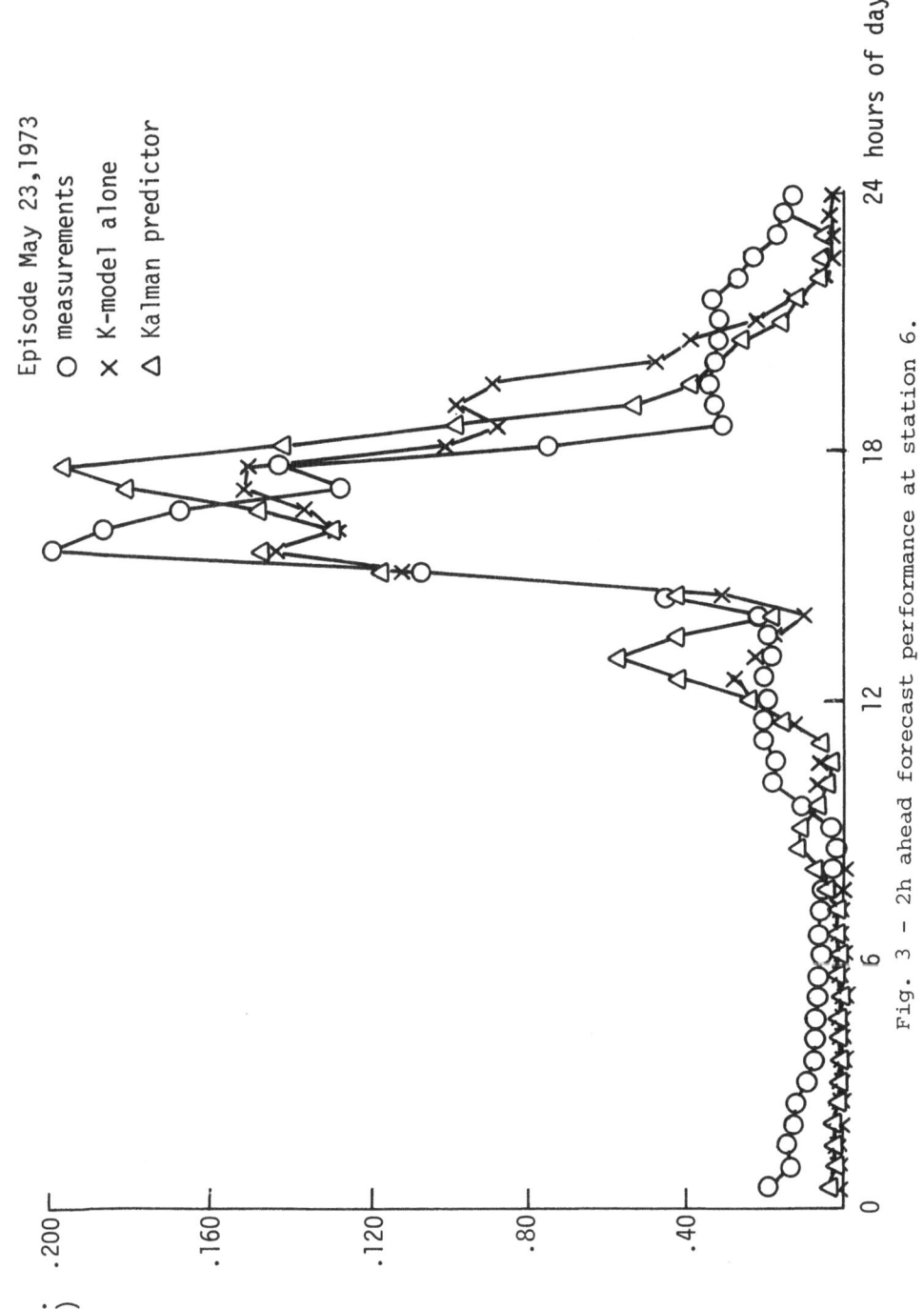

Fig. 3 - 2h ahead forecast performance at station 6.

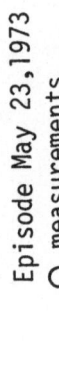
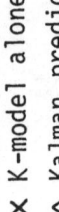

Episode May 23, 1973

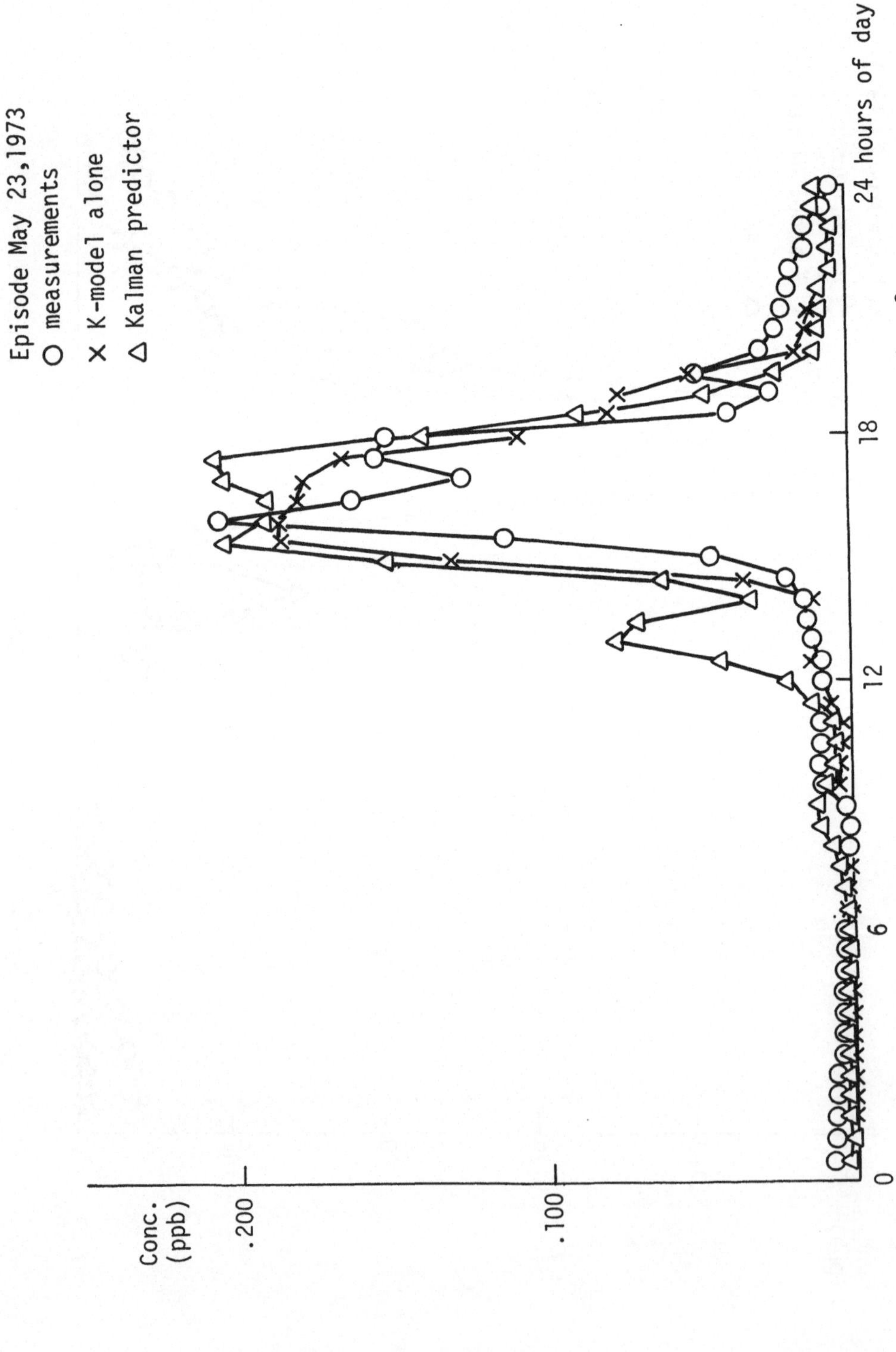

O measurements

X K-model alone

△ Kalman predictor

Conc.
(ppb)

.200

.100

0   6   12   18   24 hours of day

Fig. 4 — 2h ahead forecast performance at station 8.

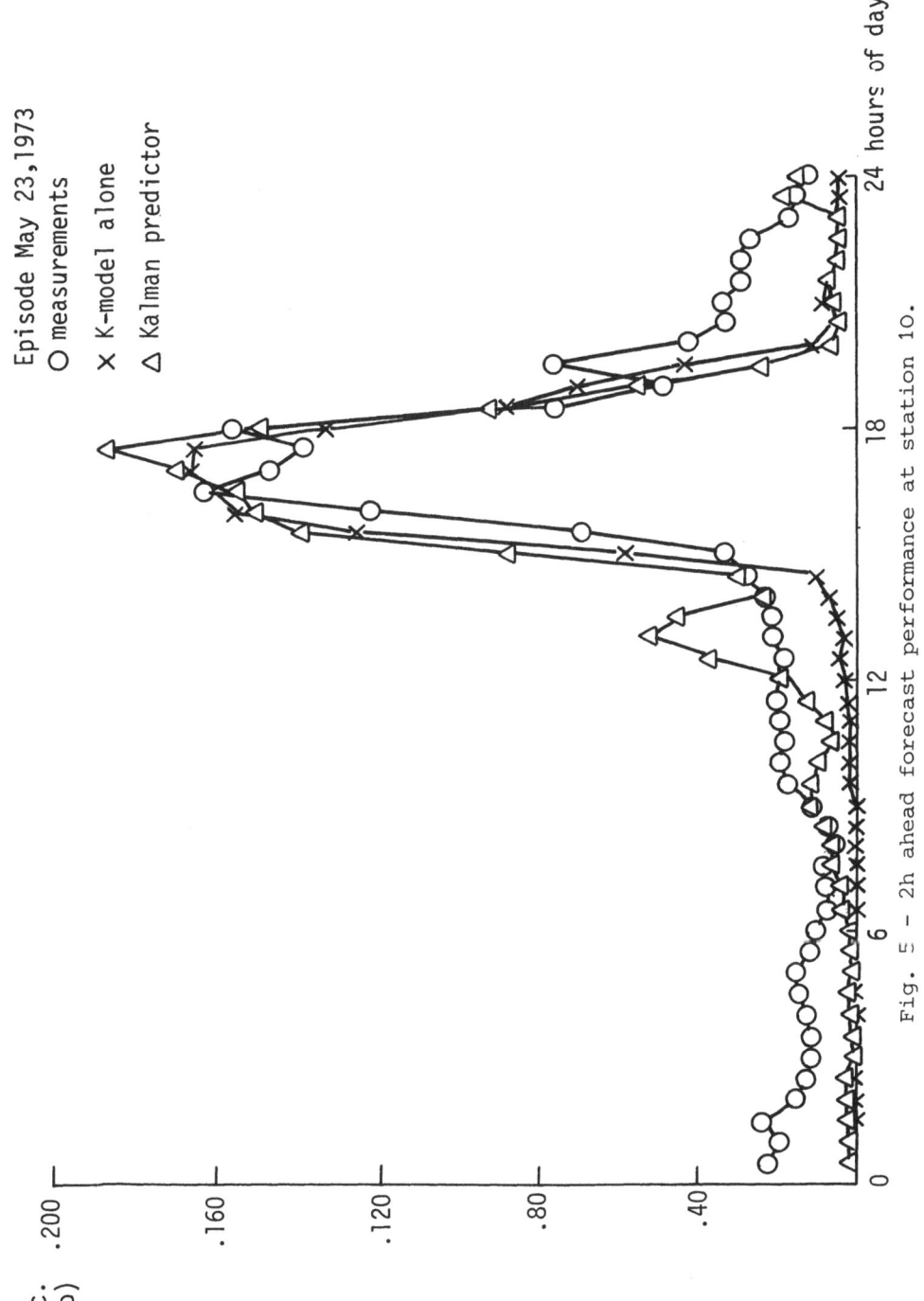

Fig. 5 – 2h ahead forecast performance at station 10.

Table 2 - K -model forecast performances (table entries: correlation between observations and predictions)

Station	K model	Kalman predictor
3	.59	.66
6	.89	.89
8	.92	.90
10	.95	.91

assigning K_z a profile in a given class of functions and by searching the function in that class which guarantees the best model performance.

ACKNOWLEDGMENT

The authors thank Dr. P. Bacci from ENEL for his useful suggestions and criticism.

REFERENCES

Anfossi D., Bacci P., and Longhetto A., 1976, Forecasting of vertical temperature profiles in the atmosphere during nocturnal radiation inversions from air temperature trend at screen height, Quart. Journ. Roy. Met. Soc., 102:173.

Bacci P., Longhetto A., and Anfossi D., 1975, Models of temperature profiles under convective and inversion conditions connected to air pollution, Proc. of 6th NATO-CCMS. Int. Techn. Meet. on Air Poll. Model., Frankfurt/Main, 313.

Bacci P., Bolzern P., and Fronza G., 1981, A stochastic predictor of air pollution based on short-term meteorological forecast, Journ. Appl. Met. (in press)

Bankoff S.G., and Hanzevack E.L., 1975, The adaptive filtering transport model for prediction and control of pollutant concentration in an urban airshed, Atm. Env., 9:793

Crank J., and Nicolson P., 1947, A practical method for the numerical solution of partial differential equations of the heat conduction type. Proc. Cambridge Philos. Soc., 43;50.

Desalu A.A., Gould L.A., and Schweppe F.C., 1974, Dynamic estimation of air pollution, IEEE Trans. on AC, AC-19:904.

Finzi G., Fronza G., and Rinaldi S., 1978, Stochastic modeling
    and forecast of the dosage area product, Atm. Env., 12:831.

Fronza G., Spirito A., and Tonielli A., 1979, Real-time forecasting
    of air pollution episodes in the Venetian region. Part II,
    The Kalman predictor. Appl. Math. Model., 3:409.

Jazwinski A.H., 1970, "Stochastic Processes and Filtering Theory",
    Academic Press, New York.

Kalman R.E., 1960, A new approach to linear filtering and predic-
    tion problems, Trans. ASME. Journ. Bas. Eng., 82:17

Pasquill F., 1976, "Atmospheric Diffusion", Van Nostrand, Prince-
    ton.

Randerson D., 1970, A numerical experiment in simulating the
    transport of sulphur dioxide through the atmosphere, Atm.Env.,
    4:615.

Richtmyer R.D., and Morton K.W., 1967, "Difference Methods for
    Initial Value Problems", Interscience, New York

Runca E., Melli P., and Spirito A., 1979, Real-time forecast of
    air pollution episodes in the Venetian region.
    Part I : the advection-diffusion model, Appl. Math. Model,
    3:402.

Shir C.C., and Shieh L.J., 1974, A generalized urban air pollution
    model and its application to the study of SO_2 distributions
    in St.Louis Metropolitan area, Journ. Appl. Meteor., 13:185

Slade D.H., (ed.), 1968, "Meteorology and Atomic Energy", US
    Atomic Energy Commission, TID 24190.

DISCUSSION

H. VAN DOP   I understand from your fi-
    gure 1 that the area considered is of the order
    of 10 x 10 km^2, which the gridspacing is ⌄1 km.
    I think that for that configuration you can neg-
    lect horizontal diffusion in your model, even for
    stability class A.

G. FRONZA   Yes I agree at least when
    wind speed is higher than say 1 m sec^{-1}

M.J.M. QUINAULT   Have you compared the results
            of your model with those of a more simple model
            like Anfossi's ?

G. FRONZA   No, maybe in the future there
            will be a K-theory application to the Turbigo area
            considered by Anfossi.

A. VENKATRAM   I think it is a good idea
            to use the eddy diffusivity as an empirical data
            fitting parameter rather than as a physical quan-
            tity related to the details of the micrometeorolo-
            gy. In your case you have specified the eddy dif-
            fusivity profile rather than having derived it
            from the available velocity and concentration pro-
            files. This suggests the question : Have you
            looked at the sensivity of you model results to
            the prescribed K-profile ?

G. FRONZA   In one simulation we tried
            to give the K-profile the "most suitable" from
            the viewpoint of concentration data fitting. The
            resulting K-profiles vary with time in a reasonable
            way.

FORECASTING OF POLLUTANT CONCENTRATION UNDER EPISODE CONDITIONS

J. M. Fage, G. Gallay, and J. Moussafir

BERTIN & Cie
BP n° 3
78370 PLAISIR (FRANCE)

1- INTRODUCTION

In case of stagnation conditions (low wind speed, strong temperature inversion) high pollution peaks can occur in the early morning in industrial areas (figure 1).

They reach very high levels quite simultaneously on a large horizontal scale and therefore one cannot think of advective effects but of vertical transfers.

Actually figure 2 shows that during the night preceding an episode a dramatic storage of pollutant takes place close above the inversion layer base in urban areas.

In the early morning large downdrafts due mainly to breaking unstable waves at the inversion base pull down the polluted layer to the ground.

These waves take their energy from the average vertical horizontal wind shear located close to the inversion base.

At the same time the vertical turbulent heat flux reaches high negative values close to the inversion base, that cannot be explained by convective entrainment leading to usual fumigation.

Moreover usual K diffusivity models have no chance representing such effects.

One then can think of using more sophisticated models taking into account equations for the second order correlations (see Donaldson (1968), Mellor & Yamada (1974), Lumley (1974), Wyngaard (1976)).

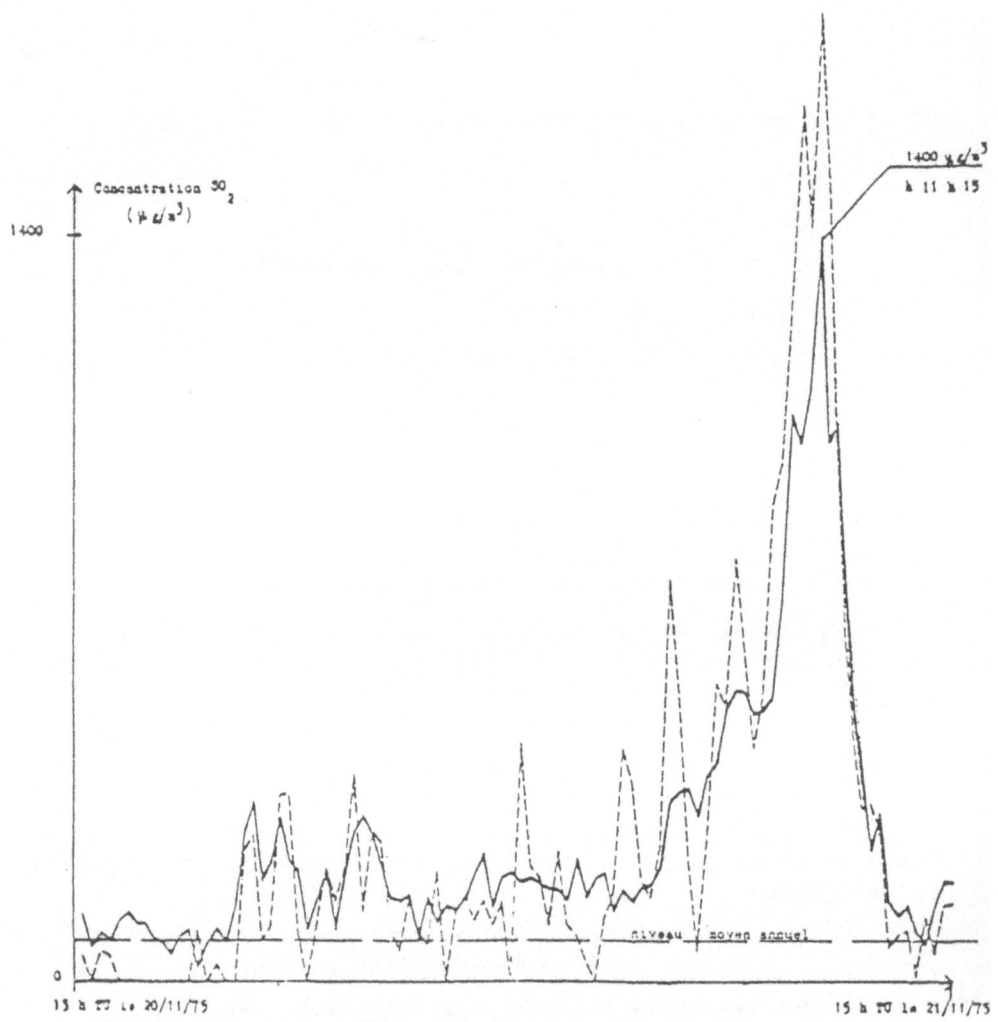

<u>FIGURE 1</u> :   Time-history of SO$_2$ concentrations during
              a pollution episode observed in PARIS.
              (EIFFEL TOWER, November 21, 1975).
              Dotted line is the first floor sensor
              (125 m).
              Full line is the ground level sensor.

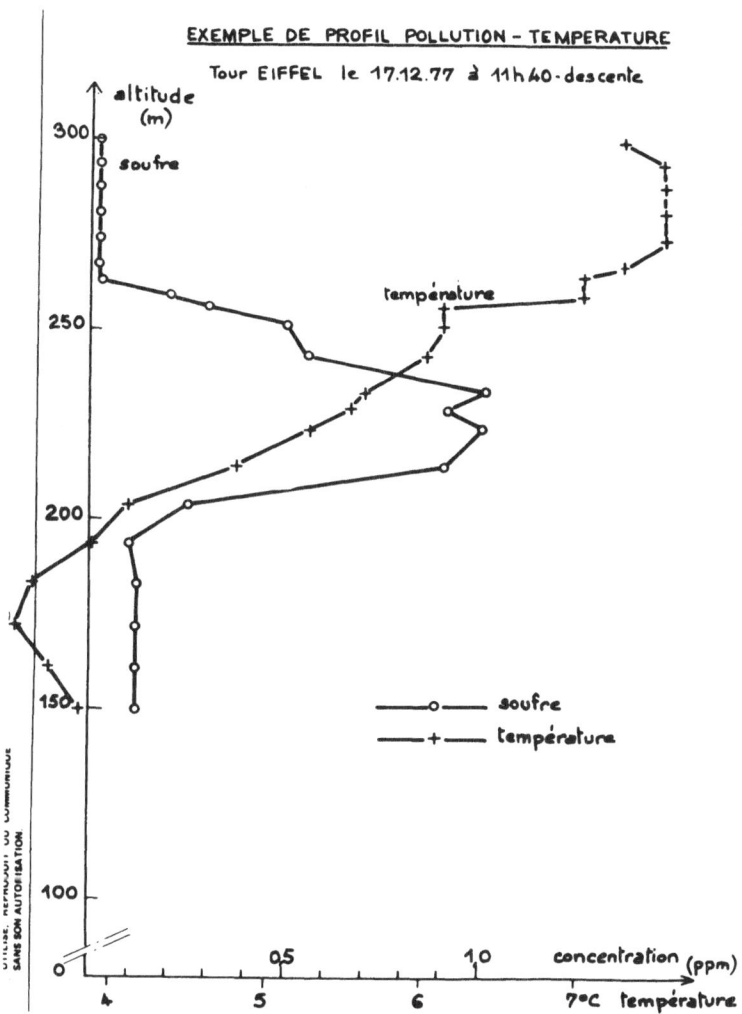

FIGURE 2 : Simultaneous profiles of potential
temperature (+) and SO_2 concentration
(o) observed in PARIS.
(EIFFEL TOWER, December 17, 1977,
11.35 A.M)

What we think is that even such models make the hypothesis
that turbulence scales (time and space) are rather small compared
to average scales.

This is not the case for what we are concerned with. For
example a downdraft due to a breaking wave can easily scale with
the depth of the mixed layer.

Another important point is that different turbulent scales
will exist at the same time due to different types of physical phenomena. Therefore second order modeling techniques should care
about dozens and dozens of equations which seems out of scope considering how boresome this type of approach is with only one mechanical scaling and one thermal scaling.

An alternate solution (Deardorff (1974)) is to use second
order closure and at the same time small space grids but this leads
to very long computer times.

This is why we tried to incorporate in our model some physics in a phenomenological way rather than working with many equations, a spectral approach being used in order to deal simultaneously with different length and time scales.

One main feature of this approach is to consider that one
Fluxmeter and one Doppler Sodar would be used in real time to give
the initial and boundary conditions. As a matter of fact we think
some problems encountered with even sophisticated models are due
to the general problem of lack of good inputs. This is why these
last past years we developed these two instruments (Fluxmeter and
Doppler Sodar) of which some data are shown on figures 3 to 6.

Up to now the model which is going to be presented deals
only with meteorological variables and this for daytime conditions.
It is planned to include a submodel taking into account passive pollutants.

2- SCALING OF THE MIXED LAYER

It has been shown by experimental studies (Lenschow (1980))
that the average distance (λ) between convective rolls scales on
the height of the mixed layer (h) and on the Monin Obukhov length
(L). We use the following formula :

$$\lambda \sim 8.8 \ h \ \left(\frac{h}{L}\right)^{-1/3}$$

Making use of the convective scaling velocity w_* this can be rewritten as :

$$\lambda \sim 12.6 \ h \ \frac{u_*}{w_*}$$

in which u_* is the usual friction velocity.

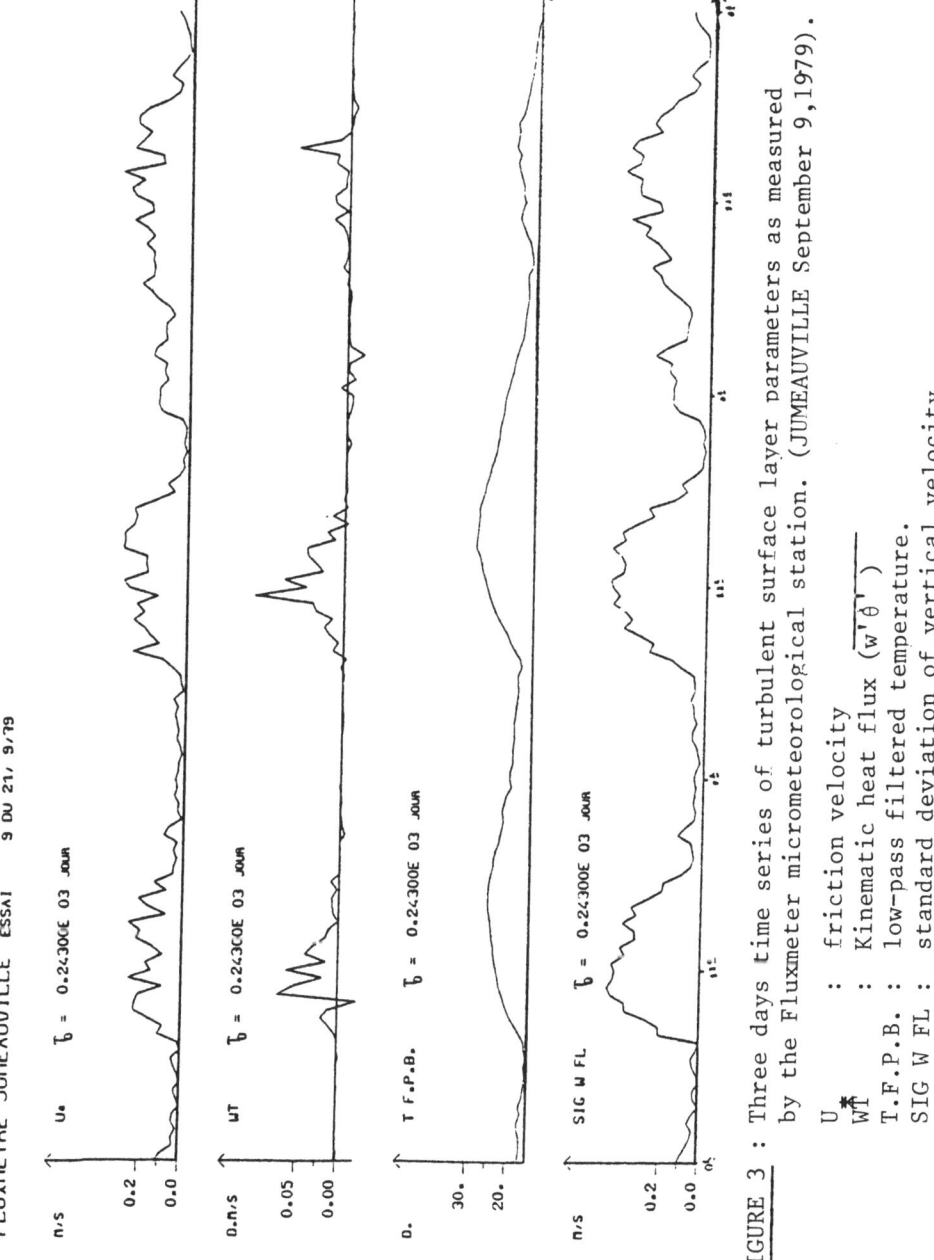

FIGURE 3 : Three days time series of turbulent surface layer parameters as measured by the Fluxmeter micrometeorological station. (JUMEAUVILLE September 9,1979).

U   : friction velocity
W*   : Kinematic heat flux ($\overline{w'\theta'}$)
T.F.P.B. : low-pass filtered temperature.
SIG W FL : standard deviation of vertical velocity.

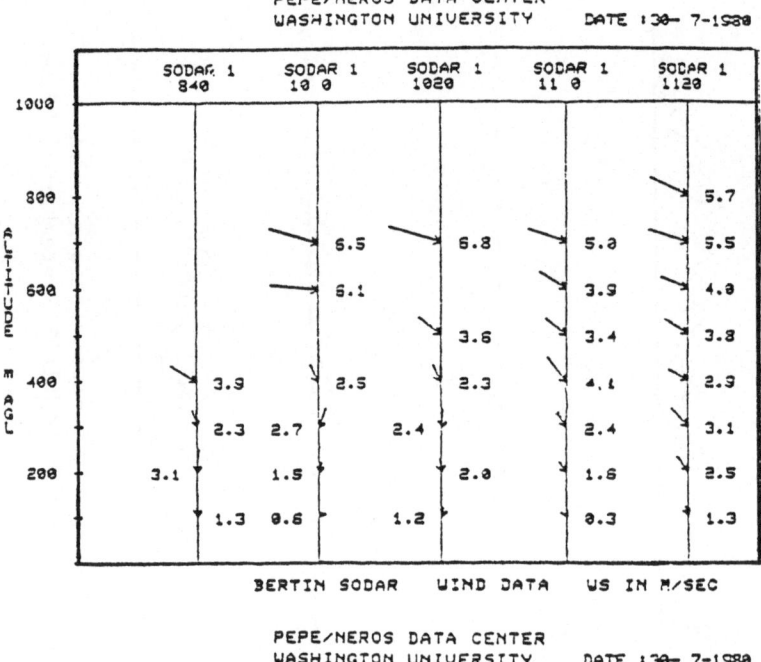

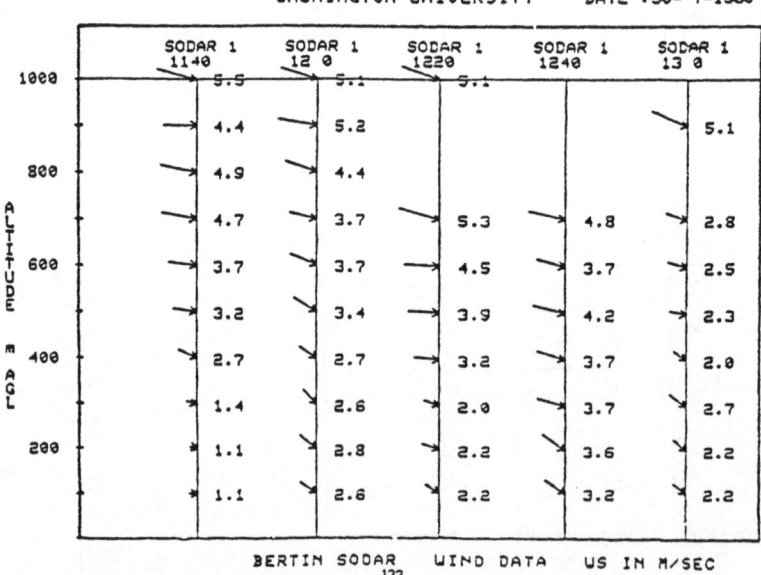

FIGURE 4 : Doppler Sodar wind velocity graphic printouts. (E.P.A.
NEROS/PEPE field study, COLUMBUS, OHIO, August 1980).
The length of the arrows at each level is proportional
to the horizontal windspeed and their position is related to wind direction. Note the dramatic wind rotation
between ground level and 600 m until 11.20 A.M.

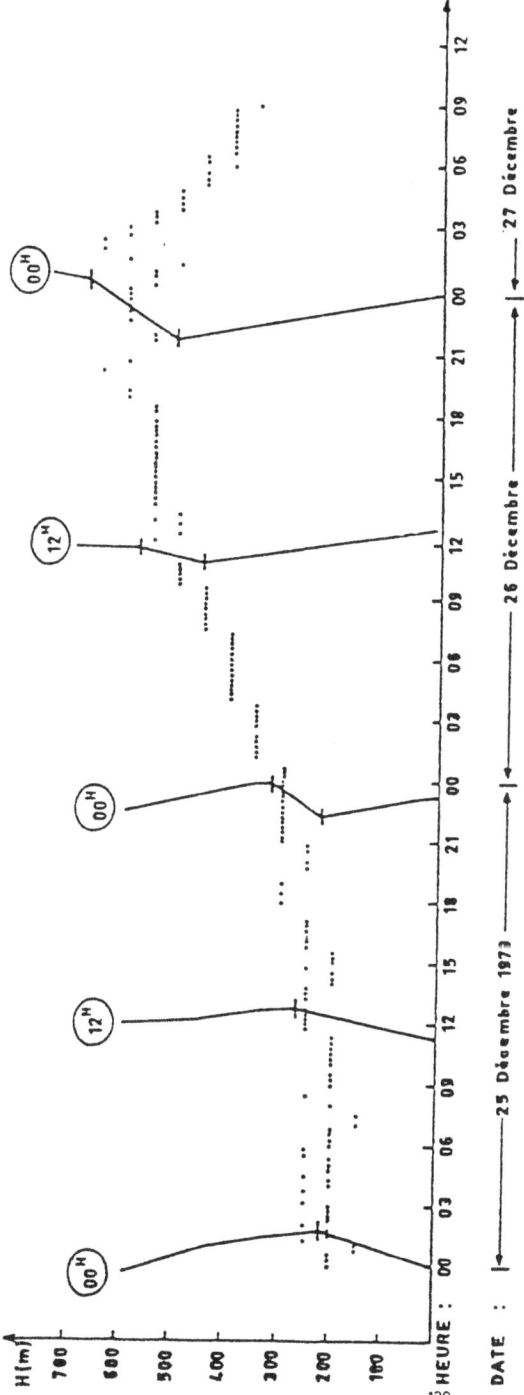

FIGURE 5 : Analysis of digital Sodar reflectivity data, compared to radiosonde temperature profiles. The horizontal separation between the Sodar site (JUMEAUVILLE) and the radiosounding station (TRAPPES) is about 20 km.

Courtesy of H. GLAND ⌐ EDF/DER/MAPA ⌐

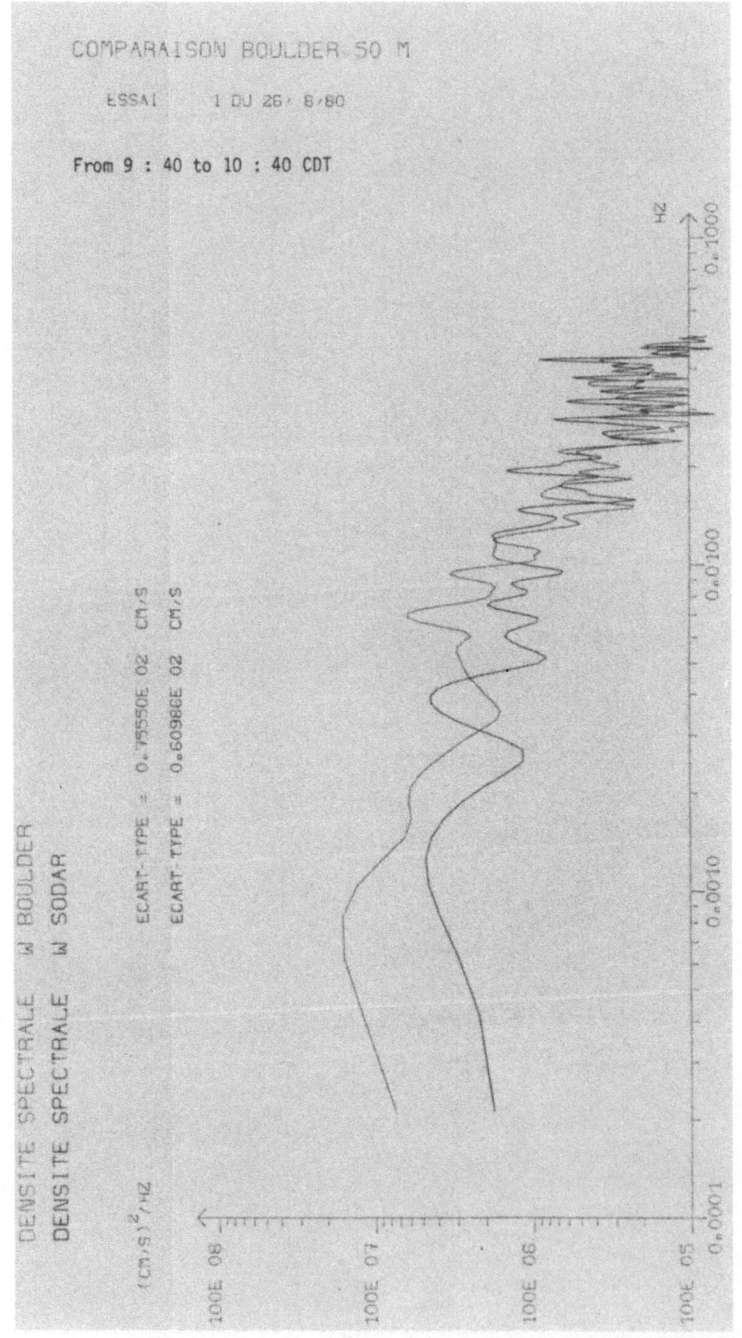

FIGURE 6 : Comparison of vertical velocity spectra as measured
by a Doppler Sodar and a Sonic Anemometer.
(Boulder Atmospheric Observatory – BOULDER, COLORADO,
August 1980). Courtesy of J.E. GAYNOR – N.O.A.A./WPL/ERL.

Now, following Manton (1975), we make the hypothesis that the size of each thermal is about one fourth of (λ).

Therefore, k_{th}, being the characteristic wavenumber associated to a convective roll, is given by :

$$k_{th} \sim 2 h^{-1} \left(\frac{w_*}{u_*} \right)$$

As we use a monodimensional model we have now to think about a time scale. If we take an Eulerian time scale, this works with the fixed frame of the model but is in contradiction with the intrinsic wave number k_{th} that we just defined and which is essentially Lagrangian.

Therefore, we make a compromise by setting the pseudotime scale T_{th} between two thermal passages in our Eulerian frame to :

$$T_{th} = \alpha \times \frac{h}{\sigma w}$$

Where :
- α is an empirical coefficient close to 1.
- σ_w is the standard deviation of vertical velocity in the mixed layer, used as a scaling velocity.

3- PHENOMENOLOGICAL DESCRIPTION OF TURBULENCE AT THE INVERSION BASE

We deal with two mechanisms of Turbulent Kinetic Energy production at the inversion base, one being called Wave Induced Turbulence (WIT) and the other being called Convection Induced Turbulence (CIT).

3.1.- W.I.T.

The main idea is that when a convective roll corrugates or impinges the inversion base an unstable gravity wave originates. Taking its energy from the mean horizontal wind shear, it finally breaks by overturning and dramatically increases the entrainment process.

First hypothesis is that the wavelength :

$$\lambda \, th = \frac{2 \pi}{k_{th}}$$

of a penetrative thermal scales the thickness : e_t of the layer in which waves can lead to turbulence.

e_t is given by :

Our model is a "three and a half" layer integral model as represented below :

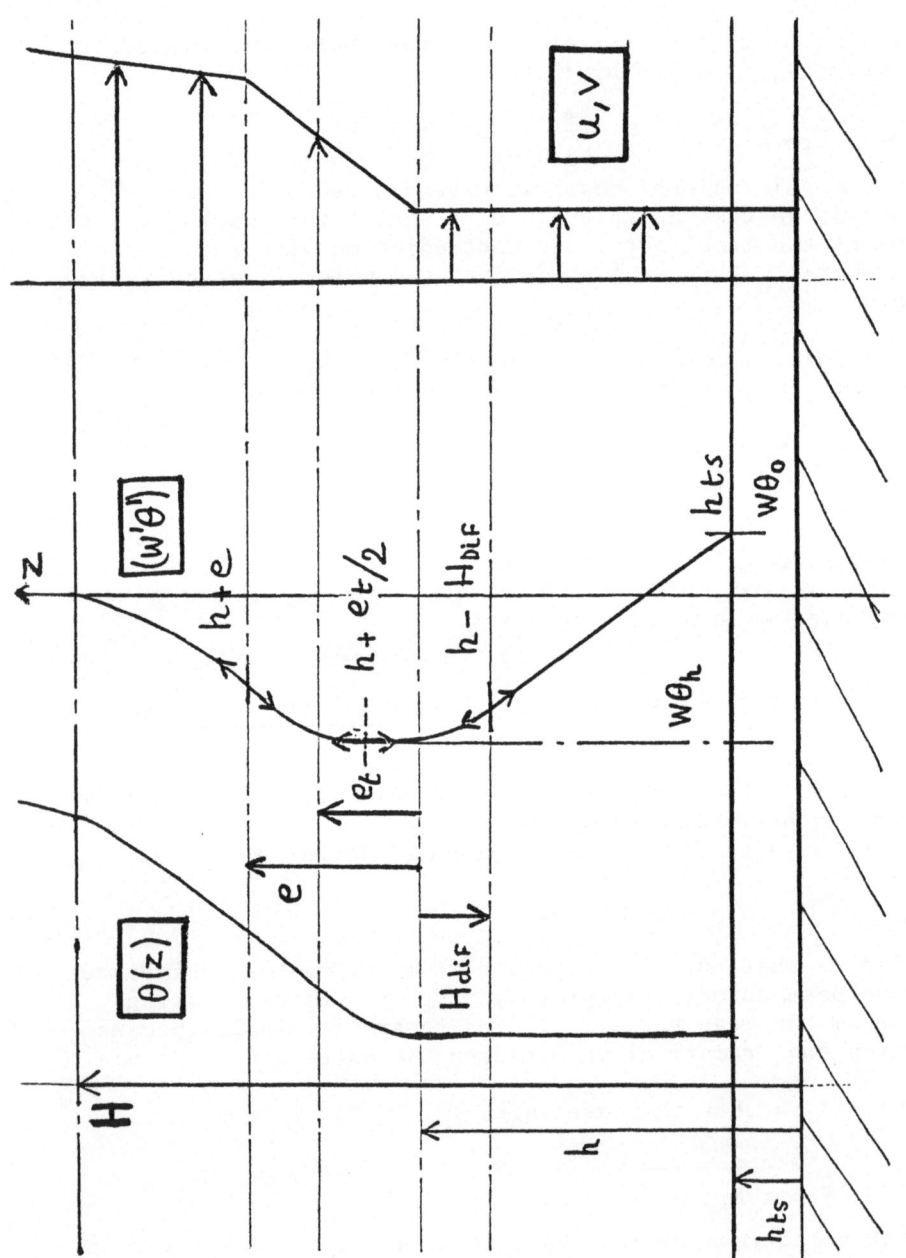

FIGURE 7 : Inversion rise model – Mean Values and Heat Flux profiles.

Figure 7. (continued)

H   : top of the stable layer

e   : thickness of the wind shear layer

e_t : thickness of the turbulent layer at the inversion base

h   : (no real physical meaning) Usually considered as the altitude of the inversion base.
        (Starting point of positive temperature gradient).

$h - H_{dif}$: average altitude of the downdrafts from the stable layer.
        In Rosset (1975) this should be called the base of the convective front.

h_{ts} : top of the surface layer.

In the shear layer we make the hypothesis that the average bulk Richardson number $Ri_g = 0.5$.

In the surface layer we use either conventional similarity based models or inputs derived from our micrometeorological "Fluxmeter" station.

$$e_t = \delta \times \lambda_{th}$$

with :

$$\delta = 1 - 0.4 \text{ Log } (-\frac{e}{H-h}---)$$

This last equation is derived from Hazel (1972), e being the thickness of the shear layer and H and h the heights of the top and the base of the inversion layer.

Before the wave diverges and then breaks the thermal hummock intrudes the inversion layer during a period of duration :

$$T_p = \pi/2 \times N$$

Where :

N : Brunt-Vaïsala frequency

And during this time the amplitude of the intrusion a_p reads :

$$a_p = \sigma_w/N \times \sin (N \times (t - t_{op}))$$

with t_{op} being the initial time of the intrusion.

We have seen that apart from divergence and breaking stages the bulk Richardson number Rig of the shear layer is kept equal to 0.5. During divergence stage e_t grows up to about $2 \times e_t$ according to a transition for Rig from 0.5 to 0.67. At the same time the wave amplitude A obeys :

$$A = a_p \times \exp ((t - t_{ep}) \times 0.14 \times N)$$

with t_{ep} being the time at which intrusion stops, and the constant 0.14 originating from theoretical considerations on the most unstable wave numbers. We can now focus on the breaking stage and especially on spatial, wind speed and time scales of the overturning billow which is created by the destruction of an unstable gravity wave.

The horizontal wave number is k_{th} while the vertical wave number k_z scales on the turbulent layer of thickness about $2 \times e_t$.

Turbulent Kinetic Energy is then produced by the overturning billow. The total energy E brought to the inversion base zone can be written:

$$E = \frac{1}{8k_{tot}^2} \times \left[1 + \left(\frac{k_{th}}{k_z}\right)^2\right] \times \left[\frac{\delta u}{\delta z}\right]^2$$

Where :

$$(k_{tot})^2 = (k_{th})^2 + (k_z)^2 \text{ is the total wave}$$
number.

$$\frac{\partial U}{\partial z}$$ is the average wind shear at the inversion base,

The breaking time scale T_b obeys :

$$T_b^{-1} = 0.2 \times \left[\frac{k_{th} * k_z}{(k_{tot})^2} \right] \times \left[\frac{\partial U}{\partial z} \right]$$

We can therefore estimate a part of the production term in the Turbulent Kinetic Energy Budget at the inversion base by writing that the production rate during the breaking stage is proportional to E/T_b.

3.2.- C.I.T.

Apart from entrainment due to unstable gravity waves there can be interfacial mixing due to small billows on the top of the thermal domes. Thes billows are supposed to scale on k_{th} with a wave number k_c of the order of $10 \times k_{th}$.

The overturning rate of these small billows is controlled at the same time, for a given spatial scale, by the average horizontal wind shear and by the vertical wind speed of the intruding thermal.

In a similar way as for WIT one can estimate the average Turbulent Kinetic Energy produced by these billows and a characteristic time scale. This latter and a time corresponding to the duration of the thermal passage lead as for WIT to an estimation of a second contribution to the production term in the Turbulent Kinetic Energy Budget at the inversion base.

4 - SPECTRAL DESCRIPTION OF TURBULENT KINETIC ENERGY AT THE INVERSION BASE

The purpose of this spectral description is to be able to hold at the same time different turbulent scales corresponding to WIT and CIT and this along time.

The Spectral Turbulent Kinetic Energy Budget (STKEB) reads in case of horizontal homogeneity :

$$\underbrace{\frac{\partial E(k,t)}{\partial t}}_{(1)} + \underbrace{\frac{\partial Q(k,t)}{\partial z}}_{(2)} = \underbrace{- S(k,t)}_{(3)} \underbrace{\frac{\partial U}{\partial z}}_{} + \underbrace{B(k,t)}_{(4)}$$

(STKEB)

$$\underbrace{- \frac{\partial}{\partial k} \epsilon(k,t)}_{(5)} - \underbrace{2 \, v \, k^2 \, E(k,t)}_{(6)}$$

Where :

$E(k,t)$: is the spectral density of Turbulent Kinetic Energy ($q^2 = 1/2 \, (u'^2 + v'^2 + w'^2)$).

$Q(k,t)$: is the spectral density of the vertical flux of Turbulent Kinetic Energy ($\overline{w'q'^2}$)

$S(k,t)$: is the spectral density of the vertical flux of momentum ($\overline{u'w'}$ and $\overline{v'w'}$).

$B(k,t)$: is the spectral density of the vertical Kinematic heat flux ($g/\theta \times \overline{w'\theta'}$).

(k,t) : is the transfer rate at wavenumber k.

v : viscosity.

The different terms in STKEB stand respectively for :
(1) temporal variation of TKE at wavenumber k.

(2) turbulent vertical transport of TKE.

(3) mechanical production through wind shear - turbulence interaction (CIT/WIT).

(4) transformation into potential energy through work against buyoancy forces.

(5) spectral energy transfer from lower to higher wave numbers k.

(6) viscous dissipation

The energy spectrum is discretized in bands according to $k_{n+1} = h \times k_n$ with $h = 1.5$ and $k_o = 0.02$ m^{-1}. The energy (En) in band number (n) is given by :

$$E_n = \frac{3}{2} Un^2 = \int_{h^{-1/2} \cdot kn}^{h^{+1/2} \cdot k'_n} E(k,t)\, dk$$

In order to have a realistic description of the negative heat flux at the inversion base we need to describe the unstationary time evolution of $E(k,t)$ according to STKEB.

The third term (source of Turbulent Kinetic Energy) is fed into the STKEB by the WIT and CIT subroutines at corresponding wave-numbers.

In order to model the fifth term $\delta\varepsilon(k,t)/\delta k$ we use a method that has been developed recently especially by Bell & Nerkin (1977). The main hypothesis is that (E_n) depends only on (E_{n-1}) and (E_{n+1}). In the same way time scale (t_n) in band (n) depends only on (t_{n-1}) and (t_{n+1}).

Under these hypothesis, we can write :

$$E_{n-1} \sim U_{n-1}{}^2 \qquad\qquad E_n \sim U_n{}^2 \qquad\qquad E_{n+1} \sim U_{n+1}{}^2$$

and :

$$t_{n-1}^{-1} \sim k_{n-1} \cdot U_{n-1} \qquad t_n^{-1} \sim k_n \cdot U_n \qquad t_{n+1}^{-1} \sim k_{n+1} \cdot U_{n+1}$$

Starting with the simplest case in which all terms are 0 but (1) and (5), we will look for a budget in band (n) of the following form :

$$\frac{\partial}{\partial t}\left(\frac{3}{2} Un^2\right) = \epsilon\, h^{-1/2} \cdot kn\, (t_{n-1},\ t_n,\ E_{n-1},\ E_n,\ \sqrt{E_{n-1}\, E_n})$$

$$-\, \epsilon\, h^{1/2} \cdot k_n\, (t_n,\ t_{n+1},\ E_n,\ E_{n+1},\ \sqrt{E_n\, E_{n+1}})$$

More we will infer by analogy with Navier-Stokes equations that only quadratic terms on wind speeds such as U_n^2 or $(U_n) \times U_{n+1}$ must be considered.

Then we can write :

$$\frac{\partial}{\partial t}(\frac{3}{2}\ u_n^2)=k_n U_n(ah^{-1}(U_{n-1}^2 - hU_n U_{n+1})+b(U_{n-1}U_n-hU_{n+1}^2))$$

Which is in slightly simplified form :

$$\frac{\partial}{\partial t}(U_n) = \beta\ k_n((U_{n-1}^2 - hU_n U_{n+1})-C(U_{n-1}U_n - h\ U_{n+1}^2))$$

We can see that C plays the role of a relaxation term but also means that energy can "go back" from high wavenumbers to lower (k). We preferred to set C = 0 and then (β) comes from Kolmogorov spectrum stationary case.

One cannot obviously work with very high wavenumbers up to the viscous dissipation range, this for computer time problems. We therefore infer that above the higher discretized wavenumber k_N, the spectrum obeys Kolmogorov law and we adapt ε (k_N) in order to match rather quickly ε_0 of the Kolmogorov spectrum. One can remark that this procedure implicity includes the viscous term (6).

We now consider terms (4) and (2) which are respectively B(k,t) and $\delta Q(k,t)/\ \delta z$.

We will model them respectively as :

$$B\ (k,t) = K_z\ (k,t)*\frac{\partial\Theta}{\partial z}$$

$$\frac{\partial Q(k,t)}{\partial z} = K_z\ (k,t) * \frac{\partial^2 q^2}{\partial z^2}$$

In these equations ($\delta\theta/\delta z$) stands for the vertical gradient of potential temperature while ($\delta q^2/\delta z$) stands for the vertical gradient of Turbulent Kinetic Energy.

For K_z (k,t) we make use of calculations made by Weinstock (1978) :

$$K_z\ (k,t) = E\ (k)\ \frac{k\ v_m}{0.8\ N^2 + k^2\ v_m^2}$$

v_m being the characteristic speed in the spectrum of vertical velocity defined as :

$$vm^2 = \frac{1}{n-1} \sum_n U_i^2$$

and N being the Brunt-Vaïsala frequency.

To estimate ($\delta q^2 / \delta z$) we need a diffusion length scale H_{dif}. This is given by making the hypothesis that H_{dif} corresponds to the difference between the middle of the turbulent layer at the inversion base and the lower level reached by the inverted thermals originated from the breaking waves.

Once terms (2) to (6) of STKEB are known or related to E(k,t), as B(k,t), we can describe the time evolution of E(k,t), which is term (1). The source terms in this evolution are the production terms defined in WIT and CIT which "breed" the spectral budget module at a small time step (dt_i). To obtain the average turbulent terms which drive the evolution of the mean values, we integrate the STKEB over all wavenumbers at each "long" time step (dt).

5 - <u>CALCULATION OF σ_w</u>

We have seen that we use this scaling velocity of the mixed layer quite extensively. Usually one would write :

$$\sigma_w^2 = w_*^2 + \eta_*^2 u_*^2$$

We made an improvement on this formulation by taking into account the turbulent Kinetic energy coming from the stable layer into the mixed layer (term (2) of STKEB integrated over all wavenumbers).

We start by writing the turbulent Kinetic Energy Budget for the mixed layer neglecting the time derivative term :

$$0 = \overline{(u'_i w)} \frac{\partial U_i}{\partial z} + \frac{g}{\theta} (\overline{w'\theta'}) - \frac{\partial}{\partial z} (\overline{w'q^2}) - \overline{\epsilon}$$

The Turbulent Kinetic Energy input from the stable layer can be written as

$$I = \int_{h-H_{dif}}^{h} (\frac{g}{\theta} (\overline{w'\theta'}) - \frac{\partial}{\partial z} (\overline{w'q^2})) \, dz$$

and :

$$I \sim \quad \frac{g}{\theta} \times \frac{H_{dif}}{2} \overline{(w'\theta')}_h - \overline{(w'q^2)}_h$$

There Hdif has been defined in last paragraph while h is the height of the inversion base.

As already noticed :

$$\overline{(w'\theta')}_h = K_z \frac{\partial \theta}{\partial z} \qquad \text{and} \qquad \overline{(w'q^2)}_h = K_z \frac{\partial q_2}{\partial z}$$

with :

$$K_z = \sum_n \frac{3/2 \times U_n^2 \times k_n \times vm}{0.8 \times N^2 + k_n \times vm^2}$$

Moreover ($\delta q_2 / \delta z$) can be written as :

$$\frac{\partial q^2}{\partial z} = \frac{1}{H_{dif}} (3/2 \times vm^2 - q^2)$$

q^2 being the average Turbulent Kinetic Energy in the mixed layer. With little calculation I can be rewritten as :

$$I = \frac{K_z}{H_{dif}} (3/2 \times vm^2 - q^2 - 1/2 \times (N \times H_{dif})^2)$$

The production term from the surface layer reads in purely convective conditions :

$$1/2 \times g/\theta \times h \times \overline{(w'\theta)}_s = 0.5 \times w_*^3$$

s being the surface layer subscript and w* the usual convective scaling velocity.

Extending this to the general case leads to :

$$0.5 \times \sigma_w^3$$

in which :

$$\sigma_w^2 = w_*^2 + \eta^2 \times u_*^2$$

Finally the dissipation term ( ) can be considered as constant in the mixed layer (Lenschow (1980)), and one can write :

$$\overline{\epsilon} \sim 0.8 \frac{q^3}{h}$$

Then the turbulent Kinetic Energy Budget in the mixed layer leads
to the following equation :

$$0.8 \, q^3 = 0.5 \, \sigma_w^{\,3} + \frac{K_z}{H_{dif}} \, (3/2 \, vm^2 - q^2 - 1/2(N \times H_{dif})^2)$$

which is non-linear and can be solved by an iterative Newton procedure.

Actually we use this scaling velocity q in our model rather than
usual σ_w and this is mostly important in the early morning stage
of inversion destruction.

6 – <u>VERTICAL HEAT FLUX PROFILE PARAMETERIZATION</u>

Up to now we have described mostly what happens at the inversion
base. One can imagine that our two layer model needs somehow a
turbulence model in the stable layer in order to control especially H which is the altitude of the top of the stable layer.

This is done through imposed conditions on vertical $(\overline{w'\theta'})$ (z)
profile as shown on figure 7, to which the reader can refer. The
conditions are the following :

(1) Between ground level (z = h_{ts}) and (h – H_{dif}) the heat
flux decreases linearly as usual.

(2) Between (h – H_{dif}) and (h + $e_t/2$) we make a parabolic fitting as well as between (h + $e_t/2$) and (h + e). The heat
flux derivative is vertical at (h + $e_t/2$).

(3) The maximum negative value of the heat flux is given by
the integrated B (k,t) term in the spectral module and
set to the height (h + $e_t/2$).

(4) Last above (h + e) the fitting is exponential with the
following constraints :

- derivative at (h + e) is the same for the parabolic
fitting up to (h + e) and the exponential fitting above.

- the exponential decay is related to N.

Our model gives for each time step (dt) a $(\overline{w'\theta'})$ (z) vertical profile allowing the calculation of the   (z) profile according to

$$\frac{\partial \theta}{\partial t} = - \frac{\partial (\overline{w'\theta'})}{\partial z}$$

With this new temperature profile (h – H_{dif}) can be determined for
the next time step by using a thin vertical grid.

7 – AVERAGE EQUATIONS

First one can write momentum conservation laws for the mixed layer :

$$\left(\frac{\partial U}{\partial z}\right)_{+} \times \frac{\partial h}{\partial t} \;-\; \frac{\partial}{\partial t}\,(U_g - U_n) \;+\; \frac{1}{h}\,\overline{(u'w')}_i \;=$$

$$\frac{1}{h}\,\overline{(u'w')}_s \;+\; f\,(V_n - V_g)$$

$$\left(\frac{\partial V}{\partial z}\right)_{+} \times \frac{\partial h}{\partial t} \;-\; \frac{\partial}{\partial t}\,(V_g - V_n) \;+\; \frac{1}{h}\,\overline{(v'w')}_i \;=$$

$$\frac{1}{h}\,\overline{(v'w')}_s \;-\; f_*(U_n - U_g)$$

and then for the shear layer in the temperature inversion :

$$\left(DU -\left(\frac{\partial U}{\partial z}\right)_{+} \times e\right)\frac{\partial h}{\partial t} \;+\; \frac{e}{2}\,\frac{\partial}{\partial t}\,(DU) \;+\; \overline{(u'w')}_i$$

$$+\;\left(DU/2 -\left(\frac{\partial U}{\partial z}\right)_{+} \times e\right)\frac{\partial e}{\partial t} \;=\; -\,f_*e\,(V_n + DV/2 - V_g)$$

and :

$$\left(DV -\left(\frac{\partial V}{\partial z}\right)_{+} \times e\right)\frac{\partial h}{\partial t} \;+\; \frac{e}{2}\,\frac{\partial}{\partial t}\,DV \;+\; \overline{(v'w')}_i$$

$$+\;\left(DV/2 -\left(\frac{\partial V}{\partial z}\right)_{+} \times e\right)\frac{\partial e}{\partial t} \;=\; +\,f_*e\,(U_n + DU/2 - U_g)$$

with : ·

$$\left(\frac{\partial U}{\partial z}\right)_+ \text{and} \left(\frac{\partial V}{\partial z}\right)_+:$$

vertical gradients of horizontal wind components above the shear layer.

DU   : $U_g - U_n$ where U_g is the first component of the geostrophic wind.

DV   : $V_g - V_n$ where V_g is the second component of the geostrophic wind.

i   : subscript for the inversion base.

s   : subscript for the surface layer.

n   : subscript for the mixed layer.

Finally the Richardson number being kept at 0.5 in the shear layer (apart from divergence and breaking stages) leads (with Ri : 0.5) to :

$$\frac{N^2 \times e}{2(H-h)} * \frac{\partial}{\partial t}(H - h) - \frac{e}{2*(H-h)} * \frac{\partial}{\partial t}(D\theta) + Ri * \frac{DU}{e} \frac{\partial}{\partial t}(DU)$$

$$+ Ri * \frac{DV}{e} * \frac{\partial}{\partial t}(DV) - N^2 * \frac{\partial e}{\partial t} = 0$$

With :

$$D\theta = \theta_H - \theta_h$$

Actually we could also write :

$$P = - \overline{(u'_i w')} \times \frac{\partial U_i}{\partial z}$$

P being the mechanical production term in the Turbulent Kinetic Energy Budget at the inversion base, integrated over all wavenumbers.

This shows that our model is overdetermined due to the very strong constraints imposed on the vertical profile of heat flux $\overline{(w'\theta')}$ (z).

We plan to use this overdetermination as an overall check of the behaviour of this model.

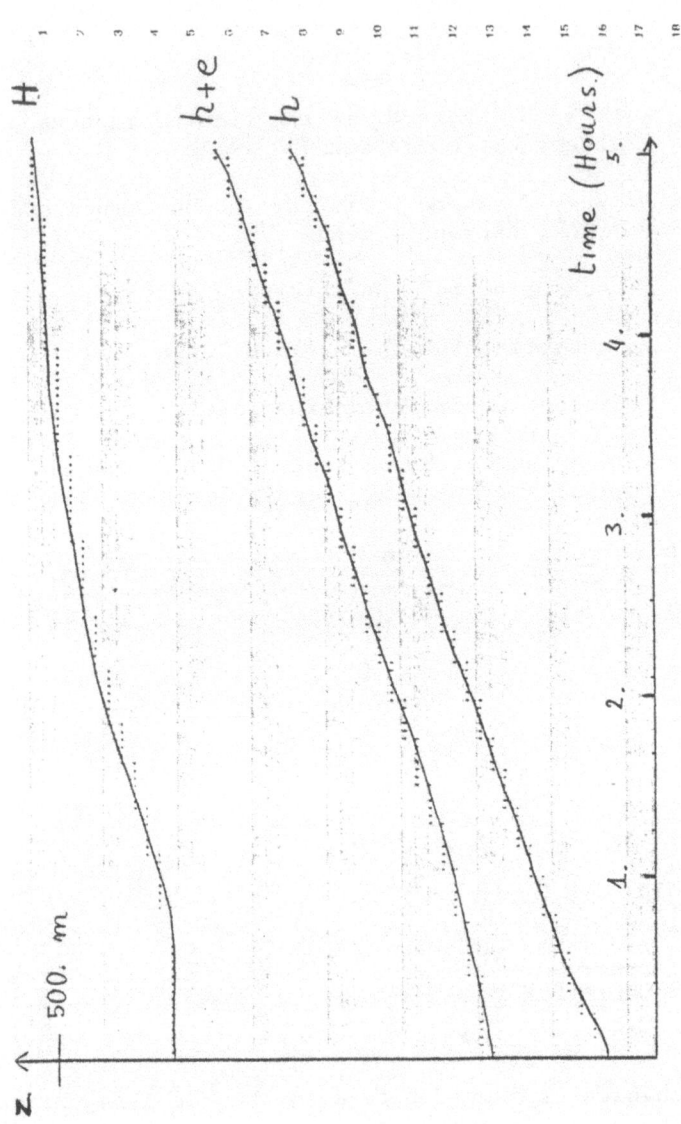

FIGURE 8 : Simulated evolution of h, h+e, and H during a morning transition period. (LIMAGNE July 9, 1977 : 6.00 A.M. to 11.00 A.M.)
Initial conditions h_0 = 20 m
                    H_0 = 400 m
                    $D_0\theta$ = 18.5 °C

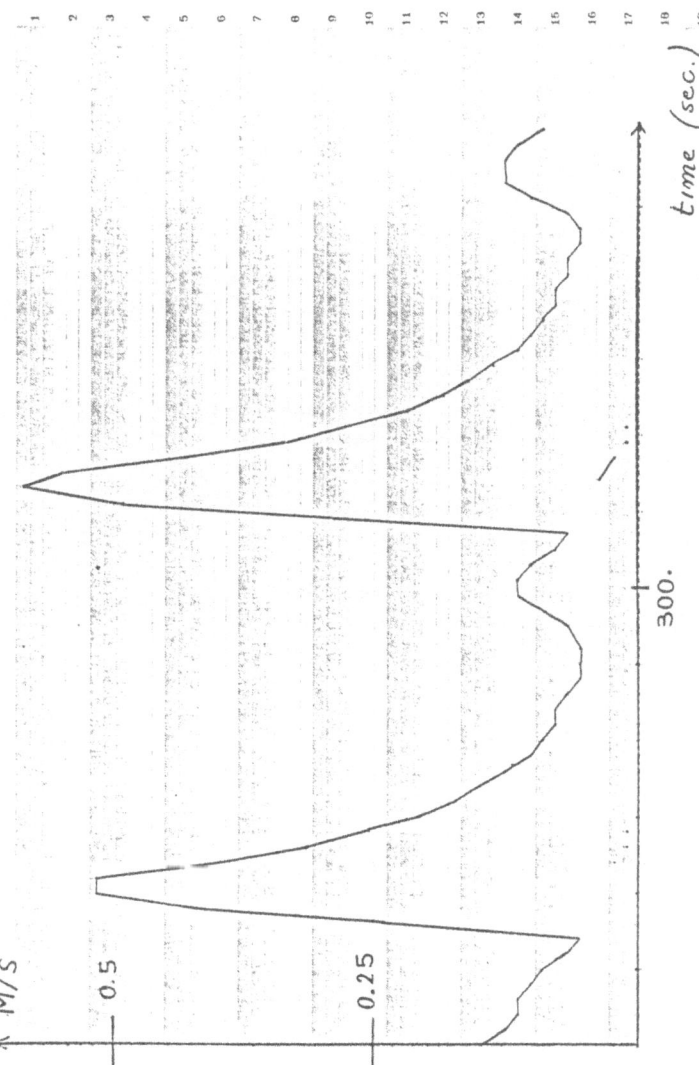

FIGURE 9 : Unstationary evolution of vm : the turbulent velocity scale
at the inversion base during a 600 sec. sample of model execution.

8   SOME SIMULATION RESULTS

We present in figure 8 a simulation result based on some data
gathered during the LIMAGNE experiment. Presently we are looking
for data of morning inversion rise with a strong wind shear in
order to test the WIT and CIT parameterizations more extensively.

REFERENCES

BELL T. et NERKIN M. (1977) : "Non linear cascade models for fully
    developed turbulence".
    Phys. of Fluids. V. 20 ; page 345.

DEARDORFF J.W. (1974) a : " Three dimensional numerical study of
    the height and mean structure of a heated planetary boundary
    layer".
    Boundary Layer Meteorology, 1, 81 - 106.

DEARDORFF J.W. (1974) b : "Three dimensional numerical study of tur-
    bulence in an entraining mixed layer".
    Boundary Layer Meteorology, 1, 199 - 226.

DONALDSON, C du P (1973) : "Construction of a dynamic model of the
    production of atmospheric turbulence and the dispersal of
    of atmospheric pollutants".
    Workshop on Micrometeorology. BOSTON.
    American Meteorological Society, 313 - 390.

HAZEL P. (1972) : "Numerical studies of the stability of inviscid
    stratified shear flows".
    Journal of Fluids Mechanics, V. 51, Part. 1 39 - 61.

LENSCHOW D.H. (1980) : "The role of Thermals in the convective
    boundary layer".
    Boundary Layer Meteorology, 19, 509 - 531.

MANTON M.J. (1977) : "On the structure of convection".
    Boundary Layer Meteorology 13, 491 - 503.

MANTON M.J. (1975) : "Penetrative convection due to a field of
    thermals".
    J.A.S. 32, 2272 - 2277.

MELLOR and YAMADA (1974) : "A hierarchy of turbulence closure mo-
    dels for planetary boundary layers".
    J.A.S. 31, 1791 - 1806.

LUMLEY J.L. and KHAJEH-NOURI B. (1974) : "Modeling homogeneous de-
    formation of turbulence".
    Advances in Geophysics, Vol. 17, NEW YORK.

ROSSET R. (1975) : "Mécanismes, rôle et paramétrisation de l'en-
    trainement au sommet de la couche limite planétaire convec-

tive".
Thèse univ. CLERMONT FERRAND.

WEINSTOCK J. (1978) : "On the theory of turbulence in the Buoyancy
    Subrange of Stably Stratified Flows".
    J.A.S., V. 35, 634 - 648.

WYNGAARD J.S. and COTE O.R. (1974) : "The evolution of a convec-
    tive planetary boundary layer : a higher order closure model
    study".
    Boundary Layer Meteorology, 7, 289 - 308.

5: REGULATORY APPLICATIONS

Chairman: N. D. Van Egmond   Rapporteur: J. G. Kretzschmar

AN APPLICATION OF THE EMPIRICAL KINETIC MODELING APPROACH (EKMA)

TO THE COLOGNE AREA

Rainer Stern and Bernhard Scherer

Freie Universität Berlin, Institut für Geophysikalische
Wissenschaften – Fachrichtung Meteorologie –
1000 Berlin 33, Thielallee 50, West-Germany

ABSTRACT

The Empirical Kinetic Modeling Approach (EKMA), recently proposed by the EPA, is a method for deriving ozone-percursor relationships that can be used to design a control strategy for urban ozone
reduction. The technique is based on an isopleth diagram generated
by computer simulations of the chemical reactions occuring among
pollutants in a well-mixed box subjected to dilution, time varying
light intensity, diurnal dependence of the emissions, and background concentration levels. The performance of EKMA is discussed
in an application to the Cologne area of the FRG. The lateral and
vertical spatial distribution of the precursor emissions in this
area strongly violates the assumption of homogeneous mixing underlying EKMA. Therefore, the applicability of EKMA for developing
control strategies in the Cologne area is limited.

INTRODUCTION

For some years it has been known that solar intensity in the
higher latitudes is sufficient to produce substantial concentrations
of oxidants in the presence of organic compounds and oxides of nitrogen[1]. Measurements of oxidant levels have shown that it is also
necessary for Central Europe to develop emission control strategies
that could be applied during photochemical smog episodes in order
to improve air quality[2].

As oxidants are secondary products, resulting from a series of
chemical reactions initiated by irradiation of nitrogen dioxide, any
control strategy must be based on the precursor emissions which are

681

oxides of nitrogen and reactive hydrocarbons. At present there
exist several methods for relating photochemical oxidants to their
precursors, but every method has its shortcomings. The greatest potential for assessing the effectivness of oxidant control strategies
involves the use of deterministic photochemical dispersion models.
These models compute ambient oxidant concentrations resulting from
chemical transformations and atmospheric dispersion[3],[4]. However, a
key limitation in using such complex photochemical models is the
need for large computer facilities and a very extensive data base,
which is often not available for many regions. The simplest approach
involves the use of a rollback method, which assumes that improvement
in air quality is incrementally related to reduction of emissions.
This principle is used in the Appendix J-method[5], where a single
curve relates percentage reduction in hydrocarbon emissions required
in a given area to attain a 80ppb standard to the maximum one-hour-
average oxidant concentration monitored in that area. The Appendix
J-curve derived from data from several US-cities is based on the
comparison of 6-9 a.m. average nonmethane hydrocarbon (NMHC) concentrations to the subsequent peak one-hour-average oxidant concentration observed at the same station. While the method is easy to apply,
it suffers from several severe limitations, including the failure to
consider pollutant transport or the role of NO_x in oxidant formation.

A relationship between ozone levels and both the organic compounds and the nitrogen oxides (NO_x) can be constructed in the form
of an ozone diagram. Such a diagram can be extracted from aerometric
data[6], smog chamber results[7],[8] or computer simulations of the chemical
processes occuring in the polluted atmosphere. The latter technique
was used in the development of the Empirical Kinetic Modeling Approach (EKMA) proposed by the EPA[9] to replace the Appendix J relationship. The model underlying EKMA computes the chemical reactions
occuring within a well-mixed box, subjected to dilution and time
varying light intensity. Based on these modeling results, ozone
isopleths are constructed for a wide range of initial precursor conditions. Using ambient air quality data in the region of interest,
the percentage change in emissions of NMHC and/or NO_x required to
attain an oxidant standard can be assessed from the isopleth diagram.
At present, it seems that EKMA may be the most reasonable compromise
between very complex photochemical dispersion models and the simpler
rollback procedures.

In this paper the performance of EKMA is discussed in an application to the Cologne area of the Federal Republic of Germany.

THE MODEL UNDERLYING EKMA

The model underlying EKMA, OZIPP[10], simulates the photochemical
reactions which take place in a column of air moving with the ambient wind. The column extends from the ground to the base of the
inversion layer and uniform mixing is assumed within the reaction

volume. Emissions of hydrocarbons and NO_x enter the column according to the emission density of the area being traversed. The reaction volume changes due to the rise of the mixing height, and this in turn causes dilution of the material within the box and entrainment of additional pollutants from above the column.

The 75-step chemical mechanism contains two hydrocarbons as surrogates for the total NMHC mixture of a real urban atmosphere : Propylene for the reactive components, and n-butane for the less reactive hydrocarbons. The mechanism was first calibrated using the result of smog chamber experiments with irradiated auto exhaust and then adapted to more realistic atmospheric conditions[11]. Additional assumptions include :
- Clear sky
- Temperature independent reaction rates
- No ground removal of pollutants
- Representation of emissions prior to the time the model simulation begins (8 a.m.) by initial precursor concentrations.

The isopleths are constructed from the maximum one-hour average ozone concentrations calculated as a function of a wide range of initial concentrations of NMHC and NO_x, subsequent NMHC and NO_x emissions, sunlight intensity, dilution, reactivity of the precursor mixture, and transport of ozone and precursors from upwind areas. To apply the EKMA isopleth diagram the following information is needed
- The design value of ozone (the highest of the second highest hourly concentrations observed at all monitors during the base period).
- The prevailing NMHC/NO_x ratio occuring between 6-9 a.m. within the area covered by the air column before the simulation begins.

DESCRIPTION OF THE COLOGNE AREA

Emissions

The Cologne basin is a highly industrialized region in the western part of the FRG (Fig. 1). The emissions in that region are predominantly from chemical and petrochemical industries situated primarily north and south of the Central Business District (CBD) of Cologne. The next adjoining large urban-industrial area is Düsseldorf situated about 40 km to the north, and which itself lies on the southern edge of the Ruhr area, the largest industrial area in Europe.

Table 1 presents the total precursor emissions in the Cologne area as derived from the inventory of the "Rheinschiene Süd"[12]. This inventory is for the region framed by the dashed-dotted line in Fig. 1. As previously mentioned, the main sources of nitrogen oxides and organic compounds are the industrial complexes around the city of

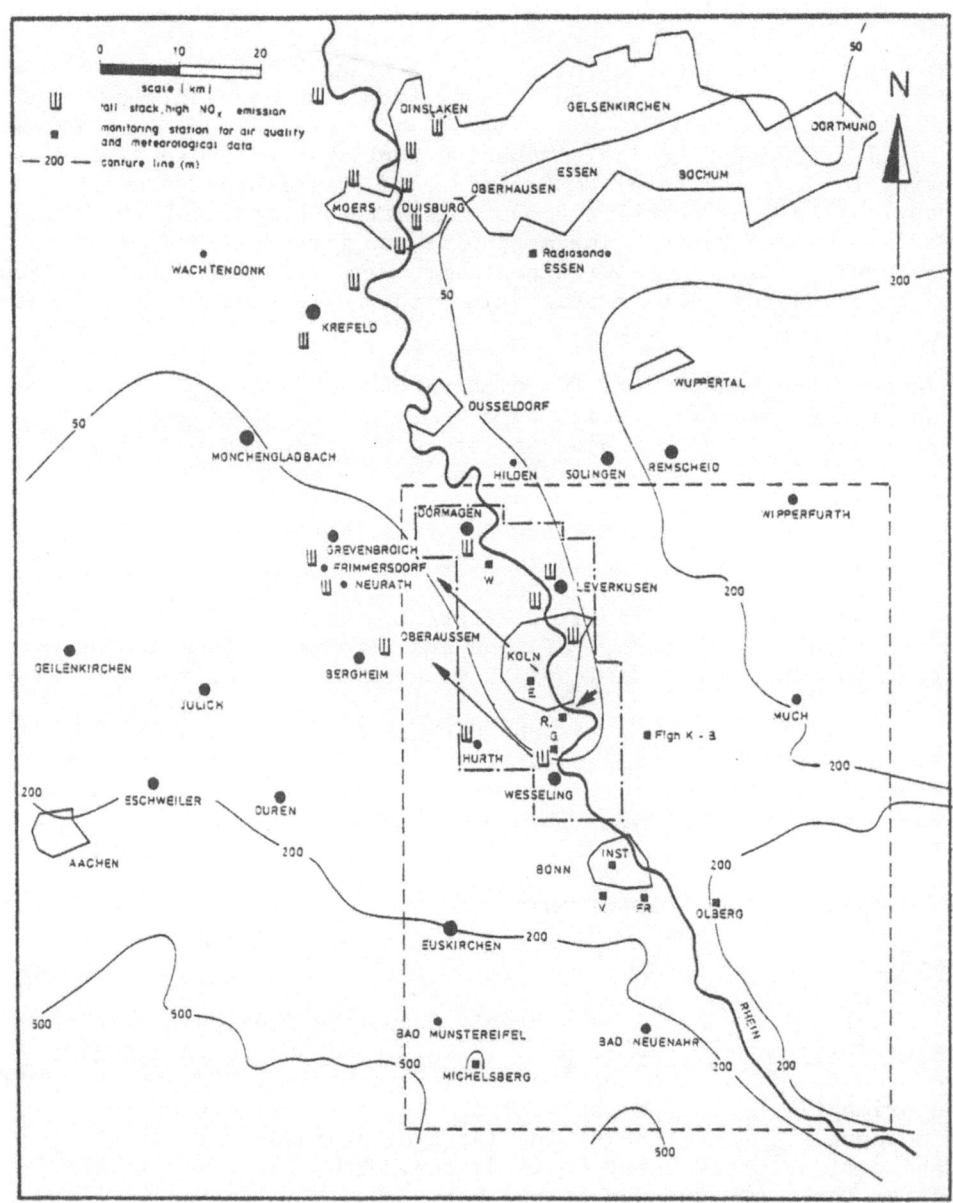

Figure 1. : The Rhine-Ruhr area. The dash-dotted line encloses
the area of the emission inventory "Rheinschiene Süd"
The dashed line encloses the region of monitoring.
Additional symbols are described in the text.

Table 1. Total precursor emissions (in metric tons per year) in the
Cologne area[12]

	Industry	Domestic heating Small Industries	Traffic
Oxides of Nitrogen as NO_2	75761	4499	9734
Volatile Organic Compounds	84117[x]	6322[xx]	11500[xxx]

[x] approx. 62 % NMHC   [xx] approx. 91 % NMHC   [xxx] corrected,see Ref.13

Cologne. Fig. 2 illustrates the lateral and vertical spatial distributions of the precursor emissions. It is particulary interesting
to note that :
- 95 % of the hydrocarbons are emitted below 30 m, while
- 85 % of the NO_x is emitted above 30 m.

Air Quality Data

        Commencing in 1975, half-hour-average air quality levels
have been monitored[2] at several stations within the region shown
in Fig. 1. Two rural stations, Michelsberg and Ölberg, are situated
in the southwest and the southeast of Bonn on top of hills 580 m and
450 m high, respectively. Two urban monitoring stations are located
inside Bonn, Institut für Physikalische Chemie (Inst) and Venusberg
(V) and another two inside Cologne, Eifelwall(E) and Rodenkirchen(R).
Two additional stations are situated in rather industrialized areas
to the north and south of Cologne. Godorf (G), to the south, is
located near an oil refinery and a petrochemical plant. The northern
station, Worringen (W), is located in the neighbourhood of a chemical plant. All stations provide ozone data and most of them NO_x data
as well; however, NMHC data were available for only one year (1979)
and only from the Rodenkirchen station. This station, accented by
an arrowhead in Fig. 1, is located south of the CBD and reflects
the influence of both traffic and industry.

        Table 2 presents some significant statistics of the data[14]
measured at Rodenkirchen, while Table 3 shows the number of days
during the years 1976 to 1979 for which the observed ozone concentrations exceeded 75 ppb. While it is obvious that oxidant levels
in 1979 were not very high, this is the only year for which NMHC
measurements were available; thus, this year was used as a base.

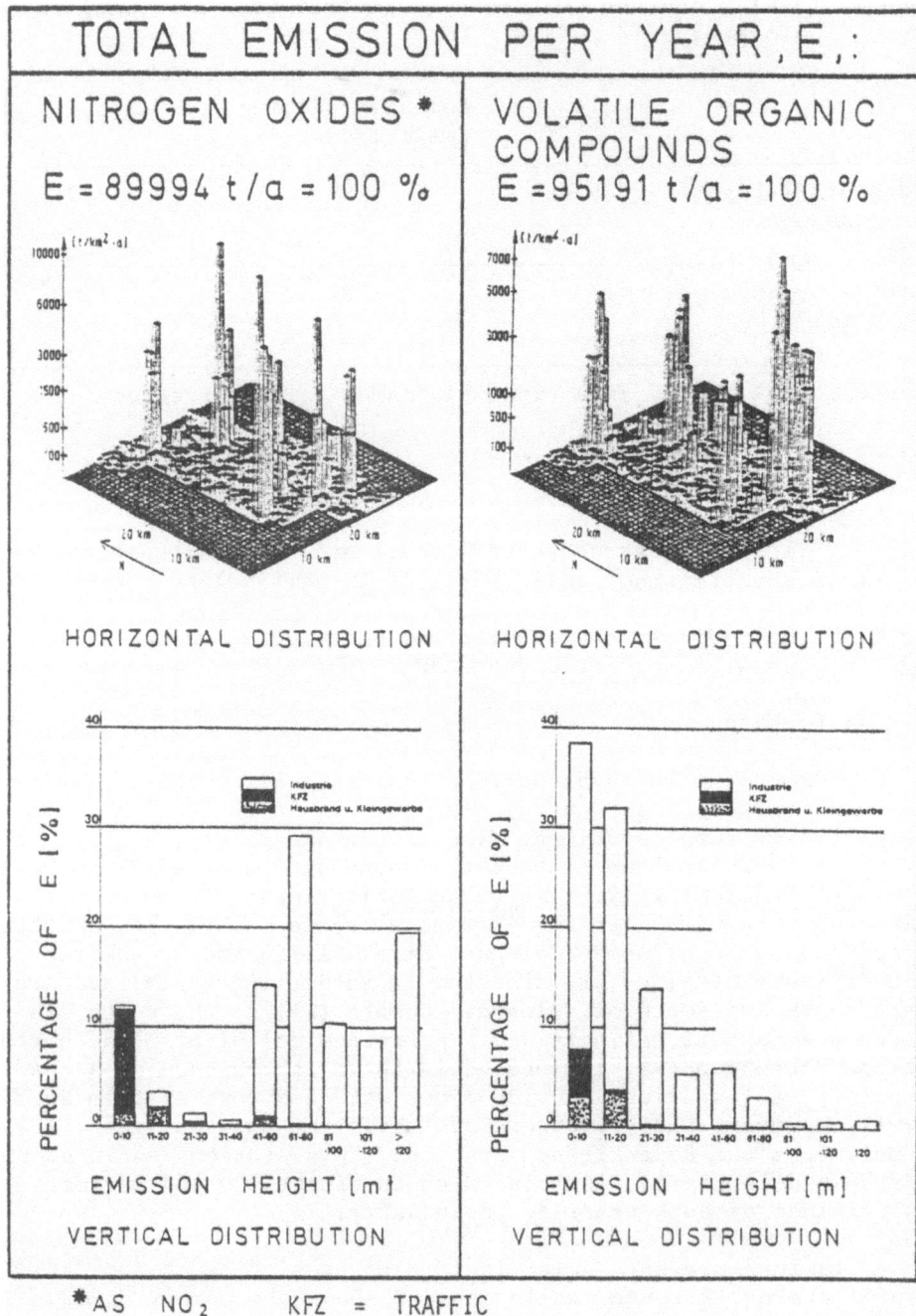

Figure 2. : Spatial distribution of precursor emissions in the
Cologne area[12].

Table 2. Average O_3, NMHC and NO_x concentrations measured at Rodenkirchen for the period 14.5. - 13.9. 1979
source : Ref. 14

	O_3 ppb	(NMHC)6-9 ppbC	(NO_x) 6-9 ppb
Median	23	490	40
95percentile	68	930	81
peak value	122		

Table 3. Number of days for which the observed ozone concentration
exceeded 75, 100, 120, 150 ppb, respectively.
Site : Rodenkirchen, source : Ref. 14

	No. of days O_3 exceeded				O_{3max}
	75	100	120	150	ppb
1976	30	16	9	6	185
1977	15	3	2	1	205
1978	20	6	3	3	163
1979	19	4	1	0	122

APPLICATION OF EKMA

To develop area specific isopleths, two separate meteorological scenarios were considered. The first case represents a
day characterized by stagnant conditions, while the second case
assumes a constant wind from the southeast, a prevailing wind direction during photochemical smog episodes in the Cologne area[2].

Stagnation case

  To simulate a stagnation case, June 21, 1979 has been chosen.
The weather situation was determined by a high pressure system over
Central Europe. The airport Köln/Bonn reported a noon temperature
of 28°C, 1/8 cloud cover and calm wind conditions. At Rodenkirchen,
the peak regional ozone concentration of 106 ppb was measured. Due
to the stagnant conditions, it is likely that this peak ozone concentration at Rodenkirchen is, as assumed in EKMA, related to the
early morning precursor concentrations in the area, i.e., the reaction volume remains within the city throughout the day. To generate
the ozone isopleths diagram, the following input parameters for
OZIPP were determined :
Dilution : Derived from mean mixing height conditions observed during photochemical smog episodes, the mixing height is assumed to
be 400 m at 8 a.m., rising to a maximum of 1900 m at 4 p.m..
Transport : Precursor concentrations transported within the surface
layer are assumed to be negligible. To allow for the ozone already
generated by 8 a.m. from local precursor emissions, an initial
"transport related" value of 14 ppb, as measured at Rodenkirchen,
was chosen (n.b., it is necessary in OZIPP to consider non-zero
initial ozone levels as being transport related). Transport aloft
of ozone was estimated from surface measurements at the Ölberg site,
where 60 ppb was monitored at the time of the inversion break up.
Transport aloft of NMHC and NO_x was derived from aircraft measurements during typical smog episodes and set 15 ppb and 5 ppb, respectively.
Emissions : Hourly emissions of the precursor pollutants occuring
after 8 a.m. are expressed as the fractions $Q_i/(c_o hA)$, where Q_i is
the emission over the are A for the ith hour after 8 a.m., c_o is the
measured 8 a.m. mass concentration and h is the initial mixing height.
For this stagnation case, c_o is the concentration measured at Rodenkirchen and A is the area of the inventory.
Reactivity : The EPA recommended specification of the reactivity of
the initial NMHC mixture consists of 25 % propylene and 75 % n-butane.
Smog chamber results[8] with irradiated outdoor air from the Rhine-Ruhr
area provide evidence that the reactivity should be lowered. Therefore, these simulations assume 15 % propylene.

  Table 4 summarizes the OZIPP input parameters for the stagnation case. The isopleth diagram resulting from using the input data
of Table 4 is shown in Fig.3. The most obvious feature of this diagram is the strong inclination of the isopleths in the upper part,
reflecting the fact that high NO_x emissions, assumed to be 90 % NO
and 10 % NO_2 inhibit the formation of ozone. Thus, these calculations suggest that a NMHC/NO_x ratio of about 20 is required for
maximum ozone formation. This contrasts sharply with the optimal
ratio of 5-7 derived from smog chamber experiments[8] and with the
optimal ratio of 8-9 observed from measurements in the Netherlands[6].
However, assuming a NMHC/NO_x ratio of 6, the model fails under any

Table 4. Input parameters for the stagnation case (June 21, 1979)

Dilution : Minimum mixing height	400 m
Maximum mixing height	1900 m
Time period for rise	8 a.m. – 4 p.m.

Emission : Expressed as fractions of initial concentrations
Measured initial concentration NO_x = 44 ppb
NMHC= 350 ppbC

Emission fractions versus time

hour	8-9	9-10	10-11	11-12	12-13	13-14	14-15	15-16	16-17	17-18
NMHC	.14	.15	.15	.15	.15	.14	.15	.15	.16	.16
NO_x	.41	.41	.45	.46	.46	.48	.48	.49	.50	.50

Surface layer transport : NHMC = 0.
NO_x = 0.
O_3 = 14 ppb

Transport aloft :   NMHC = 15 ppbC (10 % C_3H_6, 90 % nC_4H_{10})
NO_x = 5 ppb   (100 % NO_2)
O_3 = 60 ppb

Reactivity : Fraction of NMHC as propylene = .15
Fraction of NMHC added as aldehydes = .05
NO_2/NO_x = .59

initial conditions to predict ozone maxima above 80 ppb. This ano-
maly results from the fact that, in the model, high initial NMHC
levels are required to prevent the destruction of ozone by the large
quantities of NO_x subsequently emitted by elevated point sources.

Further insight in this apparent contradiction can be obtained
through examination of the modeling assumptions. The box model
approach implies that all pollutants released into the box are instantaneously well mixed. This is certainly not true for an urban
area having markedly different vertical emission profiles for the
NMHC and NO_x precursors. Differences in vertical emission profiles
lead to two problems :
- Different dispersion processes for the two precursors violate the
  uniform mixing and stationary box assumption.
- The observed NMHC/NO_x ratio at ground level reflects the actual vertical emission profiles and not the vertically averaged profiles as

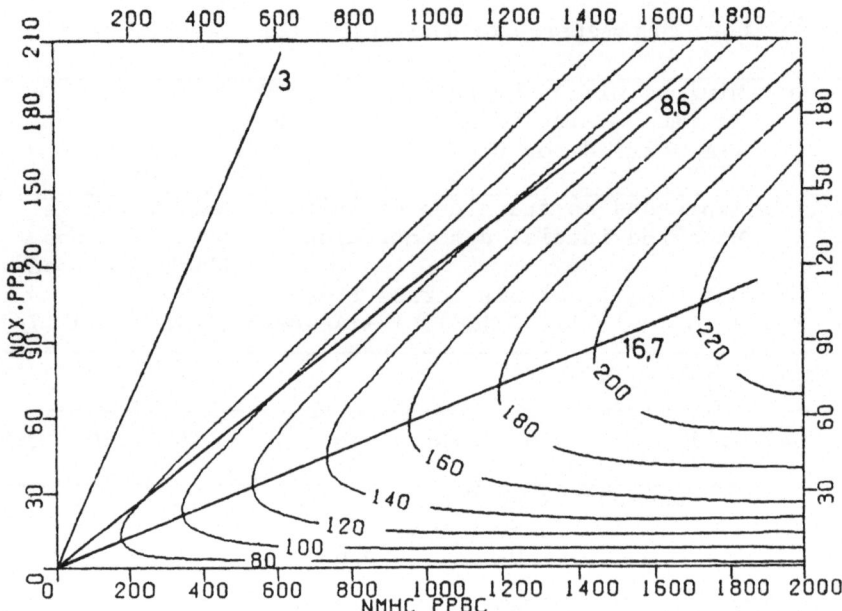

Figure 3. : Ozone isoplets based on the input parameters of Table 4.
The straight lines indicate the different 6 - 9 a.m.
NMHC/NO$_x$ ratios discussed in the text.

assumed in OZIPP; thus, measured and calculated ratios are not comparable in such cases.
This latter point is particulary important for deriving any control
strategy from the isopleth diagram. Peak ozone concentrations calculated by OZIPP are functions of the initial,box averaged precursor
concentrations. If measured and box averaged initial precursor concentrations are not comparable, then the use of measured NMHC/NO$_x$
ratios is not consistent with the assumptions underlying the isopleths. On the other hand, use of model consistent ratios is also
questionable, as the model does not consider the complicated transport and mixing processes occuring in the real atmosphere, i.e.,
that the two precursors may spatially separated and hence incapable
of interacting chemically.

To illustrate this point NMHC/NO$_x$ ratios for different subsets
of the emission inventory are presented in Table 5. Comparing these
ratios with those derived from measurements (Table 6), one notes that
corresponding ratios can be found only for the CBD. The CBD is assumed to be characterized by measured early morning ratios of about 8,

Table 5. Approximate NMHC/NO_x volumetric concentrations ratios derived from the inventory using 6-9 a.m. emissions

	(NMHC/NO_x)6-9
All emissions	3
All NMHC emissions, NO_x reduced by 80 % of industrial emission (sources above 60 m)	9
All emissions of the CBD	8
All emissions of the industrial area Wesseling	3
All NMHC emissions of the industrial area Wesseling, NO_x reduced by 80 %	15

Table 6. Measured 6-9 a.m. NMHC/NO_x ratios in the Cologne area

	(NMHC/NO_x)6-9
Rodenkirchen (influenced by traffic and industry), all measurements	13
Rodenkirchen, averaged from the 5 days having the highest ozone concentrations[xx]	16.7
Rodenkirchen, averaged from 25 days when T_{max} above 25°C	14
Rodenkirchen, June 21, 1979	8.6
Wesseling (site G), industrial area[x]	15 – 21

x incomplete data   xx as recommended by EPA

similar to that observed on June 21 at Rodenkirchen, a station which under stagnant conditions is predominantly influenced by sources similar to those found in the CBD. For the industrial area of Wesseling, to the south of Cologne, as well as for the total inventory, measured and calculated ratios agree approximately only if 80 % of the industrial NO_x emissions (i.e. sources above 60 m) are excluded from the calculation.

To investigate the effect of this NO_x exclusion, OZIPP was used to calculate the diurnal variation of the concentration levels for June 21 under both emission scenarios. Fig. 4 illustrates the diurnal dependence of the calculated and measured concentrations. Two

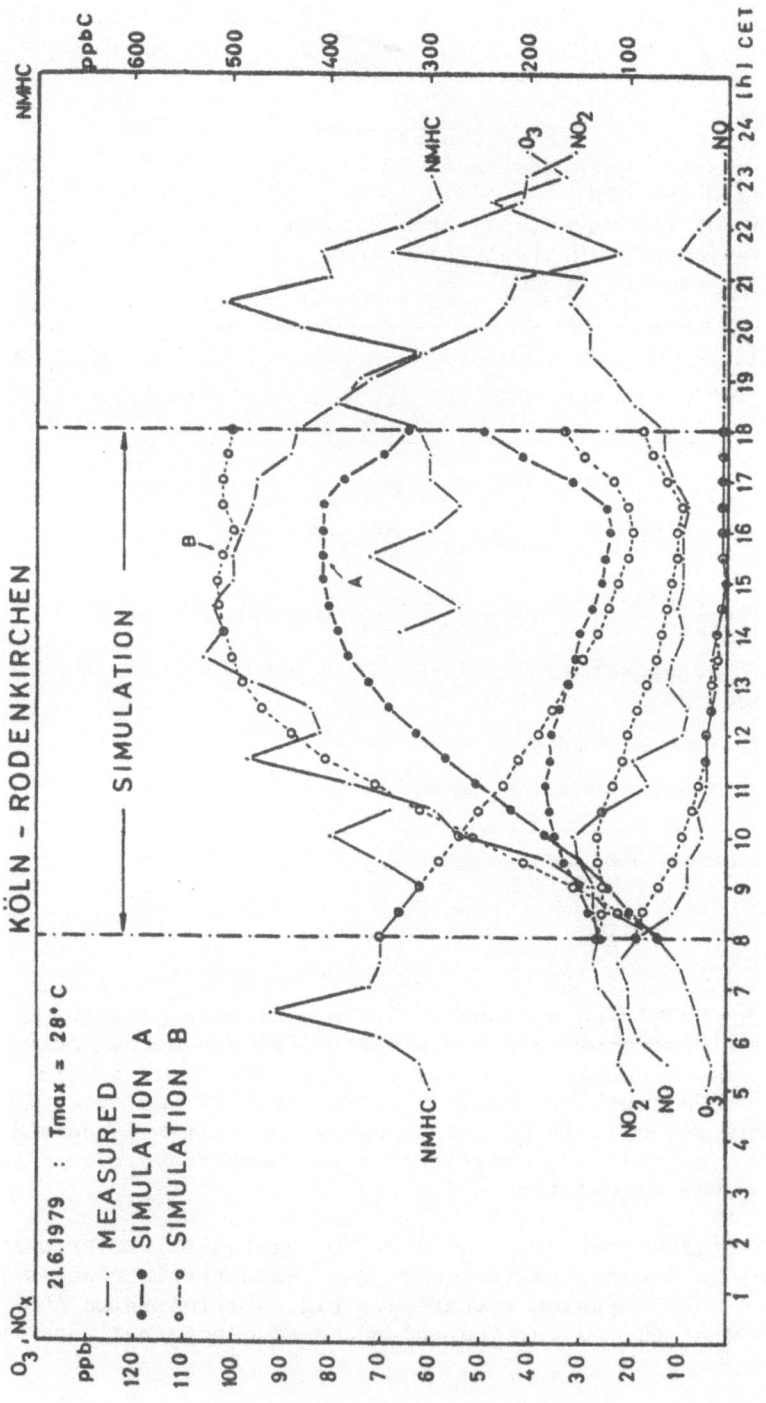

Figure 4. : Diurnal variations of measured and calculated concentration levels. Simulation A is based on the input parameters of Table 4. Simulation B differs in that it is based on only 20 % of the industrial NO_x emissions (i.e.,sources below 60 m)and an increased value of 20 ppb NO_x transported aloft. NMHC and NO curves for simulation A are not shown since they differ only slightly from those of simulation B.

simulations are shown : Simulation A is based on the input parameters
presented in Table 4, while simulation B is based on only 20 % of the
industrial NO_x emissions and an increased value of 20 ppb NO_x transported aloft. The increase in the NO_x transported aloft represents
an attempt to simulate within OZIPP the downward mixing of NO_x from
elevated plumes during the day. Both simulations are started using
the 8 a.m. concentrations measured at Rodenkirchen. Simulation A
predicts a peak ozone concentration well below that observed, while
the NO_2 levels are nearly double those measured. For simulation B
the calculated and observed pollutant levels agree quite well. This
supports the direction of the hypothesis that measured concentrations
near ground level and close to the emission area are not impacted by
elevated point sources. While not a critical factor in the present
comparison, the difference between measured and calculated NMHC
levels in both simulations probably can be related to the sensitivity of measured concentrations to the detailed spatial distribution
of ground level hydrocarbon sources.

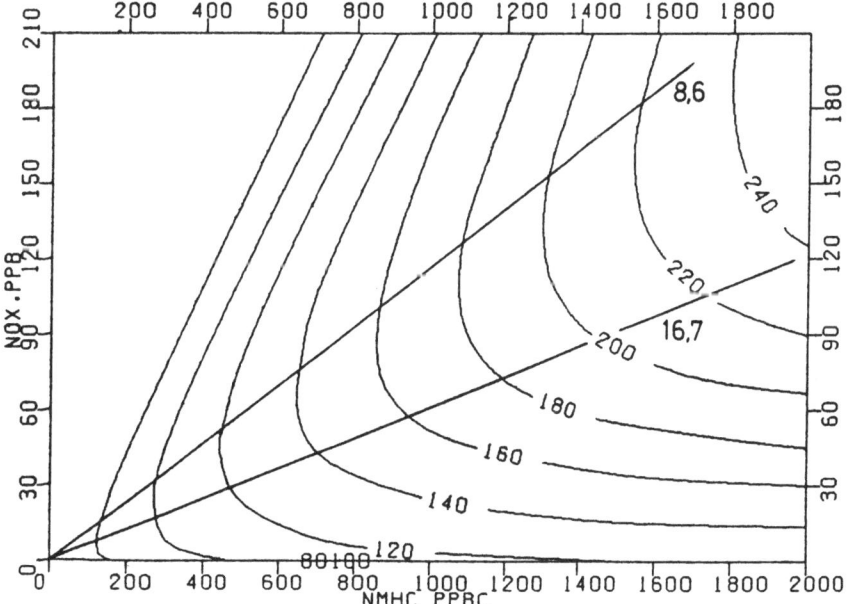

Figure 5. : Ozone isopleths based on input parameters of Table 4
            excepting that only 20 % of industrial NO_x emissions
            (i.e. below 60 m) and an increased value of 20 ppb NO_x
            aloft are assumed.

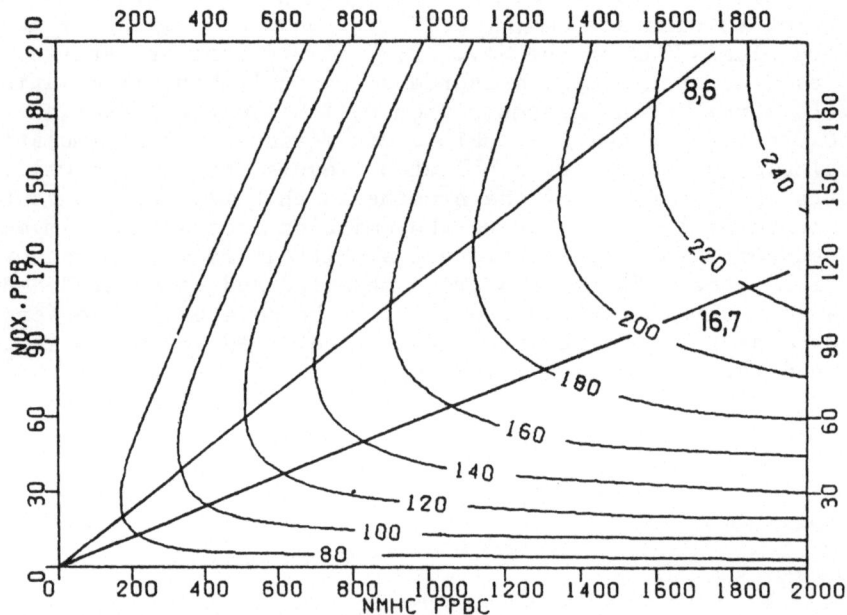

Figure 6. : Ozone isopleths based on the input parameters of Table 4
           excepting that only 20 % of industrial NO$_x$ emissions
           (i.e. below 60 m) are assumed.

        Isopleths are derived based on the input parameters of simula-
tion B. This diagram(Fig.5) differs dramatically from the diagram
of simulation A (Fig.3) and clearly illustrates the effect of lowe-
ring the NO$_x$ emissions.One also notes that the increased level of
NO$_x$ leads to greater ozone under a zero NO$_x$ emission scenario;how-
ever this artifact results from having the incremental NO$_x$ aloft
(i.e. the excess over 5 ppb) remain constant rather than proportional to the NO$_x$ emissions. However, this does imply that control
strategies derived from Fig. 5 should apply only to the near ground
level NO$_x$ sources.

        Isopleths generated under the assumption that the excluded,
elevated NO$_x$ sources are not compensated for by an increase in the
NO$_x$ aloft are shown in Fig. 6.Comparison of Fig.5 and Fig.6 further
allows assessment of the influence of transported NO$_x$ aloft on ozone
control strategies.

        Table 7 illustrates the percentage precursor reductions required
to meet a 80 ppb ozone limit for both measured and calculated early

Table 7. Reduction requirements to attain a 80 ppb ozone level derived from the isopleth diagrams of Figs. 3,5,6 using calculated and measured 6-9 a.m. $NMHC/NO_x$ ratios.
A design value of 106 ppb ozone is assumed.

$(NMHC/NO_x)6-9$		3^x	16.7^x	8.6^x
NMHC reduction NO_x constant (%)	3^{xx}	+	42	16
	5^{xx}	−	61	42
	6^{xx}	−	63	51
NO_x reduction NMHC constant (%)	3^{xx}	+	86	97
	5^{xx}	−	0	0
	6^{xx}	−	79	84

x   ratios explained in Tables 5 and 6
xx   Fig. number used for strategy
+   no reduction can be derived
−   not applicable
0   ozone limit can not be attained even through complete elemina-
tion of NO_x emissions as background NO_x concentration is too high.

morning precursor ratios, given an ozone design value of 106 ppb.It
is obvious how strong an influence this choice of initial precursor
ratio has on the proposed ozone reduction strategy.

The following conclusions can be drawn from this stagnation
case :
- Using the box averaged (i.e. total emission inventory) $NMHC/NO_x$
  ratio of 3, no reduction strategy can be developed as this ratio
  line has no intersection with relevant ozone isopleths (see Fig.3).
- Measured and calculated $NMHC/NO_x$ ratios are not directly comparable
  unless careful consideration is given to the differences in vertical emission profiles of the precursors.
- Reduction strategies based on measured early morning ratios are
  not consistent with model assumption of total mixing and should
  be considered with some care.
- Upon exclusion of the elevated NO_x sources, the calculated $NMHC/NO_x$
  ratios approach those measured.

Transport Case

In the transport case, a constant wind of 2.5 m/sec from the
southwest was considered. Starting in the morning two air masses
having different morning origins move downwind as indicated by the
long arrows in Fig. 1. One air column was assumed to be initially

Table 8. Input parameters for the transport case

Dilution :	Minimum mixing height	400 m
	Maximum mixing height	1900 m
	Time period of rise	8 a.m. - 4 p.m.

Emission : Expressed as fractions of the total emissions between
7 and 8 a.m. given from the inventory for the particular
area.
Emission fractions versus time[x]

hour	8 - 9		9 - 10	
	A	B	A	B
NMHC	.41	.46	.01	.06
NO_x	.38	.29	.02	.02

Surface layer transport : NMHC = 50 ppb (10 % C_3H_6, 90 % nC_4H_{10})
NO_x = 10 ppb (100 % NO_2)
O_3 = 20 ppb

Transport aloft : NMHC = 50 ppb (10 % C_3H_6, 90 % nC_4H_{10})
NO_x = 5 ppb (100 % NO_2)
O_3 = 80 ppb

Reactivity : Fraction of NMHC as propylene = .25
Fraction of NMHC added as aldehydes = .05
NO_2/NO_x = .33

[x] A : Air column starting in the CBD
[x] B : Air column starting in the industrial area of Wesseling

located over the CBD, the other over the industrialized area of
Wesseling. Input parameters for both situations are given in
Table 8. Background concentrations were based on data observed
during a severe smog episode in the Cologne area[2,15].

The results of the calculations are shown in Figs. 7 and 8;
Shape and location of the isopleths are quite similar in the two dia-
grams.This can be attributed to the fact that,first,although the absolute emissions in the CBD and Wesseling differ dramatically, the
emission fractions for the subsequent hours are quite similar,and second,the background concentration levels are chosen to be identical.
To utilize the diagrams the peak ozone concentration observed downwind and the prevailing NMHC/NO_x ration in the starting area of the
air parcels are needed. A reasonable estimate for the ozone de-
signe value is 160 ppb, as observed downwind of Cologne. For the

urban core the calculated NMHC/NO$_x$ ratio is 8 (see Table 5), which
is comparable with measured early morning values. As previously
described, measured and calculated ratios for the industrial area
of Wesseling differ by a factor of 5-6, indicating that the assumption of a well mixed box is strongly violated. This is particulary
true for well defined, finite wind conditions where aircraft measurements show distinct NO$_x$ plumes several hours of transport downwind of industrial emission areas. It is clear, however, that in
the transport case it is not possible to exclude the elevated NO$_x$
sources from the calculations as this NO$_x$ mixes downward during
transport, partakes in the chemical processes, and, thus, affects
the peak ozone observed at the downwind receptor. Nevertheless,
as in the stagnant case, this elevated NO$_x$ does not impact the region of the air column's origin. Thus, the discrepancy between
the measured and box-averaged NMHC/NO$_x$ ratio persists.

     Table 9 presents the precursor control requirement based
on the relevant NMHC/NO$_x$ ratios. As with the stagnant case, the
control strategy is a strong function of the assumed initial precursor ratio; thus casting uncertainty upon EKMA results, particulary
for the industrial area.

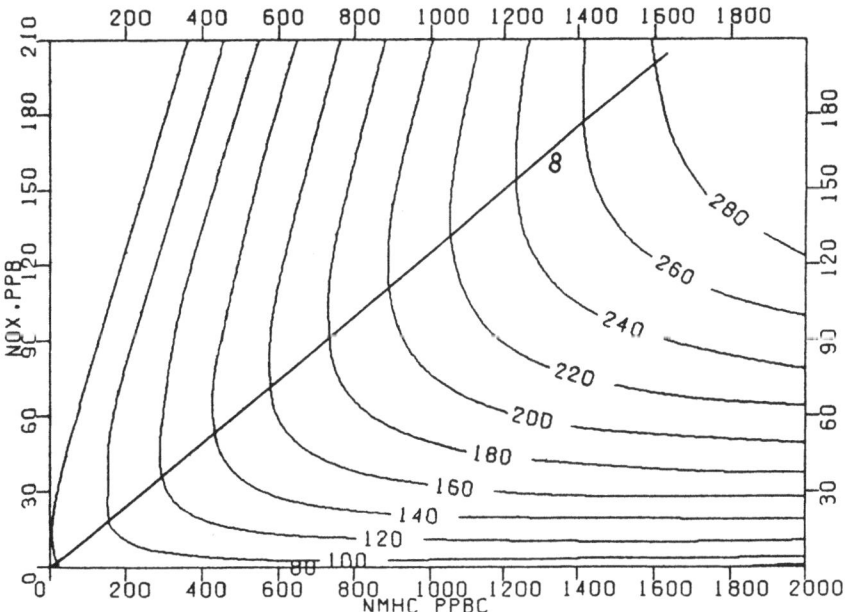

Figure 7. : Ozone isopleths based on the input parameters of Table
            8. Air column starting in the CBD.

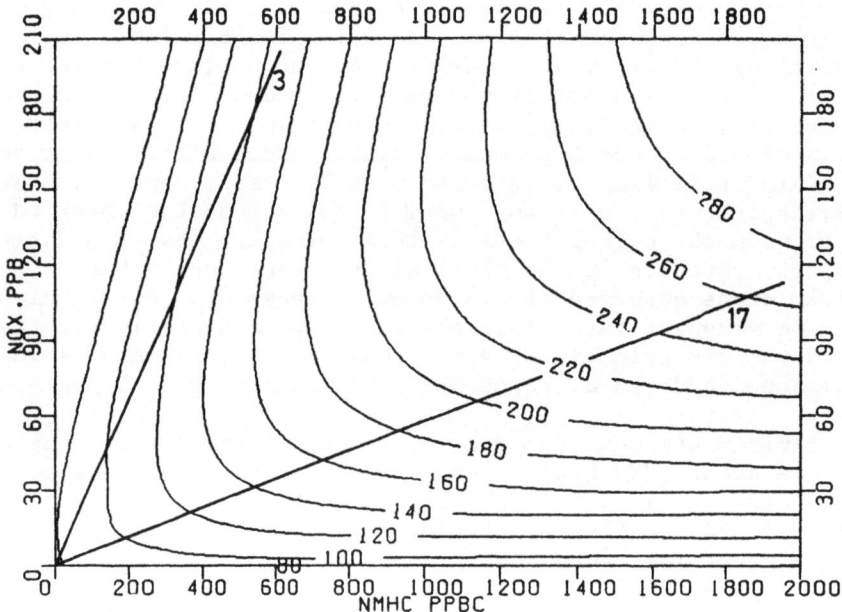

Figure 8. : Ozone isopleths based on the input parameters of Table 8.
          Air column starting in the industrial area of Wesseling.

CONCLUSION AND DISCUSSION

        The Empirical Kinetic Model Approach was discussed in an
application to the Cologne area of the FRG. Two separate cases,
representing days with different meteorological conditions, were
selected.

        The results show that the performance of EKMA is quite de-
pendent upon the emission pattern in the region of interest. A main
assumption in OZIPP, the instantaneous mixing of all pollutants within
the reaction volume, is strongly violated in areas with inhomogeneous
emission conditions. The vertical separation of NO_x and NHMC emissions in the Cologne area leads to large differences between near
ground measured $NMHC/NO_x$ ratios and the box-averaged ratios used in
OZIPP. This problem is visiually depicted in Fig. 9. Therefore control strategies derived from the isopleths should be interpreted with
some care. Only for air parcels starting at the urban core of Cologne
may this box approach have some justification, as the NO_x and NMHC
precursors are emitted predominantly near ground level, so that the
total amount of precursors are available both for interacting
chemically and ground level measurement.

Table 9. Reduction requirements to attain a 80 ppb ozone level derived from the isopleth diagrams of Figs. 7 and 8 using calculated and measured 6-9 a.m. $NMHC/NO_x$ ratios
A design value of 160 ppb ozone is assumed.

$(NMHC/NO_x)$ 6-9		3^x	17^x	8^x
NMHC reduction	7^{xx}	–	–	89
NO_x constant	8^{xx}	51	88	–
(%)				
NO_x reduction	7^{xx}	–	–	0
NMHC constant	8^{xx}	0	0	–
(%)				

x ratios explained in Table 5 and 6 (17:chosen from measurements in Wesseling)
xx Fig. number used for strategy

– not applicable
0 ozone limit can not be attained even through complete elimination of NO_x emissions as background NO_x concentration is too high.

This paper highlights only the problem associated with using EKMA in areas having different vertical precursor emission profiles. Problems can also arise in connection with the lateral extent of the reaction volume in source region having large lateral inhomogeneties. If the column area is large and includes both very polluted and relatively unpolluted sub-regions, incorrect estimation of the smog forming potential within the box may result. On the other hand, if the column area is too small, horizontal pollutant exchange with air outside the column may considerably alter concentrations inside the reaction volume.

Uncertainties in the area specific input parameters are not discussed but can also lead to large differences in the calculated ozone concentrations. In particular, the specification of the mixing height diurnal dependence is a critical factor in the OZIPP calculation.

Chemical aspects related to the mechanism used in OZIPP were also not considered in this paper. Calibration of the mechanism using auto exhaust may lead to large uncertainties in areas not dominated by traffic emissions.

In conclusion, the limited horizontal and vertical resolution of OZIPP can lead to erroneous control strategies in regions having inhomogeneous emission patterns.In such regions,detailed control strategies should be based on more sophisticated modeling approaches.

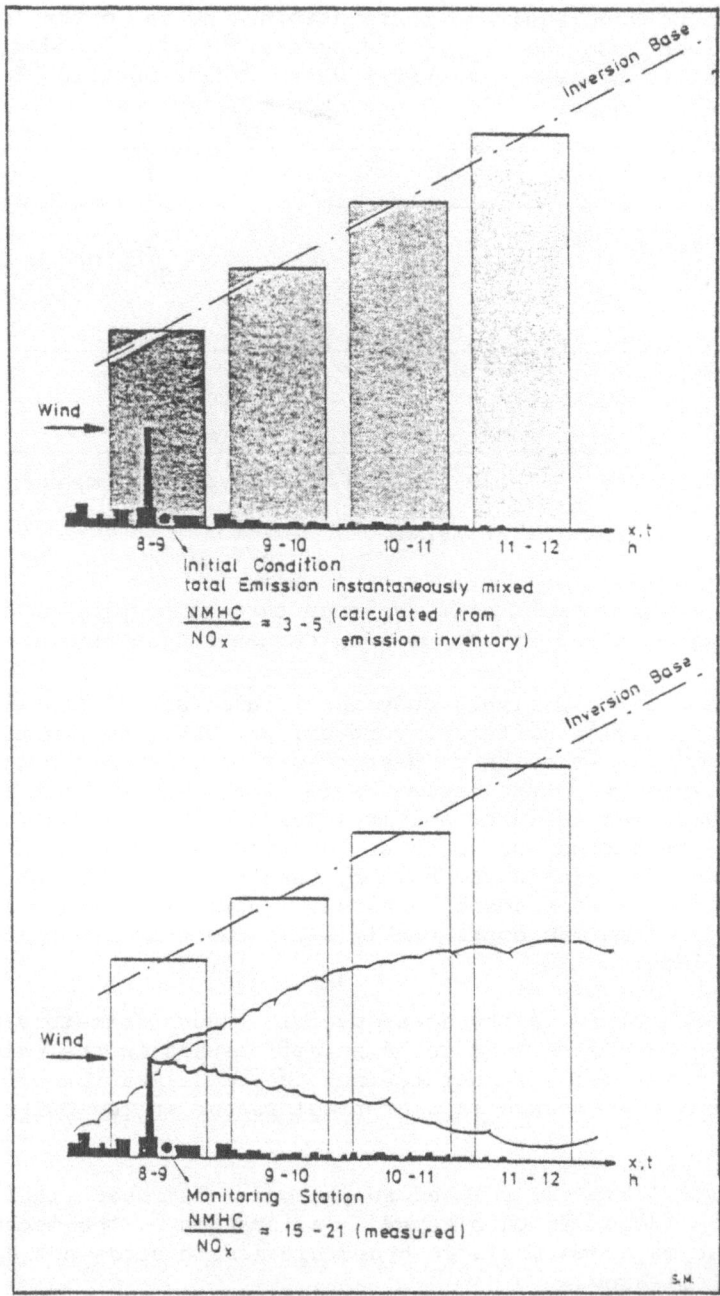

Figure 9. : Precursor dispersion near an industrialized area dominated by elevated NO_x point sources and near ground level
NMHC sources. The upper sketch represents the situation
assumed in OZIPP, while the lower sketch approximates
reality. The resulting NMHC/NO_x ratio is seen to be
quite different in the two cases.

REFERENCES

1. R. Guicherit and H. van Dop, Photochemical Production of Ozone
   in Western Europe (1971-1975) and Its Relationship to Meteorology Atmos. Environm. 11 (1977)
2. K.H. Becker,U.Schurath,H.W.Georgii and M. Deimel, Untersuchungen
   über Smogbildungen insbesondere über die Ausbildung von Oxidan-
   tien als Folge der Luftverunreinigungen in der Bundesrepublik
   Deutschland,Forschungsbericht 79/104 02 502/03/04 Umweltbundesamt, August 1979
3. S. Reynolds,P.Roth and J.Seinfeld, Mathematic Modeling of Photochemical Air Pollution, Atmos. Environm. 7 (1973)
4. M.McCracken, D.Wuebbles,J.Walton,W.Duewer,K.Grant, The Livermore
   Regional Air Quality Model : Concept and Development, J. of Appl.
   Met. Vol 17 (1978)
5. Federal Register 36,84, Part II,8186-8201,April 30, 1971
6. R. Guicherit, P.J.Blokzijl and C.J.Plasse, Some Notes on the Abatement of Photochemical Ozone Production, in : Photochemical
   Smog Formation in the Netherlands, TNO 1978
7. B.Dimitriades, Oxidant Control Strategies : Urban Oxidant Control
   Strategies Derived from Existing Chamber Data, Env. Sci. and
   Techn. 11,1 (1977)
8. P. Bruckmann, M. Buck and P.Eynck, Modelluntersuchungen über den
   Zusammenhang zwischen Vorläufer und Photooxidantienkonzentra-
   tionen, Staub, Sept. 1980
9. EPA, Uses, Limitations and Technical Basis of Procedures for
   Quantifying Relationships between Photochemical Oxidants and
   Precursors, EPA-450/2-77-21 a, Nov. 1977
10. G.Z.Whitten and H.Hugo, User's Manual for a Kinetic Model and
    Ozone Isopleth Plotting Package, EPA-600/8-78-014 a, July 1978
11. M.C. Dodge, Combined Use of Modeling Techniques and Smog Chamber
    Data to Derive Ozone-Precursor Relationships, International Con-
    ference on Photochemical Oxidant Pollution and Its Control,
    Vol. II EPA 600/3-77-001 b, Jan. 1977
12. Luftreinhalteplan Rheinschiene Süd, Minister für Arbeit, Gesund-
    heit und Soziales des Landes NRW, 1976
13. F.J. Dreyhaupt, Vergleich der Emissionssituation in verschiedenen
    Ballungsgebieten, in : Abgasimmissionsbelastungen durch den
    KFZ-Verkehr, TüV Rheinland 1978
14. P. Bruckmann, and P.Eynck, Messungen von Photooxidantien im
    Rhein-Ruhr-Gebiet, in : Photochemische Luftverunreinigungen
    in der Bundesrepublik Deutschland, VDI Düsseldorf 1980
15. B. Scherer and R. Stern, Untersuchung einer Photochemischen
    Smogepisode im Raum Köln-Bonn, Institut für Geophysikalische
    Wissenschaften, Freie Universität Berlin, 1980.

DISCUSSION

M. WILLIAMS   There has been some criticism of the EKMA
                   method in the past because of the simple way it

treats the chemistry involved, and therefore does
not easily allow for differences between areas.
Do you think that, in using EKMA, on is really
using the $(NMHC/NO_x)$ ratio as an empirical tuning
parameter to fit the model with the result that
unreliable conclusions regarding control strategies may be produced?

R. STERN   I think, the chemistry in EKMA is not simple
as the mechanism uses 75 separate reactions invol-
ving more than 30 species. A print of criticism
may be the calibration of the chemical mechanism
using irradiated auto exhaust. This may create
problems in areas not dominated by traffic emissions. In such cases the composition of the
NMHC's could be different. The $NMHC/NO_x$ ratio is
not a tuning parameter in the model calculations.
The model uses a wide range of initial precursor
ratios to generate the diagram.
In deriving control strategies EKMA demands that
the measured precursor ratio be used and that
is the problem we have addressed.

AIR POLLUTION IMPACT CALCULATED AND MEASURED DURING LICENSING PROCEDURES

Lothar Kropp

Tüv Rheinland

Köln, Fed. Rep. of Germany

ABSTRACT

The licensing of industrial plants is subject to investigations of the ambient air quality and the expected change after plants have come into operation. Regulations dealing with licensing procedures ask for measurements as well as for forecasts of expected air quality. Measurements are mostly performed during day time and a one year period. However, a reduction of measuring time is possible.

Monthly and seasonal variations of meteorological data imply deviations in the calculated or measured air pollution impact. Investigations employing an ensemble of low sources yielded lower calculated concentrations for the winter months than for the whole year. This is opposite to results received for high sources. Further to this a comparison of calculated and measured concentration data is given thus validating the model results.

The significance of these effects in view of licensing procedures is discussed.

INTRODUCTION

The licensing of new industrial plants or plants to be expanded in Germany is subject of the expected air pollution impact. The expected change due to expanded plants may result in a higher or lower impact depending on the measures implied on the already existing plants.

The German legislation for licensing procedures ask as well for measurements as for air quality simulation calculations. Both are normally for time periods of one year or more. However, this time may be reduced if the influences of time variations of the emissions and of seasonal changes in the meteorological conditions are foreseeable.

This paper deals with such a problem :

The plant operator intended to receive the license for a capacity expansion as soon as possible and therefore asked for an investigation with the goal of extrapolating measurement results of a three months winter period to the annual air pollution impact.

Calculations were therefore performed using the dispersion model as it is prescribed for the German Land Nordrhein-Westfalen [1].

It was expected on the background of former investigations [2] that model calculations for this source ensemble using the actual joint meteorological statistics for the winter period would lead to a higher impact than employing the representative long term joint meteorological statistic for the whole year.

If this could be confirmed and if the measurements taken at the same time (actual winter) would not exceed air quality standards the conclusion should be that the air quality standards are met at all time.

MODEL CALCULATIONS

The investigation was performed employing the Gaussian plume model with the following parameters:

- dispersion parameters as cited in [1], derived from those of Mc Elroy and Pooler (St. Louis Study)
- plume rise formula for neutral turbulence conditions of Moses and Carson, as cited in [1]
- turbulence classification of Klug and Manier[3]
- local joint meteorological statistics of wind direction, wind speed and turbulence classes

and the source data of 13 sources:

- source heights between 11 and 38 m
- source strenghts between 0,1 and 6 kg/h, with a total of 22 kg/h
- waste gas flows up to 100000 m^3/h
- waste gas temperatures up to 45 °C.

The air quality calculations were done for 6 sites and these are combined to be representative of 2 areas (I and II) of a total size of about 0,6 km^2 of populated area.

The results are given as average values, 95- and 99-percentiles for

- different actual months (1979/1980)
- the actual winter (1979/1980)
- representative winter months (averages of 1951 - 70)
- representative year (average of 1951 - 70).

MODEL RESULTS

The results of the calculations are given in the figures 1 to 5.

The figures 1 to 3 refer to comparisons of the calculated air pollution impact for the meteorological statistics of a) the representative months, December, January and February and b) of the actual months December 1979, January 1980 and February 1980. The results are displayed as sets of columns for each site 1 to 6 and the two areas I and II.

In detail the results are:

The calculated average concentrations are up to 50 % higher for the representative winter months in comparison to the corresponding values for the actual winter months (refer to fig. 1 site 3) or even 100 % higher (refer to fig. 1 site 1). Looking to the concentration values of area II, the deviations are about 20 % between actual and representative winter months. For area I the deviation is larger, i.e. up to 70 %.

The calculated 95-percentiles (fig. 2) show in general differences between the representative and actual winter months of about 20 %. In one extreme case, however, there is a deviation by a factor of 2,5 (refer to fig. 2 site 1).

The calculated 99-percentiles for the representative winter months are by about 50 to 80 % higher than the values for the actual winter months (refer to fig. 3 areas II and I), respectively.

The figures 4 and 5 refer to a comparison of calculations for meteorological statistics of the representative winter, the representative year and the actual winter and display also the measurement during three months in winter. In general it can be stated that the differences are smaller than those for the comparison of the winter months.

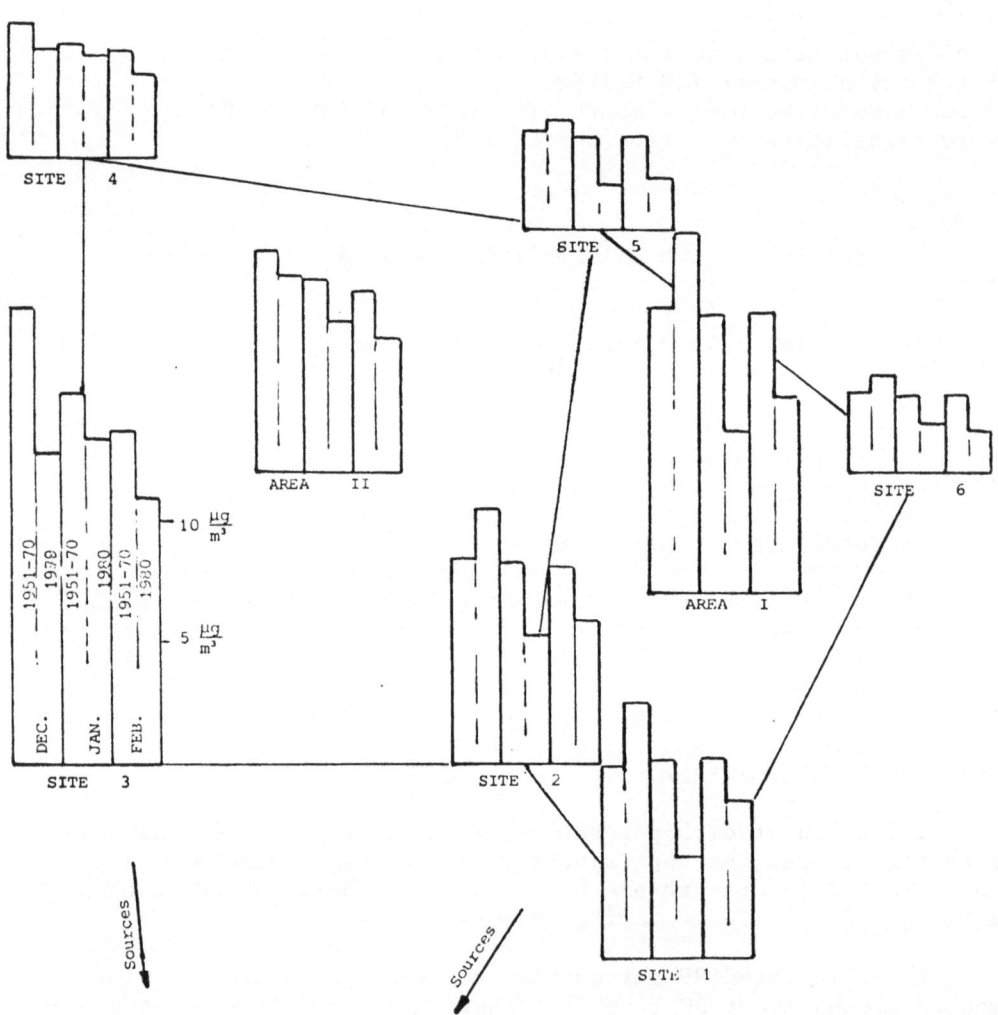

Fig. 1 : Calculated average concentrations for representative and actual winter months

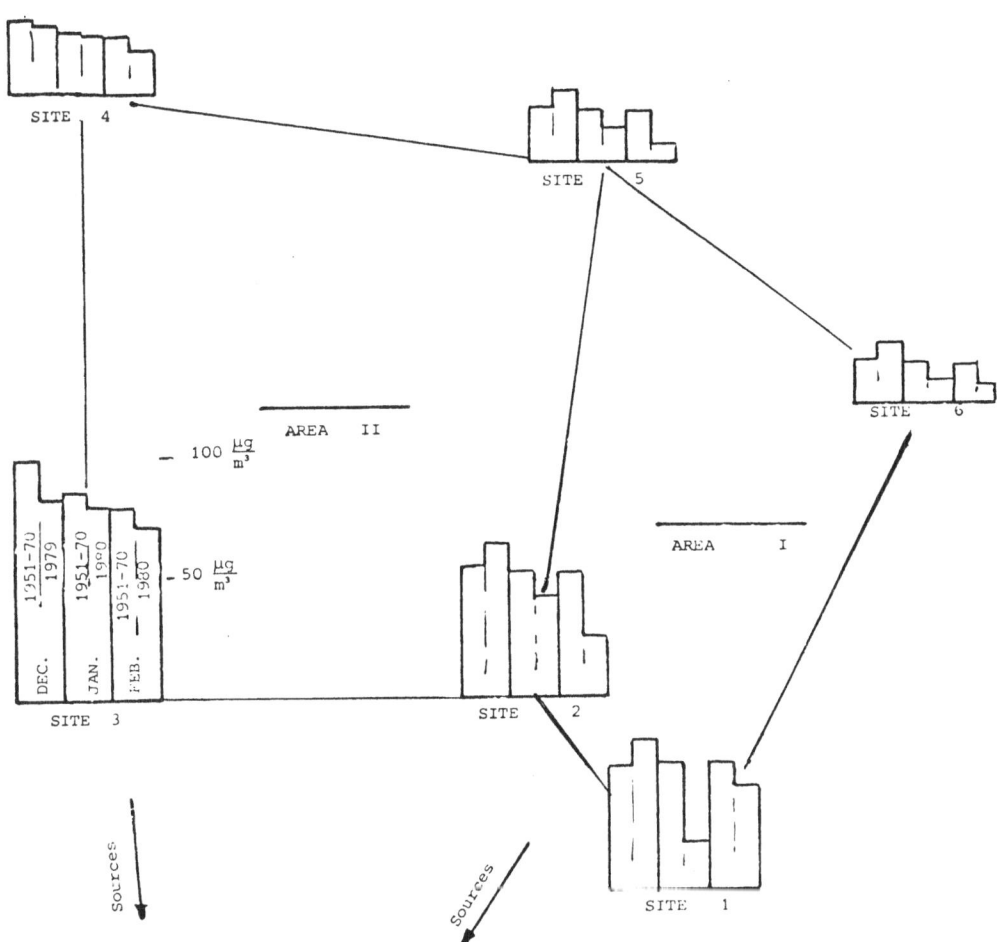

Fig. 2 : Calculated 95-percentiles for representative and
actual winter months

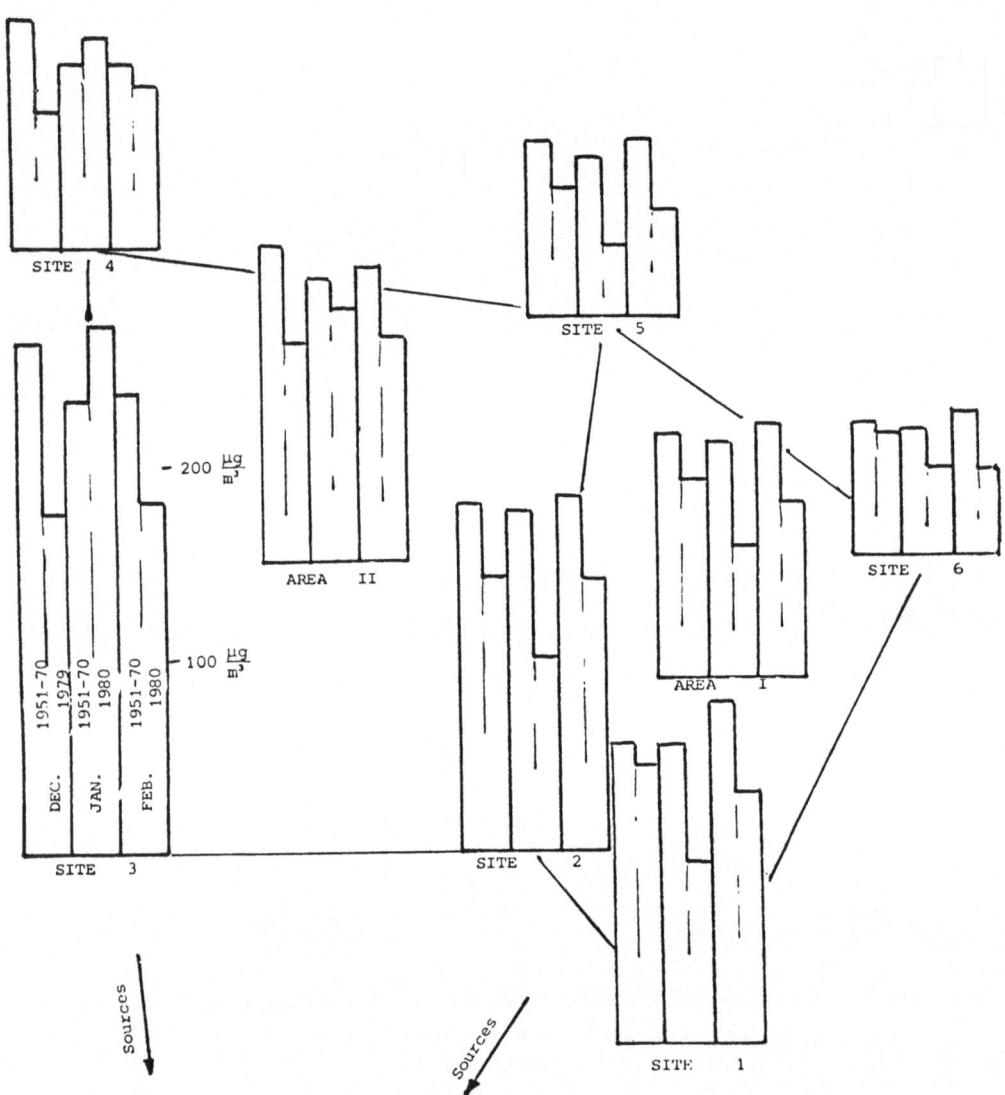

Fig. 3 : Calculated 99-percentiles for representative and
          actual winter months

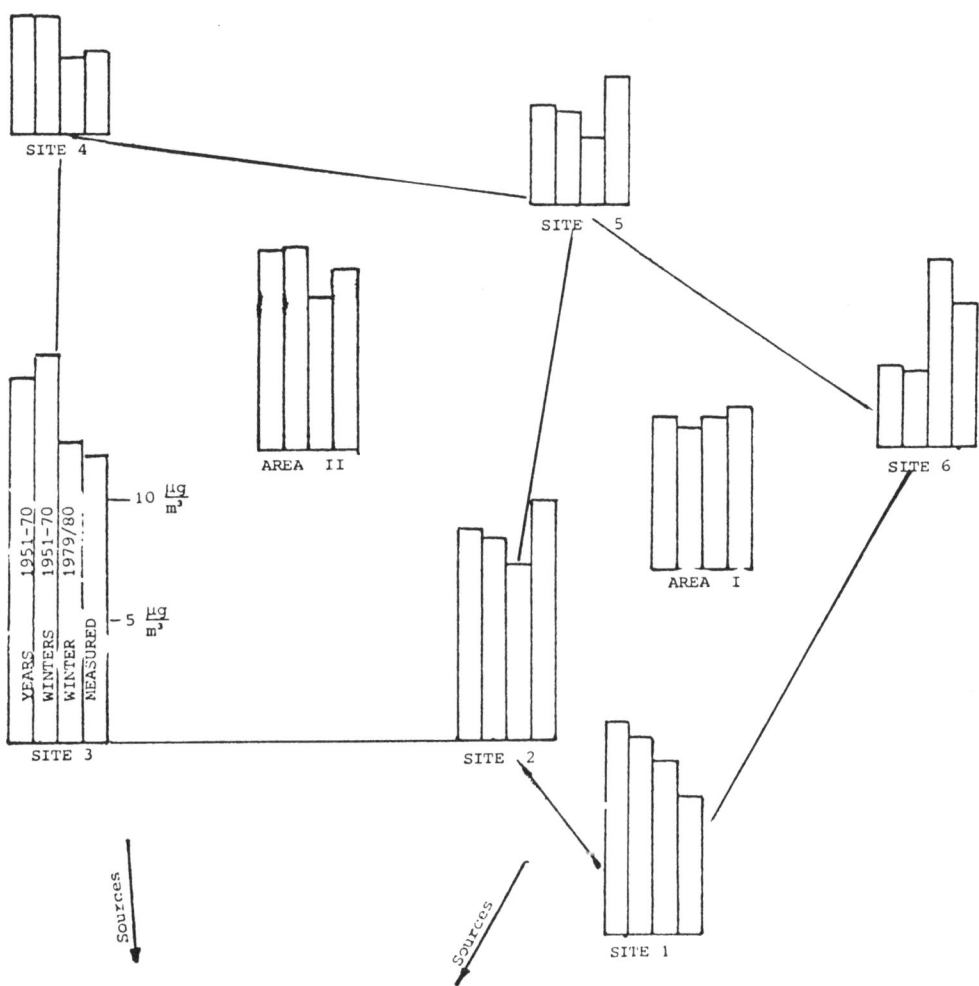

Fig. 4 : Calculated average concentrations for representative year, representative and actual winter and measurements of 3 winter months

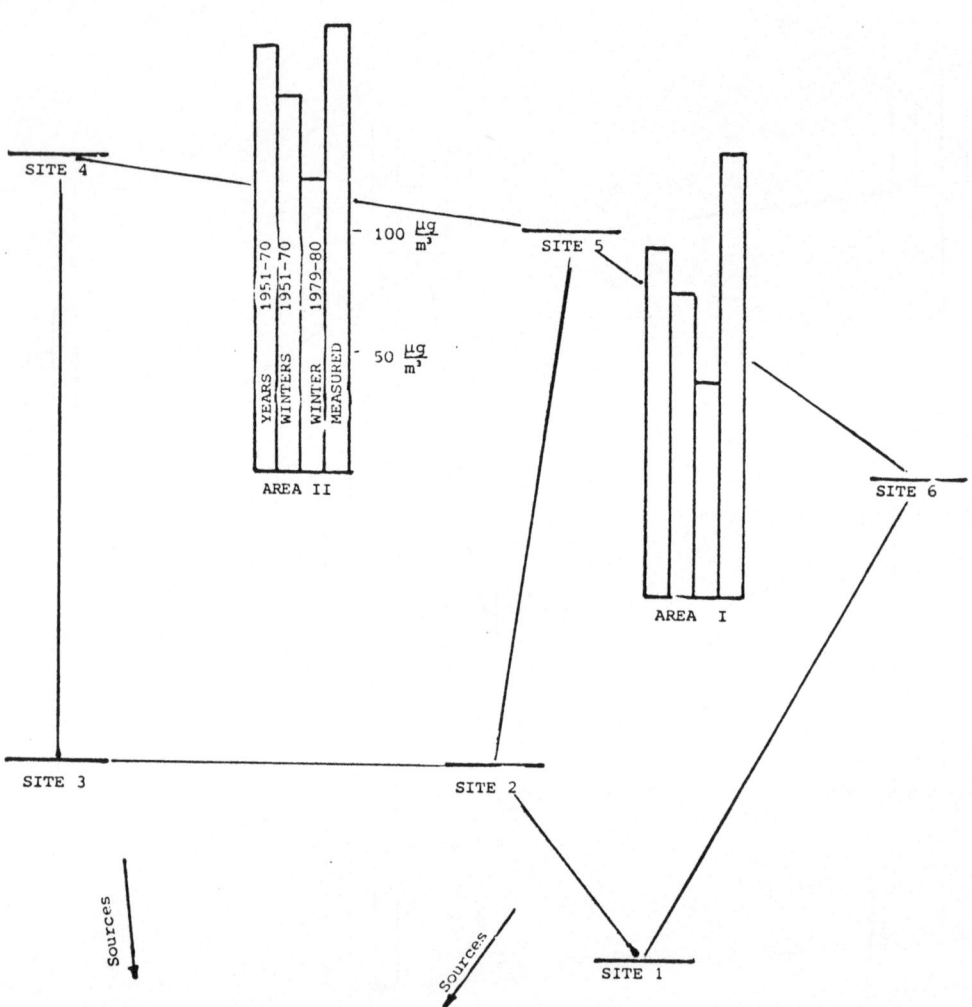

Fig. 5 : Calculated 99-percentiles for representative year,
representative and actual winter and measurements
of 3 winter months

The calculated average concentrations for area I (refer to fig. 4) show deviations of the different calculations of less than 10 %. Even the measurement is of the same size. The average values of the actual winter for area II is approximately 20 % lower than the representative values and the measured value, too. The deviations are not systematic as can be seen (the average concentration of the actual winter for site 3 is about 20 % lower than the representative winter value whereas the average concentration of site 6 is about 100 % higher).

The 99-percentiles (refer to fig. 5) at both areas under investigation are higher by a factor of about 1,5 comparing the representative winter value with the actual winter value.

A comparison of measured and calculated air pollution impact during the same time periods shows the following:

The measured and calculated averages for the areas fit rather well whereas the 99-percentiles of the measurements and calculations deviate strongly with an underestimation by the calculation: The ratio between measurements and calculations during the actual winter is 2 for area I and 1,5 for area II (refer to table 1).

DISCUSSION

A direct explanation for the differences in the results of the calculations is not possible on the basis of the one-dimensional meteorological statistics of wind speed, wind direction or turbulence situation as they are given in fig. 6 to 8.

Too many factors, as e.g. the combination of the three meteorological parameters and its frequency distribution influence the results of the model calculations strongly.

SUMMARY AND CONCLUSION

Calculations of the air pollution impact due to an ensemble of 13 sources were performed with different joint meteorological statistics for actual and representative winter months, for the actual and representative winter and for the whole year.

The comparison of the results show that an extrapolation of the calculated average concentration and especially the 99-percentile-value from the actual winter data of 1979/1980 to the expected long term values of average concentrations and 99-percentiles cannot be done, with confidence.

A comparison of calculated and measured average concentration values for the winter 1979/1980 show rather good agreement while the 99-percentiles deviate by a factor of up to 2.

Table 1 : Measured and calculated concentrations (µg/m3)

Time period.	Area I		Area II	
	Average	99-percentile	Average	99-percentile
calculations				
years (1951–70)	6,3	144	8,2	175
winters (1951–70)	5,9	126	8,3	155
winter (1979/80)	6,3	88	6,4	120
measurement				
winter (1979/80)	6,7	183	7,5	183
winter/spring (1979/80)	9,0	183	8,3	183

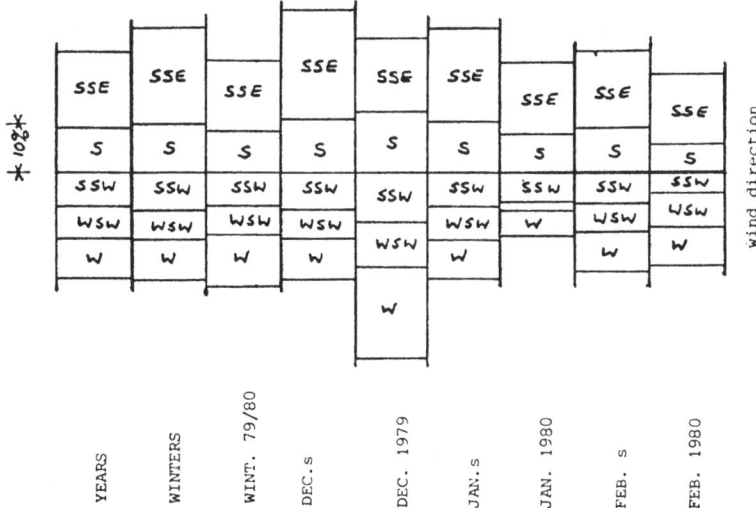

Fig. 6 : Frequency distribution (%) of selected wind directions (of 30 degree sectors)

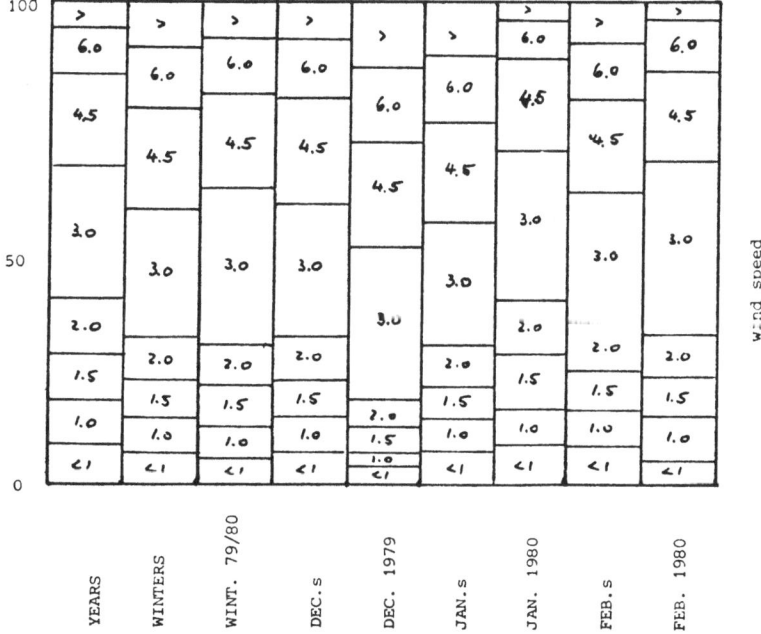

Fig. 7 : Frequency distributions (%) of representative windspeeds (m/s)

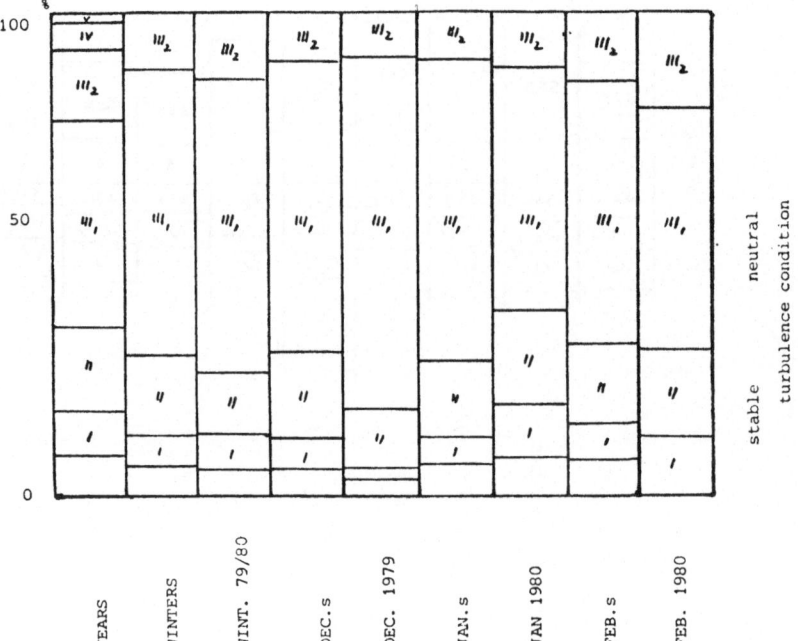

Fig. 8 : Frequency distributions (%) of turbulence classes

The measurements meanwhile were extended from a quarter of a year to half a year and yielded higher average concentrations by about 10 to 30 % but no change of the 99-percentile.

The practical consequences of this investigation on model calculations were drawn by the plant operator as well as the licensing authority as follows:

- there was not sufficient confidence in the extrapolation procedure from calculations for the winter time to the average whole year in order to give the license

- if this extrapolation would have been performed the air pollution impact would raise to about 80 % of the air quality standard, a further expansion of the plant would therefore yield an excess of the air quality standard.

- due to these thoughts measurements had to be carried on in order to receive a broader statistical basis for the measurement data of the actual air pollution impact.

REFERENCES

1. Minister für Arbeit, Gesundheit und Soziales (MAGS)
   RdErl. III B4/III B6 - 8856.4 (III Nr. 13/1975) vom 14.04.1975

2. Kropp, L. and Bahmann, W. Sensitivity of Model Calculations on the Meteorological Data and its Relevance to Regulatory Aspects, 10. ITM on Air Pollution Modeling and
   its Application, Rome 1979

3. Manier, G. Vergleich zwischen Ausbreitungsklasse und Tem-
   peraturgradienten Met. RdSch. 28 (1975) 6

DISCUSSION

H. VAN DOP   Do you expect that a 99-
           percentile-value can be calculated with suf-
           ficient accuracy in the near future?
           If not, do you think that it is useful to in-
           clude 99-percentile-value in regulation and
           license procedures.

L. KROPP   The confidence in high per-
           centiles as e.g. 99-percentiles is lesser than
           for lower percentiles as 95-percentile. The
           accuracy may become better but I can't state
           that it will be enough. Reality is that we do
           have regulations with a request of such high
           percentiles. As there is a safety margin
           "pro environmental protection" we must
           accept this within the licensing procedures.
           It is of course not satisfactory for model
           validations.

THE REGULATORY IMPLICATIONS OF USING AIRPORT METEOROLOGICAL DATA INSTEAD OF ONSITE DATA IN AIR QUALITY MODELING

Patrick T. Brennan and Mark L. Kramer

Meteorological Evaluation Services, Inc.
134 Broadway
Amityville, New York 11701   U.S.A.

SUMMARY

There has been much discussion in the United States in recent years over selection of the appropriate meteorological input data for diffusion modeling with regulatory applications. The procedure normally employed is the use of data from nearby National Weather Service (NWS) stations, with low-level (7 m) wind measurements, and atmospheric stability classifications according to the method of Turner (1964). Alternate suggestions have ranged from a simple substitution of onsite wind data to more elaborate schemes incorporating concepts of the Monin-Obukhov similarity theory.

This paper presents the results of a comparison study using the U.S. Environmental Protection Agency's CRSTER model with varying meteorological inputs. A hypothetical power plant, typical of many coal-fired units in operation in the United States today, has been modeled for SO_2 using one year of National Weather Service data as input. These "reference" calculations have been compared with calculations made substituting onsite wind data into the NWS data base, and calculations made implementing the concepts of similarity theory with onsite data as described by Irwin (1979a).

The calculated concentrations were averaged for the 3-hour, 24-hour and annual time scales used to assess compliance with the U.S. National Ambient Air Quality Standards for SO_2. The results show that for the 3-hour averaging period, the standard CRSTER-NWS combination produced higher concentrations than either of the alternative scenarios utilizing onsite meteorological data. However, for the 24-hour and annual time scales, the calculations based upon similarity theory produced the highest values.

BACKGROUND

The Clean Air Act of 1970 and its 1977 amendments have mandated the use of diffusion modeling in the U.S. EPA regulatory
review processes for large sources. The determination of compliance with the U.S. National Ambient Air Quality Standards is often
based upon a very few hours at the upper end of the frequency distribution of calculated rather than observed hourly concentrations. Selection of the appropriate input values, including meteorology, can have a significant effect upon concentration estimates at the extreme upper end of the distribution.

In the Guideline on Air Quality Models (U.S. EPA 1978), the
U.S. EPA has made provisions for the inclusion of onsite meteorological data into the regulatory modeling process. However, experience has shown that for a variety of reasons, this is most often
not used. Problems involving length of record, data recovery, and
the large amount of processing necessary to convert onsite data
into a form acceptable to the approved EPA models, often preclude
its use.

Recent workshops and conferences bringing together many of
the noted experts in the field of diffusion have recommended that
onsite data be incorporated into diffusion modeling. The Workshop
on Stability Classification Schemes and Sigma Curves (Hanna, et
al. 1977) sponsored by the American Meteorological Society (AMS),
recommended the use of meteorological data and dispersion coefficients which are representative of both the site location and
release elevation, and the introduction of surface layer theory
into routine modeling. In addition, a recent EPA sponsored workshop on onsite instrumentation needs has recommended low-level
(~10 m) instrumentation to determine the Monin-Obukhov length necessary to utilize similarity theory. Draxler (1979) has proposed
a technique which allows estimation of the Monin-Obukhov length
from towers ranging in height up to 50 m, through the use of the
bulk Richardson number. This technique, which Draxler showed to
compare well with bivane data, uses measurements of temperature
lapse rate and wind speed, which are routinely available at many
power plants throughout the United States.

In order to assess the effect of the inclusion of onsite data
into diffusion modeling for regulatory compliance, a comparative
study has been done using the U.S. EPA single-source CRSTER model.
A hypothetical power plant has been assumed to exist at the
location of a well-maintained meteorological tower, similar to
that used in the Draxler study. This plant has been modeled using
the standard EPA CRSTER model, with three different meteorological
inputs. The results, though based only upon one year of data,
give an indication as to what one might expect if the various expert panel recommendations do become a part of U.S. regulatory
practice.

PLANT DESCRIPTION

The hypothetical power plant was assumed to be located in eastern Pennsylvania. The plant simulates a single 600 MW coal-fired unit, typical of many plants existing in the United States today which are subject to regulatory review and diffusion modeling. The plant operational parameters are listed below in Table 1.

As is the case with all diffusion modeling for regulatory application in the U.S., the plant has been assumed to be operating at full load during all hours of the year. However, in certain instances 75% and 50% load are considered. For the purposes of this demonstration, only SO_2 emissions have been modeled.

METEOROLOGICAL DATA SOURCES

The hypothetical plant has been modeled using meteorological data from three primary sources. Data from the period May 1972 through April 1973 were used. The data were rearranged to simulate a calendar year, since the CRSTER model output is expressed in Julian days.

Philadelphia, Pennsylvania, National Weather Service

The Philadelphia NWS station is located at Philadelphia International Airport, approximately 50 km southeast of the hypothetical plant site. The standard synoptic hourly observations of wind speed and direction, sky cover and ambient temperature were used for each hour. Radiosonde observations are not taken at the Philadelphia station.

Table 1.   Hypothetical Plant Full-Load Operating Parameters

Stack Height	252.4 m
Stack Diameter	6.7 m
Exit Velocity	23.2 m/sec
Stack Temperature	425.0 °K
Emission Rate SO_2	3000.0 g/sec

Dulles Airport - Sterling, Virginia

Dulles International Airport near Washington, D.C. is the NWS
radiosonde station most representative of the hypothetical plant
site. These radiosonde data, combined with Philadelphia surface
temperatures, were used to compute twice daily mixing heights, according to the technique of Holzworth (1972).

The other upper air station in the site region, JFK International Airport in New York City, was not used because of its
coastal location.

Onsite Measurements

Meteorological data at the site of the hypothetical plant
have been obtained from an instrumented tower, typical of those
constructed at many power plant sites during the past decade. The
tower is instrumented for wind speed and direction at elevations
of 9, 53 and 82 m, and temperature differences are measured between the 81-8 and 52-8 m levels. Ambient temperature is also measured at the 8-m level.

The period of record chosen has been studied extensively and
found to be climatologically representative of the site. A more
complete description of the tower site and climatology has been
given by Brennan, et al. (1977).

MODELING METHODOLOGY

The plant was modeled using the United States Environmental
Protection Agency's single-source CRSTER model. This model has
been designated by the EPA as its "benchmark" model for regulatory
application on isolated sources in rural environments. CRSTER is
an hourly sequential Gaussian plume model, producing concentration
estimates necessary to assess compliance with the United States
National Ambient Air Quality Standards. The highest and second
highest 3-hour and 24-hour concentrations, as well as the annual
averages, are summarized for receptors located on concentric rings
around the source. The CRSTER model is designed to operate primarily with National Weather Service data. This includes low-
level (7 m) wind measurments, ambient temperature and atmospheric
stability classifications according to the method of Turner (1964).
The meteorological data are preprocessed through a program which
performs the Turner stability calculations, unit conversions, and
randomizes the last integer of the wind directions, which are normally read to the nearest 10° azimuth. In addition, this preproces-
sor calculates a mixing height for each hour, using an interpolation
scheme based upon the twice daily values computed from the Holzworth
technique.

The CRSTER model has been described extensively in the literature, and there is no need to review it here. A complete description may be found in the CRSTER model user's guide (U.S. EPA 1977).

Following the 1977 AMS Workshop, there have been many suggestions regarding ways to either partially or fully implement its
recommendations. These have ranged from a simple substitution of
onsite wind data to more elaborate schemes incorporating the concepts of similarity theory and site specific dispersion coefficients into regulatory modeling.

To assess the effect of these suggestions upon the calculated
ground-level concentrations, the hypothetical plant was modeled
with the CRSTER model under three different scenarios. The model
was run for a series of receptors on 15 concentric rings located
at distances ranging from 0.2 to 10.0 km. Each ring contained 36
receptors, separated at 10° azimuth intervals. In order to focus
only on the differences caused by varying meteorological inputs,
the hypothetical plant was modeled assuming flat terrain.

CRSTER - Airport Meteorology

The plant was first modeled using the CRSTER model as it is
now employed for regulatory assessments. Wind data and Turner stability classifications from the Philadelphia NWS Station were used,
along with mixing heights derived from the Dulles Airport radiosonde data and Philadelphia surface temperatures. The standard
Pasquill-Gifford dispersion coefficients were employed. These calculations provided a reference point against which to compare the
effects of the other two modeling scenarios.

CRSTER - Onsite Wind

The second modeling scenario substituted onsite wind data for
the airport winds. All other meteorological parameters were identical to those used in the CRSTER-Airport calculations. Wind speed
and direction from the 82-m level of the tower were used, following
the AMS recommendation that wind data be obtained as close as possible to the release height. The onsite wind directions were ran-
domized with the same random number seed used in the CRSTER prepro-
cessor program. In the case of calm hours, wind speeds were set
equal to one-half the starting speed of the sensor.

CRSTER - Onsite Wind and Stability

The third modeling scenario used onsite wind, temperature and
temperature lapse rate data. This technique incorporates a continuum of stability based upon the Monin-Obukhov length scale, and
allows the dispersion coefficients to change as a function of

source elevation and surface roughness. This was accomplished using the interim scheme of Irwin (1979a), which is based upon a reanalysis of some of the well-known field experiments, as well as new tank data. This interim scheme has been refined by Irwin (1979b), into an algorithm compatible with the CRSTER model.

The interim scheme requires hourly values of the Monin-Obukhov length, L, the friction velocity, u_*, and the standard deviations of the horizontal and vertical wind direction fluctuations, σ_a and σ_e, respectively. These values were obtained through relationships with the bulk Richardson number, using the methodology described by Draxler (1979).

The bulk Richardson number, B, was determined for each hour using the relationship:

$$B = \frac{g\,z\,\Delta\theta}{\bar{T}\,\bar{u}^2} \tag{1}$$

where: g = acceleration of gravity, m/sec^2
  $\bar{T}$ = ambient temperature, $°K$
  $\Delta\theta$ = potential temperature difference between the 52- and 8-m tower levels
  $\bar{u}$ = wind speed, m/sec

The values of $\bar{T}$ and $\bar{u}$ were determined at the computational height z, which is the median height through which the temperature gradient was measured.

The hourly values of the bulk Richardson number calculated from equation (1) were then used to determine the Richardson number for each hour using the relationship:

$$B = R_i \left[\frac{\ln\,(z/z_o) - \psi_m}{\phi_m} \right]^{-2} \tag{2}$$

where: R_i = Richardson number
  z_o = roughness length, m
  ϕ_m = dimensionless wind shear
  ψ_m = integral of ϕ_m, from z_o to z/L

The value of z_o was determined to be 0.3 m through the land use technique described by Counihan (1975). ϕ_m and ψ_m are both functions of R_i, requiring equation (2) to be solved using a numerical iteration technique. The exact expressions for ϕ_m and ψ_m are found in Draxler (1979).

The resulting values of R_i were then used to solve for the Monin-Obukhov length using the following relations :

For unstable conditions, $R_i < 0$

$$z/L = R_i \tag{3}$$

For neutral and stable conditions, $R_i \geq 0$

$$z/L = R_i (1-5 R_i)^{-1} \tag{4}$$

The final solution for ψ_m was used to determine the value of the friction velocity, u_x, from the equation.

$$u_x = k \bar{u} (\ln(^z/z_o) - \psi_m)^{-1} \tag{5}$$

where : k = the von Karman constant, 0.4.

The calculation of hourly values of L and u_x allows the atmospheric stability to be treated as a continuum, rather than one of seven discrete classes. However, for some portions of the CRSTER model, such as the selection of a proper plume rise equation, it is still necessary to make a determination of the proper Pasquill stability class. For this purpose, the nomograms of Golder (1972) were used to determine the Pasquill stability class for each hour, as a function of L and z_o.

Following the determination of L and u_x, hourly values of σ_e were determined using the relationships of Panofsky et al. (1977) during unstable conditions and Panofsky (1973) during stable and neutral conditions. Values of σ_a were determined according to Panofsky et al. (1977) during unstable conditions and using the suggestions of Irwin (1979a) during stable and neutral conditions. These estimates of σ_a and σ_e are both a function of $\bar{u}$, and are therefore valid at the reference elevation where L was determined. They were then corrected to the plume elevation, h_e, using the suggestion of Cramer (1976) where :

$$\sigma_a(h_e) = \sigma_a(z)\left(\frac{h_e}{z}\right)^{-p} \tag{6}$$

$$\sigma_e(h_e) = \sigma_e(z)\left(\frac{h_e}{z}\right)^{q} \tag{7}$$

The values of p are the stability dependent wind speed power law exponents. In this evaluation, site specific values of p determined by Brennan et al. (1977) were used. The values of q are a function of p, and are as follows :

Unstable   q =   (0.33 - p)
Neutral   q =   0
Stable   q =   -(p + .1)

The values of σ_a and σ_e were only allowed to vary in the low-est 25% of the mixed layer during unstable conditions.

The last hourly variable needed to run the interim scheme is an estimate of the depth of the mixed layer. During daylight hours the hourly mixing heights determined from the Dulles upper air sound-ings were used. During nighttime stable conditions, the nocturnal boundary layer depth was assumed to be 250 m. This is admittedly an assumption, but based upon the findings of Yu (1978), it appears rea-sonable. This substitution for the nocturnal boundary layer depth was only used for the calculation of dispersion coefficients. The Dulles derived rural mixing heights were used unaltered in the CRSTER dif-fusion algorithm.

DISCUSSION OF RESULTS

The National Ambient Air Quality Standards for SO_2 are spec-ified for 3-hour, 24-hour and annual averaging periods. The deter-mination of whether a source will be in violation of these stan-dards is based upon the highest estimated concentration for annual averages, and upon the highest, second-highest concentration at any receptor for periods of 24 hours or less.

The concentration estimates which would be used to determine com-pliance at the hypothetical plant are compared in Table 2. Past experience with the CRSTER model has shown that in many instances the entire contribution to the controlling 3-hour or even 24-hour average is the result of one single hour. For this reason hourly concentra-tions have also been included in Table 2. Since the plant emissions are fictitious, the actual magnitudes of the concentrations in Table 2 are meaningless. The true value of the table lies in the differences between the different CRSTER versions within each time frame.

Three-Hour Averages

The highest concentrations for the 3-hour period were calculated by CRSTER-Airport. The highest, second-highest value of 425 ug/m^3 is 23% higher than that calculated by CRSTER-Onsite or the CRSTER- Interim Scheme, which both coincidentally calculated a limiting concentration of 345 ug/m^3.

The highest, second-highest values calculated by the CRSTER-Airport and CRSTER-Onsite were caused by nearly identical meteorological conditions which occurred on separate days. In each case, the limiting 3-hour average was the result of one hour of sta-bility A, low wind speed conditions.

The limiting 3-hour concentration predicted by the CRSTER-Interim Scheme was the result of day 25, which was characterized by very unstable lapse rates and moderate wind speeds. The hours responsible for both the highest and highest, second-highest 3-hour concentrations were classified as stability A according to the Golder (1972) nomograms. It should be noted that wind speeds at the 9-m level of the tower were all greater than 5 knots, indicating that these hours would have been assigned to a more stable category if stability were typed according to the Turner technique.

Table 2.

SUMMARY OF CALCULATED SO_2 CONCENTRATIONS
(ug/m^3)

	CRSTER Airport Meteorology		CRSTER Onsite Winds		CRSTER Interim Scheme	
	Highest	Highest/2nd Highest	Highest	Highest/2nd Highest	Highest	Highest/2nd Highest
1-Hr. Conc.	1,353	1,254	1,484	1,034	722	501
Day	200	190	199	190	25	25
Hour	11	12	11	12	12	19
Dist. (km)	1.5	1.5	1.5	1.5	1.9	1.3
3-Hr. Conc.	661	425	495	345	519	345
Day	200	198	199	190	25	25
Hours	10-12	10-12	10-12	10-12	16-18	10-12
Dist. (km)	1.5	1.5	1.5	1.5	1.3	1.3
24-Hr Conc.	82	79	62	44	121	108
Day	200	198	199	205	25	162
Dist. (km)	1.5	1.5	1.5	7.0	1.0	1.0
Annual Conc.	2.8	–	1.6	–	5.5	–
Dist. (km)	7.0	–	7.0	–	1.5	–

Twenty-Four-Hour Averages

        The highest concentrations for the 24-hour period were cal-
culated by the CRSTER-Interim Scheme, followed by CRSTER-Airport
and CRSTER-Onsite. Though the highest 24-hour concentration from
the interim scheme was also on day 25, the highest, second-highest
concentration of 108 ug/m^3, was calculated from day 162. Day 162
was also a case of unstable lapse rates accompanied by persistent
winds with moderate speeds, with several hours determined to be
class A by the Golder nomogram. The usual experience with the
CRSTER model has been that as the length of the averaging period
increased, the location of the maximum concentration moved away
from the source. However, in the case of the CRSTER-Interim
Scheme, the 24-hour maximums were calculated to be 1.0 km from the
source, as compared to 1.3 km for the 3-hour maximums.

        The highest, second-highest 24-hour concentration from CRSTER-
Airport was a result of the same day (198), and primarily the same
stability A hour, which caused the highest, second-highest 3-hour
value. However, the highest, second-highest 24-hour concentration
from CRSTER-Onsite was on day 205 at a receptor 7.0 km downwind,
and was the result of ten hours of persistent wind directions with
stability classes C and D.

Annual Concentrations

        The annual concentrations show the same general trends as the
24-hour values. The CRSTER-Interim Scheme predicted the highest
annual average concentration of 5.5 ug/m^3, almost double the maximum of 2.8 ug/m^3 predicted by CRSTER-Airport. The maximum annual
concentration predicted by CRSTER-Onsite was 1.6 ug/m^3. As expected, the maximum annual concentrations predicted by CRSTER-Airport and CRSTER-Onsite were not close to the plant. The maximum
value from the CRSTER-Interim Scheme continued to remain close in at
1.5 km.

CONCLUSIONS

        Though a study based upon only one year of data is by no means
conclusive, the results of this exercise do provide insight as to
the possible consequences of incorporating onsite meteorological
data into air quality modeling for regulatory applications. The
simple substitution of onsite wind data into the otherwise standard
CRSTER evaluation reduced the critical concentrations by 19% on the
3-hour time scale and 44% on the 24-hour time scale. The further
introduction of turbulence typing according to the Monin-Obukhov
length scale, and the introduction of more site and source specific
dispersion parameters, tends to eliminate some of the extreme values
in the distribution of hourly concentrations. However, these

alterations tend to generally increase concentrations for the vast majority of hours not associated with peak concentrations. The interim scheme also confronts the user with situations such as higher than expected wind speeds during extremely unstable conditions, which are not allowed to coexist in the Turner typing scheme. The effects of the introduction of this and other unusual but real meteorological situations into the benchmark model are worthy of further consideration. This would hopefully be done at a real site with a quality SO_2 monitoring network, and allow comparison of the various modeling scenarios with measured concentrations.

It should also be noted that the effort required to prepare onsite data into a format suitable for the EPA approved models is considerable. In this particular instance, the site selected had a fully instrumented backup tower from which missing data could be substituted. At a site not so fortunate, this could prove to be an insurmountable problem. One must be very cautious in the selection of replacement data when working within a regulatory framework where the entire emission limitation could depend upon the calculated concentration from one single hour.

ACKNOWLEDGEMENTS

The authors would like to thank John Irwin of the U.S. EPA for his comments and suggestions regarding implementation of the interim scheme. We would also like to thank Frank J. Lawrence of Meteorological Evaluation Services for his assistance in program development, and Ruth Giacoma for typing this manuscript.

REFERENCES

Brennan, P. T., Castelli, F. P. and Smith, M. E., 1977, Comparative Measurements on Two Adjacent Meteorological Towers, Preprints of the Joint Conference on Applications of Air Pollution Meteorology, Salt Lake City, Utah, Am. Met Soc., pp. 154-159.

Counihan, J., 1975, Adiabatic Atmospheric Boundary Layers: A Review and Analysis of Data from the Period 1880-1972, Atmos. Environ., Vol. 9, pp. 871-905.

Cramer, H. E., 1976, Improved Techniques for Estimating Dispersion from Tall Stacks, Proceedings of Seventh International Technical Meeting on Air Pollution Modeling and its Application, Arlie, Virginia, pp. 731-780.

Draxler, R. R., 1979, Estimating Vertical Diffusion From Routine Meteorological Tower Measurements, Atmos. Environ., Vol. 13, pp. 1559-1564.

Golder, D., 1972, Relations Among Stability Parameters in the Surface Layer, Boundary Layer Met., Vol. 3, pp. 47-58.

Hanna, S. R., Briggs, G. A., Deardorff, J., Egan, B. A., Gifford, F. A. and Pasquill, F., 1977, AMS Workshop on Stability Classification Schemes and Sigma Curves-Summary of Recommendations, Bull. Am. Met. Soc., Vol. 53, No. 12, pp. 1305-1309.

Holzworth, G. C., 1972, Mixing Heights, Wind Speeds, and Potential for Urban Air Pollution Throughout the Contiguous United States, U.S. EPA Office of Air Programs Publication AP-101.

Irwin, J. S., 1979a, Estimating Plume Dispersion - A Recommended Generalized Scheme, Preprints of the Fourth Symposium on Turbulence, Diffusion, and Air Pollution, Reno, Nevada, Am. Met. Soc., pp. 62-69.

Irwin, J. S., 1979b, Scheme for Estimating Dispersion Parameters as a Function of Release Height, U.S. EPA, EPA-600/4-79-062.

Panofsky, H. A., 1973, Tower Micrometeorology, in "Workshop on Micrometeorology", D. A. Haugen, Ed., Am. Met. Soc., pp. 151-176.

Panofsky, H. A., Tennekes, H., Lenschow, D. H. and Wyngaard, J. C., 1977, The Characteristics of Turbulent Velocity Components in the Surface Layer Under Convective Conditions, Boundary Layer Met., Vol. 11, pp. 355-361.

Turner, D. B., 1964, A Diffusion Model for an Urban Area, Journ. Appl. Met., Vol. 3, pp. 83-91.

U.S. EPA, 1977, User's Manual for Single Source (CRSTER) Model, EPA-450/2-77-013.

U.S. EPA, 1978, Guideline on Air Quality Models, OAQPS, EPA-450/2-78-027.

Yu, T.W., 1978, Determining Height of the Nocturnal Boundary Layer, Journ. Appl. Met., Vol. 17, pp. 28-33.

DISCUSSION

P. SAMSON   The difference between the on-site and airport data may be due, at least in part, to your choise of modifying the calm cases to one-half the starting speed of the anemometer rather than the default value of CRSTER. The lower wind speed input may allow a greater computed plume rise and thus lower concentrations. Why weren't wind speeds treated the same way for both techniques?

P. BRENNAN   The use of a default wind speed of one half the anemometer starting speed was necessary to insure a representative calculation of the bulk Richardson number when using 9 meter winds. The wind speeds used in the diffusion calculations in CRSTER were from the 82 meter tower level, where the percentage of calm was very small.

List of participants at the 11th NATO/CCMS International Technical Meeting on Air Pollution Modelling and its Application
Amsterdam November 24-27, 1980   The Netherlands.

AUSTRIA

Kolb H.   University of Vienna
                                          38,Hohe Warte
                                          A-1190 Vienna

Runca E.   IIASA
                                          Schloss Laxenburg
                                          A-2361 Laxenburg

BELGIUM

Berger A.   UCL - Institut d'Astronomie
                                          et de Géophysiaue
                                          2, Chemin du Cyclotron
                                          B-1348 Louvain-La-Neuve

Cieslik A.   SCK/CEN
                                          200, Boeretang
                                          B-2400 Mol

Condé F.   INIEX - Institut National
                                          des Industries Extractives
                                          200, Rue du Chera
                                          B-4000 Liège

Cosemans G.   SCK/CEN
                                          200, Boeretang
                                          B-2400 Mol

Cravatte M.   UCL - Institut d'Astronomie
                                          et de Géophysique
                                          2, Chemin du Cyclotron
                                          B-1348 Louvain-La-Neuve

Dermonne C.   INIEX - Institut National
                                          des Industries Extractives
                                          200, Rue du Chera
                                          B-4000 Liège

Guillot P.   Commission of the
                                          European Communities
                                          200, Rue de la Loi
                                          B-1049 Brussels

Hecq P.	UCL - Institut d'Astronomie et de Géophysique 2, Chemin du Cyclotron B-1348 Louvain-La-Neuve
Kretzschmar J.	SCK/CEN - Studiecentrum voor Kernenergie 200, Boeretang B-2400 Mol
Schayes G.	UCL - Institut d'Astronomie et de Géophysique 2, Chemin du Cyclotron B-1348 Louvain-La-Neuve
Vandendriessche St.	KUL - Laboratorium voor Ana-lytische en anorganische Schei-kunde 200 F, Celestijnenlaan B-3030 Heverlee
Wispelaere C., De	Programmatie van het Weten-schapsbeleid - Nationaal R-D Programma Leefmilieu 8, Wetenschapsstraat B-1040 Brussels

CANADA

Davidson A.	University of Waterloo Waterloo, Ontario N2L 3G1
Ferguson H.L.	Atmospheric Environment Service 4905, Dufferin Street City of North York Downsview, Ontario M3H 5T4
Portelli R.	Atmospheric Environment Service Fisheries and Environment Canada 4905, Dufferin Street Downsview, Ontario M3H 5T4
Reid J.D.	Atmospheric Environment Service 4905, Dufferin Street Downsview, Ontario M3H 5T4
Slawson P.R.	Mechanical Engineering Dept. University of Waterloo Waterloo, Ontario N2L 3G1
Turner H.E.	Atmospheric Environment Service 4905, Dufferin Street Downsview, Ontario M3H 5T4

Venkatram A.   Ontario Ministry of Environment
                                      880 Bay Str. 4 FLR
                                      Toronto, Ontario M5S   1Z8

DENMARK
Berkowicz R.   Danish Meterological Institute
                                      100, Lyngbyvey
                                      DK-2100 Copenhagen

Jensen N.O.   Riso National Laboratory
                                      Physics Department
                                      DK-4000 Roskilde

Nielsen L.B.   National Agency of Envir. Pro-
                                      tection-Air Pollution Lab.
                                      Riso National Laboratory
                                      DK-4000 Roskilde

Markvorsen J.   Cowiconsult
                                      Consulting Engineers and
                                      Planners AS
                                      45, Teknikerbyen
                                      DK-2830 Virum

Prahm L.P.   National Agency for Environmen-
                                      tal Protection, Air Pollution
                                      Laboratory
                                      RISO National Laboratory
                                      DK-4000 Roskilde

FRANCE
Blondin M.C.   Direction de la Météorologie
                                      EERM-GMA
                                      73-77, Rue de Sèvres
                                      F-92100 Boulogne-Billancourt

Camps R.   Soc. Elf-Aquitaine
                                      CIRN Laboratoire Pollution
                                      B.P. N23
                                      F-64170 Lacq

Cariou J.   Centre d'Etudes Techniques
                                      de l'Equipement de Rouen
                                      Chemin de la Pardière -B.P. 247
                                      F-76120 Grand-Quevilly

Fage J.M.   Bertin & Cie
                                      Zone Industrielle B.P. 3
                                      F-78370 Plaisir

Fulleringer D.	AIRPARIF 152, rue de Picpus F-75012 Paris
Gabel J.G.	Embassy of Israel Paris France
Menard T.	Institut National de Recherche Chimique Appliquée Boîte Postale 1 F-91710 Vert Le Petit
Moussafir H.	Bertin & Cie Zone Industrielle B.P. 3 F-78370 Plaisir
Oppeneau J.C.	Min. de la Cult. et de l'Environ. Serv. des Affaires Scientifiques 14, BLVD. du General Leclerc F-92521 Neuilly-Seine
Quinault J.M.	Commissariat à l'Energie Atomique 19, Tour d'Aygosi F-13100 Aix-en-Provence
Racher I.	LAMP University of Clermont II France
Rosset R.	LAMP University of Clermont II France
Saab A.E.	Electricité de France 6, Ouai Watier F- CHATOU
Starkand Y.S.	Embassy of Israel Paris France
Willefert J.M.	Electricité de France 3, rue de Messine F-75008 Paris

FEDERAL REPUBLIC OF GERMANY

Axenfield F.	Dornier System GMBH Postfach 1360 D-7990 Friedrichshafen

Chamier J., Von	VDI-Kommission Reinhaltung der Luft Geschäftsstelle Postfach 1139 D-4000 Dusseldorf
Eppel D.P.	Institut für Physik GKSS Forschungszent. Geesthacht GMBH D-2054 Geesthacht
Giebel J.	Landsanstalt für Imissionsschutz Wallneyer Str. 6 D-4300 Essen
Heinz H.D.	University of Berlin D-Berlin
Herberg G.	Abteilung für Praktische Mathematik - Hoechst AG. Postfach 800320 D-6230 Frankfurt-Main 80
Janicke L.	Dornier System GMBH Postfach 1360 D-7990 Friedrichshafen
Kämmer Kl.	CAREMA Rudiger Str. 20 D-4000 Dusseldorf
Klug W.	Institut für Meteorologie Technische Hochschule Darmstadt D-6100 Darmstadt
Kropp L.	TÜV RHEINLAND Institut für Umweltschutz Postfach 101750 D-5000 Köln 1
Lehmann Kl.	Deutsches Institut für Normung Burggrafenstr. 4-10 D-1000 Berlin 30
Löbel J.	VDI-Kommission Reinhaltung der Luft Geschäftstelle Postfach 1139 D-4000 Dusseldorf
Lohmeyer A.	Institut Wasserbau III Sonderforschungsbereich 80 University of Karlsruhe D-7500 Karlsruhe, 1

Ludwig C.   Umweltbundesamt
                                             Bismarckplatz 1
                                             D-1000 Berlin 33

Meinl H.   Dornier System GMBH
                                             Postfach 1360
                                             D-7990 Friedrichshafen

Meroney R.N.   Institut Wasserbau III
                                             Universität Karlsruhe
                                             Kaiserstrasse 12
                                             Postfach 6380
                                             D-7500 Karlsruhe 1

Müller H.G.   Dornier System GMBH
                                             Postfach 1360
                                             D-7990 Friedrichshafen

Schultz H.   Universität Hannover
                                             Theodor Storm Str. 5.
                                             D-3007 Gelwelen

Stern R.   Freie Universität Berlin
                                             Inst. für Geophysikalische
                                             Wissenschaften
                                             Fachrichtung Meteorologie
                                             Thielallee 50
                                             1000 Berlin 33

Weber E.   Bundesministerium des Innern
                                             Graurheindorfer Str. 198
                                             D-5300 Bonn 1

Yamartino R.J.   Geomet. GMBH
                                             D-Berlin

GREECE

Aravantinou S.A.   Ministry of Social Services -
                                             Environmental Pollution Control
                                             Project - Athens
                                             147, 28 Octovrioustreet
                                             Athens 814
                                             Greece

ITALY

Anfossi D.   Laboratorio di Cosmo-Geofisica
                                             CNR
                                             Corso Fiume 4
                                             I-TORINO

Bonino G.	Laboratorio di Cosmo—Geofisica CNR Universita Degli Studi di Torino Corso Fiume 4 I-TORINO
Borghi S.	Osservatorio Meteorologico di Brera Via Rera 28 I-20121 Milano
Colacino M.	Instituto di Fisica Dell Atmos- fera, CNR Piazza Sturzo 31 I-Roma
Ferrara V.	CNEN - Casaccia Nuclear Centre Via Anguillarese I-00060 ROMA
Giovannini I.	Tecneco Societa Per Azioni Pian di Rose I-61040 S. Ippolto-Pesaro
Longhetto A.	ENEL Bastioni Porta Volta 10 I-Milano
Melli P.	Centro Scientifico IBM Via Del Giorgione 129, I-ROMA
Ruspolini F.	INAIL-Centro Tecnico Accerta- mento Rischi Professionali Via Nomentana 74 I-00161 ROMA

THE NETHERLANDS

Aronds C.A.	Hoogovens Ijmuiden BV Milieubeheer, 3010 PO. BOX 10000 NL-1970 Ijmuiden
Booij C.G.	Landbouwhogeschool Van Huevenstraat, 46 NL-WAGENINGEN

Bovenkerk M.	Ministerie van Volksgezondheid en Milieuhygiene Postbus 439 NL-2260 A.K. Leidschendam
Bremer S.	R.I.V. Postbus 1 NL-Bilthoven
Brull N.A.	Prov. Waterstraat Limburg Parkweg 32 NL-6212 XN Maastricht
Builtjes P.	MT-T.N.O. Postbus 342 NL-7300 AH Apeldoorn
Buytenen C.J.P., Van	Prins Maurits Laboratorium-TNO Postbus 45 NL-2280 AA Rijswijk (ZH)
Colenbrander G.W.	Koninkl. Shell Laboratorium Afd. EE Badhuisweg 3 NL-1031 CM Amsterdam
Dauwer R.M.	Dow Chemical Netherlands NL-Terneuzen
Dop H., Van	Koninklijk Nederlands Meteorologisch Instituut P.O. BOX 201 NL-3730 AE De Bilt
Egmond K., Van	Dutch National Institute of Public Health P.O. BOX 1 NL-Bilthoven
Guicherit R.	IMG-TNO Postbus 214 NL-Delft
Guldemond C.P.	DSM- Central Laboratory P.O. BOX 18 NL-6160 MD Geleen
Haan B.J., De	Koninklijk Nederlands Meteorologisch Instituut P.O. BOX 201 NL-3730 AE De Bilt
Ham J., Van	SCMO-TNO Postbus 186 NL-2600 AD Delft

Harssema H.	Department of Environmental Sciences - Air Pollution Agricultural University Wageningen Binnenhaven 12 NL-6709 PD Wageningen
Heida H.	Gemeentelijk Centraal Milieu-laboratorium Amstelveense weg 90 NL-1057 XJ Amsterdam
Herzberg S.	Shell International Research Mij B.V. P.O. BOX 162 NL-2501 AN Den Haag
Holtslag B.	Koninklijk Nederlands Meteorologisch Instituut P.O. BOX 201 NL-3730 AE De Bilt
Hout K.D., Van Den	IMG- TNO Postbus 214 NL-Delft
Huygen C.	IMG-TNO Postbus 214 NL-Delft
Janssens L.A.	CTI-TNO P.O. BOX 342 NL-Apeldoorn
Leeuwen, Van	Prins Maurits Laboratorium-TNO Postbox 45 NL-2280 AA Rijswijk (ZH)
Mahieu A.P.	Koninklijk Shell Laboratorium Afd. EE - Badhuisweg 3 NL- 1031 CM Amsterdam
Muskens P.J.	Hfd. Bur. Dataverwerking Dienst Cent. Milieubeheer Rijnmond S-Gravelandsesteenweg 565 NL-3119 XT Schiedam
Nieuwstadt F.	Koninklijk Nederlands Meteorologisch Instituut P.O. BOX 201 NL- 3730 AE De Bilt

Quast F.J.   Dienst Centraal Milieubeheer
                                  Rijnmond
                                  NL-Schiedam

Römer F.G.   NV KEMA
                                  Utrechtseweg , 310
                                  P.O. BOX 9035
                                  NL-6800   Arnhem

Schneider T.   National Institute of Public
                                  Health
                                  P.O. BOX 1
                                  NL- 3720   BA Bilthoven

Steenkist R.   NV KEMA
                                  Utrechtseweg   310
                                  P.O. BOX 9035
                                  NL-6800   ET Arnhem

Ulden A., Van   Royal Netherlands Meteo-
                                  rological Institute
                                  P.O. BOX 201
                                  NL-3730   AE   De Bilt

Veldhuizen J.G., Van   Inspectie Milieuhygiëne
                                  NL-Heerlen

Wieringa J.   Koninklijk Nederlands Meteoro-
                                  logisch Instituut
                                  P.O. BOX 201
                                  NL-3730   AE   De Bilt

Zonneveld S.C.   Zonegge 18-23
                                  NL-6903   GS   Zevenaar

Zwerver S.   Ministerie van Volksgezondheid
                                  en Milieuhygiëne
                                  Postbus   439
                                  NL-2260   AK Leidschendam

NORWAY

Gram F.   NILU
                                  Norwegian Institute for Air
                                  P.O. BOX 130
                                  N-2001 Lillestrøm

Grønskei K.E.   NILU
                                  Norwegian Institute for Air
                                  Research
                                  P.O. BOX 130
                                  N-2001 Lillestrøm

SWITZERLAND

Broder B.   Atmospheric Physics
                                       CH-Zürich

UNITED KINGDOM

Benjamin S.F.   B.L. Technology Ltd.
                                       Brown Lane,
                                       UK-Coventry

Fisher B.E.A.   Central Electricity
                                       Research Laboratory
                                       Kelvin Avenue
                                       UK-Leatherhead-Surrey K622   7SE

Foster F.O.   Esso Engineering Europe Ltd.
                                       UK-London

Fryer L.S.   Safety and Reliability
                                       Directorate
                                       Wigshaw Lane
                                       Culcheth
                                       UK-Warrington WA 3   4NE

Henderson-Sellers A.   University of Liverpool
                                       Dept. Geography
                                       UK-Liverpool

Henderson-Sellers B.   University of Salford
                                       Dept. Civil Engineering
                                       UK-Salford MS 4 WT

Jagger S.   Safety and Reliability
                                       Directorate
                                       Wigshaw Lane
                                       UK- Warrington WA3   4NE

Puttock G.F.   Shell Research Ltd.
                                       UK-Chester

Selby K.   Shell Research Ltd.
                                       UK-Chester

Smith F.B.   Meteorological Office
                                       London Road
                                       Bracknell
                                       UK-Berkshire   R612   2SZ

Williams M.   Warren Spring Laboratory
                                       Department of Trace and In-
                                       dustry
                                       Gunnels Wood Road
                                       Stevenage SG1   2BX
                                       UK-Herts

UNITED STATES OF AMERICA

Brennan P.T.   Meteorological Evaluation
                                        Services Inc.
                                        134 Broadway
                                        Amityville - New York
                                        U.S.A.

Chan M.W.   AeroVironment Inc.
                                        145 Vista Avenue
                                        Pasadena - California 91107
                                        U.S.A.

Hanna S.R.   Atmosph. Turbulence and Diff.
                                        Lab.
                                        Nat. Oceanic and Atmosph.
                                        Administrat.
                                        P.O. Box E, Oak Ridge
                                        Tennessee   37880
                                        U.S.A.

Johnson W.B.   SRI International
                                        333 , Ravenwood Ave.
                                        Menlo Park , California 94025
                                        U.S.A.

Kramer M.L.   Meteorological Evaluation Serv.
                                        New York
                                        U.S.A.

Liu Mei-Kao   Systems Applications Inc.
                                        950 Northgate Drive
                                        San Rafael, California 94903
                                        U.S.A.

Meroney R.N.   Engineering Research Center
                                        Colorado State University
                                        Fort Collins, Colorado 80523
                                        U.S.A.

Mills M.   Teknekron Research Inc.
                                        60 , Hickory Drive
                                        Waltham Massachusetts - 02154
                                        U.S.A.

Niemeyer L.E.   Meteorology and Assessment
                                        Division EPA (MD 80)
                                        Research Triangle Park
                                        North Carolina   27711
                                        U.S.A.

Samson P.J.   Dept. of Atmospheric and
                                         Oceanic Science
                                         University of Michigan
                                         2218 Space Research BLDG.
                                         Ann Arbor, MI 48109
                                         U.S.A.

Tikvart J.A.   Office of Air Quality Plan-
                                         ning and Standards
                                         EPA (MD- 14)
                                         Research Triangle Park -
                                         North Carolina 27711
                                         U.S.A.

Zannetti P.   AeroVironment Inc.
                                         145 ,Vista Avenue
                                         Pasadena, California 91107
                                         U.S.A.

Woodward J.L.   Exxon Res. and Eng.
                                         Florham Park
                                         U.S.A.

AUTHORS INDEX

Aalst R., Van, 621
Allard, D.W., 133
Anfossi, D., 595

Bacci P., 595
Bacon J.W., 223
Basting W., 621
Berkowicz R., 267, 385
Bonino G., 509
Borghi S., 235
Blondin Ch., 191
Brennan P.T., 717
Briatore L., 509
Bruin H.A.R., De, 401
Builtjes P.J.H., 621
Bultynck H., 369

Caneill Y., 201
Cats G.J., 181
Chan M.W., 133
Cieslik S., 369
Conradsen K., 385
Cravatte M., 251, 609

Davidson G.A., 409
Demuth Cl., 609
Dop H., Van, 181
Duuren H., Van, 575

Egmond N.D., Van, 111
Elisei G., 509

Fage J.M., 553
Fisher B.E.A., 99
Fronza G., 639

Gallay G.J., 653

Giebel J., 319
Grønskei K.E., 551
Guicherit R., 575

Haan B.J., De, 181, 411
Hanna S.R., 337
Hecq P., 609
Henderson-Sellers A., 223, 309
Henderson-Sellers B., 223, 309
Hirata A., 147
Holtslag A.A.M., 401
Horn A., Van, 147
Hout K.D., Van Den, 575, 621
Huldy H.J., 621
Hulshoff J., 621
Huygen C., 575

Jagger S.F., 463
Jensen N.O., 477
Johnson W.B., 3

Kaiser G.D., 463
Kesseboom H., 111
Kramer M.L., 717
Kretzschmar J.G., 369
Kropp L., 703

Liu Mei-Kao, 55
Lohmeyer A., 433
Longhetto A., 509, 595

Melli P., 639
Meroney R.N., 433, 449, 489
Mills M.T., 147
Moody J.L., 43
Moussafir J., 653

Neff D.E., 449
Nielsen L.B., 385
Nieuwstadt F.T.M., 357

Plate E.J., 433
Prahm L.P., 267, 385

Racher P., 201
Reid J.D., 297
Rolin C., 213
Rosset R., 201
Roth P.M., 55
Römer F.G., 575
Runca E., 509

Saab A.E., 213
Samson P.J., 43
Schayes G., 251, 609
Scherer B., 681

Sivertsen B., 551
Slawson P.R., 417
Smith, L.F., 147
Steenkist R., 357
Stern, R., 681

Tong E.Y., 147

Ulden A.P. Van, 401

Veldt C., 621
Venkatram A., 169
Viljeer J.W., 575
Villouvier V., 213

Wieringa J., 279

Zannetti P., 537

SUBJECT INDEX

Acid rain, 61
Advection, 411, 642
Albedo, 252, 386

Boundary
  conditions, 21
  layer, 169, 213, 251, 267,
     297, 343, 369, 385
     609
Box model, 464
Buoyancy, 309

Calm conditions, 544
Cloud
  cover, 317
  flammable-, 469
Complex terrain, 489, 509,
     537, 551
Convection, 267
Convective layer, 513
Covective velocity scale, 270
Conversion
  chemical-, 46
  rate, 575
Cycles, 317

Data base ,417
Dense gas, 450, 463
Deposition
  coefficient, 26
  dry-, 27, 46, 57, 100, 113
  wet-, 27, 57, 67, 100
Diffusion, 3, 468, 609, 717
Diffusivity, 182, 251
Dispersion, 169, 297, 357, 417,
     450, 551
  calculations, 369
  model, 251, 621

Doppler sodar, 656
Downdrafts, 663
Drag coefficient, 281

Elevated sources, 575
Emission projection, 148
Energy
  balance, 311, 385
  budget, 401, 481
Entrainment, 464, 481
Episode, 117, 589, 595, 639
     653
Eulerian
  grid, 6
  time scales, 663

Flashing, 464, 483
Bluxmeter, 656
Föhn, 213
Fumigation, 595

Gaussian
  climatological model, 509
  method, 537
Gravity, 463
Ground
  concentration, 357
  observations, 369

Heat
  budget, 394
  flux, 385, 401
  island, 213
  turbulence, 267
Heavy gases, 433, 477
Hydraulic model, 509
Hydrodynamic tank, 516

Impact, 551, 703
Interregional, 3
Inversion, 186, 309
  nocturnal-, 513
Isopleth, 681

Kalman filter, 644
K. Theory, 251, 538

Lagrangian, 663
  trajectory, 6
Land breeze, 513
Level
  high air pollution-, 609
  ground -, 171, 575
Licensing, 703
Liquid Natural Gas, 449
Long range transport, 3, 99,
    111, 134, 235

Markov, 99, 223
Meso-scale, 111, 191, 201,
    213
Meteorological
  data, 717
  observations, 176
Mixed layer, 44, 57, 169,
    214
Monin-Obukhov length, 170,
    251, 401, 483, 663,
    718

Nitrogen oxides, 575

Orographic effects, 191
Ozone, 51, 134, 681

Pasquill, 579
Photochemical episodes, 589
Photochemical conditions, 591
Photochemistry, 57
Plume rise, 309
Poisson distribution, 101
Precipitation, 46, 99, 219
  scavenging, 99
  sulfate-, 99
Propylene, 464
Puff model, 465, 537

Radiation, 385, 402
Rainfall, 67, 99

Regional modelling, 55, 161
Regulatory modelling, 718
Regulatory problems, 537
Removal processes, 99
Removal rate, 575
Residence time, 576
Richardson number, 466, 665,
    718
Roughness, 379

Scaling effect, 470
Sea breeze, 513
Similarity, 551
  theory, 267, 717
Stability, 172, 251, 332,
    357, 401
  classes, 579
Standards
  air quality-, 147, 181, 703
Statistical diffusion model,
    99, 397
Statistical trajectory, 6
Stochastic model, 223
Sulphate, 55, 99
Surface layer, 251, 267, 657
Synoptic data, 251

Tall stack, 337
Tracer experiments, 551
Trajectory, 99, 235
  analysis, 43
  patterns, 191
Transport
  air pollution-, 191, 223
Turbulence, 267, 664
Turbulent
  kinetic energy, 481
  diffusion, 267, 470
  entrainment, 464

Urban
  heat island, 311
  mixing layer, 309
  ozone, 681

Valley, 201
  drainage, 489
Volocity
  friction-, 169, 254
  scales, 169, 270
Visibility

Wake effects, 551
Wind
  field, 181, 191, 201
  profile, 135, 279
  rose, 135
  speed, 223, 267, 317
  velocity fluctuations, 267